AA007709

11th International Scientific Conference on Aeronautics, Automotive and Railway Engineering and Technologies (BulTrans-2019)

IOP Conference Series: Materials Science and Engineering
Volume 664

Sozopol, Bulgaria
10 - 12 September 2019

Editors:

S. Stoilova
M. Todorov
N. Nikolov
I. Kralov
V. Stoilov
B. Gigov

ISBN: 978-1-5108-9852-3
ISSN: 1757-8981

Printed from e-media with permission by:

Curran Associates, Inc.
57 Morehouse Lane
Red Hook, NY 12571

Some format issues inherent in the e-media version may also appear in this print version.

This work is licensed under a Creative Commons Attribution 3.0 International Licence.
Licence details: http://creativecommons.org/licenses/by/3.0/.

No changes have been made to the content of these proceedings. There may be changes to pagination and minor adjustments for aesthetics.

Printed with permission by Curran Associates, Inc. (2020)

For permission requests, please contact the Institute of Physics at the address below.

Institute of Physics
Dirac House, Temple Back
Bristol BS1 6BE UK

Phone: 44 1 17 929 7481
Fax: 44 1 17 920 0979

techtracking@iop.org

Additional copies of this publication are available from:

Curran Associates, Inc.
57 Morehouse Lane
Red Hook, NY 12571 USA
Phone: 845-758-0400
Fax: 845-758-2633
Email: curran@proceedings.com
Web: www.proceedings.com

Table of contents

Volume 664

11th International Scientific Conference on Aeronautics, Automotive and Railway Engineering and Technologies (BulTrans-2019)

10–12 September 2019, Sozopol, Bulgaria

Accepted papers received: 8 October 2019
Published online: 29 October 2019

Preface

Papers

Aeronautics

Dynamics, Strength and Reliability of Vehicles

Internal Combustion Engines

Railway Engineering and Technologies

Transport Management and Logistics

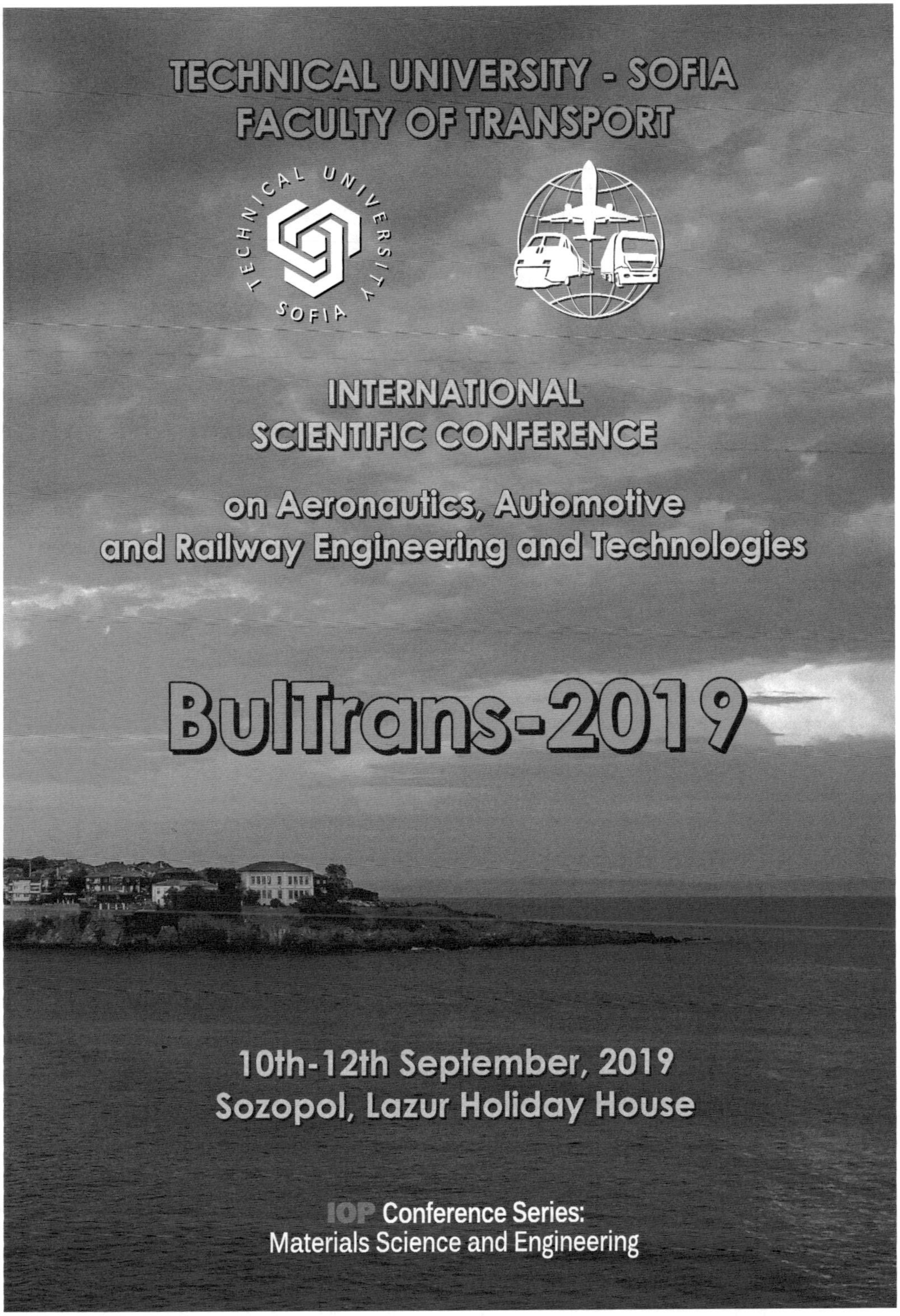

Content from this work may be used under the terms of the Creative Commons Attribution 3.0 licence. Any further distribution of this work must maintain attribution to the author(s) and the title of the work, journal citation and DOI.
Published under licence by IOP Publishing Ltd

IOP Conf. Series: Materials Science and Engineering **664** (2019) 011001 doi:10.1088/1757-899X/664/1/011001

Scientific Editorial Board

Prof. Todorov M., PhD
BulTrans Chairman, Dean, Faculty of Transport, Technical University – Sofia

Assoc. Prof. Nikolov N., PhD
BulTrans Vice-Chairman and Editor-in-Chief, Vice-Dean on R&D, Faculty of Transport, TU – Sofia

Assoc. Prof. Stoilova S., DSc
Editor-In-Chief for IOP Proceedings, Faculty of Transport, Technical University – Sofia

Prof. Kralov I., DSc
Vice-Rector on Research and Development, Technical University – Sofia

Prof. Stoilov V., PhD
Vice-Director of Research and Development Sector, Technical University – Sofia

Assoc. Prof. Gigov B., PhD
Faculty of Transport, Technical University – Sofia

Reviewers

Assoc. Prof. Asenov A., PhD,
University of Ruse, Bulgaria

Assoc. Prof. Barta D., PhD,
University of Zilina, Slovakia

Assoc. Prof. Camaj J., PhD,
University of Zilina, Slovakia

Eng. Dobrev I., PhD,
Arts et Métiers ParisTech, Paris, France

Assoc. Prof. Dolinayova A., PhD,
University of Zilina, Slovakia

Prof. Fomin O., Dsc,
State University of Infrastructure and Technology, Kiev, Ukraine

Assoc. Prof. Gigov B, PhD,
Technical University – Sofia, Bulgaria

Assoc. Prof. Grenčík J., PhD,
University of Zilina, Slovakia

Prof. Hunicz J., DSc,
Politechnika Lubelska,Lublin, Poland

Prof. Ivanov R., Dsc,
University of Ruse, Bulgaria

Assoc. Prof. Kendra M., PhD,
University of Zilina, Slovakia

Prof. Kralov I., DSc,
Technical University – Sofia, Bulgaria

Prof. Kunchev L., PhD,
Technical University – Sofia, Bulgaria

Assoc. Prof. Marcheva Y., PhD,
Technical University – Sofia, Bulgaria

Assoc. Prof. Marinkovic D., PhD,
University of Belgrade, Serbia

Prof. Massouh F., PhD,
Arts et Métiers, ParisTech, Paris, France

Assoc. Prof. Nikolov N., PhD,
Technical University – Sofia, Bulgaria

Prof. Rotaru C., PhD,
Air Force Academy,Brashov, Romania

Assoc. Prof. Stoilova S., Dsc,
Technical University – Sofia, Bulgaria

Prof. Todorov M., PhD,
Technical University – Sofia, Bulgaria

Prof. Vatulia G., DSc,
State University of Railway Transport, Kharkiv,Ukraine

Statement of Peer Review

All papers published in this volume of *IOP Conference Series: Materials Science and Engineering* have been peer reviewed through processes administered by the Editors. Reviews were conducted by expert referees to the professional and scientific standards expected of a proceedings journal published by IOP Publishing.

IOP Conf. Series: Materials Science and Engineering **664** (2019) 011001 doi:10.1088/1757-899X/664/1/011001

PREFACE

BulTrans International Scientific Conference on Aeronautics, Automotive and Railway Engineering and Technologies is held for the first time in 2009, as part of *The Science Days* of Technical University of Sofia. Till now, eleven editions of the Conference have been organized, ten in USOB "Lazur" and a jubilee forum in Sofia.

BulTrans-2019 was the 11th edition of the Conference. It was attended by over 120 participants from 10 countries.

Thorough scientific achievements in the field of transport are being presented at the conference. **BulTrans** is the most prestigious scientific forum for road, railway and air transport in Bulgaria.

Our sponsors have significant contribution for the realization of the event. Among them are: Siemens Mobility, Knorr-Bremse, Automotor Corporation, Sofia France Auto, Sofia Airport.

OPENING ADDRESS

Prof. M. Todorov, PhD,
Chairman of the Organizing Committee (on the right) presenting a gold medal to Prof. F. Massouh, PhD, France

Ladies and Gentlemen,

Dear Colleagues and Friends,

I have the pleasure and honor, on behalf of the Organizing and Scientific Committees, to congratulate you with the opening of the Scientific Conference on Aeronautics, Automotive and Railway Engineering and Technologies – BulTrans-2019.

The main objective of this conference is to bring together academic and industry professionals from the universities, companies and government agencies to present their achievements in the transport science and technology.

The wide variety of more than 50 scientific reports included in the program, as well as the participation of foreign and Bulgarian renowned lecturers, is really impressive.

I am particularly grateful to our sponsors for their financial support.

I hope you find this conference both interesting and stimulating and you take advantage of this opportunity to meet old friends and make new friendships.

Finally, I ask you to forgive eventual faults of ours – believe me, they are nor deliberate.

I look forward to meeting you again at the next BulTrans-2020.

IOP Conf. Series: Materials Science and Engineering **664** (2019) 011002 doi:10.1088/1757-899X/664/1/011002

Peer review statement

All papers published in this volume of *IOP Conference Series: Materials Science and Engineering* have been peer reviewed through processes administered by the proceedings Editors. Reviews were conducted by expert referees to the professional and scientific standards expected of a proceedings journal published by IOP Publishing.

Content from this work may be used under the terms of the Creative Commons Attribution 3.0 licence. Any further distribution of this work must maintain attribution to the author(s) and the title of the work, journal citation and DOI.
Published under licence by IOP Publishing Ltd

IOP Conf. Series: Materials Science and Engineering **664** (2019) 012001 doi:10.1088/1757-899X/664/1/012001

Transonic flow hysteresis in divergent bent channels

A Kuzmin

Department of Fluid Dynamics, St. Petersburg State University, St. Petersburg, 198504 Russia

E-mail: a.kuzmin@spbu.ru

Abstract. This paper presents a numerical study of the 2D and 3D turbulent airflows in divergent 9°-bent channels. The incoming flow is supersonic, whereas the exit flow may be either supersonic or subsonic. Solutions of the Reynolds-averaged Navier-Stokes equations are obtained with a finite-volume solver ANSYS CFX using the Spalart-Allmaras and Shear Stress Transport k-ω turbulence models. The solutions reveal a significant hysteresis of the flow field under variations of the free-stream Mach number or exit pressure. At the ends of hysteresis bands, the flow pattern changes crucially due to instability of a shock wave formed near the bend of channel. The instability is caused by the shock foot interaction with an expansion flow developed over the convex wall of channel. Boundary conditions, in which the flow admits a double hysteresis, are figured out. The occurrence of non-unique flow regimes must be taken into account in the advanced intake design, as different losses of the total pressure may cause different trusts of an air breathing engine.

1. Introduction

Supersonic air intakes are of major importance for the efficient operation of aircraft engines. At on-design conditions, a train of oblique shocks forms in a conventional convergent-divergent intake [1]. However, this shock system is very sensitive to small perturbations. A small reduction of the free-stream Mach number M_∞ triggers an expulsion of the oblique shocks from the channel and formation of a nearly normal shock in front of it. This results in losses of the total pressure and reduction of the engine efficiency. Numerical studies of the supersonic flow in convergent-divergent channels demonstrated a significant hysteresis under gradual variations of M_∞ [2, 3].

Supersonic flow in bent channels can be sensitive to small perturbations as well [4]. In this case, instability of the shock wave arising ahead of the concave wall is caused by its interaction with a flow acceleration region over the opposite (convex) wall. Numerical simulations of the flow in channels with a short upper wall (cowl) demonstrated a significant hysteresis under variations of M_∞ or angle of attack [4-6].

Guo *et al.* [7] performed a numerical modelling of supersonic flow in a curved channel of nearly constant cross-section. The existence of a flow hysteresis under variations of the pressure given at the outlet was demonstrated.

In practice, supersonic flow in bent channels is utilized, e.g., in Y-shaped intakes of battle-plane and fighter aircraft engines. Kotteda and Mittal [8] studied the effect of sideslip angle on the performance of a slightly divergent Y-intake. Different flow regimes under variations of the outlet pressure were identified, and an effect of the initial condition was examined.

Feng *et al.* [9, 10] performed a numerical and experimental investigation of supersonic flow in a bent channel which models a variable geometry dual mode combustor. The study indicated that the static

Content from this work may be used under the terms of the Creative Commons Attribution 3.0 licence. Any further distribution of this work must maintain attribution to the author(s) and the title of the work, journal citation and DOI.
Published under licence by IOP Publishing Ltd

pressure distribution on a wall had an obvious hysteresis phenomenon under continuous variations of the outlet cross section area.

In [11] we studied shock wave hysteresis in a bent divergent channel with a longer upper wall than that considered in [4-6]. It was shown that flow features depend essentially on the thickness δ of leading edges of walls, $0.1 \leq \delta$, mm ≤ 1.0. In the present paper, we focus on the case $\delta = 0.1$ mm and study in detail the behaviour of shock waves. Sections 2 and 3 are concerned with the problem formulation for 2D flow and description of a numerical method. Section 4 addresses locations of the shock wave (SW) generated by the concave upper wall; the SW position as a function of the free-stream Mach number M_∞ is investigated at zero angle of attack and supersonic exit condition. In Section 5 we study the behaviour of SW under variations of the pressure given at the subsonic exit. In Section 6, preliminary simulations of the 3D flow are discussed.

2. 2D problem formulation

Let the inner part of the lower wall be a convex corner with a vertex at $x = x_{corner}$, $y = 0$:

$$y=0 \text{ at } -60 \leq x \leq x_{corner};$$

$$y = (x - x_{corner}) \tan(-9°) \text{ at } x_{corner} < x \leq x_{corner} + 27, \tag{1}$$

whereas the inner part of the upper wall be a concave corner with a vertex at $x = 0$, $y = 30$:

$$y= 30 + x \tan 1° \text{ at } -60 \leq x \leq 0;$$

$$y= 30 + x \tan(-8°) \text{ at } 0 \leq x \leq 55. \tag{2}$$

Evidently, the walls are 1° divergent before and after the bends. In expressions (1), (2) and further in the paper, the Cartesian coordinates (x, y) are dimensional and given in millimetres. The outer parts of the lower and upper walls are a segment and corner, respectively, see figure 1. The thickness of the leading edges is set to 0.1 in order to attenuate the bow shocks and reduce perturbations they generate.

The above specific geometry is chosen to provide a clear demonstration of the hysteresis phenomenon in bent channels. From a practical point of view, this geometry can be of interest, e.g., for the Y-type intake design [8]. The small dimensions of the channel should be convenient for future experimental verification in a midsize wind tunnel.

The inflow boundary Γ_{in} of the computational domain is constituted by a vertical segment $x = -90$, $-30 \leq y \leq 60$ and two inclined segments with endpoints $x = 0$, $y = \pm 200$, which are remote enough to eliminate an interaction of Γ_{in} with the bow shocks produced by the channel. We prescribe the *x*-, *y*- and *z*-components of the flow velocity on Γ_{in} as follows:

$$U_\infty = M_\infty a_\infty,\ V_\infty = 0,\ W_\infty = 0. \tag{3}$$

In addition, we impose on Γ_{in} the static pressure p_∞, the turbulence level of 1%, and static temperature $T_\infty = 250$ K which determines the sound speed $a_\infty = 317.02$ m/s. The pressure p_∞ is set to 10^5 N/m^2 throughout the paper, except for figure 6 in which we compare results obtained for different values of p_∞.

The outflow boundary Γ_{out} of the computational domain is constituted by two segments which begin at $x = 0$, $y = \pm 200$ and end at $x = 20$ on the outer surfaces of walls. We prescribe the supersonic outflow condition $M > 1$ on Γ_{out}. In the exit section Γ_{exit} of the channel we impose either the supersonic condition $M > 1$ or subsonic one and the static pressure p_{exit}.

The vanishing heat flux and no-slip condition are prescribed on the walls. The air is treated as a perfect gas whose specific heat at constant pressure is 1004.4 J/(kg K) and the ratio of specific heats is 1.4. We use the Sutherland formula for the molecular dynamic viscosity and adopt the value of 28.96 kg/kmol for the molar mass. The Reynolds number based on $p_\infty = 10^5$ N/m^2, $M_\infty = 1.35$, and height of the channel is 1.1×10^6. Initial data are either parameters of the free stream or a flow field calculated for other values of M_∞, p_{exit}.

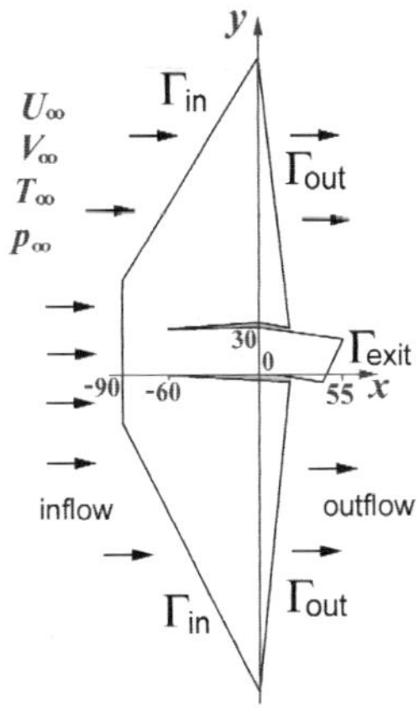

Figure 1. Sketch of the 2D computational domain.

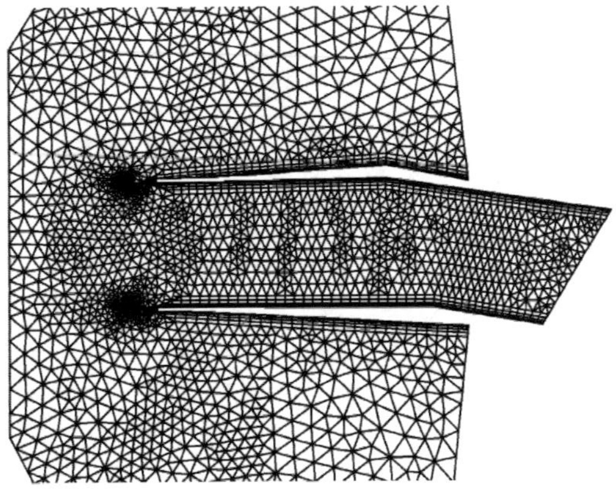

Figure 2. Schematic of the computational mesh.

3. Numerical method

Solutions of the unsteady Reynolds-averaged Navier-Stokes equations were obtained with a commercial ANSYS-18.2 CFX finite-volume solver of second-order accuracy. Computations showed a convergence of the mean parameters of turbulent flow to steady states in less than 0.1 s of physical time. We used the Spalart-Allmaras [12] and Shear Stress Transport *k*-ω [13] turbulence models, which are known to reasonably predict aerodynamic flows with boundary layer separations. Hybrid unstructured meshes were constituted by quadrangles in 38 layers on the walls and by triangles in the remaining region, see figure 2. The non-dimensional thickness y^+ of the first mesh layer on the walls was less than 1. The sizes of triangles were essentially decreased in the channel for an accurate resolution of shock waves. Test solutions obtained on uniformly refined meshes of approximately 2.5×10^5, 5×10^5, and 10^6 cells showed that a discrepancy between shock wave coordinates calculated on the second and third meshes did not exceed 1%. Global time steps of 5×10^{-7} s and 10^{-6} s yielded indistinguishable solutions. For this reason, the time step of 10^{-6} s and mesh of 5×10^5 cells were employed for the study of flow behavior at various M_∞ and p_{exit}.

The solver was validated by computation of several benchmark supersonic and transonic flow problems. For instance, we performed numerical simulations for a channel with a bump [4], a supersonic intake [14] and Busemann biplane [11]. The obtained solutions showed good agreement with numerical and experimental data available in the literature.

4. Shock location versus M_∞ at $p_\infty = 10^5$ N/m^2 and supersonic exit condition

First, we solved the problem at $M_\infty = 1.35$, $x_{corner} = 7.5$ using the free stream (3) for initialization of the solution. The obtained flow field exhibits a train of weak shocks which propagate from the entrance downstream, being multiply reflected from the walls. A stronger shock wave (SW) begins in front of the upper wall bend and hits the lower wall at a coordinate $x > x_{corner}$. Such a flow pattern is called hereafter the flow regime with a swallowed SW.

A reduction of M_∞ step-by-step (at each step, initial conditions are the steady flow parameters obtained for the previous value of M_∞) from 1.35 to 1.27 produces a decrease of Mach numbers in the channel and formation of a small subsonic region downstream of SW, see figure 3(a). To analyse streamwise positions of SW, we will trace its coordinate x_{sh} at the height $y = 10$. Computations showed that, when M_∞ decreases from 1.35 to 1.27, the flow pattern persists while x_{sh} slightly decreases, see the upper branches of curves 1 and 2 in figure 4(a). Further reduction of M_∞ to 1.268 triggers a jump of SW upstream (and, hence, drop of x_{sh}) due to instability of the shock foot interaction with the flow acceleration region over the lower wall bend. The drop of x_{sh} means a transition from the upper branches of curves 1 and 2 to lower ones. The oblique SW becomes a nearly normal shock located at $x_{sh} = -41.5$. Such a flow pattern is called hereafter the flow regime with a partially expelled SW.

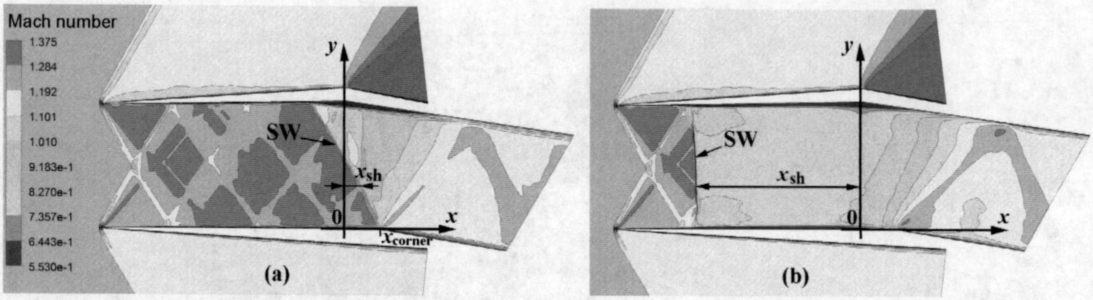

Figure 3. Mach number contours at $M_\infty = 1.27$, $M_{exit} > 1$, $p_\infty = 10^5$ N/m^2, $x_{corner} = 7.5$, SST k-ω turbulence model. The flow field is obtained by: (a) decreasing M_∞ from 1.35 to 1.27, (b) increasing M_∞ from 1.24 to 1.27.

Further decrease of M_∞ to 1.24 yields a gradual shift of SW in the channel to the entrance. A comparison of curves 1 and 2 in figure 4(a) shows a negligible distinction between SW coordinates obtained using SST k-ω and Spalart-Allmaras turbulence models.

Inversely, if M_∞ increases step-by-step from 1.24 to 1.318, then the expelled SW gradually shifts downstream and x_{sh} rises, as illustrated by the lower branches of curves 1 and 2. In particular, isoMachlines obtained at $M_\infty = 1.27$ are depicted in figure 3(b). An increase of M_∞ beyond 1.318 triggers a jump of SW downstream and relaxation to a flow pattern with a swallowed SW. This corresponds to a transition from the lower branches of curves 1 and 2 to upper ones. As seen, the coordinate x_{sh} exhibits a hysteresis in the band $1.27 \leq M_\infty \leq 1.318$ in which the realization of a certain flow field depends on the history of free-stream Mach number variation.

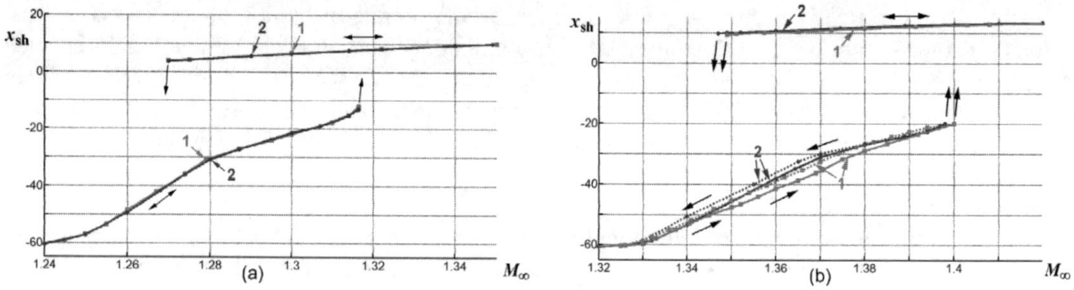

Figure 4. Coordinate x_{sh} of SW versus M_∞ at $p_\infty = 10^5$ N/m^2, $M_{exit} > 1$; turbulence models: 1 – SST k-ω, 2 – Spalart-Allmaras; (a) $x_{corner} = 7.5$, (b) $x_{corner} = 11.5$.

In a similar way we obtained a plot $x_{sh}(M_\infty)$ for $x_{corner} = 11.5$, see figure 4(b). In this case there exist two lower branches of the curves, which illustrate an extra hysteresis caused by the separated boundary layer like a hysteresis in straight nozzles [15]. The dashed (solid) curves are obtained at decreasing (increasing) free-stream Mach number, as indicated by arrows next to the curves.

Figure 5 illustrates the relative total pressure $P_{total} = p_{total}/p_{total,\infty}$ (where $p_{total,\infty} = 296{,}761$ N/m^2) in the channel and exit section at $x_{corner} = 11.5$, $M_\infty = 1.35$. As seen, pressure losses are different in the flow regimes which correspond to the upper and lower branches of curves in figure 4(b).

We notice that the thickness δ of the leading edges of walls crucially influences the location of SW. When δ increases, the bow shocks become stronger, and flow velocities in the channel decrease. Therefore the swallowing and expulsion of SW occur at larger Mach numbers M_∞. Computations showed that the hysteresis band shifts from $1.35 < M_\infty < 1.40$ to $1.41 < M_\infty < 1.473$ when δ increases from 0.1 to 0.5 in the case $M_{exit} > 1$, $x_{corner} = 11.5$.

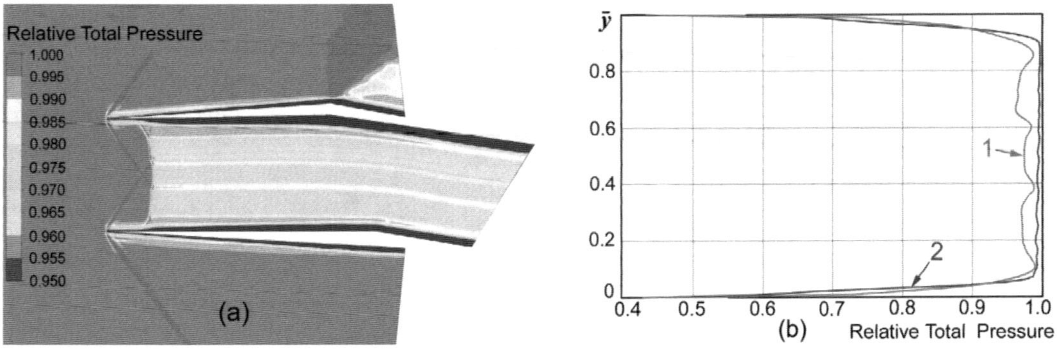

Figure 5. The relative total pressure P_{total} at $M_\infty = 1.35$, $x_{corner} = 11.5$, $M_{exit} > 1$, SST k-ω turbulence model: (a) contours of P_{total} in the regime that corresponds to the lower branch of curve 1 in figure 4(b); (b) plots of P_{total} in the exit section vs the non-dimensional height $\bar{y}$, $0 < \bar{y} < 1$, where curve 1 (curve 2) corresponds to the lower (upper) branch of curve 1 in figure 4(b).

5. Variations of the pressure given on Γ_{in} and Γ_{exit}

A doubling of the free-stream pressure p_∞ (and, consequently, the Reynolds number) produces an insignificant effect on the lower branches of plot $x_{sh}(M_\infty)$, see figure 6. Meanwhile, the upper branch essentially shifts to smaller values of M_∞; therefore, the hysteresis band expands. Further increase of p_∞ to 4×10^5 N/m^2 virtually does not influence the shock location.

The calculated pressure p_{exit} is less than $1.28p_\infty$ when the supersonic condition $M_{exit} > 1$ is used on Γ_{exit}. If, instead of $M_{exit} > 1$, we impose the subsonic condition $M_{exit} < 1$ along with $p_{exit} = 1.65p_\infty$, then an extra shock appears and terminates the supersonic region near the exit. This shock gradually moves upstream as p_{exit} increases. Figure 7 illustrates the flow field obtained at $p_{exit}/p_\infty = 1.88$, $M_\infty = 1.40$. Further increase of p_{exit}/p_∞ triggers a jump of the coordinate x_{sh} from 2 to −30 and a transition from the upper branch of curve 2 in figure 8 to lower one. Inversely, a decrease of p_{exit}/p_∞ to 1.83 produces a gradual increase of x_{sh} to −18. After that a decrease of p_{exit}/p_∞ to 1.825 triggers a jump of x_{sh} to 9 and a transition to the upper branch of curve 2.

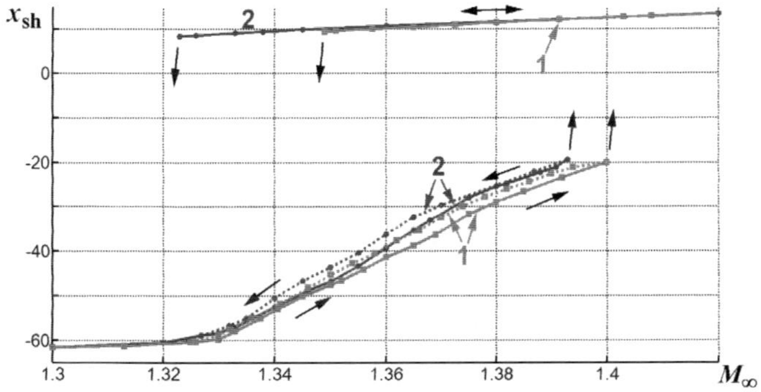

Figure 6. Coordinate x_{sh} of SW versus M_∞ at x_{corner}=11.5, $M_{exit} > 1$ and SST k-ω turbulence model: 1 – $p_\infty = 10^5$ N/m^2, 2 – $p_\infty = 2\times10^5$ N/m^2.

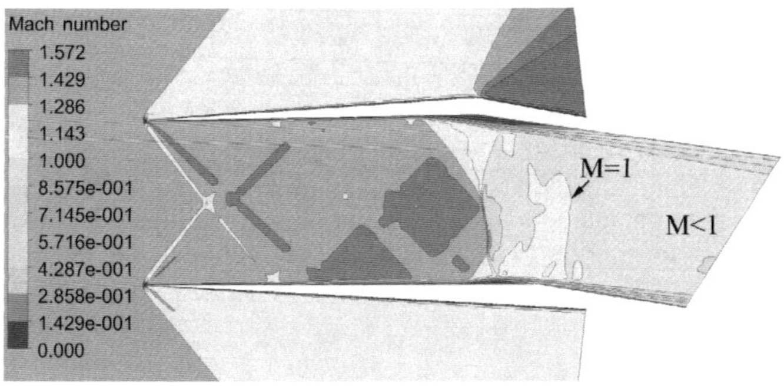

Figure 7. Mach number contours in the flow at $x_{corner} = 11.5$, $M_\infty = 1.40$, $M_{exit} < 1$, $p_{exit} = 188000$ N/m^2, SST k-ω turbulence model.

At the larger free-stream Mach number $M_\infty = 1.41$, computations demonstrated a similar dependence of x_{sh} on p_{exit}/p_∞, see curve 3 in figure 8. Meanwhile, at the smaller Mach number $M_\infty = 1.39$, a transition from the upper to lower branch of curve 1 is irreversible, since any decrease of p_{exit}/p_∞ fails to produce a return to the upper branch, i.e., to the flow regime with a swallowed SW.

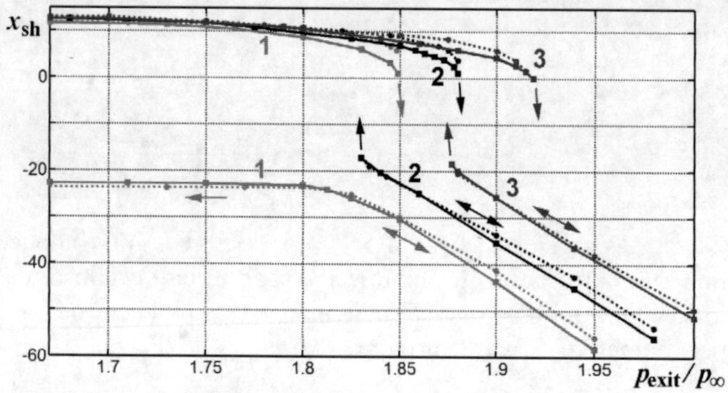

Figure 8. Coordinate x_{sh} of SW versus p_{exit} at $M_{exit} < 1$, $x_{corner} = 11.5$, $p_\infty = 10^5$ N/m^2: 1 – $M_\infty = 1.39$, 2 – $M_\infty = 1.40$, 3 – $M_\infty = 1.41$. Solid curves – SST k-ω turbulence model, dashed curves – Spalart-Allmaras model.

6. 3D flow simulation

A three-dimensional channel was created by an extrusion of the profile (1), (2) with $\delta = 0.5$, $x_{corner} = 11.5$ from the plane $z = 0$ to $z = \pm 45$. The sidewall thickness was set to 0.5, so that the outer surfaces of sidewalls were located at $z = \pm 45.5$. The boundaries Γ_{in} and Γ_{out} of the 2D computational domain were extruded from $z = 0$ to $z = \pm 120$. For CPU savings, we assumed the flow to be symmetric about the plane $z = 0$ and solved the problem in a half of the domain, at $-120 < z < 0$. On the side boundary $z = -120$ we imposed the free-slip condition. A hybrid 3D mesh was constituted by about 11.3×10^6 tetrahedrons, 11.1×10^6 prisms, and by 2.1×10^6 hexahedrons in 38 layers on the walls.

Computations showed that, due to a sidewall effect, flow velocities in the 3D channel are noticeably smaller than those in 2D one at the same M_∞. As a consequence, the expulsion of SW takes place when M_∞ becomes smaller than 1.50 instead of 1.41 in 2D flow. The 3D flow hysteresis is realized in the band $1.50 \le M_\infty \le 1.52$ instead of $1.41 \le M_\infty \le 1.473$ pointed out in the end of Section 4.

Figure 9 displays the bow shocks and partially expelled SW at $M_\infty = 1.49$. Figure 10 exposes Mach number contours in the plane $x = -30$. As seen, a dependence of flow parameters on the spanwise coordinate z is insignificant, except for a vicinity of the sidewall. The boundary layer separation from the walls is larger than that in 2D flow because of the larger M_∞ and, as a consequence, stronger shocks.

Thus, 3D flow simulations confirm the occurrence of flow hysteresis in the channel at hand. With increasing span or Reynolds number, the sidewall effect is expected to decrease.

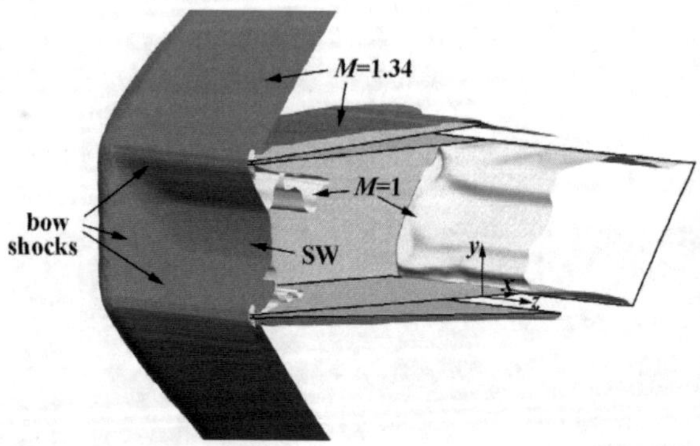

Figure 9. Isosurfaces $M(x,y,z) = 1.34$ and $M(x,y,z) = 1$ in the 3D flow at $M_\infty = 1.49$, $M_{exit} > 1$, $x_{corner} = 11.5$.

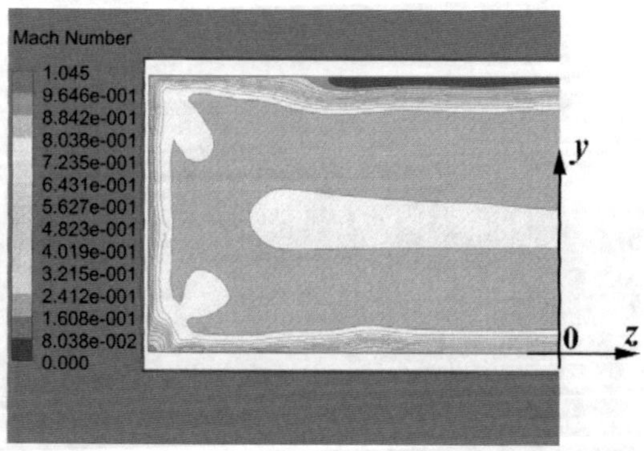

Figure 10. Mach number contours in the cross section $x = -30$ of 3D flow at $M_\infty = 1.49$, $M_{exit} > 1$, $x_{corner} = 11.5$.

IOP Conf. Series: Materials Science and Engineering **664** (2019) 012001 doi:10.1088/1757-899X/664/1/012001

7. Conclusion

Transonic flow in divergent bent channels at zero angle of attack is studied numerically. Simulations of the 2D flow revealed a shock wave in front of the bend of channel and significant hysteresis in the shock location under variations of the free-stream Mach number or exit pressure. At the ends of hysteresis bands, there are abrupt changes of the flow patterns, whereas within the bands there exist up to three flow regimes. This must be taken into account in the design of air breathing engines, as different losses of the total pressure in different regimes may essentially influence the engine thrust. The 3D flow simulations confirmed the findings, though demonstrated a noticeable sidewall effect.

Acknowledgements
This research was performed using computational resources provided by the Computational Center of St. Petersburg State University (http://cc.spbu.ru). The work was partially supported by the Russian Foundation for Basic Research under grant no. 19-01-00242.

References

[1] Sforza P M 2017 *Theory of Aerospace Propulsion* (Waltham: Butterworth-Heinemann)
[2] Kusunose K, Matsushima K and Maruyama D 2011 Supersonic biplane—A review *Progress in Aerospace Sciences* **47(1)** 53–87
[3] Chang J, Li N, Xu K, Bao W and Yu D 2017 Recent research progress on unstart mechanism, detection and control of hypersonic inlet *Progress in Aerospace Sciences* **89** 1–22
[4] Kuzmin A 2016 Shock wave bifurcation in channels with a bend *Archive of Applied Mechanics* **86(5)** 787–95
[5] Kuzmin A 2019 Instability of shock waves in bent channels of rectangular cross section *Proc. 1st Int. Conf. on Numerical Modelling in Engineering (Ghent) (Singapore: Springer)* **2** 142
[6] Kuzmin A 2019 Shock wave instability in a bent channel with subsonic/supersonic exit *Advances in Aircraft and Space Sci.* **6**(1) 19–30
[7] Guo S, Wang Z and Zhao Y 2014 *J. National Univ. Defense Technology* **36** 10
[8] Kotteda V M K and Mittal S 2016 *J. Propulsion and Power* **32** 171
[9] Feng S, Chang J, Zhang Ch, Wang Y, Ma J and Bao W 2017 *Aerospace Sci. and Technology* **67** 96
[10] Feng S, Chang J, Zhang Y, Zhang Ch, Wang Y and Bao W 2017 *Aerospace Sci. and Technology* **68** 320
[11] Kuzmin A 2019 *Computers and Fluids* **182** 52
[12] Spalart P R and Allmaras S R 1992 *AIAA Paper* 92-0439 1
[13] Menter F R 2009 *Int. J. Comput. Fluid Dynamics* **23** 305
[14] Kuzmin A and Babarykin K 2018 *Advances in Aircraft and Spacecraft Sci.* **5** 445
[15] Setoguchi T, Matsuo S, Alam M M A, Nagao J and Kim H D 2010 *J. Thermal Science* **19** 526

Calculation of the maximum endurance of a small unmanned aerial vehicle in a hover

J Hnidka[1] **and D Rozehnal**

Department of Aircraft Technology, University of Defence in Brno, 602 00 Brno, Czech Republic

[1] E-mail: jakub.hnidka@unob.cz

Abstract. Optimizing the flight performance, chiefly maximum endurance, of small unmanned aerial vehicles (SUAVs) relying on propellers powered by an electric motor/s as a form of propulsion is key challenge in SUAV development. Choosing the correct combination of propellers, electric motors and batteries is necessary and knowledge of characteristics of all three factors is of a great help. The paper presents a simple blade element model (BEM), which can be used to calculate the propeller performance characteristics. This model is further validated by measurements in the wind tunnel. The paper also presents measurements of electric motor and battery discharge characteristics. Finally, the paper presents a method of calculation of maximum endurance in hover.

1. Introduction

During the past decades, the amount of both military and civilian applications of small unmanned aerial vehicles (SUAVs) increased, and the focus of the development of SUAVs has shifted from development of the drivetrain to increasing its efficiency. Due to the limited amount of energy on board, often the main parameter to tune is the maximum endurance of the SUAV [1]. This paper presents a method of calculating the maximum endurance in hover of an SUAV of a multirotor type, equipped with a low-Reynolds fixed-pitch propeller, brushless DC electric motor (BLDC motor) with a regulator and a battery. A modified blade-element model with Joukowsky theorem is used to calculate the propeller performance characteristics (PPCH) and the paper shortly discusses necessary corrections for this type of calculations. The performance characteristics of the BLDC motor with a regulator and discharge characteristics of the battery were measured prior to the calculation. The calculated results were then compared with measurement of the maximum endurance of a Yuneec Typhoon H hexacopter drone.

2. Blade element model (BEM)

There are several different models, which can be used to calculate the PPCH, with each having its pros and cons, as described in [2]. BEM allows for rapid determination of PPCH, which allows e.g. optimization of the propeller geometry based on the flight configuration of the SUAV. Unfortunately, BEM models all assume quasi-static inflow conditions and cannot compensate for the dynamic behaviour of the propeller mounted on the SUAV [3, 4]. This problem can be overcome by an implementation of the vortex models described in [5] or [6]. However, for purposes of this paper, in a hover under semi-ideal ambient conditions, it can be assumed that the inflow velocity is equal to zero (with no disturbances). Thus, only relatively small deviations of revolutions of each propeller can be expected, and the dynamics of the rotor do not significantly affect the results. Under these assumptions, the PPCH calculated by BEM can be used in the model.

Content from this work may be used under the terms of the Creative Commons Attribution 3.0 licence. Any further distribution of this work must maintain attribution to the author(s) and the title of the work, journal citation and DOI.
Published under licence by IOP Publishing Ltd

2.1. Blade element model with Joukowsky modification

Principle of all BEM models is similar. The rotor is represented as a finite sum of elements with constant geometry, as shown in the figure 1. The element on the radius r with the length of dr has a constant chord length c, thickness t and aerofoil. In a general form, if the propeller is subjected to a non-axial inflow, the forces acting on the blade element depend further on azimuthal position of the blade ψ and thus discretisation along the azimuthal angle is needed as well – it is assumed that the forces acting on the blade element do not change during the increment of the azimuthal angle $d\psi$. Furthermore, the airflow over each element is considered two-dimensional with non-interference between neighbouring elements or different blades of the propeller is assumed. The validity of this assumption was proven e.g. in [7]. Each blade element is, as per Joukowsky theorem, represented as a system of vortices. The fundamental parameter for the calculation is circulation. The speed of the flow, induced by the propeller, can be calculated from the value of the circulation, i.e. circulation defines the speed of the flow. The relation of the local inflow velocity W_l is shown in the figure 2. The inflow velocity V is acting on all blade sections with the same magnitude. However, the circumferential component $U = r\Omega$ changes with the position of the blade element along the blade span and the velocity triangle alters as well.

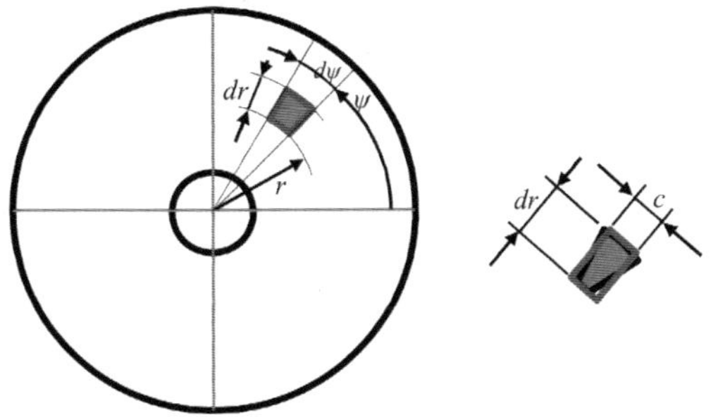

Figure 1. Division of a rotor into finite number of blade elements [8].

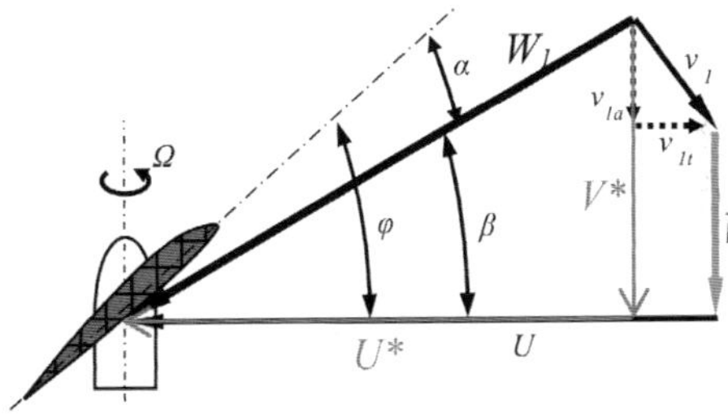

Figure 2. Velocity triangle of Joukowsky theorem acting on a blade element [9].

The local inflow angle β is defined as a subtraction of the pitch angle φ and the local angle of attack (AoA) α. It can be shown, that the local inflow angle can be calculated as:

$$\beta = \tan^{-1}(V^*/U^*), \tag{1}$$

where V^* is a sum of inflow velocity V and axial component of the induced velocity v_{1a} and U^* is a sum of circumferential velocity and a tangential component of the induced velocity v_{1t}. Even though the inflow velocity in hover is equal to zero, the axial component of the induced velocity can be non-zero, resulting in a non-zero local inflow angle β. Furthermore, it can be shown that both axial and rotational component of the local inflow velocity W_l are a function of local circulation Γ. The value of local circulation can be derived as:

$$\Gamma = \frac{1}{2} c_L c \frac{U^*}{\cos(\beta)} \, . \tag{2}$$

The local circulation is directly proportional to the local chord length c and local value of lift coefficient c_L. The local value of the lift coefficient is then depended on the local value of AoA and, therefore, an iterative solution to the equations is required. Also, the lift curve for each aerofoil is required as well.

Result of the BEM calculation is propeller performance characteristics – dependence of thrust coefficient c_T, power coefficient c_N and efficiency η on the advance ratio λ. Thrust and power coefficients can be defined as follows:

$$c_T = \frac{T}{\rho n^2 D^2}; \tag{3}$$

$$c_N = \frac{N}{\rho n^3 D^5}. \tag{4}$$

T and *N* are thrust or power consumed by the propeller, while *n* is its revolution and *D* is the diameter of the propeller. Advance ratio is a ratio between tangential speed of the tip of the blade and the inflow velocity. It can be expressed as:

$$\lambda = \frac{V}{nD}. \tag{5}$$

The efficiency is, in this case, a ratio between thrust and power coefficient, multiplied by the advance ratio:

$$\eta = \frac{c_T}{c_N}\lambda. \tag{6}$$

BEM with Joukowsky modification is further detailed in [10].

2.2. *Impact of Reynolds number to input aerofoil data for BEM calculation*

The non-constant distribution of both aerofoil thickness on a propeller with general geometry and local inflow velocity along the blade span, the value of local Reynolds number can change significantly. This phenomenon is demonstrated in figure 3. As shown in the figure 4, for large Reynolds numbers (500000+) this is not a problem. However, for Reynolds numbers below 500000 the drag of the aerofoil starts considerably increasing and the aerofoil polars can change significantly even with a relatively small change of the Reynolds number. At Reynolds number of 50000, a distortion can be observable. At some point, the drag of the aerofoil increases significantly, indicating stall. However, after further increase of the AoA, the flow over the aerofoil fully recovers and then the aerofoil polar behaves as expected. This highly atypical behaviour could be attributed to convergence problems. At these very low Reynolds numbers, similar behaviour was observed on e.g. Hofsaess Espada aerofoil [11]. Design of aerofoil, which can operate efficiently at these low Reynolds numbers is described in [12].

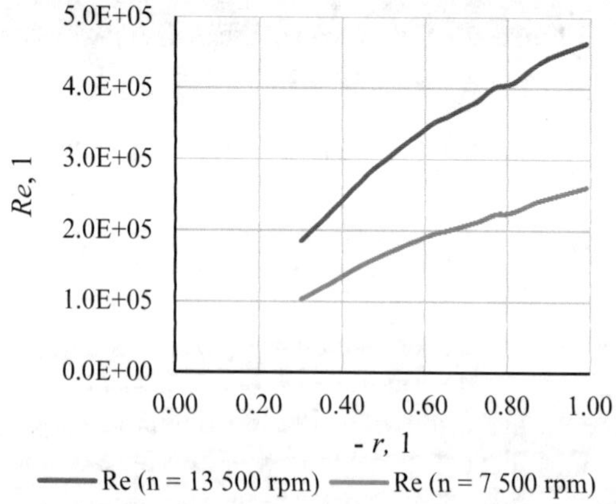

Figure 3. Calculated distribution of Reynolds numbers along the blade span for two different rotational speeds.

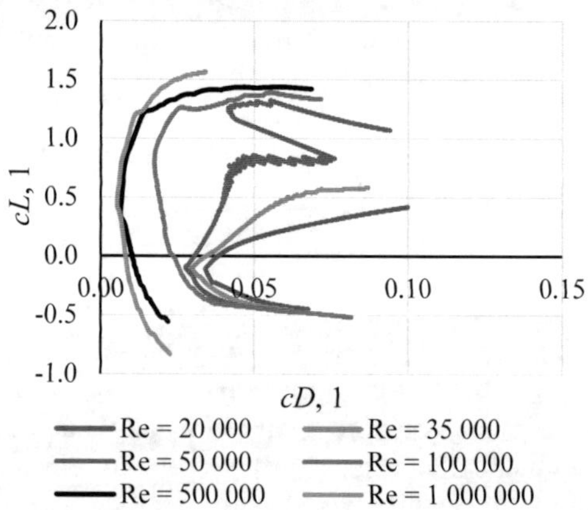

Figure 4. Aerofoil polars of Clark-Y as calculated by XFOIL.

2.3. *Calculation of input aerofoil polars*

Input Aerofoil data, namely aerofoil lift curve and polar are critical for the correctness of the BEM calculation. However, as demonstrated in previous chapter, obtaining them is perhaps the most difficult step of the calculation. While it is possible to measure each polar, it is impractical and expensive due to the sheer amount of polars required. As demonstrated in [9] or in [13], XFOIL or CFD can be used to approximate the aerofoil polars well. Comparison is shown in figure 5, where both XFOIL and CFD fail to correctly estimate the behaviour close to the stall conditions, otherwise the fit seems to be good enough.

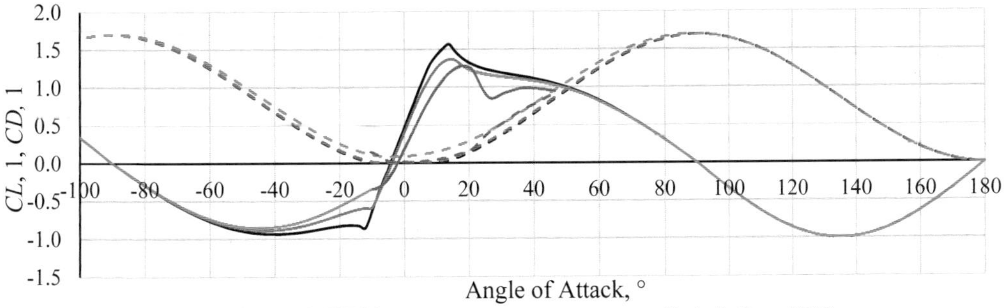

Figure 5. Extended data of NACA 4412 aerofoil to ±180°, $AR = \infty$ and $Re < 10^6$.

Often the input polars need to be extended to include post-stall data. While it is possible to obtain the data numerically, typically the extrapolation is accurate enough. Different methods of extrapolation are discussed in [14]. All calculations presented in this paper included the Viterna-Corrigan extrapolation [15].

2.4. *Corrections to the calculation*

Different corrections to the BEM model are described in [8] or [16]. Typically, Mach number correction, stall delay corrections, rotational augmentation corrections and surface roughness corrections should be performed before running the BEM calculation. As a general rule of thumb, it is always advisable to measure the propeller performance at least in a static regime, because even with the corrections, the BEM typically overestimates the propeller performance in static regime (advance ratio λ equal to 0) by about 10-15%. This overestimation can be improved by introducing a tip-loss factor to the calculation, as discussed in [2].

2.5. *PPCH as an input for maximum endurance calculations*

To decrease the calculation time of the maximum endurance, the PPCH can be inputted as a map of thrust and power coefficient for different revolutions to compensate for effects of different Reynolds numbers. The thrust and power coefficient for required rpm can be calculated as an interpolation between different maps. The example of such map is shown in the figure 6. The thrust coefficient is plotted against advance ratio λ and incidence angle θ. In hover, both the incidence angle and advance ratio are equal to zero. In regimes, where thrust coefficient is negative, the propeller no longer produces thrust, but is braking. This regime is not suitable for propeller operation.

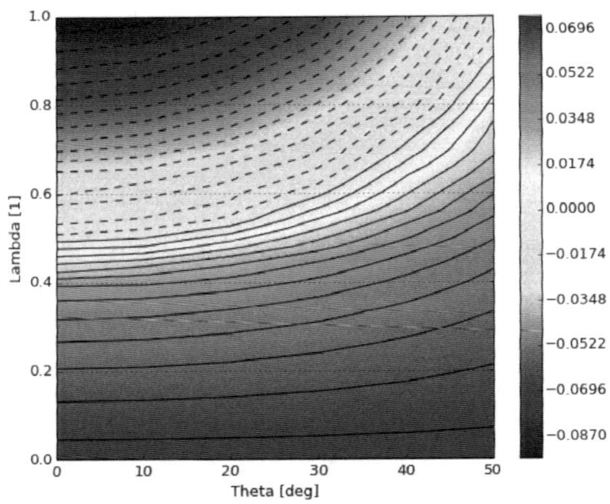

Figure 6. Map of thrust coefficient.

3. Motor performance characteristics

For purposes of this paper, an SUAV powered with an BLDC electric motor is calculated. For SUAV to sustain hover under aforementioned assumptions, the propeller needs to maintain its revolutions. However, as the battery discharges, the voltage in the system drops and the controller needs to increase the throttle (or PWM) on all motors. Furthermore, the propeller is directly mounted to the electric motor and its revolution are equal with no losses in transmission. The efficiency of the electric motor η_{mot} is defined as a ratio between output mechanical power P_{OUT} and consumed electrical power P_{IN}. Measured motor efficiency for different input voltages is shown in figure 7. The initial drops in the efficiency are caused by the heating effects. The measurement started with cold motor and constant load. As the motor heated up, the internal resistance increased as well and at the same PWM, both the efficiency and speed dropped. Then, the load on the motor was decreased and the typical motor efficiency was measured.

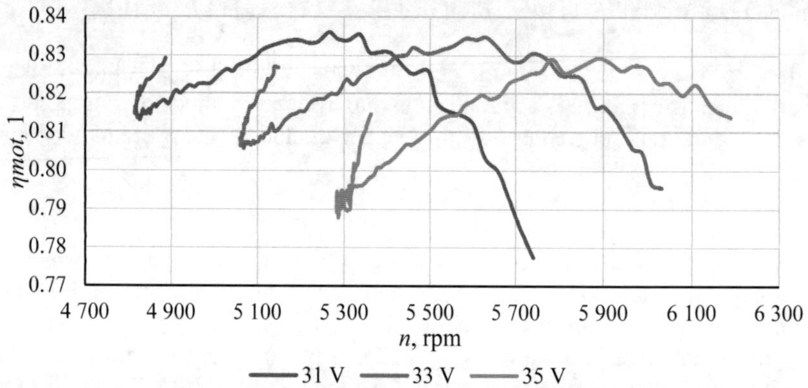

Figure 7. Measured efficiency of an electric motor.

4. Battery discharge characteristics

For purposes of this paper, a nickel-manganese-cobalt (NMC) Li-Ion battery was assumed. During the discharge process, the voltage on the battery drops. The rate at which it drops is shown in the figure 8 in battery discharge characteristics.

The maximum voltage on commercially available battery cells is 4.25 V. Also, if the voltage on cell drops below 2 V, the battery suffers a permanent damage. Therefore, a cut-off voltage is typically selected, at which the battery is indicated as empty. There are two factors contributing to the value of cut-off voltage: battery protection and minimal revolutions, at which the SUAV can sustain flight. Choosing a correct value of cut-off voltage can significantly affect the maximum endurance of an SUAV. E.g. if the SUAV was equipped with a battery presented in figure 8 and the cut-off voltage was set to 3.1 V because of e.g. motor limitations, at larger loads it would not be possible to drain all the available capacity of the battery. I.e. not only the battery has limited energy available, this amount can be proportionally decreased by incorrect value of cut-off voltage.

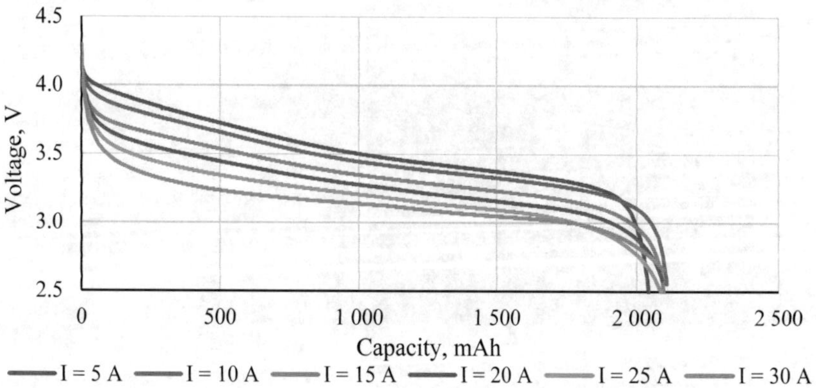

Figure 8. Measured battery discharge characteristics.

Other factors, which can affect the battery performance are ambient conditions (temperature), investigated e.g. in [17], or number of discharge cycles of each cell during their life cycle, as described in [18].

5. Maximum endurance in hover

The hover conditions can be solved by solving the equations of motion (SUVAT) simplified for a single axis problem (no disturbances and perfect symmetry assumptions). The scheme of the calculation for a given time step with a cut-off voltage of 3.2 V is shown in figure 9.

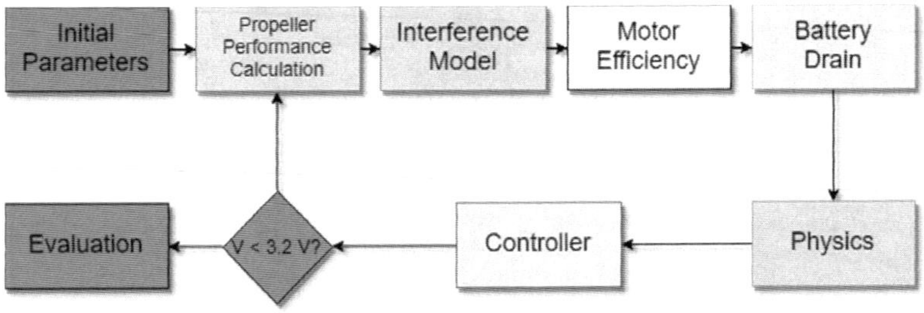

Figure 9. Algorithm of the calculation.

At first, the initial parameters need to be set (number of propellers, propeller performance input maps, weight of the SUAV, initial rpm, etc.). For given parameters, thrust and power coefficient is calculated. Then, the thrust of the propeller is adjusted for the obstacles in the downwash (mounting legs of the SUAV). After that, the consumed power by the propeller is corrected for motor efficiency at given voltage and PWM. If the motor cannot produce the required power, the thrust of the propeller is recalculated for the maximum available power. Then, the battery voltage drop is calculated. "Physics" represents the calculation of SUVAT. Furthermore, while it is possible to find the required revolutions of the propeller by directly solving this system, it is much easier to use controller with multiple PID controllers to stabilize the system. Because of the high instability at the beginning of the calculation, which depends on the "quality" of the initial guess of the revolutions, the maximum endurance is calculated only after the SUAV stabilization. The result of such calculation is shown in figure 10.

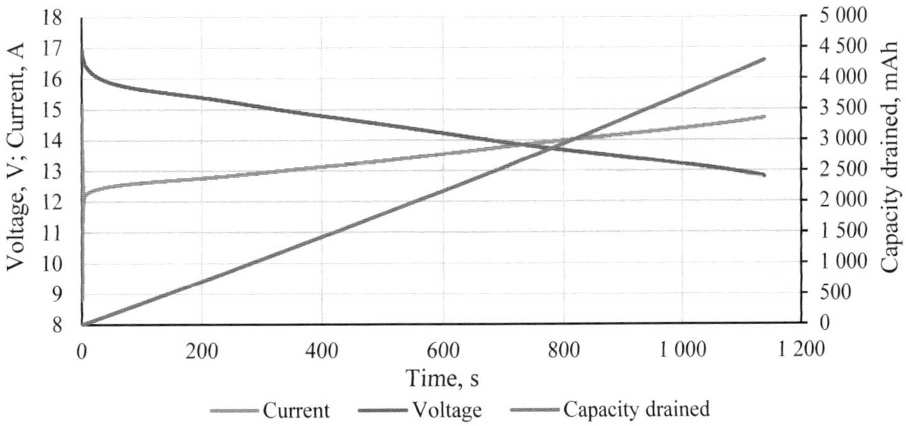

Figure 10. Algorithm of the calculation.

In the first phase, the stabilization can be seen. Then the current in the system proportionally increases, as the voltage decreases. The overall amount of capacity drained is non-linear and as the battery is drained more and more, the efficiency of the motor decreases, causing the increase rate of the consumed energy.

6. Comparison of measured and calculated results

The measurement of maximum endurance of Yuneec Typhoon H, presented in [19] and showed in figure 11, served as the source for the validation of the computational program. Yuneec Typhoon H is a hexarotor with fixed-pitch propellers. The measurement was performed at different altitudes with variable weight. The results are presented in the table 1.

Figure 11. Yuneec Typhoon H [19].

Table 1. Comparison of Measured and Calculated Maximum Endurance.

Altitude, m.a.s.l.	Weight, g	Measured, mm:ss	Calculated, mm:ss
120	1,950	19:30	19:33
183	1,950	19:00	19:20
1,000	1,950	18:00	18:47
370	2,165	18:10	17:33

In most cases, the calculation overpredicted the overall maximum endurance. However, with the increased maximum weight caused the calculation to significantly lower the maximum endurance, more than it did in real life. Furthermore, the drop of the maximum endurance was more significant with the increased altitude as the calculation suggested. The resulting maximum endurance values seems to be within margin of error considering the simplifications of the model.

7. Conclusion

The paper presents a simple method of combining measured or calculated characteristics of different components of an SUAV of a multirotor type to calculate its maximum endurance. The propeller performance characteristics calculation via Joukowsky BEM method was presented, along with references to necessary corrections. Also, the motor power curves for different voltages at different RPM were discussed. Finally, battery discharge characteristics were introduced as well, with highlight of the importance of cut-off voltage.

The model itself is simple, though despite all simplifications, it produces relatively good results. However, it is computationally expensive. Furthermore, any dynamical changes in the inflow velocity vector are outside of the scope of this model. Corrections for incident inflow to BEM calculation are highly advisable.

Comparison of measured and calculated results was promising. In most scenarios, the calculation slightly overestimates the propeller performance, but the results are within expected range.

Acknowledgment
This work was completed thanks to the successful cooperation with company „TECHNOFIBER“.

References

[1] Hnidka J, Rozehnal D and Nguyen B V 2019 Impact of non-zero incident angle to propeller performance characteristics and its BEM model *Engineering Mechanics* **25** 149–152

[2] Hnidka J 2018 *Determination of the Flight Performance of a Hybrid-Wing UAV* (Brno: University of Defence)
[3] Moriarty P J and Hansen A C 2005 AeroDyn Theory Manual *Office of Scientific and Technical Information* (OSTI)
[4] Suzuki A 2000 *Application of Dynamic Inflow Theory to Wind Turbine Rotors* (Salt Lake City: D. Bell & Howell Information and Learning)
[5] Panayotov F, Dobrev I, Massouh F and Todorov M 2017 Studying the wake contraction of the flow-field of a rotor in hover *MATEC Web of Conferences* **133** 01009
[6] Panayotov F, Dobrev I, Massouh F and Todorov F 2017 Numerical model for rapid computatuon of the flow-field of a rotor in hover *Int. Conf. Military Technologies* 570–577
[7] Lock C N H, Bateman H and Townend H C H 1924 Experiments to Verify the Independence of the Elements of an Airscrew Blade *Aeronautical Research Committee: Reports and memoranda* (London: H.M. Stationery Office)
[8] Hnidka J Rozehnal D and Nguyen V B 2019 Evaluation of BEM calculation method used for low-reynolds propellers *Int. Conf. Military Technology* 520–525
[9] Rozehnal D, Nguyen V B and Hnidka J 2018 Hysteresis assumptions for propeller in yaw in close stall conditions *Engineering Mechanics* **24** 721–724
[10] Okulov VL, Sørensen JN and Wood DH 2018 The rotor theories by Professor Joukowsky: Vortex theories *Progress in Aerospace Sciences* **73** 19–46
[11] Koike M and Wuerz W 2003 Aerodynamic Characteristics of the 'Hofsass Espada' *The National Free Flight Society Symposium*
[12] Selig M 2003 Low Reynolds Number Airfoil Design Lecture Notes *VKI Lecture Series* 24-28
[13] MacNeill R and Verstraete D 2017 Blade element momentum theory extended to model low Reynolds number propeller performance *The Aeronautical J.* **121(1240)** 835–57
[14] Nguyen V B and Rozehnal D 2016 Determine 360° polar for calculation of wind turbines *Transfer* **16** 1–7
[15] Viterna L A and Corrigan R D 1982 Fixed pitch rotor performance of large horizontal axis wind turbines *Energy Production and Conversion* 69–85
[16] Branlard E 2011 *Wind Turbine Tip-Loss Corrections* (Riso: DTU)
[17] Aris A M and Shabani B 2017 An experimental study of a lithium ion cell operation at low temperature conditions *Energy Procedia* **110** 128–35
[18] Marques P, Garcia R, Kulay L and Freire F 2019 Comparative life cycle assessment of lithium-ion batteries for electric vehicles addressing capacity fade *J. Cleaner Production* **229** 787–94
[19] Žárský P 2017 *Využití multikoptér v podmínkach AČR* (Brno: University of Defence)

IOP Conf. Series: Materials Science and Engineering **664** (2019) 012003 doi:10.1088/1757-899X/664/1/012003

Numerical study of a shock-wave boundary-layer interaction: compression ramp flow-fields

O A Elsayed

International University of Rabat, School of aerospace engineering, LERMA Lab, Campus UIR Parc Technopolis, Rocade, Rabat-Sale, 11100 – Sala Al Jadida, Maroc

E-mail: omer.almatbagi@uir.ac.ma

Abstract. Shock-wave boundary-layer interaction (SBLI) causes a serious aerodynamic and structural problems, therefore understanding its complex flow phenomenon is essential. Flow properties at the interaction region were numerically predicted via solving steady, compressible, 2D Navier-Stokes Equations with Roe-FDS flux type and a second order upwind flow discretization. Two compression ramps with 12° and 24° angles were investigated. The flow over a 12° ramp were fully attached with a λ-shock type and a limited upstream influence. Wall temperature increases to 2.55 times the edge temperature value and affected region increases with the downstream distance. The pressure field for the 24° ramp case reveals a substantial increase in the upstream influence due to the strong shock. A larger shock-leg at the wall provoke a significant flow separation and the wall temperature increases to 2 times the edge temperature value. Separation bubble is generated at the corner, with a separation length of 3.25 times the boundary layer thickness. This leads to a noticeable increase in the total pressure drag. Simulated results were fairly agreeing with the experimental result at the separation region.

1. Introduction

Shock wave turbulent boundary layer interaction and shock reflection results in a complex flow structure. Interaction has an adverse consequence on the transonic, supersonic and hypersonic flights and must be considered at the early vehicles design stages. SBLI influence aircraft external structure via losses of controllability and extreme heat and mechanical surface loading in an unsteady manner. Also, it increases the pressure losses and distortion in the internal flow. Viscous-inviscid interactions will take place at this flight regime and basically leads to flow separation, vorticial flow structures, turbulence intensification and structure alterations. Research of the SBLI started in the mid of the twentieth century.

Inside the boundary layer shock bends as it meets a relatively low Mach numbers, breaks into compression fan and shock reflection may be generated [1]. Further if the shock is strong enough boundary layer may separate. The compression corner flow has a single major shock associated with the ramp. In the mid-1990s extra experimental measurements were conducted to provide validation data for numerical codes [2].

The standards chosen were that the data be turbulent, on relatively simple geometries, applicable in the sense that they are good enough to explore turbulence modelling, those studies shows that experimental boundary conditions and error bounds, were self-consistent and adequately resolved. Similar researches of surface pressure, field Pitot survey, skin friction and heat transfer at Mach 8.3 on impinging shock, single-fin and double fin interactions were conducted [3-4]. In the same research area [5-6] published skin friction coefficient results for the double fin. These data were useful to predict

Content from this work may be used under the terms of the Creative Commons Attribution 3.0 licence. Any further distribution of this work must maintain attribution to the author(s) and the title of the work, journal citation and DOI.
Published under licence by IOP Publishing Ltd

surface properties using turbulence modelling aspects. [7] compile RANS result from simulations for numerous 2D and 3D interactions.

The models started from algebraic to full Reynolds Stress Equation formulations for the single fin, double fin and the hollow cylinder flare. They concluded that skin friction and heat transfer predictions were poor, with up to 100 percent difference for the strong interactions showing a substantial region of separation. There are still significant quantities that cannot be predicted correctly, such as peak heating in strong interactions [8].

Direct numerical simulation codes were used by [9] to study the SBLI on a flow at Mach 2.25 with an impinging oblique shock wave with a shock angle of 33.2°. The study reveals that due to the vortex-shedding near separation point, along with the vortical structures diffusion in the mixing layer, a pressure waves was produced due to shock interaction at the foot of impinging shock, and upstream propagation of acoustic frequencies of cavity tones, caused the low-frequency unsteadiness. [10] evaluated SBLI unsteadiness by studding a shock reflection using plate angles of incidence from 7° to 9.5° with a free stream flow of Mach 2.3. Experiments reveals fluctuation shock motion produces a relatively lower frequency than the incoming boundary layers characteristic frequencies. The shockwave acted as a low pass filter, which selected only the lower frequency excitation. Velocity fields was measured in case of shock impinging boundary layer at Mach 2.3 on two wedges with angles 8° and 9.5° using particle image velocimetry [11]. It reveals that flow reattachment downstream the shack wave depends on the fluid properties in the mixing layer. The low frequencies of the separation shock were associated with the separation bubble dilatations and contractions.

The dominant frequencies for the 8° and 9.5° configuration were close to the theoretical value. The calculated Strouhal number agreed with the values obtained in experiments conducted by [12] and [13] for different configurations. [14] in his review paper declared that, correlations and coherent structures in the separation bubble, were successfully described on nominally 2D interactions. Heat transfer can be measured and, in several cases, predicted, accurately for axisymmetric laminar situations. [15] simulated Mach 2 turbulent compression corner flows with 22.5° and 37.5° sweep angle and found no indications of a low-frequency unsteadiness. They argue that compared to unswept interactions for swept interactions the reverse flow intensity is not a good indicator for the onset of instability since disturbances are convected outward in the spanwise direction. [16] carried out simulations of a hybrid turbulence model with 1-equation renormalization group turbulence model as well as LES with wall-adapting local eddy-viscosity model of turbulent swept and unswept interactions. The simulations revealed ripples of shock separation for the unswept case that appeared weakened in the presence of sweep.

In this work a steady 2D numerical simulation of SBLI flow-field phenomenon at supersonic flow regime was conducted. Shock-wave boundary-layer interaction flow-filed features, such as pressure loads (reflecting the shape and criteria of the SBLI), velocity profiles, wall shear stress distribution and heat transfer were predicted. All those considerations are vital for supersonic engine inlet design, control surfaces functionality and understanding of the internal supersonic nozzles flow.

2. Mathematical Model

Navier-Stokes equations were solved numerically. Spalart-Allmaras turbulence model was selected to account for turbulence. The equations for conservation of mass, momentum and energy are given below:

$$\frac{\partial \rho}{\partial t} + \nabla (\rho \vec{v}) = 0; \quad (1)$$

$$\frac{\partial}{\partial t}(\rho \vec{v}) + \nabla (\rho \vec{v} \vec{v}) = -\nabla p + \nabla (\bar{\bar{\tau}}) + \vec{F}, \quad (2)$$

where , $\bar{\bar{\tau}}$ and F are static pressure, stress tensor and surface forces, respectively. The stress tensor $\bar{\bar{\tau}}$ is below shown:

$$\bar{\bar{\tau}} = \mu \left[(\nabla \vec{v} + \nabla \vec{v}^{T}) - \frac{2}{3} \nabla \vec{v} I\right], \quad (3)$$

where μ and I are the molecular viscosity and unit tensor respectively. Energy transport equation is:

$$\frac{\partial(\rho E)}{\partial t} + \nabla\left[\vec{v}(\rho E + p)\right] = \nabla\left[k_{eff}\nabla T + \left(\bar{\bar{\tau}}_{eff}\vec{V}\right)\right], \quad (4)$$

where $k_{eff}\nabla T$ is the conduction and $\bar{\bar{\tau}}_{eff}\vec{V}$ is the viscous dissipation terms. Energy E per unit mass is defined as:

$$E = h - \frac{p}{\rho} + \frac{v^2}{2}, \quad (5)$$

where h is enthalpy. The viscous stress tensor τ_{ij} is related to the Reynolds stresses through the eddy viscosity μ_t, effectively modelling the momentum transfer by turbulent eddies, written as

$$\tau_{ij} = 2(\mu_1 + \mu_t)(s_{ij} - \frac{1}{3}\frac{\partial u_k}{\partial x_k}\delta_{ij}), \quad (6)$$

where δ_{ij} is Kronecker delta and s_{ij} is the strain rate tensor given by

$$s_{ij} = \frac{1}{2}(\frac{\partial u_i}{\partial x_j} + \frac{\partial u_j}{\partial x_i}). \quad (7)$$

The Spalart-Allmaras turbulence model is effective for flows at high Reynolds number and shows good results for boundary layers subject to adverse pressure gradients. Spalart-Allmaras model was used in this study to consider the effect of turbulence. The SA model utilizes one equation to solve the transport variable where $\tilde{v}$ is identical to the turbulent kinematic viscosity except in the near wall region. The transport equation for $\tilde{v}$ is:

$$\frac{\partial}{\partial t}(\rho\tilde{v}) + \frac{\partial}{\partial x_i}(\rho\tilde{v}u_i) = G_v + \frac{1}{\sigma\tilde{v}}\left\{\frac{\partial}{\partial x_j}\left[(\mu + \rho\tilde{v})\frac{\partial\tilde{v}}{\partial x_j}\right] + C_{b2}\rho\left(\frac{\partial\tilde{v}}{\partial x_j}\right)^2\right\} - Y_v + S_{\tilde{v}}, \quad (8)$$

where G_v is turbulent viscosity, and Y_v is the near wall turbulent viscosity destruction, $\sigma_{\tilde{v}}$ and C_{b2} are the constants and v is the molecular kinematic viscosity. $S_{\tilde{v}}$ is the source term.

3. Numerical Setup

Shock wave originates at a discontinuity in the wall direction, through which the supersonic flow experiences a deflection equal to the corner angle. The flow structure depends on the corner angle, therefore two ramps with a 12° and 24° angles were selected for this numerical study. The structure meshes generated over the two-dimensional domain of total length L = 0.6 m and ramp located at L = 0.2 m from leading edge. The mesh for 12° and 24° ramp cases consisted of 8891 nodes and 17790 faces as shown in figure 1.

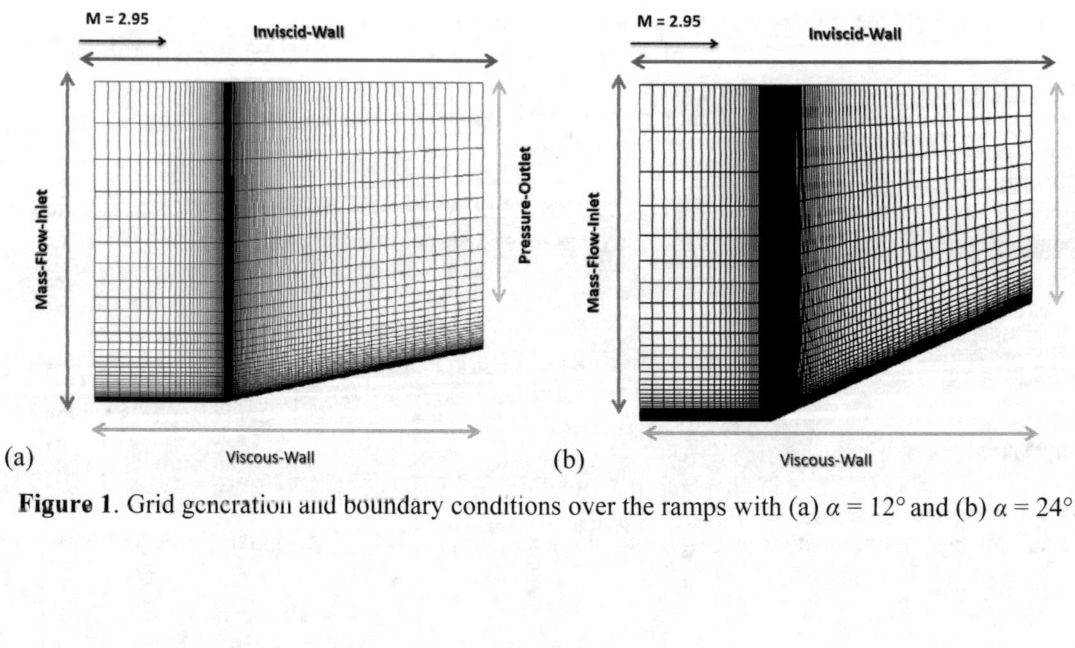

Figure 1. Grid generation and boundary conditions over the ramps with (a) α = 12° and (b) α = 24°.

Grids were clustered within the boundary layer and at the expected shock regions to ensure a proper flow resolution. The values of *Y* plus, $y+ = \frac{y}{\mu}\sqrt{\rho\tau_w}$ in the wall-adjacent cells ranges between 0.3-0.6 and 0.15-0.35 for the 12° and 24° corner angles respectively. Those ranges indicate a laminar sublayer resolution was achieved. Flow domain boundary conditions are illustrated in figure 1.

4. Results and discussions

Results of a supersonic flow at free-stream Mach number $M = 2.95$, Reynolds number based on total length equal to $R_{eL} = 3.9\times10^7$ and boundary layer thickness at the edge corner $\delta_{i\,(L=0.2)} = 2.79$ mm are discussed below.

4.1. Wall Pressure Distribution

Pressure distribution over the 12° and 24° ramps are shown in figure 2 (a) and (b) respectively. For the case of 12° the pressure at the wall starts to rise by the upstream propagation mechanism reach a value of 2.35 times the free stream value. This result shows a very good agreement with the theoretical value which found to be equal to 2.3136. This trend indicates a weak shock-boundary layer interaction.

Figure 2 (b) shows the pressure distribution of the 24° ramp. The pressure curve exhibit three inflection points. Those respectively associated with separation, the onset of reattachment and the reattachment compression. This result indicates a good agreement with experimental results of [17-18] and excellent agreement with inviscid solution outside the interaction region.

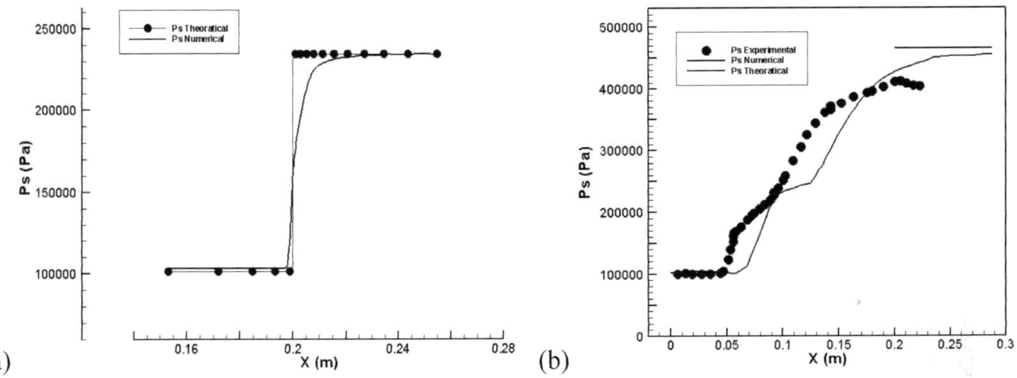

Figure 2. Pressure distribution over a 12° and 24° ramps.

4.2. Pressure contours over the flow field

When a shock-wave propagates through a boundary-layer, it sees an upstream flow of a decreasing Mach number as it approaches the wall. The shock adapt itself and becomes vanishingly weak when it reaches regions of sonic flow. The pressure is transmitted in the upstream direction through the subsonic inner part of the boundary-layer. Thus, a pressure rise is felt upstream of the point where the shock would meet the surface.

Figure 3 (a) show the pressure distributions over the 12° ramp. It reveals that a distinct shock-wave arise from the corner location with the shock wave angle of $\mu = 31.5°$, this value is 6% higher than the theoretical value. Within the boundary layer the shock is curved due to its propagation and the entropy changes from one streamline to the other. Outside the boundary-layer, the shock is linear. The upstream influence is evidence but relatively take place over a small region in front of the ramp.

Pressure distribution for 24° corner angles case is shown in figure 3 (b). Again, a single major λ shock associated with the ramp is evidenced. A substantial increase of the upstream influence length due to an intensified shock strength is noticed along with a large spreading of the shock near to the wall is visible. This is due to the compression waves induced by the thickening of the low velocity portion of the boundary-layer. In this case the pressure rise is strong enough to encourage significant separation with a length of 9 mm, leads to a high-pressure rise.

IOP Conf. Series: Materials Science and Engineering **664** (2019) 012003 doi:10.1088/1757-899X/664/1/012003

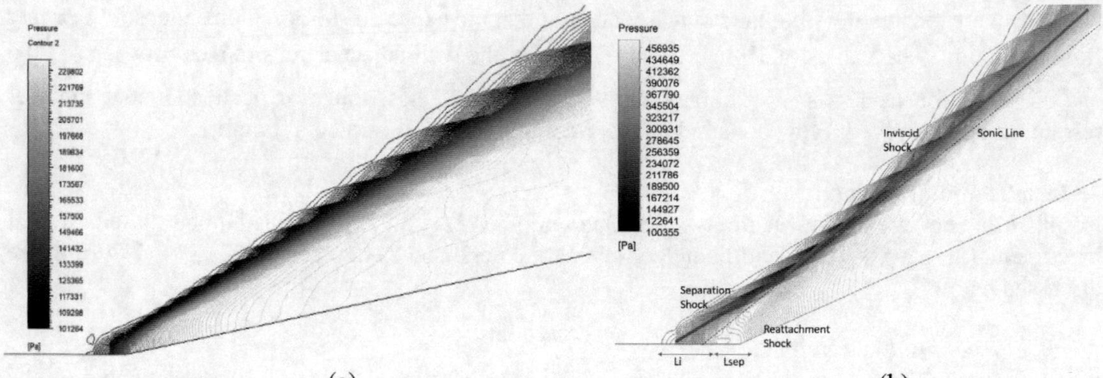

Figure 3. Static pressure contours flow over (a) 12° and (b) 24° ramps.

4.3. Velocity profile and streamlines distribution

Velocity profiles for the 12° and 24° ramps are shown in figure 4. The velocity profile of the incoming flow for both cases shows a trends of a typical turbulent velocity profile. The transition to zero velocity at the wall takes place over a very short normal distance, hence the subsonic layer is extremely thin, and the shock originates from a region very close to the wall. Incoming turbulent boundary layer, δ_i is found to be 2.77 mm. Right behind the shock, the boundary layer thickness increases to 3.3 and almost an order of magnitude for the 12° and 24° ramps, respectively. which indicates a large region of separation, especially for the 24° ramp. For the latter case, the pressure rise is strong enough to provoke significant flow separation clearly shown by the reverse flow portion near the ramp leading edge as illustrated in figure 5 (a). Streamlines distribution of the flow shown in figure 5 (b) illustrates clearly formation of separation bubble stating at point *S*, with a length of L_{sep} equal to 9 mm, generated at the corner. Flow reattachment is taking place right behind this bubble at point *R*.

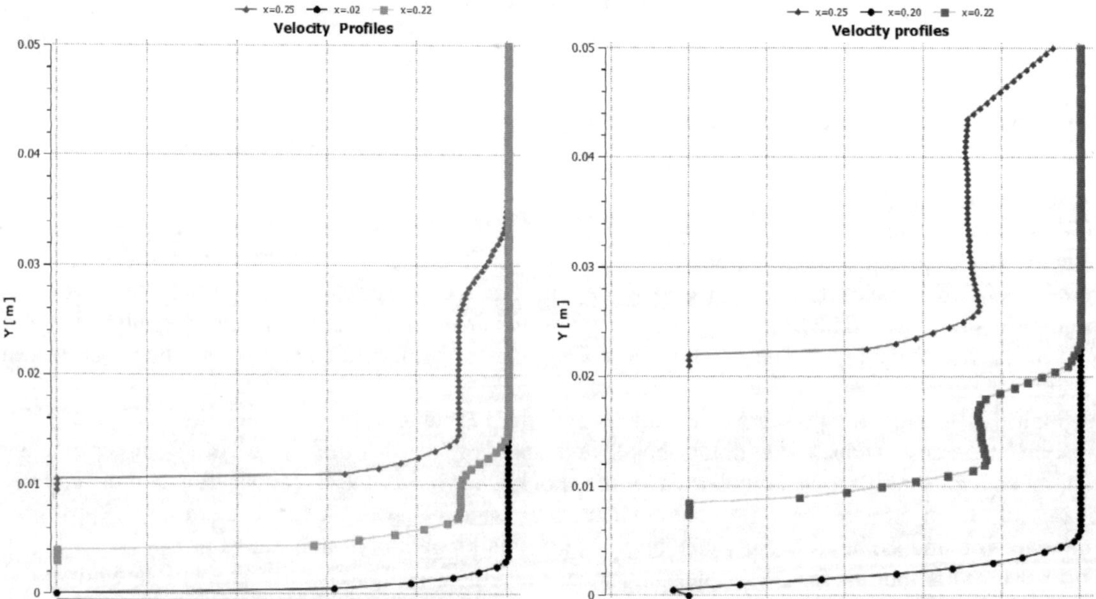

Figure 4. Velocity profiles over the 12° and 24° ramps.

IOP Conf. Series: Materials Science and Engineering **664** (2019) 012003 doi:10.1088/1757-899X/664/1/012003

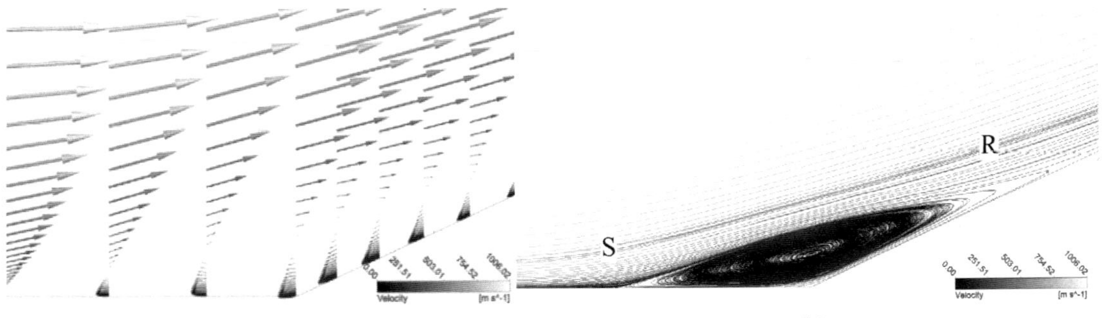

(a) (b)

Figure 5. (a) Velocity profiles and (b) streamlines at the ramp leading edge for the case of 24°.

4.4. *Heat transfer and near wall temperature distributions*

Wall adjacent temperature profiles at the corner region is presented for 12° and 24° ramps in figure 6. The near wall temperature is directly reflected into heat transfer to the wall as well as the skin friction at the wall. Temperature profiles at the corner region for 12° ramp shows an increment at an order of 2.55. Affected region is increasing with the downstream distance. Same trend is evidence for 24° ramp case with smaller order of increment due to flow separation.

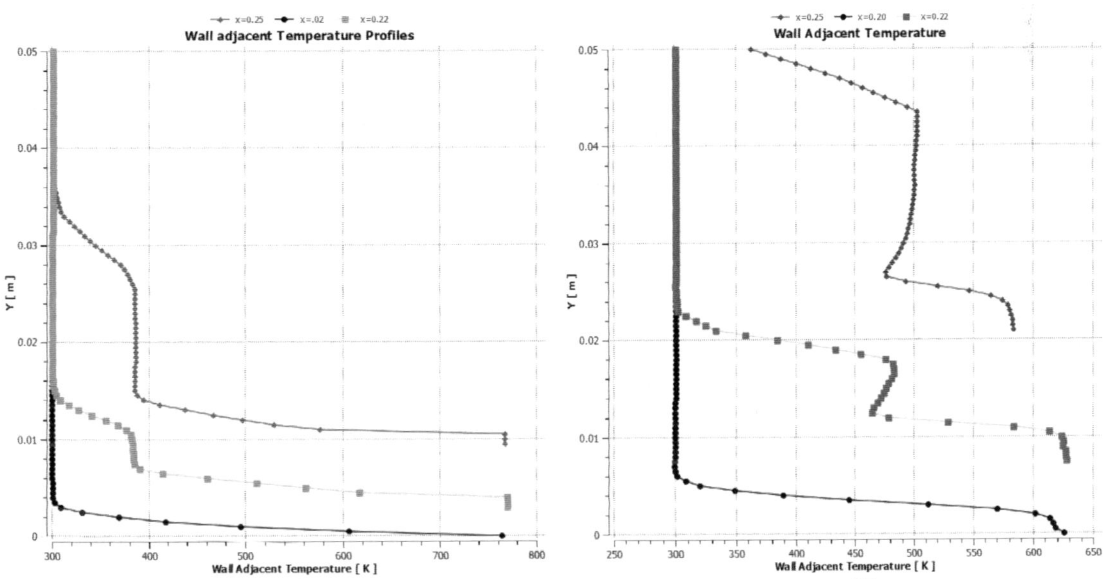

(a) (b)

Figure 6. Wall adjacent temperature for (a) 12° and (b) 24° angles ramps.

4.5. *Skin friction coefficient*

Skin friction coefficient of the 12° and 24° ramps are shown in figure 7 (a) and (b) respectively. For the case of 12° ramp the skin friction coefficient decreases to its minimum value in front of the corner where $C_f = 0.00015$ at the ramp corner. This value grows sharply up to 0.0025 at the shock region at the corner leading edge and remains nearly constant over the rest of the ramp. This skin friction coefficient distribution reveals attached flow at the corner. While for 24° ramp the skin friction coefficient plummets to the value of $C_f = 0.0005$ near separation point *S*, then reverse its direction in a rather complex variation and further decreases to zero indicating reattachment of the flow *R*. This complex variation indicates flow separation at the corner. Further skin friction coefficient recovers to a value 0.003 reattachment point.

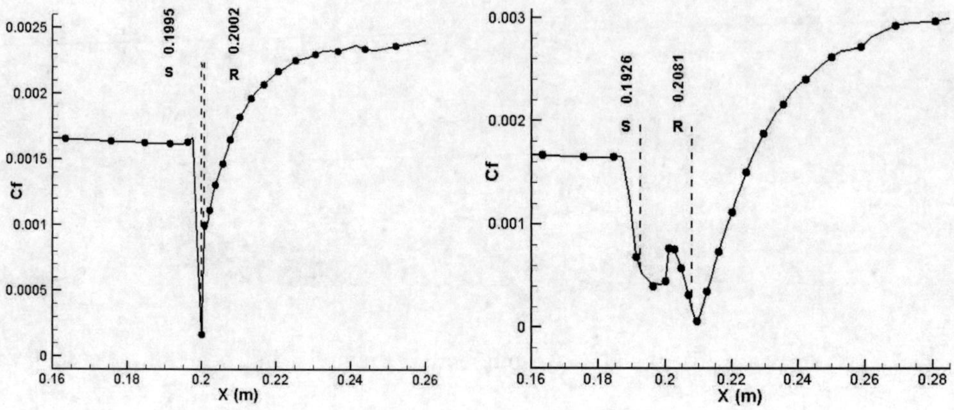

(a) (b)

Figure 7. Surface shear stress distributions turbulent flow over a 12° ramp at $M = 2.95$.

5. Conclusion

Turbulence, compressibility and viscous-inviscid interaction phenomena are present in the SWBLI flow. For all that, numerical simulation is providing a further insight of those phenomena. SWBLI is numerically simulated for a compression ramp flow with corner angles of 12° and 24°. Propagation through a boundary layer results in a λ shape curved shock for the case of 12° corner with fully attached flow at the corner. Temperature profiles shows an increment of 2.55 times the edge temperature value. Affected region is increasing with the downstream distance. For the 24° ramp the upstream influence is relatively higher due to the increased shock strength with a wide spreading near the wall that provoke significant flow separation. The streamlines of the flow reveal that a separation bubble is generated at the corner, which leads to the boundary layer separation. Same wall temperature trend is evidence as for 12° ramp but, with a smaller increment due to flow separation in this case. Simulation and experimental results in the shock impinging region are found to be in a fair agreement.

References

[1] Adamson T and Messiter A 1980 Analysis of two-dimensional interactions between shock waves and boundary layers *Annu Rev Fluid Mech* **12** 103–38
[2] Settles G and Dodson L 1994 Supersonic and hypersonic shock/boundary-layer inter- action database *AIAA J.* **32(7)** 1377–83
[3] Kussoy M and Horstman K 1991 Documentation of two-and three-dimensional shock-wave/turbulent-boundary-layer interaction flows *Technical Report 103838 (NASA, Ames Research Center, Moffet Field, California 94035-1000)*
[4] Kussoy M, Horstman K and Horstman C 1993 Hypersonic crossing shock-wave/turbulent-boundary-layer interactions *AIAA J.* **31(12)** 2197–203
[5] Garrison T, Settles G and Horstman C 1996 Measurements of the triple shock/wave turbulent boundary-layer interaction *AIAA J.* **34(1)** 57–64
[6] Garrison T and Settles G 1993 Laser interferometer skin-friction measurements of crossing-shock wave/turbulent boundary-layer interactions *AIAA J.* 93-3072
[7] Knight D and Degrez G 1997 Shock wave boundary layer interactions in high Mach number flows – a critical survey of current CFD prediction capabilities *Tech. Rep. AR-319* **2** AGARD
[8] Dolling D 2001 Fifty years of shock wave/boundary layer interaction research: what next? *AIAA J.* **39(8)**
[9] Pirozzoli S and Grasso F 2006 Direct numerical simulation of impinging shock wave/turbulent boundary layer interaction at M=2.25 *Physics of Fluids* **18**
[10] Dussauge J P, Dupont P and Debieve J F 2006 Unsteadiness in shock wave boundary layer interactions with separation *Aerospace Science and Technology* **10** 85–91

[11] Piponniau S, Dussauge J P Debieve J F and Dupont P 2009 A simple model for low-frequency unsteadiness in shock-induced separation *J. Fluid Mechanics* **629** 87–108
[12] Erengil M E and Dolling D S 1991 Unsteady wave structure near separation in a Mach 5 compression ramp interaction *AIAA J.* **29(5)**
[13] Wu M and Martin P 2008 Analysis of the shock motion in shockwave and turbulent boundary layer interaction using direct numerical simulation data *J. Fluid Mechanics* **594** 71–83
[14] Datta V G 2015 Progress in shock wave/boundary layer interactions *Progress in Aerospace Sciences* **72** 80–99
[15] Adler M C and Gaitonde D V 2017 Unsteadiness in swept-compression-ramp shock/turbulent-boundary-layer interactions *55th AIAA Aerospace Sciences Meeting* 987
[16] Gross A, Little J C and Fasel HF 2018 Numerical investigation of shock wave turbulent boundary layer interactions *AIAA* 1807
[17] Settles G S, Vas I E and Bogdonoff S M 1976 Details of a shock-separated turbulent boundary layer at a compression corner *AIAA J.* **14** 1709–15
[18] Settles G S, Fitzpatrick T J and Bogdonoff S M 1979 Detailed study of attached and separated compression corner flow fields in high Reynolds number supersonic flow *AIAA J.* **17(6)**

IOP Conf. Series: Materials Science and Engineering **664** (2019) 012004 doi:10.1088/1757-899X/664/1/012004

Prediction of the impact of friction's coefficient in cylindrical deep drawing for AA2090 Al-Li alloy using FEM and Taguchi approach

K Bouchaâla[1,2,4], M F Ghanameh[1,3], M Faqir[1], M Mada[2] and E Essadiqi[1]

[1] International University of Rabat, AERO school of engineering, LERMA lab. Rocade Rabat-Salé, 11 100 Sala el Jadida, Morocco

[2] Mohammed V University, Mohammadia School of Engineers, ITACS lab. Rabat, Morocco

[3] Department of mechanical conception-faculty of mechanical and electrical engineering-Damascus University, Damascus, Syria

[4] E-mail: kenza.bouchaala@uir.ac.ma

Abstract. To produce different cups from sheet metal, deep drawing is a vital process demanded by different industries such as automobile, aeronautics and so on. The comprehension of the process's mechanics led to the correct determination of the factors which affect the deep drawing results. Experimental tests and numerical simulations are two significant ways for analysing the deep drawing process. Experimental tests are beneficial in studying the process to control the defect of the final product. However, to gain time and substantially reduce costs the numerical modelling became a common method which could be investigate the deep drawing technique through all steps of deformation. The Coulomb friction in contact regions is one of the most influencing parameters during deep drawing. This research emphasizes the influence of friction coefficient on deep drawing in three different regions: between blank surface and punch, blank surface and die, and blank surface and blank holder. For this purpose, a 3D Finite Element Model of AA2090 Al-Li alloy was carried out to simulate the process using ABAQUS software with A combination of the statistical approach based on Taguchi Techniques in order to predict the impact of friction coefficient Thickness variation have been studied, and the thinning region have been analysing for a cylindrical deep drawing process the results obtained were used as imputes of Taguchi method and the most appropriate combination of the Coulomb friction in the three regions concerning AA2090 Al-Li alloy was found.

1. Introduction

Aluminium is one of the lightest sheet metals which are very demanded in the aeronautic industry to satisfy the demands of environmental protection and fuel economy. In recent years, the Al-Li alloys have attracted increasing interest worldwide for large-scale applications in the aerospace industry, mainly due to soaring fuel prices and the development of a new generation of civil and military aircraft. Lithium is the lightest metallic element, it's the only metal when it's alloyed to aluminium the density decreases and the elastic modulus improves, the addition of 1% of Li could reduce 3% in density, and nearly raises 6% in elastic modulus [1,2]. Moreover, adding 2% lithium, the density is decreased by 10%, and the modulus of elasticity is increased by 25% to 35% [3], furthermore, the aluminium-lithium alloy has high specific strength, specific stiffness, high modulus of elasticity, instead of the conventional

Content from this work may be used under the terms of the Creative Commons Attribution 3.0 licence. Any further distribution of this work must maintain attribution to the author(s) and the title of the work, journal citation and DOI.
Published under licence by IOP Publishing Ltd

high-strength aluminium alloy, the structural quality can be reduced by 10% to 20%, and the rigidity is increased by 15% to 20% [4].

Three generations of Al-Li alloys were obtained [5]. The third-generation alloys began in the late 1980s and is ongoing, and is considered to be the main structural material of the 21st century aircraft [6-8].

An important part of manufacturing is covered by sheet metal forming industry from kitchen to aircraft. Deep drawing process is one of the most widespread sheet metals forming techniques, it contains a great pact of plastic bending, sliding and unbending phenomena of the held blank between die and blank holder [9]. The deep drawing process is categorized by high efficiency and excellent surface quality of the deep drawn part.

Several factors affect the quality of the metal sheet forming during the deep drawing process such as yield criteria, the geometry of the design, the process parameters and the improper definition of these parameters led to failures like earring [10], wrinkling [11] and thinning [12]. A strong affectation of strain during formability appears due to the big contact surface with the tool compared to the sheet cross sectional region [13]. Therefore, the deep drawing manufacture of sheet metal is extremely affected by the friction condition. [14] analysed a combination between finite element method and experience to evaluate the coefficient of friction between the blank and tools. [15] studied a friction model by the friction condition, that was a generalization of non-associated sliding rule was defined using isotropic Coulomb's frictional contact law to anisotropic friction conditions. [16] used Finite element method to investigate a high frequency vibration to decrease friction. [17] analysed a theoretical approach on friction limit surfaces and sliding rules. [18] investigated the impacts of friction coefficient alteration on the required punch load for forming and thickness reduction. [19] investigated the effects of forming parameters, including lubrication type and lubrication position, on the drawing formability. [20] carried out FEM simulation and investigated the contact situation between the die and the lubricant coated steel sheet through varying the friction coefficient between the die and the lubricant layer.

The friction phenomenon and its impact on thinning in deep drawing has been studied under different coefficient of friction between blank and die, blank and holder and blank and punch.
For this purpose, finite element model has been carried out using a 3D modelled by ABAQUS software with a combination of Taguchi orthogonal test design method in order to optimize the coefficient of friction in the three different regions.

2. Numerical modelling of deep drawing

In the present paper the modelling of the deep drawing process was carried out by ABAQUS which was used as finite element software. The 3D quarter model of cylindrical deep drawing was modelled with the corresponding symmetry boundary conditions as it is shown on figure 1, the mass scaling as well was used in order to reduce the CPU time and save computer memory.

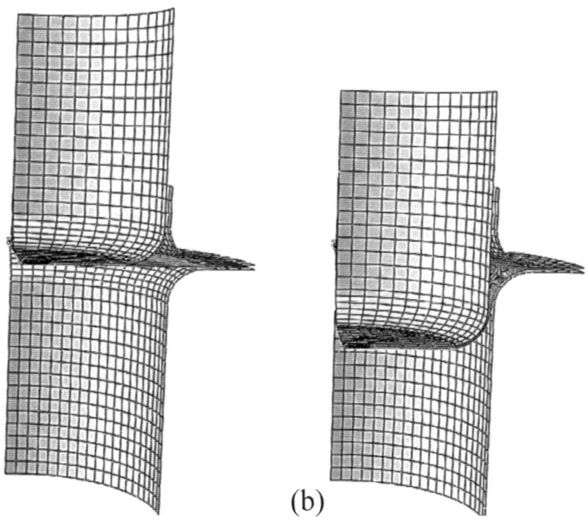

(a) (b)

Figure 1. Finite element model of deep drawing process: (a) After deformation; b) During deformation.

The problem studied on this paper is the effect of friction condition on the thinning of deep drawing of AA2090 Al-Li material. The mechanical properties of AA2090 Al-Li alloy are shown on the table 1. The material is used with a thickness of 1.5 mm, the blank holder force is 22.2 kN, the blank diameter is 158.76 mm, the die diameter is 101.48 mm, the punch diameter is 97.46 mm, die and punch shoulder radius are 12.70 mm both [21].

Table 1. Mechanical properties of AA2090 Al-Li Alloy.

Density (Kg/dm^3)	Young's modulus E (GPa)	Poisson's ratio υ	Hardening law
2.89	78.6	0.3	$\sigma = 646(0.025 + \varepsilon)^{0,227}$

The die, punch and holder were modelled as rigid bodies whose motion was governed by the reference node and were meshed with (R3D4) surface elements, while the blank was modelled as a deformable body, using reduced integration shell element (S4R) [22].

3. Taguchi design

A combination between FEA simulation and the statistical Taguchi approach was analysed to study the effects of contact between the sheet blank and the three rigid tools. The orthogonal Taguchi array adjacent the design such a method that parameter levels are similarly weighted. Due to this, each factor can be evaluated separately of all the other factors, therefore, the impact of one factor does not influence the estimation of another factor.

Under three levels of each factor as illustrated in table 2. The values of chosen control parameters were based on the results of [23] the models were conducted so as to predict the thinning on cylindrical deep drawn process using relevant ranges of friction coefficient, L9 orthogonal array had used to conduct the only nine simulations. The assignment of parameters along with the OA matrix is shown in table 3. Taguchi approach is most powerful Design of experiments (DOE) tool for parametric optimization of engineering process in the tool the concept of Signal to Noise ratio (S/N ratio) is used for improvement of quality through variability reduction and improvement of measurement [24].

Table 2. Process parameters and their levels.

Process parameter	Levels		
	1	2	3
Coefficient of friction between blank and die μd	0.14	0.16	0.18
Coefficient of friction between blank and holder μh	0.14	0.16	0.18
Coefficient of friction between blank and punch μp	0.1	0.15	0.20

Table 3. Orthogonal array (L9) of Taguchi method.

Treat No	Coefficient of friction between blank and die μd	Coefficient of friction between blank and holder μh	Coefficient of friction between blank and punch μp
1	0.14	0.14	0.1
2	0.14	0.16	0.15
3	0.14	0.18	0.2
4	0.16	0.14	0.15
5	0.16	0.16	0.2
6	0.16	0.18	0.1
7	0.18	0.14	0.2
8	0.18	0.16	0.1
9	0.18	0.18	0.15

4. Results and discussion

According to the test factor and the number of levels. Orthogonal test L9 (3^3) is used to develop orthogonal array (OA). S/N ratio had determined for every Finite Element Analysis simulation and the values of thinning obtained from each model is listed in table 4.

Table 4. Thinning results for each model.

Treat No	1	2	3	4	5	6	7	8	9
Thinning (mm)	1.335	1.422	1.421	1.418	1.442	1.325	1.4394	1.319	1.308

After the nine tests. Taguchi analysis was done and in order to investigate the optimal level of the coefficient of friction on the deep drawing process as shown in table 5. The optimum levels of the three factors given above. In order to obtain ameliorate the thinning in the final drawn cup and the percentage of the contribution of each factor on the quality of the drawn shape as shown in table 6.

Table 5. Signal to Noise Ratios.

Level	μd	μh	μp
1	2.877	2.904	2.455
2	2.887	2.883	2.812
3	2.636	2.613	3.133
Delta	0.251	0.291	0.678
Rang	3	2	1

Table 6. Optimum Condition and the contribution of each factor.

Process parameters	Level	Contribution %
μd	2	27
μh	1	33
μp	3	40
Thinning (mm)	1.4445	

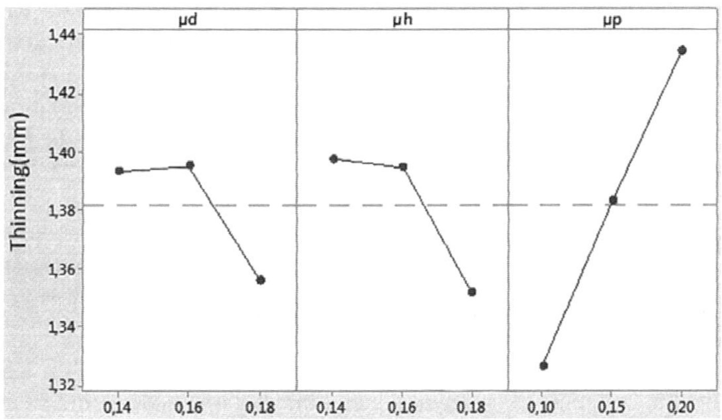

Figure 2. Variation of thinning with the variation of control parameters.

From figure 2 it is detected that the optimum condition to reduce thinning is μd at Level2, μh at Level1 and μp at Level3. An extra run was done again by FEA simulation to confirm the results obtained by Taguchi analysis and it was found that the combination of the levels that the value of thinning is 1.4445 mm which is the bigger values compared to the results of tests in table 4 which confirmed the validation of Taguchi method.

5. Conclusion

Friction is one of the pertinent parameters which affect the quality of final drawn shape. Therefore, a numerical simulation for the prediction of friction effect of AA2090 Al-Li alloy sheet were achieved using ABAQUS/EXPLICIT software. In this study a combination between finite element analysis and the statistic Taguchi method was used. As a result. the best combination of friction coefficients to enhance the quality of the final drawn shape of AA2090 is 0.16, 0.14 and 0.20 concerning µd. µh and µp respectively, as well the friction between punch and blank is the most contributed parameter with a percentage of 44% followed by the coefficient of friction between the blank and holder with a percentage of 33% and the less contributed is the coefficient of friction between die and blank with a percentage of 27%.

References

[1] Sankaran K K and Grant N J 1980 The structure and properties of splat-quenched aluminium alloy 2024 containing lithium additions *Mater. Sci. Eng.* **44** 213–27

[2] Wanhill R and Bray G 2014 *Aluminum-Lithium Alloys: Processing, Properties, and Applications* (Elsivier)

[3] Rioja J R 1998 Fabrication methods to manufacture isotropic Al-Li Fabrication methods to manufacture isotropic Al-Li *Mater. Sci. Eng.* **257** 100–7

[4] Yu-wei X, Yi-yuan Z, Wen-feng M and Jain-zhong C 1997 Superplastic forming technology of aircraft structures for Al-Li alloy and high-strength Al alloy *J. Mater. Proc. Tech.* **72** 183–7

[5] Prasad N E, Gokhale A A and Wanhill R J H 2014 *Aluminum-Lithium Alloys: Processing, Properties, and Applications* (Amsterdam: Butterworth-Heinemann/Elsevier)

[6] Lavernia E J and Grant N J 1987 Aluminium-lithium alloys *J. Mater. Sci.* **22(5)** 101–19

[7] Abd El-Aty A, Xu Y, Guo X, Zhang S, Ma Y and Chen D 2018 Strengthening mechanisms, deformation behavior, and anisotropic mechanical properties of Al-Li alloys: A review *J. Adv. Research* **10** 49–67

[8] Jian Li Y, Li Y, Huang M, Xiang Y and Liao Y 2018 Improvement of aluminum lithium alloy adhesion performance based on sandblasting techniques *Int. J. Adhesion and Adhesives* **84** 307–16

[9] Chen F and Tseng P 2005 DeformationAnalysis of side –wall curl in the sheet metal forming of flanged channels *Manuf. Sci. Eng*. 369–75

[10] Bouchaâla K, Essadiqi E, Mada M, Ghanameh M F, Faqir M and Meziane M 2018 Prediction of earing in cylindrical deep drawing of aluminum alloys using finite element *Int. J. Mechanical and Production Engineering* **6** 63–69

[11] Bouchaâla K, Ghanameh M F Faqir M, Chouiyakh H, Mada M and Essadiqi E 2019 Wrinkling defect prediction and formability improvement in cylindrical deep *14th Congress of Mechanics (Rabat, Morocco)*

[12] Bouchaâla K, Ghanameh M F, Essadiqi E, Mada M, Faqir M and Meziane M 2018 Modeling of anisotropy influence on thickness distribution of deep drawing sheet *Proc. 2018 Int. Conference on Robotics, Control and Automation Engineering (ACM)* 142–6

[13] B,pınar 2011 *Modeling and Simulation of Friction in Deep Drawing* (M.Sc. thesis, Middle East Technical University)

[14] Jayahari L, Sasidhar P V, Reddy P P, BaluNaik B, Gupta A K and Singh S K 2014 Formability studies of ASS 304 and evaluation of friction for Al in deep drawing setup at elevated temperatures using LS-DYNA *J. King Saud University-Engineering Sciences* **26** 21-31

[15] Hjijaj M, Feng Z Q, de Saxc G and Mróz Z 2004 On the modelling of complex anisotropic frictional contact laws *Int. J. Eng. Sci.* **42(10)** 1013–34

[16] Ashida Y and Aoyama H 2007 Press forming using ultrasonic vibration *J. Materials Processing Technology* **187-188** 118–22

[17] Mróz Z and Stupkiewicz S 1994 An anisotropic friction and wear model *Int. J. Solids Struct.* **31(8)** 1113–31

[18] Olguner S and Bozdana A T 2016 The effect of friction coefficient on punch load and thickness reduction in deep drawing process *Int. J. Materials* **3** 64–8

[19] Browne M and Hillery M 2003 Optimising the variables when deep-drawing CR 1 cups *J. Material Processing Technology* 64–71

[20] Wang Z, Nakamura T, Dohda K and Obika T 2003 FEM analysis of contact mechanism in press-forming of lubricant pre-coated steel sheet *J. Materials Processing Technology* **140** 514–9

[21] Rabahallah M, Bouvier S, Balan T and Bacroix B 2009 Numerical simulation of sheet metal forming using anisotropic strain-rate potentials *Materials Science and Engineering (Elsevier)* **517** 261–75

[22] Abaqus Analysis User's Guide Manual **IV** Elements

[23] Bouchaâla K, Ghanameh M F, Faqir M, Mada M and Essadiqi E 2019 Friction behavior of aluminum-lithium sheet thought deep drawing analysis *Int. Conf. Comp. Sci. and Renewable Energies ICCSRE'2019 (Agadir, Morocco)*

[24] Reddy A C S, Rajesham S, Reddy P R, Kumar T P and Goverdhan J 2015 An experimental study on effect of process parameters in deep drawing using Taguchi technique *Int. J. Engineering, Science and Technology* **7(1)** 21–32

IOP Conf. Series: Materials Science and Engineering **664** (2019) 012005 doi:10.1088/1757-899X/664/1/012005

Effective use of a helicopter with a Bambi bucket firefighting system in Bulgaria

D T Seyzinski, I Georgiev, H P Panayotov and S I Penchev[1]

Department of Transport and Aircraft Equipment and Technologies, Technical University – Sofia, Plovdiv Branch, 25 Tsanko Diustabanov St., 4000 Plovdiv, Bulgaria

[1] E-mail: spenchev@tu-plovdiv.bg

Abstract. The present paper studies the effectiveness of the use of the Mi-17 helicopter with a Bambi bucket firefighting system in operations against mountain and forest fires in the Republic of Bulgaria. It makes an analysis of the conditions under which it is used, on the basis of numerous publications about theoretical and experimental studies, including ones aimed at increasing the efficiency of using helicopters with other external hanging firefighting systems. The preparation and conduct of flying experiments with a Bambi bucket system have been made. They take into account the specifics of this system, the characteristics of the firefighting methods used in the Republic of Bulgaria, and the of country specific external factors. On the basis of the results of these studies, and by taking into consideration the flight safety requirements, recommendations have been made to improve the effectiveness of using a Bambi bucket firefighting system in Bulgaria.

1. Introduction

Forest fire is a violent spread of fire across the entire forest area, unmanageable by humans, and characterized by dense smoke and intense heat radiation. It reduces the protective characteristics of the forests, their water-conservation and other properties. It causes the destruction of trees and shrubs, fauna and, in some cases, impacts populated areas. Due to their social and economic costs scale, forest fires could grow to national disasters. Such an example are the fires in Greece last summer, which have been identified as the largest in Europe since the Second World War. Therefore, firefighting is a matter of national security, and all supporting means must be used, including aviation firefighting systems.

World practice shows that the extinguishing of complex combined (both low and peak) forest fires with water (currently the best extinguishing agent in the world used for this purpose) is possible only using helicopters. This is a multifarious, labour-intensive and costly process characterized with little efficiency and effectiveness. Unfortunately, there is no other fire extinguishing method for extinguishing fires in inaccessible, difficult to reach and remote areas, far from water basins.

Over the last few decades around the world, and since 2010 in our country as well, helicopters with a suspension fire extinguishing system (SFS) have been used for extinguishing mountain and forest fires. The most common firefighting method is to pour water from the system's external storage tank onto the place of fire or to create an enclosing lane. The advantages of the use of helicopters for forest fires are their operability in delivering water to the hearth, the great instantaneous efficiency of pouring a lot of water, airport independence, and the high level of safety for humans.

Content from this work may be used under the terms of the Creative Commons Attribution 3.0 licence. Any further distribution of this work must maintain attribution to the author(s) and the title of the work, journal citation and DOI.
Published under licence by IOP Publishing Ltd

In Bulgaria, during the extinguish mountain and forest fires, the use of Mi-17 helicopters with Bambi bucket system was established. In order to make it safe and effective a number of theoretical and experimental studies [1, 2], including those aimed at increasing the efficiency of this use in compliance with the requirements of flight safety, have been completed in Russia and in Europe [3]. A problem with the application of the results of these studies in our country is the fact that in Russia the SFS is VSU-5, whose characteristics vary from those of the Bambi bucket (table 1). Thus, specific adaptation of the results of this research is required and conduction of surveys in our country in order to improve the overall performance in our current conditions and environments.

Table 1. Technical characteristics of SFS Bambi bucket and VSU-5A.

Technical Characteristics	Bambi bucket	VSU-5A
Maximum volume, m^3	3.0	4.5
Diameter of the top, mm	1400	2200
Bottom diameter, mm	800	800
Length of rope, m	15	20-40
Maximum helicopter speed with empty tank, km/h	176	160
Maximum helicopter velocity with full tank, km/h	176	180
Most advantageous helicopter speed, km/h	80-100	80-100
Efficient altitude for running water above the surface, m	10-20	20-30
Required time to pour water after pushing the pilot button, s	4-5	6-8
Average water consumption, l/s	800 -1000	800-1000
Time to fill the system tank, s	75	60
Closing time of the valve, s	1-2	1-2
Mass of the empty system, kg	75	160

The present work analyses the conditions under which a Mi-17 helicopter with a Bambi bucket suspension system is used for extinguishing mountain and forest fires in Bulgaria. Descriptive and experimental materials are presented, taking into account the specifics of our conditions. The main conclusions and recommendations for providing higher firefighting efficiency in compliance with the requirements of the flight safety are formulated.

2. Fire extinguishing in Bulgaria

Helicopters could be used in different firefighting and emergency operations, performing the following tasks:

- search for and rescue victims in the case of wild fire;
- targeting and on site control of the ground fire and rescue forces;
- transport of rescue teams and fire-extinguishing means and their delivery by parachute, fast-rope and landing;
- victims' evacuation from the disaster area;
- extinguishing of wild fires from the air with water or a flame-retardant solution in areas not accessible for ground firefighting teams;
- conducting aerial fire surveillance and monitoring;
- preventing fire evolution until the ground forces arrive, preventing the spread of fire;
- establishing enclosing lanes using water, foam and chemical solutions;
- communication between firefighting teams, search and rescue teams and the on-site decision-maker.

In Bulgaria, in the case of firefighting, an aviation fire system helicopter Mi-17 is used with external suspension Bambi bucket system. Mi-17 is a military version helicopter converted to use a Bambi bucket system. It is capable of delivering up to 4000 litres of water and discarding it with great precision at the place of the fire. It is also capable of transporting firefighting teams and special equipment to the affected area. Up to now, over 14 000 units have been produced of various modifications, a world record for two-engine helicopters. These helicopters are operated in over 100 countries around the world and have a total of about 100 million flight hours.

The external fire extinguishing system is the Bambi bucket, produced by the SEI – Canadian Company, with a capacity of 3000 litres [4]. The main features of this system, compared to those of the Russian VSU-5A, are depicted in table 1.

For firefighting of mountain and forest fires by helicopters with SFS, two methods are applied – direct (active) and indirect (passive). In the direct extinguishing method, the fire is directly attacked by water or another fire-extinguishing agent. It is applied when the fire is flat and is in an early stage of development. The indirect method is used to create a restraining lane at a specific distance away from the fire. The end of these lanes must finish with natural or artificial fire barriers such as roads, rivers, and lakes or pre-prepared mineralized lanes.

The main stages of the firefighting flights are: take off; flight with SFS to the water basin to fill in the tank; hovering and filling the SFS with water; flight to the fire site; release of water into the fireplace.

Each stage of the flight is related to a threat for its safety. From the point of view of flight safety, the final stage is most dangerous due to the fact that complex air dynamic processes are developed over the fire site as result of combustion. There is a large difference in the temperature fields due to the different pyro-characteristics of solid combustible materials. This leads to the formation of winds different as speed and direction. Following the relief of the area, especially when mountain fires are extinguished, these winds create swirling and turbulent airflows. In many cases it is possible to form convective thermal pylons, the power of which is directly dependent on combustion intensity.

The intensity of combustion depends on the wind speed over the fire site, the type of the solid combustible materials and the height of the fire itself. Following the relief of the area, the flat fire can easily turn into a peak fire when the wind has a speed of over 20 m/s and the slope angle is 20-30°.

Helicopter flight over such convective pylons is especially dangerous because of the temperature of the elevated air. It could cause engine surge. It could happen when the air inlet temperature is above 70°C. Therefore, when flying over such areas, the limitations of maximum ambient air temperature limits must be taken into account. For example, using the methodology set out in [1] for high intensity fires (20000 kW/m) with diameter of 15 m, the following values of the ambient air temperature are observed at altitude H: 190°C at H = 30 m; 110°C at H = 50 m; 15°C at H = 100 m. Therefore, helicopter flight at an altitude of 30-70 m is dangerous because of the possibility of engine surge.

Turbulent flows in the area of fire also affect the work of the crew. They must constantly react to the external environment, which makes them tired much faster.

When the water from the water reservoir is released, rope backwards deflection is observed within the range of 45-60°. The angle of rope and SFS deviation is directly related to flight speed and the impact of the horizontal and vertical air masses. When the tank is full it has a streamlined shape and a low ballistic coefficient. After water is released, this coefficient increases, and the SFS creates a high drag. When moving the helicopter with an empty SFS over a vertical thermal pillar, it is possible for the airflow to impact the system for short period of time. From research [2] done so far it has been shown that even a vertical flow at 20-30 m/s cannot have a significant impact on the SFS and there is no danger of falling into the horizontal stabilizer and tail rotor.

Taking into account the conditions accompanying a fire (wind speed, high temperature, quivering) and the inability to operate the helicopter in a smoky environment due to the probability of engine surge, it should be assumed that fire extinguishing takes place at a forward speed. At high speeds and flight altitudes, the efficiency of falling water is reduced due to the fact that it covers a larger area and the amount of fallen water per 1 m^2 is less. At low speeds (less than 60 km/h), power demand is close to the available limit, and the helicopter operates in critical modes where flight safety is under threat. The analysis of these conditions indicates that it is advisable to fly horizontally in the speed range of 80-100 km/h

3. Firefighting efficiency

The effectiveness of fire extinguishing is the ratio of the water entering the fire site to the total amount of water discharged. This possible maximum is achieved by supplying a fire-extinguishing liquid in the fire site at the required concentration in accordance with the flight safety requirements.

The main factors affecting the effectiveness of helicopter usage and fire extinguishing capability are [1, 2]:

- prehistory of the flight until the moment of liquid release;
- altitude and speed of the flight when throwing the liquid;
- turbulence of the atmosphere and ascending flow from the fire site;
- induced flow of the main propeller;
- spatial position of the dropping points (the position and construction of the outlet of the SFS tank);
- initial drip rates (how the liquid is delivered to the SFS tank outlet);
- liquid flow rate per second.

These factors have been determined through the results of a number of theoretical and experimental studies [5-8]. These findings are of great importance for the development of aviation technologies and aeronautical fire extinguishing methods, which is a promising direction for achieving high firefighting performance.

When a helicopter with SFS is used, there is a loss of extinguishing liquid during transport, which reduces firefighting performance. Experiments show that up to 30% is lost due to evaporation of the low-dispersed part of the water, and about 20-30% of it remains on the branches and the tops of the trees.

In order to reduce losses and improve fire performance of liquid solutions, modern helicopters are equipped with chemical delivery systems - retarders, pigments, moisturizers or thickeners. Pure water can support fire extinguishing properties for only 5-15 minutes. Adding a moisturizing agent or retarder increases this ability for up to 2 hours. In general, special additives contribute to increasing cooling efficiency with liquid solution from 2 to 4 times.

It has been experimentally recognized and practically confirmed [9,10] that in order to achieve the necessary efficiency in fighting mountain and forest fires, it is necessary to provide the following water supply intensity J:

- in case of flat forest fire: $J = 0.1 - 0.15$ l/m^2 s, extinguishing time $t_{ff} = 30$ - 60 s;
- at peak forest fire: $J = 1.0$ l/m^2 s, $t_{ff} = 30$-60 s.

These data show that they are much higher than the intensity of fire extinguishing fluid supplied in fire in a residential building $J = 0.1$ l/m^2 s.

On releasing a fire-extinguishing liquid at the edge of the fire, the following results should be obtained:

- the edge of the fire can be extinguished when specific flow of water or other fire extinguishing solution D_{sp} reaching the burning edge is not less than the specific flow D_{need} required to extinguish the entire width and perimeter of the burning edge;
- the linear spread rate of the fire edge V_f is reduced at least twice for a period of 15 minutes if the specific extinguishing rate of the fire extinguishing solution reaching the edge of the burning edge is not less than 0.5 l/m^2;
- for extinguishing the burning edge of a forest fire, the required specific water consumption should be $D_{need} = 5$ l/m^2.

The values given for the specific water consumption are effective in extinguishing forest areas with a total density of not more than 0.5.

When setting an enclosing lane at the edge of the fire, it is necessary to provide the required specific consumption of fire-extinguishing solution or water along the entire length of the lane.

The necessary water consumption for laying an enclosing lane in the fight against forest fires, depending on the specifics of the forest vegetation, are given in table 2.

Table 2. The necessary water consumption for laying an enclosing lane depending on the specifics of the forest vegetation.

Forest conditions	Necessary specific water consumption, l/m^2
Forests with a density of planting not less than 0.2	0.7
Forests of juniper with a density of 0.2 to 0.3 with a slight pronounced overlay	1.3
Forest juniper berries with a density of 0.4 to 0.5 with a moderately developed superior cover	2.0
High density forest plantations	2.6

When creating a locking barrier with water for a period of time between laying the strip and approaching the edge of the forest fire, part of the water evaporates. This can be limited by the use of fire extinguishing solutions containing special chemical additives (humectants, thickeners).

If the primary extinguishing site is larger than planned, it is necessary to use 2-3 helicopters depending on the extinguishing area. It should be taken into account that the relative efficiency of simultaneous fire extinguishing with 3-4 helicopters is less, as the area of simultaneous extinguishing and consequently water losses increases.

To achieve the required firefighting efficiency, it is necessary to release the water consistently over the extinguishing site (5-6 minutes).

In order to achieve it, the crew needs to have the necessary skills that can be developed through training. This skills and experience could be assessed by measuring the uniformity of water distribution on the fire area and the rate of water filling of the measuring vessels. This will allow the pilot to correctly assess the altitude of the flight at the moment of water release [11, 12].

The minimum safe flight altitude at which the highest firefighting performance is obtained is determined by the following condition (figure 1):

$$H_{FL} = H_{TR} + H_{LR} + H_{SF}, \tag{1}$$

where H_{TR} is the height of the trees; H_{LR} – the length of the rope; H_{SF} – the minimum safe distance between the tops of the trees and the SFS.

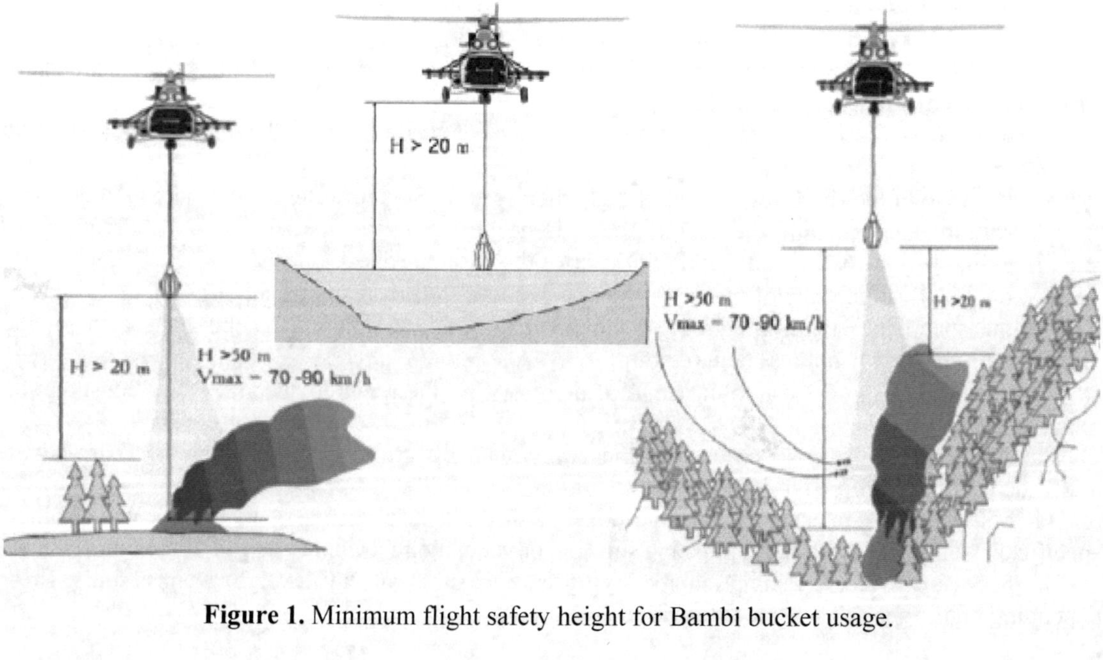

Figure 1. Minimum flight safety height for Bambi bucket usage.

In the case of open water basins, the helicopter approaches at the point of watering on a mid-slope glide. At the moment when the reservoir touches the water surface the forward speed does not exceed 5 km/h and the descent rate is about 0.5 m/s. After that, the pilot decreases the forward speed to zero and continues to decreasing the descent rate until the tank is full, then performs a hovering with a subsequent climb. This method of approach limits the generation of the vortex flow from the main rotor after SFS reservoir touches the water, which greatly facilitates the process and shortens the time of filling the tank with water.

Using the helicopter manoeuvrability higher effectiveness of the firefighting can be achieved while the required flight safety level is ensured.

To increase the fire extinguishing capability, a cycling technology is proposed in the fire area using a pair of helicopters [13]. One helicopter fills the SFS with water from the water basin, while the other delivers water to the site of the fire. The first one is then directed to pour the water on the fire and the second to the water basin.

The poor quality of interaction and coordination between ground and air forces adversely affects the effectiveness by not ensuring that the extinguishing agent is properly disposed of at the designated fire site.

4. Flight experiments

Taking into account the specific features of the Bambi bucket SFS in firefighting process in our country, and in order to increase the efficiency of its use while maintaining a high level of flight safety, flight experiments were conducted. They aim to determine the following parameters at certain flight speeds and altitudes:

- the groundwater concentration discharged from the SFS tank;
- the distribution of the water concentration in the area;
- deviation of the suspension SFS rope.

Table 3 presents theoretically determined data for the width and length of the lanes on which the experiments have been conducted to measure the fallen water from the SFS Bambi bucket. The selected speeds for the experimental study are in the range of 80-120 km/h.

Table 3. Theoretically determined data for the experiments.

Speed (km/h)	80	90	100	120
Time to release the water, s	4.0	3.43	3.6	3.7
Lane width, m	8.4	8.6	9.0	9.4
Lane length, m	89	86	100	123
Rope deviation after release, °	52	55	58	60

Using calculations based on a mathematical model of the relationship between the specific water consumption, the volume of the ejected from the SFS water and the area of a fire [9], it was determined that by reducing the volume of the water in the tank from 2500 litres to 2000-2200 litres (which is done when extinguishing mountain and forest fires) the extinguishing area can be 60 m^2 and the radius of the dropped water is 4.5 m at hover mode. Taking these values into account, the time required is 333 s. Fire extinguishing using hovering can only be applied to small fires at an early stage of evolution.

Due to financial constraints, only 2 flight experiments (8 water-release legs) were conducted with a Mi-17 helicopter Bambi bucket. They were held at Krumovo Airport in helicopter hovering areas. The track size is 8m x 80m. On it plastic containers with a radius of 15 cm and a capacity of 12 l to measure the water falling from the SFS were situated. The step between containers is 4 m wide and 5 m in longitudinal direction. Additionally, entry and exit doors (indicated by red flags) are fixed and the flight direction corresponds to 322° (figure 2).

Figure 2. Measuring lane of the amount of water after pouring from the Bambi bucket.

As a result of the conducted study, the concentration of the water fallen into the predetermined ground lane (in l/m^2) is measured. The water concentration is highest at a speed V_{fl} of 80 km/h and an altitude H_{fl} of 50 m. The data for it is shown in figure 3. It can be seen that the average amount of fallen water between the thirty and sixty meters is 4.2 l/m^2. The covered area where the cessation of pyrolysis in hard combustible materials is possible [9, 10] is 240 m^2. For the complete termination of the fire spread on such an area, it is necessary to consecutively throw water from several helicopters within 5-10 minutes. Three helicopters can be used when the water source is up to 5 km away from the fire. For a distance of 10 km – 4 helicopters may be used.

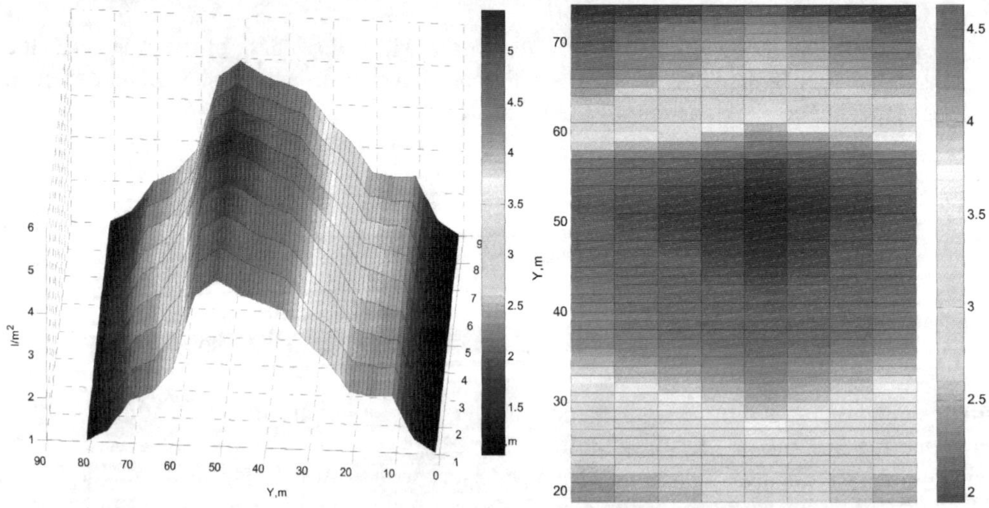

Figure 3. Distribution of fallen water over l/m^2 at V_{fl} = 80 km/h and H_{fl} = 50 m.

5. Recommendations for efficiency enhancement

Mountain and forest fires fighting efficiency could be increased if these recommendations are followed:

- aerial intelligence by drones and determining the location and characteristics of the fire, water area and landing sites of helicopters;
- calculation of the fuel charge in advance by the helicopter crew taking into consideration the helicopter load with passengers and cargo and distance to the fire;
- performing a run-over flight to determine the fire characteristics, the location of ground crews and obstacles endangering flight safety in the area;

- accurate assessment of the fire type and selection of proper firefighting tactics by the crew;
- flights at low speed and minimum safe altitude are a prerequisite for delivering more liquid per m^2;
- use of chemical additives in the water contributes to a faster extinguishing of the fire or to a greater resistance to the landing lanes;
- use of multiple helicopters for work on a single target together with ground crews is a prerequisite for rapid fire suppression.

References

[1] Asovsky V P 2009 Peculiarities of forest firefighting by helicopters with spillway devices on the external sling *Scientific Bulletin of MSTUCA "Aeromechanika i prochnost"* **138** 143–9 (In Russian)

[2] Asovsky V P 2007 Researches of sailing cargo dynamics on the helicopter external sling at the gust *Scientific Bulletin of MSTUCA "Aeromehanika i prochnost"* **111** 140–6 (In Russian)

[3] Marchi E, Neri F, Tesi E, Fabiano F and Brachetti Montorselli N 2014 Analysis of helicopter activities in forest fire-fighting *Croat. J. of Forest Eng.* **35(2)** 233–43

[4] Anon 2013 *Bambi Bucket Operation Manual Version G* (SEI Industries Ltd.)

[5] Borisov I V and Tsipenko A V 2012 Computing experiment for fire fighting helicopter operation analysis *Trudy MAI* **53** (In Russian)

[6] Eritsov A M and Gusev V G 2016 Technological improvements in using control lines in fighting forest fires in areal application zones *Vestnik of Volga State Technological University Ser.: Forest, Ecology, Nature Management* **1(29)** 42–56 (In Russian)

[7] Efimov V V 2007 The analisys of flight safety and servise efficiency interaction of helicopters with cargo on the external sling *Scientific Bulletin of MSTUCA "Aeromehanika i prochnost"* **111** 115–20 (In Russian)

[8] Parshentsev S A and Prusov N A 2010 Evaluation of field of concentration of liquid and solid particles at their distribution from helicopter external sling *Scientific Bulletin of MSTUCA "Aeromehanika i prochnost"* **154** 71–78 (In Russian)

[9] Abduragimov I M, Govorov V Yu and Makarov V Ye 1980 *Physico-Chemical Basis of Development and Extinguishing of Fires* (Moscow: VIPTSh MVD USSR) (In Russian)

[10] Abduragimov I M 2012 The problem of extinguishing forest and peat fires *Fire and explosion safety* **21** 66–76 (In Russian)

[11] Borisov I V and Tsipenko A V 2014 *Proc. of 29th Congress of ICAS (7-12 September St. Petersburg)*

[12] Tsipenko A V , Borisov I V and Parshentsev S A 2009 Simulation of helicopter flight with taking into account cargo onto external suspension and lifting rotor jet *J. Proc. of South Federal University* **1(90)** 171–177 (In Russian)

[13] Lebedev A A 2014 Recommendations of piloting a helicopter and fighting against external goods oscillations during the conduct of urgent aerial works with external suspension *Scientific Bulletin of MSTUCA "Aeromehanika i prochnost"* **199** 157–60 (In Russian)

IOP Conf. Series: Materials Science and Engineering **664** (2019) 012006 doi:10.1088/1757-899X/664/1/012006

Assessment of the fuel efficiency of unmanned cargo aircraft, based on general aviation aircraft

V S Serbezov

Department of Aeronautics, Technical University of Sofia, 8 Kl.Ohridski Blvd., 1000 Sofia, Bulgaria

E-mail: vlados@aero.tu-sofia.bg

Abstract. This study was motivated by the globally increasing interest in unmanned cargo drones. It was focused particularly on cargo drones based on existing conventional general aviation airplanes and it should be regarded as a preliminary step towards the complex assessment of unmanned cargo aircraft transport systems. The aim was to estimate the fuel efficiency of such drones and to outline the optimums of some of their key design characteristics. A sample of 26 very light and light aircraft, and motorgliders was examined. The data was taken from open sources. The results outline that for best fuel economy the cargo drone should be a composite structure, piston engine airplane with wing aspect ratio of 10 to 12. Fuel efficiency estimation at distances of 500 to 2500 km shows that such cargo drones would be competitive with large piloted commercial cargo airplanes as well as with the road transport.

1. Introduction

Nowadays the introduction of unmanned autonomous transportation systems is expected to bring significant cost and environment benefits. These expectations cover also the air cargo, which is reflected in many strategic documents such as the IATA Cargo Strategy [1] and the European Commission Flightpath 2050 [2]. The initial interest by currier and e-commerce companies was focused on so called "last mile" delivery drones for small items. At the same time many start-ups emerged aimed at the creation of more heavy unmanned cargo aircraft (UCA) with longer range. This activity led eventually to the creation of organizations like the Platform for Unmanned Cargo Aircraft (PUCA).Most of the projects are for aircraft with payloads of 1 to 10 tons. Much practical work was conducted in China, creating cargo drones based on existing general aviation light aircraft like Staruas AT200.

An important step towards the realization of the cargo drone concepts is the estimation of their energy and transport efficiencies. A review of the problem of Unmanned Aerial Vehicles(UAV) efficiency, although not focused exactly on cargo drones is made by Amoiralis et.al. [3]. In [4] Valerdi reviews the cost metrics of UAV that can be used also for efficiency estimates.

Particularly for the case of "last mile" delivery drones efficiency analyses were performed by researchers funded by the U.S. Department of Energy [5] and RAND [6]. Regarding the larger UCA research has been undertaken under the auspices of PUCA [7, 8], trying to cover a broad area of the UCA commercial implementation aspects. At the same time there are no studies dedicated to the segment of UCA with range and payload corresponding to these of the very light class of general aviation aircraft (VLA), although there are start-ups aimed at the development of such UCA, including the Bulgarian company Dronamics.

Content from this work may be used under the terms of the Creative Commons Attribution 3.0 licence. Any further distribution of this work must maintain attribution to the author(s) and the title of the work, journal citation and DOI.
Published under licence by IOP Publishing Ltd

The aim of this study is to estimate the fuel efficiency of such UCA, and to serve as a basis for further studies on their complex performance and efficiency as a transportation system. The approach used for this assessment is common with that of similar studies concerning the passenger aviation [9-12]. The difference is that the data used is for existing light and very light aircraft and some assumptions regarding the transformation of piloted aircraft to unmanned are applied. Also the study looks at the effect on fuel efficiency of some of the key design features of the aircraft such as the type of the materials used in the airframe and the wing aspect ratio. Finally a comparison with the efficiency of a contemporary mid-size air freighter and a panel van road vehicle is performed.

2. Assessment basis

It should be stated that the real performance of the future UCA will depend from a great number of their design features. The complexity of the UCA design challenge can be seen in works like [13], where different types of aircraft and rotorcraft layouts are compared, meeting different requirements. The aim of this paper is not to assess the variety of all the possible UCA, but to estimate the energy efficiency of UCA that can be easily developed from existing general aviation aircraft, using the same airframe and engine technologies.

2.1. Assumptions for the conversion of very light aircraft to UCA

The four main characteristics that have large effect on the energy efficiency of an aircraft are the weight, the aerodynamic properties, the power plant efficiency and the power consumption of the airplane systems. To assess correctly the UCA energy efficiency, using data for piloted aircraft it is essential to properly estimate how these characteristics will be affected by the UCA design requirements. In this paper the following changes in the design are suggested:

- The absence of a cockpit in the UCA will allow a better aerodynamic shape of the nose section of the body, and in the case of a single engine with tractor propeller – higher placement of the engine, resulting in shorter landing gear legs, which are lighter and create less drag;
- The absence of pilots and passengers equipment and furnishing as well as mechanical flight control system will save weight;
- The need of relatively large cargo volume, enforced cabin floor and suitable cargo door will have negative effect on the airframe weight and aerodynamic drag, compared to piloted VLA;
- The absence of cockpit flight instruments will probably compensate the addition of sensors and autopilot systems in terms of weight and energy consumption;
- The use of control surfaces actuators will add weight and energy consumption.

All of the considerations made above are very sensitive to the particular design and need separate investigations. For this reason in the present paper a ‘neutral’ scenario was suggested, assuming that the conversion of the airplanes to UCA will not affect significantly the empty weight and all available useful load will be divided between the fuel and the payload, and the aerodynamic properties will stay essentially the same.

2.2. Airplanes included in the survey

A sample of 26 very light and light aircraft, and motor gliders was examined, based on the performance data provided in Internet by the manufacturers and data from the corresponding Type Certificate Data Sheets. The airplanes with their manufacturers are listed in table 1. They can be divided in the following groups:

- VLA with aluminium alloy airframe, Rotax 912/914 gasoline engine (10 airplanes);
- VLA with composite airframe, Rotax 912/914 gasoline engine(9 airplanes);
- Motor gliders with composite airframe, Rotax 912/914 gasoline engine (3 airplanes);
- Motor gliders with aluminium alloy airframe, Rotax 912/914 gasoline engine (2 airplanes);
- Diesel powered light aircraft with composite airframe (2 airplanes);
- The AT200 turboprop UCA, based on PAC-750 utility aircraft was taken as a reference.

Table 1. Airplanes included in the survey.

	No.	Airplane	Manufacturer
VLA with aluminium or mixed structures	1	Harmony LSA	Evector
	2	EuroStar SL	Evector
	3	DV-1	DOVA Aircraft
	4	PS-28	Czech Sport Aircraft
	5	P92js	Tecnam
	6	Skyleader 600	ZALL JIHLAVAN
	7	AC 4	Lightwing
	8	AT 3	Aircraft technologies
	9	SD 4	TOMARK, s.r.o.
	10	P2008js	Tecnam
VLA with composite structures	11	Sting S4	TL-ULTRALIGHT
	12	TL-3000	TL-ULTRALIGHT
	13	ATEC 321 Faeta NG	ATEC
	14	DA20i	Diamond aircraft
	15	KR-030	Ekolot
	16	Virus SW	Pipistrel
	17	CTLS	Flight Design
	18	WT 9	Aerospool
	19	APM 30 Lion	APM
Motor gliders with composite structures	20	Europa XS	Europa Aircraft
	21	HK36	Diamond aircraft
	22	Sinus	Pipistrel
Motor gliders with aluminium structures	23	SF 25 C Falke	Scheibe Aircraft
	24	Samburo AVO-68r	M+D Flugzeugbau
Single Diesel engine, composite light airplanes	25	Da40NG	Diamond aircraft
	26	DA50	Diamond aircraft
UCA derived from utility turbopropeller	27	AT200	Star UAV System

The VLA are the most popular nowadays category of general aviation aircraft that are widely used for pilot training and leisure flights. Most of these airplanes are relatively new designs and are equipped with Rotax 912/914 family engines, using automotive gasoline instead of more expensive aviation gasoline.

The motor gliders with thick airfoil, high aspect ratio wing are added in order to investigate the influence of the wing aspect ratio on the energy efficiency.

Diesel powered aircraft are interesting for their low specific fuel consumption. Their application in the general aviation was limited by the successful introduction of aviation piston engines certified for automotive gasoline, but diesel engines find applications also for military UAV.

3. Fuel efficiency assessment methodology

3.1. Metrics for cargo aircraft fuel efficiency

Publications [9-12] use a variety of different metrics to measure the environmental performance of commercial aircraft. In present paper as a metric, that is independent from the engine and fuel types the energy intensity is chosen. In the case of cargo aircraft the energy intensity is expressed in equation (1):

$$Ei = \frac{m_{fuel} H_u}{1000 m_{pld} L_{flt}}, \qquad (1)$$

where m_{fuel} is the mass of the fuel spent for the flight in kg, H_u– the net heat of combustion of the fuel that is used in MJ/kg, m_{pld}– the payload in kg and L_{flt}– the distance of the flight in km. The dimension of Ei could be either MJ.(ton.km)$^{-1}$ or kJ.(kg.km)$^{-1}$.

3.2. Calculation of the fuel consumption

In most of the cases the available performance data of the examined airplanes is limited to the parameters described in table 2.

Table 2. Airplane performance parameters, available for the fuel - range calculation.

No.	Parameter	Remarks
1.	Empty weight of the airplane, kg	
2.	Maximum allowed takeoff weight, kg	
3.	Capacity of the fuel tanks, litters	Assumed specific fuel weight is 0.74 kg/l for gasoline and 0.82 kg/l for jet fuel.
4.	Maximum range with 30 minutes flight reserve, km	
5.	Cruise speed, km/h	If the long range cruise is not known it is assumed to be 80% of the maximum horizontal speed.

To overcome the absence of detailed performance data for the fuel-range calculations, an approach similar to that of Burzlaff [14] was applied. Instead of using the specific fuel consumption (TSFC) and the lift to drag ratio (L/D) in the Breguet equation:

$$L_{flt} = \frac{V_{cr}\, L/D}{TSFC\, g} \ln\left(\frac{m_{TO}}{m_{TO} - m_{fuel}}\right), \tag{2}$$

where m_{TO} is the take-off mass of the airplane and V_{cr} is the cruising speed, a Breguet factor is defined:

$$BF = \frac{V_{cr}\, L/D}{TSFC\, g}. \tag{3}$$

The Breguet factor for each airplane now can be calculated using eq. (2) and the data for the maximum range and maximum usable fuel capacity:

$$BF = L_{flt.\max}\left[\ln\left(\frac{m_{TO}}{m_{TO} - m_{fuel.\max}}\right)\right]^{-1}. \tag{4}$$

If the available data is for the maximum range with 30 minutes reserve ($t_R = 0.5$ h), than the maximum range without reserve is calculated using the assumption that the airplane will fly with its cruise speed:

$$L_{flt.\max} = L_{flt.\max R} + V_{cr}\, t_R. \tag{5}$$

3.3. Calculation of energy intensity of the UCA at given flight distance

The energy efficiency at given distance is calculated at three steps:

- Calculation of the mass of the required fuel:

$$m_{fuel} = m_{TO}\left(1 - e^{-\frac{L_{flt}}{BF}}\right). \tag{6}$$

- Calculation of the mass of the reserve fuel:

$$m_{fuelR} = (m_{TO} - m_{fuel})\left(1 - e^{-\frac{V_{cr}.t_R}{BF}}\right). \tag{7}$$

- Calculation of the available payload and the energy efficiency:

$$m_{payload} = m_{TO} - m_{fuel} - m_{fuelR}. \tag{8}$$

Energy efficiency is calculated using equation (1).

4. Results and discussion

The fuel efficiency assessment was divided in two tasks. First the Energy intensity of the airplanes was evaluated at 1000 km flight distance and the influence of the wing aspect ratio and the type of airframe structure on it was analysed. Secondly the Energy efficiencies of the best performing airplanes of each group described in p. 2.2 were calculated for distances between 500 and 2500 km and were compared with a narrow body cargo airplane and a hypothetical road cargo vehicle.

4.1. UCA fuel efficiency at 1000 km flight distance

The results of the calculations of the energy intensity at 1000 km as a function of the wing aspect ratio of the airplanes are shown in corresponding groups in figure 1.

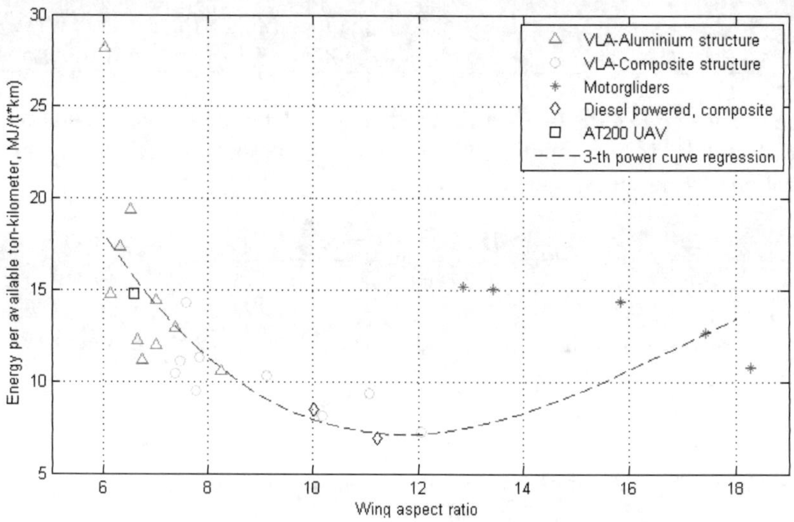

Figure 1. Energy intensity of UCA based on VLA and light aircraft at 1000 km flight distance.

Figure 1 shows good relation between the energy intensity (fuel efficiency) and the wing aspect ratio of the airplanes. Only the older generation motor gliders with aluminium airframe stay out. The relation was approximated with third power regression curve, also shown on figure 1 (The two older motor gliders are excluded from the regression data). The results indicate that best fuel efficiency can be expected for airplanes with composite structure and wing aspect ratio in the order of 11 to 12. The poorer energy efficiency at lower wing aspect ratios is explained by the increased induced drag of such wings. At high wing aspect ratios the transport efficiency of the airplanes also is poorer because their wings are optimised for low speed soaring and not for covering long distances.

The energy intensity of the AT200 turboprop UCA is also shown on fig.1. Despite the larger size of the airplane and the difference in the power plant, it fits very well with the general tendency.

4.2. Comparison of UCA fuel efficiency with other types of cargo aircraft and ground transport

On figure 2 the "best performing" airplanes of each group are compared with Boeing 737-800 freighter and with a hypothetical panel van road vehicle. The data for Boeing 737-800 was taken from [10]. The energy intensity of the panel van is calculated with the assumption that it carries 400 kg of cargo with fuel consumption of 5 litters of Diesel fuel per 100 km.

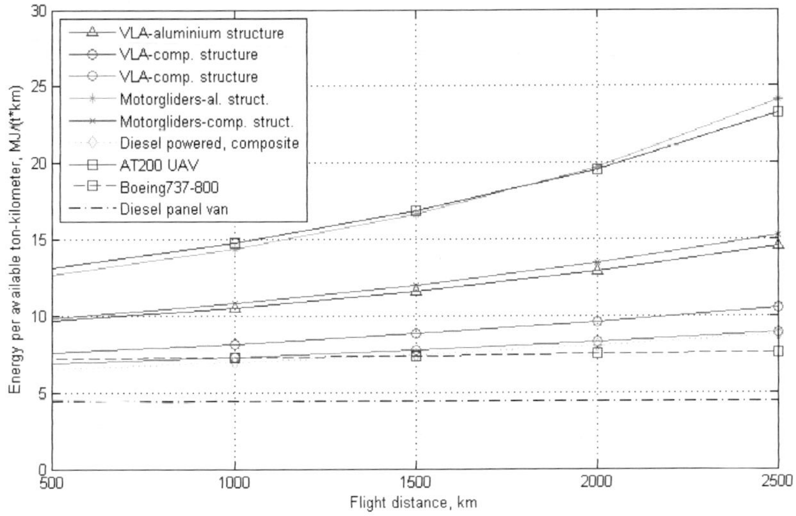

Figure 2. Energy intensity of UCA based on VLA and light aircraft as a function of the flight distance.

The results show that at shorter distances up to 1000 km the best-performing VLA-based UCA will be comparable with the present generation of large piloted cargo aircraft. At longer distances the advantage of jet cargo airplanes is due to their greater cruising speed. At the same time the VLA-based UCA will be twice more efficient then the larger turboprop UCA (AT200), because of the much better specific fuel consumption of the small piston engines compared to the turbo propeller engines.

5. Conclusions

The results outline that for best fuel economy the VLA-based cargo drone with conventional airplane layout should have composite structure and wing aspect ratio in the order 10 to 12. Fuel efficiency estimation at distances of 500 to 2500 km shows that such UCA will be competitive with the present generation of large piloted commercial cargo airplanes, achieving energy efficiency in the order of 7-8 MJ/(t.km). This is a much better figure than 14-24 MJ/(t.km) of the utility turboprop-based UCA.
To some extent and in particular cases such UCA will be competitive also with the road transport, which energy intensity is estimated around 4-5 MJ/(t.km).

This study should be regarded as a preliminary step towards the complex assessment and optimisation of UCA transport systems, including the airplane itself, the ground infrastructure, the operations, etc.

Acknowledgments

The author would like to thank the Research and Development Sector at the Technical University – Sofia for the financial support.

References

[1] International Air Transport Association 2018 IATA cargo strategy (Retrieved June, 3, 2019)
[2] Kallas S, Geoghegan-Quinn M, Darecki M, Edelstenne C, Enders T, Fernandez E and Hartman P 2011 Flightpath 2050 europe's vision for aviation *Report of the high level group on aviation research, European commission, Brussels, Belgium, Report No. EUR 98*
[3] Amoiralis E I, Tsili M A, Spathopoulos V and Hatziefremidis A 2014 Energy efficiency optimization in uavs: a review. *In Materials Science Forum* **792** 281–6
[4] Valerdi R 2005 Cost metrics for unmanned aerial vehicles *Infotech@ Aerospace* 7102
[5] Stolaroff J K, Samaras C, O'Neill E R, Lubers A, Mitchell A S and Ceperley D 2018 Energy use and life cycle greenhouse gas emissions of drones for commercial package delivery *Nature communications* **9(1)** 409

[6] Gulden T R 2017 The Energy Implications of Drones for Package Delivery: A Geographic Information System Comparison

[7] Collins M P 2017 The future market for Large Unmanned Cargo Aircraft in the national airspace system *Faculty of Lewis University Aviation & Transportation Illinois USA*

[8] Van Groningen R 2017 Cost Benefit Analysis Unmanned Cargo Aircraft: Case Study Stuttgart–Urumqi/Shenzen *Erasmus University Rotterdam, Rotterdam*

[9] Lee J J, Lukachko S P, Waitz I A and Schafer A 2001 Historical and future trends in aircraft performance, cost, and emissions *Annual Rev. Energy and the Environment* **26(1)** 167–200

[10] Peeters P M, Middel J and Hoolhorst A 2005 Fuel efficiency of commercial aircraft: an overview of historical and future trends *NLR-CR-2005-669*

[11] Hileman J, Katz J B, Mantilla J G and Fleming G 2008, September Payload fuel energy efficiency as a metric for aviation environmental performance *Int. Cong. Aeronautical Sciences*

[12] Kharina A and Rutherford D 2015 Fuel efficiency trends for new commercial jet aircraft: 1960 to 2014

[13] Hasan Y J, Sachs F and Dauer J C 2018 Preliminary design study for a future unmanned cargo aircraft configuration. *CEAS Aeronautical Journal* **9(4)** 571–86

[14] Burzlaff M 2017 *Aircraft Fuel Consumption–Estimation and Visualization* (Hamburg: Aircraft Design and Systems Group (AERO), Department of Automotive and Aeronautical Engineering, Hamburg University of Applied Sciences)

IOP Conf. Series: Materials Science and Engineering **664** (2019) 012007 doi:10.1088/1757-899X/664/1/012007

Noise sensitivity of a bio-inspired echolocation model

Krasin Georgiev[1] and Radina Nikolova

Department of Aeronautics, Technical University – Sofia, 8 Kl. Ohridski Blvd., Sofia, Bulgaria

[1] E-mail: krasin@tu-sofia.bg

Abstract. The Spectrogram Correlation And Transformation (SCAT) receiver is often cited as a model of bat auditory system that is responsible for echolocation signal processing and super resolution in range. We explore the response of a simplified baseband version of SCAT (BSCT) for two targets in white Gaussian noise. Knowing the loss of signal-to-noise ratio (SNR) relative to the matched filter (MF) is important for any practical applications of the algorithm. We show that this loss is 6~dB and increases if a critical noise level is surpassed.

1. Introduction

Bats' ability to resolve close targets has been attributed to a unique perspective of the world in which at local scale a group of reflectors is represented by the spacing between them [1]. Multiple behavioural experiments have demonstrated better resolution of two close targets than the conventional matched filter [2-5], and models of bat auditory system have been developed to reproduce such perceptual capabilities [6, 7]. These studies are limited to high to medium signal to noise ratios (SNR). The noise properties of the corresponding models are largely unexplored.

The Spectrogram Correlation And Transformation (SCAT) receiver is a model of the auditory system that was proposed by Saillant et al. for the bat Eptesicus fuscus [6] and further developed in [8]. It is a physiologically plausible model of the peripheral and the central auditory system of frequency modulating bats. Different aspects of this model are clarified and analysed in [9, 10]. The interference pattern arising from the return of two close targets is produced in a bank of filters and further processed to evaluate the distance between the targets.

A baseband version of the SCAT model (BSCT) was developed as a tool to study the SCAT model [11]. The noise-free response of the BSCT receiver was found analytically for two closely spaced scatterers and a waveform with a flat spectrum (Linear Frequency Modulated chirp) in [1]. It has been shown that such a receiver works in ultrasound and radio frequency experiments [1, 12].

Similar techniques that rely on interference of two signals to produce modulation of the spectral density have been introduced in the early ages of radar studies for range estimation in noise radars [13, 14]. The so called two-stage spectral analysis (TSSA) consists in cascading two physical spectrum analysers and processing the sum of the transmitted (zero delay) signal and the received signal. The first stage introduces frequency modulation of the spectrum related to the target delay and the second stage transforms the modulation into a range profile.

A fall-off in BSCT-like models performance at reduced SNR is expected as the noisy signal is correlated with a copy of itself rather than with a "clean" reference. The loss of Signal to Interference ratio for TSSA processing of slowly moving targets was derived in [15] and shown to be approximately 6 dB as compared to the cross correlation processing for selected example. A key paper

Content from this work may be used under the terms of the Creative Commons Attribution 3.0 licence. Any further distribution of this work must maintain attribution to the author(s) and the title of the work, journal citation and DOI.
Published under licence by IOP Publishing Ltd

on the subject of noisy references is [16] where the advantages for detection in passive MIMO radar networks are developed. Bat performance in noisy environment has been studied in [17]. Bats were trained to discriminate the closer of two targets. The background noise level was increased in 10~dB steps. The discrimination performance degraded slightly until 10~dB SNR was reached. After this threshold discrimination dropped sharply.

How big is the loss of SNR relative to the matched Filter (MF) is of major concern for any practical application of a new algorithm. The purpose of this paper is to analyse the loss of SNR for a BSCT receiver. Analytical expressions are derived and then confirmed through simulations.

2. The BSCT receiver

The BSCT receiver is described in details in [1]. This section summarizes the key results and relationship defining the model. For practical applications the BSCT procedure should be applied after signal compression, e.g. matched filtering.

The frequency profile $E[f_i]$ of the BSCT receiver is defined for $M+1$ frequencies f_i, $i = -M/2, \ldots, M/2$:

$$E[f_i] = \int_{f_i-\frac{B}{2}}^{f_i+\frac{B}{2}} |X(f)|^2\, df, \tag{1}$$

where B is a parameter of the model, $x(t) \leftrightarrow X(f)$ is the input signal in time and frequency domain.

The spacing profile is an inverse Fourier transform of the frequency profile with subtracted mean value:

$$e(t) = \mathcal{F}^{-1}\{E(f) - \mathrm{mean}[E(f)]\}. \tag{2}$$

For two targets convolved with a waveform with complex envelope $x_0(t)$ and central frequency f_0:

$$x(t) = \left[\delta(t-t_1)\mathrm{e}^{-j2\pi f_0 t_1} + \delta(t-t_2)\mathrm{e}^{-j2\pi f_0 t_2}\right] x_0(t); \tag{3}$$

$$\begin{aligned} X(f) &= X_0(f)\mathrm{e}^{-j2\pi(f+f_0)t_1} + X_0(f)\mathrm{e}^{-j2\pi(f+f_0)t_2} \\ &= X_0(f)\mathrm{e}^{-j2\pi f^I t_1}\mathrm{e}^{-j\pi f^I \tau} 2\cos\left(\pi f \tau + \frac{\psi_\tau}{2}\right); \end{aligned} \tag{4}$$

$$|X(f)|^2 = 2|X_0(f)|^2\,[\cos(2\pi f\tau + \psi_\tau) + 1], \tag{5}$$

where t_1 and t_2 are the time delays of the targets, $\tau = t_2 - t_1$ is the delay between the targets, $f^I = f + f_0$ and $\psi_\tau = 2\pi\tau f_0$.

Let's consider a waveform with flat spectrum with spectral density A and bandwidth B_C:

$$|X_0(f)| = A\,\mathrm{rect}\left(\frac{f}{B_C}\right); \tag{6}$$

$$\mathrm{rect}(f) = \begin{cases} 1, & |f| \le 1/2 \\ 0, & \text{otherwise} \end{cases} \tag{7}$$

(i.e. an approximation of a band limited impulse with amplitude A, linearly frequency modulated chirp, stepped frequency signal or white noise signal). Then

$$|X(f)|^2 = 2A^2\,[\cos(2\pi f\tau + \psi_\tau) + 1]\,\mathrm{rect}\left(\frac{f}{B_C}\right) \tag{8}$$

and equation (1) simplifies to

$$\begin{aligned} E[f_i] &= 2A^2 \int_{f_i-B/2}^{f_i+B/2} [\cos(2\pi f\tau + \psi_\tau) + 1]\,\mathrm{rect}\left(\frac{f}{B_C}\right) df \\ &= 2A^2 B\,[\mathrm{sinc}(\tau B)\,\cos(2\pi\tau f_i + \psi_\tau) + 1]), \end{aligned} \tag{9}$$

for $|f_i| \le (B_C - B)/2$, where $\mathrm{sinc}(x) = \sin(\pi x)/(\pi x)$ [1]. For $|f_i| > (B_C + B)/2$, $E_E[i] = 0$.

The spacing profile for two targets is produced by replacing (9) in (2). A conventional MF range profile and a BSCT spacing profile are compared in figure 1 for two targets at 0.122 m, radio frequency linear chirp waveform with B_C= 4 GHz, and white Gaussian noise leading to SNR of 17 dB. The abscissa is in meters and in resolution gates, where one gate is equal to $c/2\ B_C$, c is the speed of light. The single prominent peak in the spacing profile indicates the separation between the two targets, as opposed to the two peaks in the range profile that indicate the ranges to the targets.

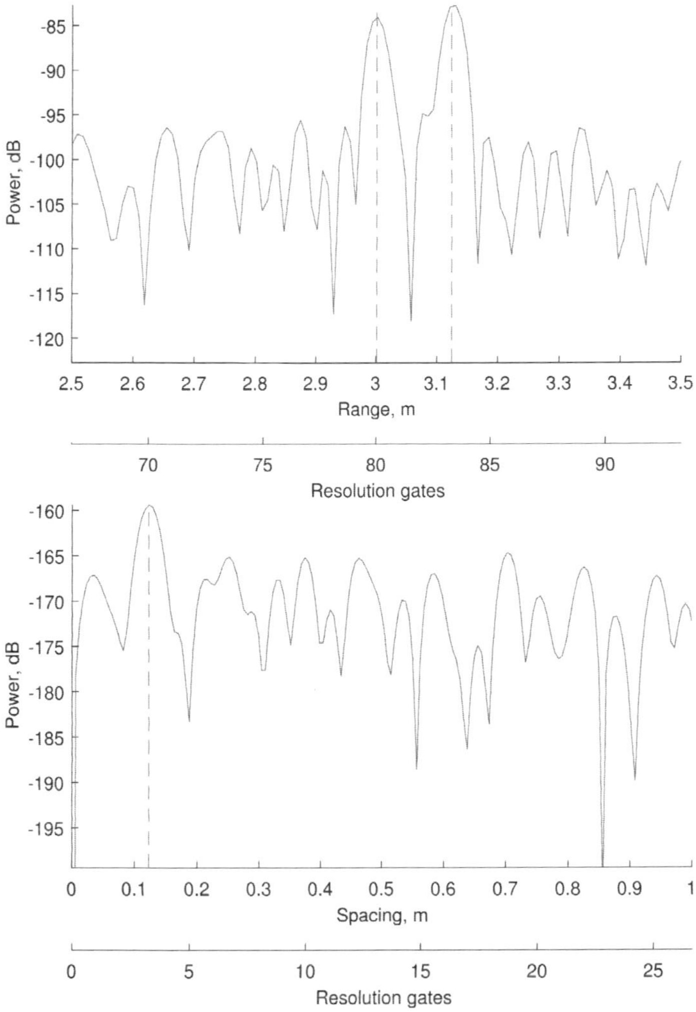

Figure 1. Matched filter range and BSCT spacing profiles for simulated targets separated by 0.122 m, 4 GHz bandwidth and 17 dB SNR.

3. SNR for a BSCT transformed signal

The SNR is defined as the ratio of the peak amplitude squared of the compressed signal and the average noise power (the noise variance).

Let's denote a noisy signal sampled at the Nyquist rate with $s(n)$ and let $w(n)$ is a band limited white Gaussian noise with variance σ_w^2:

$$s = x + w; \tag{10}$$

$$S = X + W. \tag{11}$$

First we will analyse a simplified version of BSCT without averaging (integration) in the frequency domain, i.e. frequency spacing and integration bandwidth are such that the frequency profile $E[f_i]$ (1) is equal to the frequency energy of the processed echo $|S(f_i)|^2$:

$$E[f_i] = |S(f_i)|^2 = (X + W)(X + W)^* = X X^* + X W^* + W X^* + W W^*$$
$$= |X|^2 + X W^* + (X W^*)^* + |W|^2 = |X|^2 + 2\,\mathrm{Re}(X W^*) + |W|^2 \qquad (12)$$

and for the two point target case, taking X from (4), (6) and $|X|^2$ from (8), the frequency profile (12) becomes

$$E[f_i] = 2\,A^2\,[\cos(2\pi\, f_i\, \tau\, + \psi_\tau) + 1] + 2\,\mathrm{Re}\left[2\,A\,e^{-j\,k(f,\tau)}\cos\left(\pi f_i \tau + \frac{\psi_\tau}{2}\right) W^*\right] + |W|^2, \qquad (13)$$

where $k(f,\tau)$ is a phase shift that we will ignore as it can go in the random phase of the noise. The noise energy $|W|^2$ can be decomposed into a constant and a zero-mean part $\widetilde{|W|^2}$. Therefore,

$$E = 2\,A^2\,[\cos(2\pi f_i \tau + \psi_\tau) + 1] + 4A\cos\left(\pi f_i \tau + \frac{\psi_\tau}{2}\right)\mathrm{Re}(W^*) + \widetilde{|W|^2} + \mathrm{mean}(|W|^2). \qquad (14)$$

For two targets that are not very close, i.e. $B_C\,\tau > 2$, we can simplify $\int_{-B_C/2}^{B_C/2}\cos(2\pi f \tau)\, df = 0$. Therefore, the mean removed frequency profile $\tilde{E}$ is

$$\tilde{E} = E - \mathrm{mean}(E) \approx 2\,A^2\,\cos(2\pi f_i \tau + \psi_\tau) + 4A\,cos\left(\pi f_i \tau + \frac{\psi_\tau}{2}\right)\mathrm{Re(W)} + \widetilde{|W|^2}. \qquad (15)$$

Equation (15) can be written as a sum of signal $\tilde{E}_X$ and noise $\tilde{E}_W$ components:

$$\tilde{E} = \tilde{E}_X + \tilde{E}_W, \qquad (16)$$

where the signal part is

$$\tilde{E}_X = 2\,A^2\,\cos(2\pi\, f_i\, \tau\, + \psi_\tau) \qquad (17)$$

and the noise part is

$$\tilde{E}_W = 4\,A\cos(\pi\, f_i\, \tau\, + \frac{\psi_\tau}{2})\,\mathrm{Re}(W) + \widetilde{|W|^2}. \qquad (18)$$

Inverse Fourier transform of the signal part (17) is two shifted scaled impulses (or sincs) signal e_x

$$e_x = \mathcal{F}^{-1}(\tilde{E}_X) = A^2\big(\delta(t-\tau) + \delta(t+\tau)\big) \qquad (19)$$

with peak power

$$\max|e_x|^2 = (A^2)^2. \qquad (20)$$

The power of a complex white Gaussian noise W with variance σ_W^2 has exponential distribution [18] with main statistics

$$\mathrm{mean}(|W|^2) = \sigma_W^2; \qquad (21)$$
$$\mathrm{var}(|W|^2) = \sigma_W^4. \qquad (22)$$

It is easy to switch to time domain considering that the Fourier transform of sampled in-band (complex) white noise is (complex) white noise and for M point signal [18]:

$$\sigma_W^2 = M\sigma_w^2; \qquad (23)$$

$$\mathrm{var}\,\widetilde{|W|^2} = \mathrm{var}\,|W|^2 = \sigma_W^4 = (M\,\sigma_w^2)^2; \qquad (24)$$

$$\mathrm{var}\big(\mathcal{F}^{-1}\widetilde{|W|^2}\big) = \frac{1}{M}\mathrm{var}\,\big(\widetilde{|W|^2}\big) = M\sigma_w^4. \qquad (25)$$

The Fourier transform of cosine modulated white noise is white noise with scaled power:

$$\mathcal{F}^{-1}\left(A\cos(2\,\pi\,\tau\,f)\;W(f)\right) = \frac{A}{2}\left(w(t-\tau) + w(t+\tau)\right)$$

$$\operatorname{var}\left(\mathcal{F}^{-1}\left(A\cos(2\,\pi\,\tau\,f)\;W(f)\right)\right) = \left(\frac{A}{2}\right)^2 2\,\sigma_W^2.$$

Therefore the cosine modulated white noise component in (18) has power

$$\operatorname{var}\left(\mathcal{F}^{-1}\left(4A\cos(\pi f_i\,\tau + \psi_\tau/2)\ \operatorname{Re} W(f)\right)\right) = 4A^2\sigma_W^2. \quad (26)$$

The noise power of the BSCT transformed signal follows from (18), (25), and (26), considering that the covariance of a normal random variable with its square is zero:

$$e_W = \mathcal{F}^{-1}(\tilde{E}_W);$$
$$\operatorname{var}(e_W) = 4\,A^2\,\sigma_W^2 + M\,(\sigma_W^2)^2. \quad (27)$$

The output of a filter matching s with x_0 is:

$$S_{\text{MF}} = S\,X_0^* = (X + W)X_0^* \quad (28)$$

and it can be shown that the MF SNR ratio is A^2/σ_W^2 .

The SNR after BSCT processing is

$$\text{SNR} = \frac{\max(e_x)}{\operatorname{var}(e_W)} = \frac{A^4}{4\,A^2\,\sigma_W^2 + M\,(\sigma_W^2)^2} = \frac{A^2/\sigma_W^2}{4 + M\,\sigma_W^2/A^2} = \frac{\eta}{4 + M/\eta}, \quad (29)$$

where η denotes the SNR of the compressed signal and M is the number of independent samples, i.e. characterises the time-bandwidth product for the signal, $M = T\,B_C$ (e.g. see Sec. 2.4 in [19]).

From (29) we can see that the SNR after BSCT is lower than after MF. The relationship is shown on figure 2 with continuous lines for differet values of M. For high SNR the reduction is 6 dB or the ratio of the signal energy and the noise power is reduced 4 times. For low SNR the change in dB for BSCT is two times the change in dB of the SNR. The limit curves are

$$\lim_{\eta\to\infty} \text{SNR}_{dB} = \eta_{dB} - 6; \quad (30)$$

$$\lim_{\eta\to 0} \text{SNR}_{dB} = 2\,\eta_{dB} - 10\,\log_{10} M. \quad (31)$$

The limit curves intersect at $\eta_{dB} = M_{dB} - 6$ with $\text{SNR}_{dB} = \eta_{dB} - 9$, dB. The corresponding noise level could be considered critical for BSCT as the SNR degrades faster at higher values.

Higher values of the time-bandwidth parameter M lead to worse performance of BSCT. This means that BSCT is applicable only for short sequences which is fine as far as it is used for local targets resolution.

Integration of the spectral energy over a bandwidth B will reduce proportionally the noise power. At the same time the signal e_x amplitude is scaled by $\operatorname{sinc}(\tau\,B)$, i.e. the close targets are preserved and the distant targets are attenuated. The effect of BSCT model parameters on the SNR could be subject of future research.

4. Verification of results through simulations

The BSCT SNR is evaluated by simulation of a simple scenario – a band limited impulse waveform $x_C = A\,\delta(t)$ and two point scatterer target $x_T = \delta(t - \tau) + \delta(t + \tau)$.

The following steps were implemented and run in Python (3.5.2) with Numpy (1.13.3) and Matplotlib (1.5.3):

1) generate noise waveform sequence $w[n]$ with variance σ^2, $n = 0, 1, \dots, M-1$;
2) generate signal $s[n]$ containing two impulses with amplitude A at $n = \pm p$;
3) calculate $\text{SNR} = A^2/\sigma^2$;

4) sum the noise and signal, $x[n] = w[n] + s[n]$;
5) apply BSCT transformation, $E = |\mathrm{FFT}(x[n])|^2$, $y[n] = \mathrm{IFFT}\big(E - \mathrm{mean}(E)\big)$, where (I)FFT is (Inverse) Fast Fourier Transform;
6) the BSCT peak is $A_{\mathrm{BSCT}} = A^2$, (20), at index $2p$, and for high SNR $A_{\mathrm{BSCT}} = y[2p]$;
7) calculate BSCT noise power σ^2_{BSCT} as the variance of the signal $y[n]$ with discarded values at index $\pm 2p$;
8) the SNR for BSCT is $\mathrm{SNR}_{\mathrm{BSCT}} = A^2_{\mathrm{BSCT}}/\sigma^2_{BSCT}$.

The results of the simulations are shown in figure 2 as points. Each point represents an SNR average from 1000 simulations. There is a good fit between simulated and calculated results, though some bias is observed for small values of *M*.

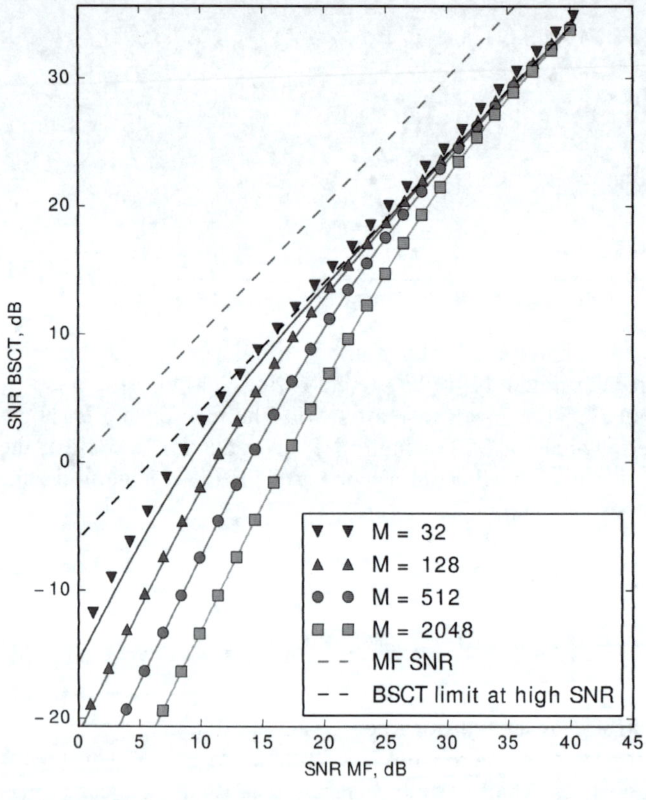

Figure 2. BSCT SNR as a function of signal SNR and the time bandwidth *M*. Continuous lines are calculated using (29). Points are averaged results from simulations with different values of *M*.

5. Conclusion

The bat inspired BSCT receiver provides 6 dB lower SNR compared to the MF when the processing window is small enough. There is a critical noise level above which the SNR of BSCT falls two times faster in dB than the SNR for MF.

The frequency integration/averaging bandwidth of the BSCT receiver can be increased to reduce the SNR loss. A side effect is attenuation of distant spacings from the spacings profile.

Topics for future research are optimisation of the BSCT parameters, exploration of the noise behaviour for target spacing close to the resolution limits, physical experiments with real targets and study of the clutter interference.

Acknowledgments

The authors would like to thank Alessio Ballery and Andy Stove for the detailed technical discussions and for their contributions to this work, and the Research and Development Sector at the Technical University – Sofia for the financial support for the publication.

References

[1] Georgiev K, Balleri A, Stove A and Holderied M 2018 Bio-inspired processing of radar target echoes *IET Radar, Sonar & Navigation* **12** 1402–09

[2] Simmons J A, Ferragamo M, Moss C F, Stevenson S B and Altes R A 1990 Discrimination of jittered sonar echoes by the echolocating bat, Eptesicus fuscus: The shape of target images in echolocation *J. Comparative Physiology* A **167** 589–616

[3] Schmidt S 1992 Perception of structured phantom targets in the echolocating bat, Megadermalyra *J. Acoustical Society of America* **91** 2203–23

[4] Simmons J A, Saillant P A, Wotton J M, Haresign T, Ferragamo M J and Moss C F 1995 Composition of biosonar images for target recognition by echolocating bats *Neural Networks* **8** 1239–61

[5] Simon R, Knornschild M, Tschapka M, Schneider A, Passauer N, Kalko E K V and von Helversen O 2014 Biosonar resolving power: echo-acoustic perception of surface structures in the submillimeter range *Frontiers in physiology* **5** 64

[6] Saillant P A, Simmons J A, Dear S P and McMullen T A 1993 A computational model of echo processing and acoustic imaging in frequency-modulated echolocating bats: The spectrogram correlation and transformation receiver *J. Acoustical Society of America* **94** 2691–712

[7] Wiegrebe L 2008 An autocorrelation model of bat sonar *Biological cybernetics* **98** 587–95

[8] Simmons J A, Gaudette J and Warnecke M 2017 *Biosonar Inspired Signal Processing and Acoustic Imaging from Echolocating Bats* (Biologically-Inspired Radar and Sonar: Lessons from Nature ed Balleri A, Grifiths H and Baker C Stevenage UK: IET) chap 2

[9] Peremans H and Hallam J 1998 The spectrogram correlation and transformation receiver, revisited *J. Acoustical Society of America* **104** 1101–10

[10] Park M and Allen R 2010 Pattern-matching analysis of fine echo delays by the spectrogram correlation and transformation receiver *J. Acoustical Society of America* **128** 1490–1500

[11] Georgiev K, Balleri A, Stove A and Holderied M W 2016 Baseband version of the bat-inspired spectrogram correlation and transformation receiver *2016 IEEE Radar Conf. (Philadelphia)*

[12] Georgiev K, Balleri A, Stove A and Holderied M W 2017 *Enhanced Range Resolution: Comparison with the Matched Filter* (Biologically-Inspired Radar and Sonar: Lessons from Nature ed Balleri A, Grifiths H and Baker C Stevenage UK: IET) chap 3

[13] Poirier J L 1968 Quasi-monochromatic scattering and some possible radar applications *Radio Science* **3** 881–886

[14] Dukat F 1969 *A simplified, CW, Random-Noise Radar System* (Monterey California U.S. Naval Postgraduate School: Calhoun)

[15] Kalinin V I and Chapursky V V 2006 Efficiency of two-stage spectral analysis for noise radar in the presence of clutter *J. Communications Technology and Electronics* **51** 286–96

[16] Hack D E, Patton L K, Himed B and Saville M A 2014 Detection in passive MIMO radar networks *2014 IEEE Transactions on Signal Processing* **62** 2999–3012

[17] Simmons J A 2017 Noise interference with echo delay discrimination in bat biosonar *J. Acoustical Society of America* **142** 2942–52

[18] Richards M A 2013 The Discrete-Time Fourier Transform and Discrete Fourier Transform of Windowed Stationary White Noise Tech. rep. https://radarsp.weebly.com/supplemental-tech-notes.html

[19] Woodward P M 1953 *Probability and Information Theory, with Applications to Radar* (London, England: Pergamon Press)

Integration of the environomic energy services for mobility and household using electric vehicle with a range extender of solid oxide fuel cell

Zlatina Dimitrova

PSA Groupe, Centre Technique de Vélizy, 78943 Vélizy- Villacoublay Cedex, France

E-mail: zlatina.dimitrova@mpsa.com

Abstract. This article studies an innovative concept for vehicle propulsion, considered to deliver mobility services and integrated energy services for a household. An innovative converter – a Solid Oxide Fuel Cell with gas turbines system (SOFC- GT) is powering an electric vehicle. The vehicle architecture is as a serial range extender used to charge the high voltage battery. The range extender vehicle is optimized according to techno-economic and environmental criteria. It is researched to deliver optimal mobility and is integrated to the extended energy system, including the household needs. When the vehicle is unused is considered to be connected to the household and to deliver part of the energy services. In this configuration, the services of extended energy system are optimized. The environomic optimization concerns the extended energy system – vehicle and household and optimal designs for the integrated system are researched. The vehicle integrates energy services for mobility and household and becomes part of the large-scale energy grid for integrated services. The vehicle is sized for both uses. The major advantage of the energy integrated dual system is the reduction of the total global warming potential (GWP) impact with around 30000 kg CO_2 eq. (60%). – from 50590 kg CO_2 eq. to 20338 kg CO_2 in comparison to separate energy services for mobility and household needs.

1. Introduction

The biggest part of the state of the art studies on energy conversion systems for vehicles are concentrated on the on-board energy storage and conversion systems during the phase where the vehicle is used, when this vehicle is driven. The holistic methods might be used to optimize not only the design and the energy systems of the vehicles, but are also enablers to evaluate new energy services of the vehicles propulsion systems, when the vehicles are not used for driving. They are related to the different energy grids-electricity and gas. In this system the vehicles are considered to be related to the energy grids. Their design is as well optimized for the additional energy services and functions in the enlarged energy integrated system.

The environomic optimization method is used in this article to explore the optimal design for vehicles with innovative propulsion system, considering the mobility and also the integrated energy service to the households. The environomic design of vehicles integrated energy systems consists in the systematic and holistic optimization for vehicle energy systems, according to technical, economic and environmental objectives. The method is developed for vehicles applications and is presented for electric vehicles in [1], for hybrid electric vehicles in [2] and for electric vehicles with innovative range extender module with SOFC in [3]. Solid oxyd fuel cells technologies are described in [4] and researched as range extenders for battery electric vehicles in [3] and [5].

Content from this work may be used under the terms of the Creative Commons Attribution 3.0 licence. Any further distribution of this work must maintain attribution to the author(s) and the title of the work, journal citation and DOI.
Published under licence by IOP Publishing Ltd

This article presents the application of the environomic design method for an extended energy system, where the vehicle is connected to the household and delivers as well the part of energy services of the household. This is the so called Vehicle-to-house configuration, part of the Vehicle-to-grid concepts are described in the literature. The idea to use that electric vehicles to supply power to the grid for stabilization and peak time supply is intersting for region where traditional forms of storage. Vehicle-to-grid scenrios have been proposed in [6]. Prototypes installations prouved that the technological means to deliver many of the vehicle-to-grid services are available. Charging electric vehicles can, however, be added to planned demand side management schemes without the need for additional capital investment. Balancing power plants operated with gas serve as stabilizers of the grid in the case of high renewable and variable electricity penetration in power systems. The smart-grid including electric vehicles is playing the grid stabilization role [7]. The results show that even a small share of the fleet of electric vehicles providing load balancing to the grid could lead to important reductions in gas use and energy excess. The integration of renewable energy in the electricity sector and as well the adoption of electric vehicles in the transportation sector have the potential to significantly reduce greenhouse gas emissions with Vehicle-to-Grid technology. The study [8] aims to evaluate the greenhouse gas emission reduction in case of intermittency coming from the introduction of wind power through. The researchers propose in [9] an optimal vehicle to grid planning and scheduling. According to [10], the strategy to reduce CO_2 emissions in the transportation sector is its renewable based electrification. However, the technical option vehicle-to-grid (V2G) requires the vehicle users to temporarily abstain from the usage of their batteries for V2G. 'Range anxiety' and the 'minimum range' are important for the acceptance of the users to participate in V2G. The litterature reflects separated optimizations of the vehciles use. Mainly vehicles-to-grid optimizations are done for electric vehicles, to support the electric grid.

This article considers the vehicle as an extended energy system (figure 1), related to the household and proposes optimal design for the energy services for mobility and for the household that the vehicle has to deliver. The optimization of the designs is done by using multi-objective environomic techniques.

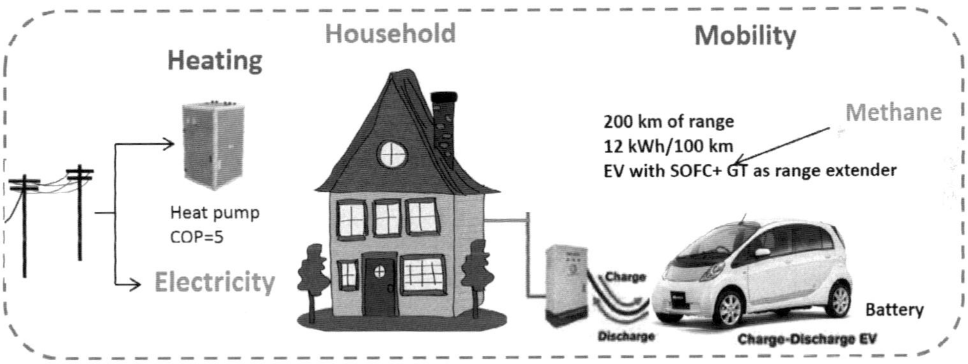

Figure 1. Definition of the extended energy system for the vehicle and the energy services.

2. Methodology for environomic design of integrated serices for the vehicle energy systems

In the environomic methods and techniques to design, the improvement and the optimizations of energy systems deal simultanously on the energy consumption, the economics and the environmental impacts.

The modelling of the SOFC GT module and the vehicle simulation models are explained in details in [3]. This article studies the environomic optimization of the extended energy system. The energy integration could be also activated in the superstructure. The energy integration method is presented in additional article [11].

3. Study definition

The study researches the dual mode of utilization of the system. This means that the driving and the residential modes are combined. In this mode, the SOFC-GT can be plugged into the house for electricity and heat co-generation.

3.1. Definition of the mobility and the residential needs
The mobility needs are estimated at 15000 km per year – 150000 km during 10 years. For the residential needs it is considered a household composed of 4 persons. The estimation of the needs of electricity consumption of this household is estimated using the average electricity consumption per capita in France in 2012. The mobility and the residential modes are summarized in the table 1.

Table 1. Definition of mobility and residential needs.

Mobility	
Annual distance, km/year	15 000
Total distance during lifetime, km	150000
Household with 4 persons and 120 m² heated surface	
Annual electricity needs, kWh	10000
Annual heat needs, kWh	7200
Annual operating hours for heating, h/year	2000

The vehicle mass characteristics are given in table 2:

Table 2. Vehicle characteristics.

Nominal mass, kg	1000
Mass of the methane tank, kg	25
Specific mass of the SOFC-GT, kg/kW	7.3

3.2. Strategy of the SOFC-GT use in dual mode
The optimization study targets the lowest possible investment cost by designing the SOFC-GT power for the vehicle. Due to its lower power, the SOFC-GT when used in residential mode will only provide a small part of the house electricity and heat needs. The rest of this needs are supplied by other energy sources (electricity for light and heat pump powering for heating). The part of the residential electricity provided by the SOFC-GT should be a design variable with the range from 0 to 1. This variable influences the operating cost and the CO_2 emissions during the usage.

3.2.1. Economic model and life cycle assessement (LCA) model. The economic and the LCA models of the SOFC-GT technology and the vehicle cost models are explained and presented in details in [3]. The study applies the economic model for SOFC-GT presented in [12].

The annualized investment cost is remained in equation (1).

$$AIC = IC\frac{i(1+i)^n}{(1+i)^n - 1} \quad (1)$$

The methane used in the SOFC-GT installation is considered produced from bio mass. We use GWP as indicator for equivalent CO_2 emissions and electricity produced in France.

4. Results: Multi-objective optimization for residential needs

4.1. Residential needs –base case scenario
In the base case scenario, electricity is provided from the grid (French electricity mix). Heating is satisfied with a gas boiler using natural gas as fuel.

4.1.1. Economic evaluation of the base case scenario. The investment cost of the energy technologies are annualized using the equation (1). The operating costs are calculated for the vehicle and the residential needs. The results are listed in table 3.

Table 3. Economic evaluation for the base mobility and residential scenario.

Driving	
Operating cost, €/year	901.5
Annualized investment cost, €/year	2328
Total driving yearly cost, €/year	3229.5
Residential	
Heating cost, €/year	492
Electricity cost, €/year	1505
Annualized boiler cost, €/year	828
Total residential cost, €/year	2825
Total operating cost, €/year	2898.5
Total annualized investment cost, €/year	3156
Total yearly cost, €/year	6054.5

4.2. Residential mode and environmental evaluation of the base case scenario
There are three sources of CO_2 emission in residential mode: methane combustion at boiler during the whole lifetime, eq. CO_2 emission of the fuel production for the total amount of fuel used in the whole lifetime and eq. CO_2 emission for the electricity production phase consumed in the house for 10 years. The CO_2 emission by combustion of methane in the boiler is computed using the CO_2 emission factor and the boiler efficiency. The results are summarized in table 4.

Table 4. Equivalent CO_2 emission for the base case scenario.

Mobility	
Car production GWP, kg eq. CO_2	3740
Gasoline production GWP, kg eq. CO_2	3775
Tank-to-Wheels CO_2 emission, kg eq. CO_2	17250
Total GWP / mobility, kg eq. CO_2	24765
Residential	
CO_2 emission by NG boiler, kg eq. CO_2	13640
GWP NG production, kg eq. CO_2	3265
GWP electric production, kg eq. CO_2	8920
Total GWP residential, kg eq. CO_2	25825
Total GWP, kg eq. CO_2	50590

The total GWP impact for the base scenario for mobility and residential is 50590 kg eq. CO_2 for the whole life duration (10 years exploitation and 150000 km). The mobility with internal combustion engine vehicle (ICEV) vehicle contributes to the half of the GWP impact. The tank-to-wheels CO_2 emission obtained by gasoline combustion represents 70% of the mobility with 17250 kg eq. CO_2.

In the residential mode the contribution of the different parts is more equilibrated. The electricity contribution based on the French mix is representing around 9000 kg eq. CO_2. This is a favourable case because the French mix is mostly based on nuclear energy. The introduction of other energy mixes majority based on coal will increase the contribution of the GWP impact of the electricity.

4.3. Vehicle as grid related system – mobility and residential mode optimization – optimal scenario
In the dual mode, the SOFC-GT is used for mobility and residential energy needs. The calculation for mobility is done on NEDC and the simulation starts with a fully charged battery. The energy for the battery charge is provided by the house charger and is produced from the solar panels that equip the house. Part of the electricity needs in residential mode is provided by the SOFC-GT module and the remaining part is supplied by the grid (French electricity mix is considered). The complement of heat is

supplied by a heat pump with a coefficient of performance (COP) of 5. The annual saving resulting from the use of SOFC-GT has to be enough to be able to amortize the total investment cost within the lifetime of the system defined at 10 years.

4.3.1. Parameters of the optimization. Two decision variables are added compared to previous optimizations – the SOFC-GT power in residential mode and the part of electricity provided by the SOFC-GT, $\beta_{el,sofcgt}$, –. The optimizer can decide the size of the SOFC-GT for driving or residential mode.

The yearly cost is an effective way to account the cost of both modes. The yearly cost includes:

- The annualized investment cost of the vehicle;
- The operating cost of methane used in driving mode for 150000 km;
- The operating cost of methane used in residential mode for electricity production during 10 years;
- The operating cost of heat production in the house via the heat pump during 10 years;
- Operating cost of pure oxygen needed for the driving and residential mode during 10 years;

The computed GWP includes:

- GWP of the vehicle production phase;
- GWP of the production of methane (wood gasification) needed for mobility and residential mode during 150000 km (driving) and 10 years (residential);
- GWP of the production of the amount of electricity used by the household during 10 years;
- CO_2 emission in both driving and residential mode during 150000 km and 10 years.

The amount of CO_2 emitted during driving and residential mode is counted in the total GWP to be minimized. The decisions variables are summarized in table 5. The 3 dimentional multi- objective optimization function is defined in equation (2):

$$\min(-autonomy\ (x),\ yearly\ cost\ (x), total\ GWP\ (x),\ x \in (Decisions\ variables). \qquad (2)$$

Table 5. Definition of the decision variables for a dual application of the range extender module.

Decision variables	
Electric machine power, kW	[15-50]
Battery capacity Q_batt, kWh	[5-80]
Power of the SOFC-GT module for driving, kW	[1-10]
Power of the SOFC-GT module for residential, kW	[1-10]
Mass of methane, kg	[5-15]
$\beta_{el,sofcgt}$, –	[0-1]

The resulting Pareto curve is shown in figure 2. The solutions converge between 6400 €/year and 7800 €/year regarding the yearly cost and between 20000 and 45000 kg CO_2 eq. regarding the total GWP during the lifetime. The yearly cost and the total GWP of the base case scenario are 6055 €/year and 50590 kg eq. CO_2. The studied integrated system cost more (from 400 to 1800 €/year) than the base case but the GWP impact for all solutions is improved. The families of solutions in the transient autonomy zone, already discussed in the previous section, are also present in this Pareto. They are represented on the figure 2 (b). The variation of the solutions is large. For example, for the autonomy of 200 km, the solutions vary between the 6900 €/year and 7300 €/year. The solutions with less cost and bigger GWP give the lower scatter band. The solutions with bigger cost and lower GWP give the high scatter band. The width of the Pareto curve increases as the autonomy increases.

The evolution of the SOFC-GT designs is plotted on the figure 3. It shows that the autonomy increases when the design power of the range extender increases, as observed in the previous optimizations. In this multi- objective optimization (MOO) the use phase is considered also, and as it contributes to the major impacts and cost, one can observed that the battery design capacity doesn't influence a lot the solutions. Figure 4 shows that the battery size between 5 and 15 kWh is selected in the most of the solutions. The efficiency of the converter and the methane production way influence on an important way the use phase CO_2 emission.

In order to identify for which application the SOFC-GT is optimally designed, a Boolean variable is created (table 5). This variable is equal to 1 when the range extender is designed for the residential needs and is equal to 0 when the range extender is designed for the mobility needs. Figure 5 shows that in most of the cases the SOFC-GT is optimized to satisfy the mobility needs and to extend the vehicle autonomy. Figure 6 shows that the repartition coefficient $\beta_{el,sofcgt}$ between the part of electricity provided by the SOFC-GT and the part provided by the electrical grid is related to the cost and the GWP.

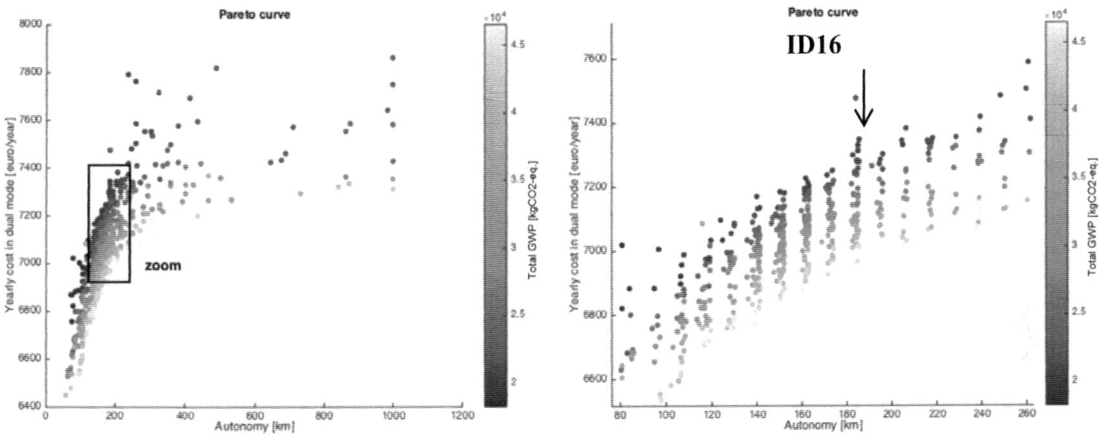

(a) (b)

Figure 2. Yearly cost/ autonomy Pareto curve in dual mode, GWP in color bar.

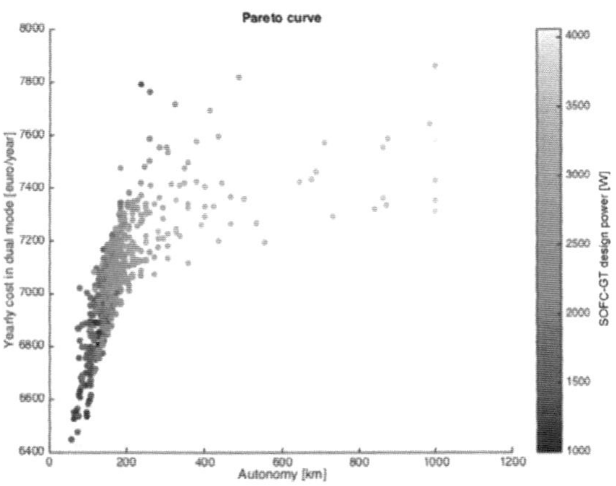

Figure 3. 3D MOO Pareto curve, SOFC-GT design in colorbar.

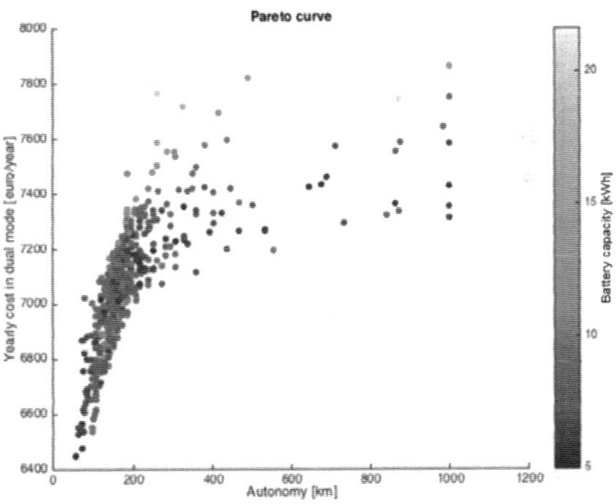

Figure 4. 3D MOO Pareto curve, battery design in color bar.

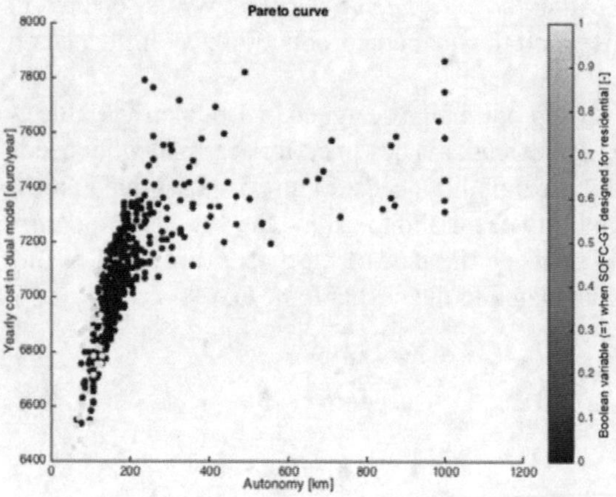

Figure 5. Boolean variable=1 when the SOFC is designed for residential use, in color bar

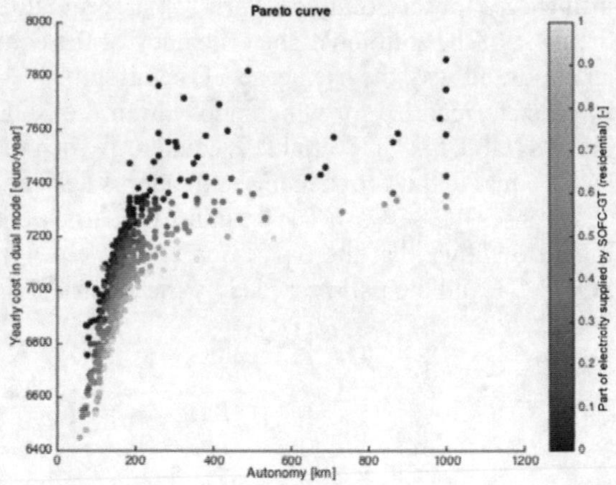

Figure 6. 3D MOO Pareto curve, with the part of the residential electricity from SOFC-GT in color bar.

If a low cost is targeted (meaning a high GWP impact), the optimizer choice is that 100% of the energy needs for the household are provided by the SOFC-GT (figure 7). The total GWP is strongly dependent on the variable $\beta_{el,sofcgt}$ and this is logical because most of the GWP comes from the use phase (the electricity use in the residential mode). The electricity needs per year is 10000 kWh and this means 100000 kWh for 10 years. Compared to that and considering a consumption of 11.8 kWh/100 km for the range extender vehicle, as observed in previous optimizations, the amount of energy needed for 150000 km is 17700 kWh. This value represents 18% of the consumption of electricity in residential mode.

The electricity mix production has a determinant impact in the optimizer choice of the $\beta_{el,sofcgt}$ and consequently on the GWP impact. The CO_2 contribution for 1 kWh is compared for different production mixes (for different countries) in table 6. The CO_2 contribution for the production of 1kWh of electricity with the SOFC and biomethane is also illustrated. One can see that the electricity produced from SOFC-GT and biomethane has more CO_2 contribution than electricity produced in France. For the case of a house supplied with electricity supplied from France, the optimal GWP will be found for solutions supplying the total electricity needs from the grid, the SOFC-GT is used at the minimum. This is the case illustrated on the figure 8.

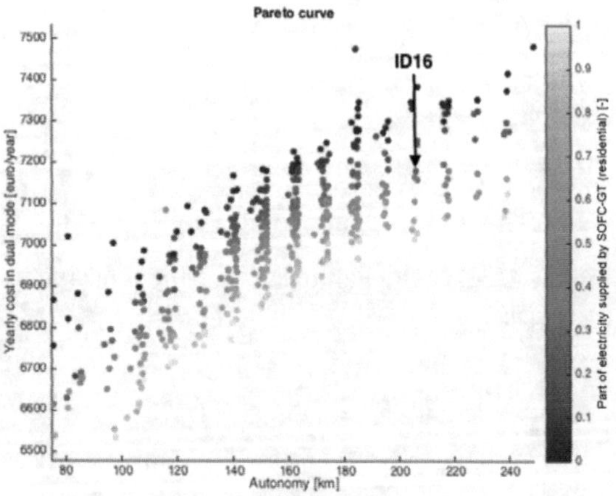

Figure 7. 3D MOO Pareto curve, with the part of the residential electricity from SOFC-GT in color bar (zoom in the autonomy transition zone).

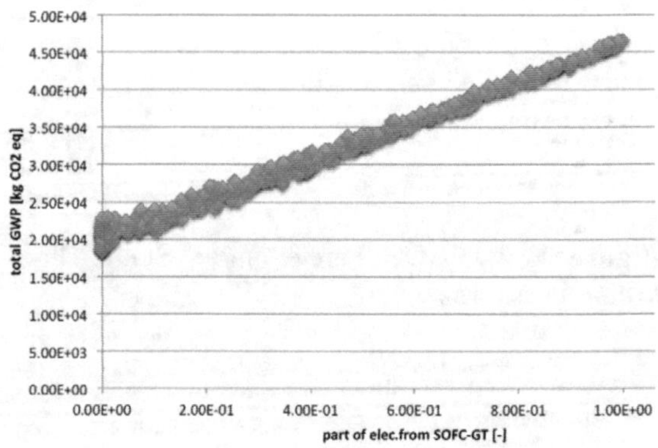

Figure 8. The total GWP as a function of the part of residential electricity from the SOFC-GT (French electricity mix).

Table 6. Specific GWP for different sources of electricity.

Source of electricity	Specific GWP of electrical production, eq. kg CO_2/kWh
Mix France	0.0892
SOFC-GT	0.2805
Mix Germany	0.638
Mix Poland	1.1

If the German or the Polish electricity production mixes were chosen then the optimizer to minimize the GWP, would surely privilege the use of SOFC-GT for the total providing of the residential electricity needs by the SOFC-GT module.

The optimal solution for a system with integrated energy services is chosen in ID 16 (figure 7). ID 16 presents autonomy of 200 km, which is the targeted autonomy for current electric vehicles. Globally, all integrated solutions present higher yearly cost than the base case. This is due to the high investment cost for the SOFC-GT installation. But all obtained solutions for the integrated services have lower total GWP impact than the base case separated house and mobility energy services. Therefore choosing a high autonomy vehicle for competitiveness with the ICEV would end up not economically competitive at all considering the high yearly cost. The vehicle with 200 km of autonomy integrated to the house energy services gives a balanced solution between cost and GWP. This solution is with integrated energy services for residential and driving mode. The details of the ID16 are presented in table 7. The comparison with the base energy scenario is also done.

The detailed contribution of the cost and the GWP of the ID 16 are compared to the base scenario. One can observe that with the chosen solution, the main outcomes are: the yearly cost 1000 € higher for the integrated solution and the GWP is with around 30000 kg CO_2 eq. smaller than in the base case.

Table 7. Comparison between ID16 at 200 km of autonomy and the base case scenario.

	ID 16	Base case
Power Electric machine, kW	32.28	
Capacity of the battery, kWh	6.28	
SOFC-GT power for driving, kW	2.80	
SOFC-GT power for residential, kW	1.60	
Mass of methane, kg	10.2	
$\beta_{el,sofcgt}$, –	0.33	
Range/Autonomy, km	206	1000
Vehicle cost, €	28430	20900
Mobility		
Operating costs, €/year	127	900
Annualized investment costs, €/year	3681	2328
Total driving yearly costs, €/year	3808	3229
Car production GWP, kg eq. CO_2	6596	3740
Fuel production GWP, kg eq. CO_2	768	3775
T-t-W CO_2 emission, kg eq. CO_2	3594	17250
Total GWP driving, kg eq. CO_2	10958	24765
Residential		
Total residential yearly costs, €/year	3258	2825
Total GWP residential, kg eq. CO_2	9380	25825
Total operating costs, €/year	1660	2898.5
Total annualized investment costs, €/year	5406	3156
Total yearly costs, €/year	7066	6054.5
Total GWP, kg eq. CO_2	20338	50590

5. Conclusions

In the study, the vehicle is considered as grid related to the house and this means that its energy service of mobility and comfort is related to the energy services for the family house. The study contributes to consider in an integrated scale mobility and energy services for a household and introduces the use of the vehicle in dual mode. SOFC-GT module is introduced as a range extender of the autonomy of the electric vehicle. A base energy scenario for the house is also defined. The total GWP impact of the base case scenario is estimated to 50590 kg CO_2 eq. The multi-objective optimization for environomic design of the new vehicle- house system is performed for the definition of the environomic design of the driving and the residential mode. In the case of the energy supplied in France the optimization sizes the SOFC-GT module for the mobility needs. The rest of the energy needs of the house are covered by the grid supply. Optimal design of the range extender vehicle in dual mode is proposed for 200 km of autonomy, investment cost of 28430 € and 3808 €/year of total driving yearly cost. The base case scenario has lower investment (20900 €) and total driving yearly cost (3229 €/year).

The total yearly cost for the integrated system in dual mode stays still higher (1000 €/year) than the base case separately delivered energy services. This is due to the high investment cost of the equipment in the vehicle and the house. The major advantage of the energy integrated dual system is the reduction of the total GWP impact with around 30000 kg CO_2 eq. (60%). – from 50590 kg CO_2 eq. to 20338 kg CO_2 eq. This study illustrates large scale economic and environmental benefits opportunities by integration of the energy services for mobility and household.

References

[1] Dimitrova Z Maréchal F 2014 Environomic design of vehicle energy systems for optimal mobility service *Energy* **76** 1019–28
[2] Dimitrova Z and Maréchal F 2015 Techno-economic design of hybrid electric vehicles using multi objective optimization techniques *Energy* **91** 630–44
[3] Dimitrova Z and Maréchal F 2017 Environomic design for electric vehicles with an integrated solid oxide fuel cell (SOFC) unit as a range extender *Renewable Energy* **112** 124–42
[4] Van Herle J , Ihringer R , Sammes N M , Tompsett G, Kendall K , Yamada K , Wen C , Kawada T , Ihara M , Mizusaki J 2000 Concept and technology of SOFC for electric vehicles *Solid State Ionics* **132** 333–42
[5] Bessekon Y , Zielke Ph , Wulff A C , Hagen 2019 A Simulation of a SOFC/Battery powered vehicle *International Journal of Hydrogen Energy* **44** 1905–18
[6] Mullan J, Harries D, Bräunl Th and Whitely S 2012 The technical, economic and commercial viability of the vehicle-to-grid concept *Energy Policy* **48** 394–406
[7] Pedro N and Brito M C 2017 Displacing natural gas with electric vehicles for grid stabilization *Energy* **141** 87–96
[8] Zhao Y, Noori M and Tatari O 2017 Boosting the adoption and the reliability of renewable energy sources: Mitigating the large-scale wind power intermittency through vehicle to grid technology *Energy* **120** 608–18
[9] Tan K M, Ramachandaramurthy V K and Yong J Y 2016 Optimal vehicle to grid planning and scheduling using double layer multi-objective algorithm *Energy* **112** 1060–73
[10] Geske J and Schumann D 2018 Willing to participate in vehicle-to-grid (V2G)? Why not! *Energy Policy* **120** 392–401
[11] Dimitrova Z and Maréchal F 2015 Energy integration on multi-periods and multi-usages for hybrid electric and thermal powertrains *Energy* **83** 539–50
[12] Akikur R K, Saidur R, Ping H W and Ullah K R 2014 Performance analysis of a co-generation system using solar energy and SOFC technology *Energy Conversion and Management* **79** 415–30

Performance and economic analysis of an organic Rankine Cycle for hybrid electric vehicles

Zlatina Dimitrova

PSA Groupe, Centre Technique de Vélizy, 78943 Vélizy- Villacoublay Cedex, France

E-mail: zlatina.dimitrova@mpsa.com

Abstract. Car manufacturers need to develop efficient technologies to reduce emissions and save fuel in order to cater of European regulations and satisfy customers' requirements. Internal combustion engines have a main limitation on efficiency due mainly to the heat losses. Processing an energy integration, it has been shown than an Organic Rankine Cycle (ORC) can improve the overall powertrain efficiency by recovering the heat into electricity. In parallel, hybrid electric vehicles attractiveness increases as it enables important energy saving. The contribution of this article is to design highly efficient vehicles, using both Organic Rankine Cycle and hybrid electric technologies. An adapted methodology based on energy integration techniques is required to choose the best points for the integrated system design. This study applies a methodology on hybrid electric vehicles, so as to define the powertrain configuration of the vehicle that is energy integrated. The energy recovery potential of a single stage Organic Rankine Cycle for a thermal engine in combination with a mild hybrid electric powertrain is studied. The assessment is done for different drive cycles. A study of economic feasibility and fuel consumption improvement is also done, in order to characterize the integrated energy system.

1. Introduction

The efficiency gains of the energy systems for vehicles are directly related to the fuel saving and the emissions reductions. The development of efficient technologies on the driving profiles and the comfort demands is then needed. There comes the idea to define a vehicle energy system with integrated energy services for mobility and comfort. With the definition of the system we need an energy integration methodology. The energy flows are integrated into the energy system to increase the efficiency. The energy integration considers the dynamic and comfort profiles and uses technologies recovering waste energy into the energy system. Organic Rankine Cycle (ORC) is selected to recover the heat waste of the internal combustion engine (ICE). Waste heat recovery technologies and hybrid electric vehicles are briefly introduced in the following section.

1.1. Waste heat recovery and hybrid electric vehicles

The vehicle is a dynamic system. The size and the efficiency of the converters depend on the dynamic driving profile. Energy integration is used to increase the energy efficiency of the vehicle energy system. An adapted methodology is proposed in [1] to select the best points for the integrated design of the system. The method clusterizes the dynamic profile on typical multi-periods where the vehicle is used. The design of the energy system is optimized for these main multi-periods. The energy integration is applied in [1] on hybrid electric vehicles. The potential of energy recovery of a single stage Organic Rankine Cycle is studied. The ORC is considered for a thermal engine in a hybrid electric powertrain.

Content from this work may be used under the terms of the Creative Commons Attribution 3.0 licence. Any further distribution of this work must maintain attribution to the author(s) and the title of the work, journal citation and DOI.
Published under licence by IOP Publishing Ltd

The contribution of the ORC is assessed for different drive cycles. After the process integration, a multi-objective optimization defines the design of a hybrid electric vehicle with an optimal waste heat recovery system. The energy balance of the internal combustion engine, for different operating points, is defined and analysed in [2]. An energy integration methodology, based on process integration techniques is discussed in [2]. It is applied on the extended energy system of the vehicle. The possibilities of an ORC to recover waste heat from a 900 kW fast passenger ferry Diesel engine are investigated in [3]. The authors analyse in [4] the feasibility of an "on-board" innovative and patented ORC recovery system. The vehicle thermal source can be either a typical diesel engine (1400cc) or a small gas turbine set (15-30kW).

Hybrid vehicles can use energy storage systems to disconnect the engine from the driving wheels of the vehicle. This option enables the engine to be run closer to its optimum operating condition. Nevertheless fuel energy is still wasted as heat through the exhaust system. The article [5] presents the model of the engine of a diesel-electric hybrid bus, coupled with a hybrid powertrain. The performances of a hybrid vehicle over a drive-cycle are analysed. This showed that including the waste heat recovery technology reduces the fuel consumption with 2.4% for a typical drive-cycle. The ORC has large applications such as agricultural biogas combined heat and power engines, biomass boilers, landfill biogas plants, small geothermal plants and concentrated solar thermal systems [6]. Most of the projects related to ORC systems, in the transport industry, discuss performance optimized solutions that include heat recovery in engine exhaust gas. The authors describe in [7] the ddevelopment and exergetic assessment of a new hybrid vehicle incorporating gas turbine as powering option. A multiple heat source supercritical ORC (Organic Rankine Cycle) for vehicle waste heat recovery is studied in [8]. A vehicle energy-supplying system is proposed to satisfy energy demand of vehicle in every season in [9]. This system is based on an ORC. The authors modeled a hybrid electric heavy duty vehicle to assess energy recovery using a thermoelectric generator in [10].

Hybrid electric vehicles have two or more prime movers as energy converters (internal combustion engine (ICE) or fuel cell and electric machine) and power sources (fuel tank, hydrogen tank and electrochemical battery) on their board. The main motivation for hybrid electric vehicles (HEV) development is the efficiency increase of the powertrain in comparison to the thermal (gasoline or Diesel) engines powertrain. HEV offer as well long ranges of autonomy, comparable to thermal vehicles. HEV vehicles recover the kinetic energy during braking. However, the ICE still has thermal losses due to the waste heat when is operating. In this case, the recovery of the waste heat is possible by using an external cycle- the Rankine cycle. The ICE energy balance is showing that there are two main sources of heat: the water jacket and the exhaust gases [1, 2]. The levels of the temperature between the water jacket and exhaust gases are different from 80°C to 900°C. To value the low heat temperature, an organic fluid is proposed to be used as working fluid.

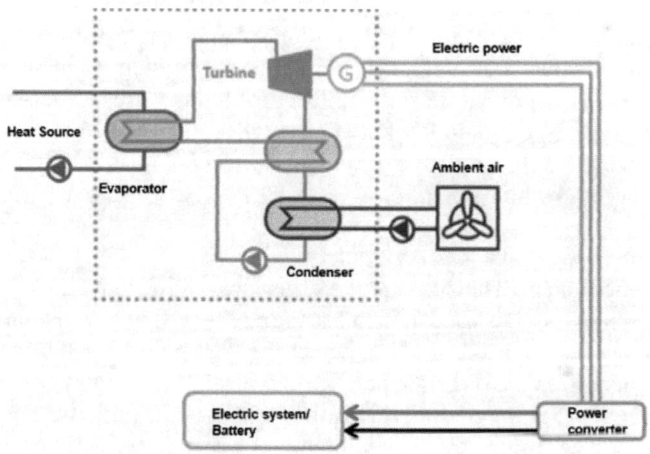

Figure 1. ORC system to recover the waste heat of the ICE.

An ORC presented in figure 1 will be investigated as principal utility for the energy integration. Its operation will be optimized for the operation points of the ICE, adapted for HEV. As shown in the figure 1 the turbine is linked to an electric generator and so an electricity is produced from the waste heat of the engine. The electricity can be then used for the battery fulling or to be sent the electrical network of the system, driving different accessories, devices around the engine or the consumers on the vehicle board (figure 2). The ORC receives the heat from the water circuit and the gases of the exhaust of the engine. The turbine out is related to the high-speed generator which produces electricity. The electricity is converted into DC and to the adapted voltage level and enters into the high voltage battery. The functional architecture is presented in a Mild hybrid electric vehicle is presented in figure 2.

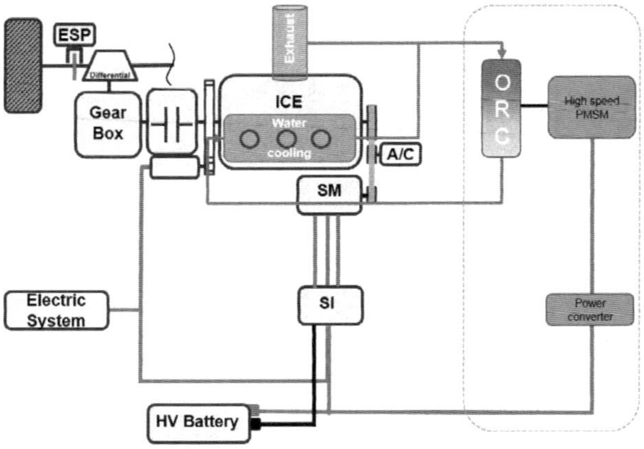

Figure 2. Functional architecture of the ORC integration in a Mild HEV Architecture.

The target of this article is to illustrate the results from a multi-constrain optimization, which take into consideration performance and economic constrains, in order to achieve an efficient design of an ORC loop for recovery of the waste heat on a hybrid electric vehicle. The presented design is adapted and evaluated to operate on hybrid electric vehicle's architecture. Once we perform the optimization, we should study if there is one design solution which is better than the other, or if there exists different solution which could correspond to different users' profiles. Finally, for the best design we should perform both a study on driving cycles, and an economic study, to determine if the Organic Rankine Cycle benefit is important enough to be implemented on those existing vehicles.

The contribution of this article is to apply energy integration methodology on mild hybrid electric vehicles for optimal powertrain design of hybrid electric vehicles. The article studies the impact of an energy recovery technology, such as the ORC on the mobility performances of HEV energy system. After integration of the energy services, a sensitivity analysis on the fuel benefits is done for different real drives representative driving cycles. The results are obtained by simulation, using vehicles models containing experimental engines and vehicles data. The ORC performances are obtained by simulation.

2. Methodology - Multi objective optimization and energy integration

The optimal size of the powertrain components is defined with multi-objective optimization. The correct sizing of the powertrain is important for the operating strategies and to reach optimal fuel consumption. In this article, the multi-objective optimization is realized with the tool presented in figure 3. The tool has already been described in [11]. The vehicle simulation model including with dynamic and thermal layouts is part of the superstructure. The cost equations are defined in the economic model. The energy integration model has as inputs the results from the dynamic flows and thermal flows calculations. The genetic algorithm optimises the final solutions. The optimization considers the following parts – a master multi objective optimization (MOO), a thermo economic simulation (TES). The energy integration is incorporated in the superstructure and is a slave optimization. The optimization process ends with the techno-economic evaluation (TEE). After the identification of the typical operating points, the energy balance is obtained from the thermal layout of the engine. The energy system to be integrated is then defined in the part "Energy Integration" of energy integration model, by listing the "cold" and "hot" streams between which a heat exchange has to occur, presented in [1] and [2]. The minimum energy requirement of the system is estimated. A utility is used to close the energy balance and estimate the energy recovery potential.

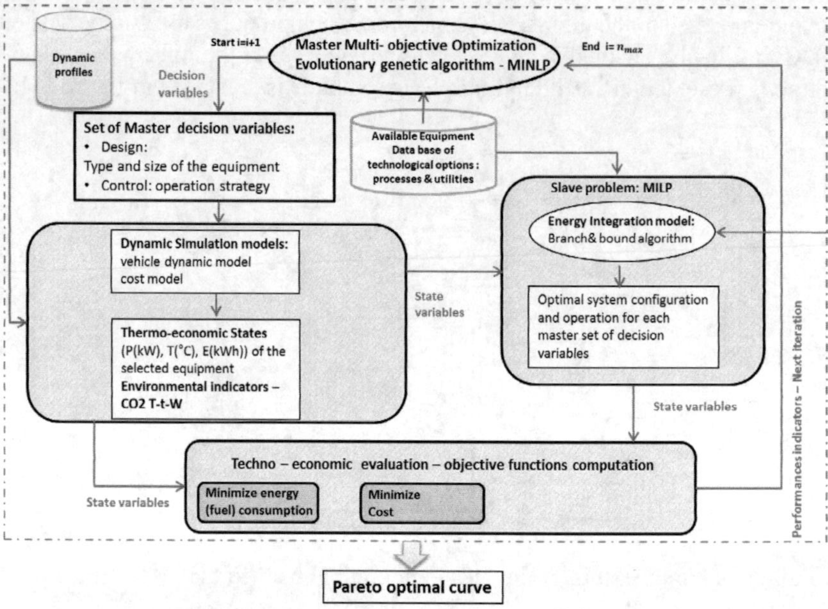

Figure 3. Framework of the energy integration and optimization program [11].

2.1. Focus on the operating points

According to [1, 2], the best ORC we can implement on a car should use two streams from the vehicle for the evaporator: the water cooling system, and the exhaust gases. Moreover, the cycle has a much better efficiency if it reaches a superheating temperature.

Usually, hybridization methods are efficient at low power of the engine. These zones allow to use an additional electric motor or to recover energy from braking. However, the ORC is more efficient at full load operations, when the total quantity of heat to recover is higher. This seems interesting to add such a system to a hybrid vehicle, as the two range are different: an electric device work mainly during urban cycles, whereas an ORC works on national roads or highway. We want to emphasize this complementariness, we try to optimize the cycle for high-loaded operating points.

2.2. Economic model

The economic model is described in [11]. We consider a serial C-Segment vehicle. The vehicle is equipped with a 1.2 l 3 cylinders gasoline engine and 60 kW of power. An automatic gearbox is as well considered.

As economic indicator, the total price of the ORC is defined. Here just the investment cost is considered. This cost is a linear function to consider the parts produced on large units' volume. The investment cost is the sum of the following equations (1-7):

$$Cost_{evaporator} = 18.84\ Area_{evaporator} + 14.49, \text{ Euro, area in m}^2; \quad (1)$$

$$Cost_{condenser} = 4.01\ Area_{condenser} + 4.96, \text{ Euro, area in m}^2; \quad (2)$$

$$Cost_{pump} = 0.083\ flowrate + 165, \text{ in Euro, flowrate in kg.s}^{-1}; \quad (3)$$

$$Cost_{turbine} = 0.015\ Power_{turbine} - 30, \text{ in Euro, power in kW}; \quad (4)$$

$$Cost_{pipe} = 26, \text{ Euro;} \tag{5}$$

$$Cost_{fluid} = 10, \text{ Euro;} \tag{6}$$

$$Cost_{ORC} = Cost_{evaporator} + Cost_{condenser} + Cost_{pump} + Cost_{turbine} + Cost_{pipe} + Cost_{fluid}, \text{ Euro.} \tag{7}$$

2.3. *Waste heat recovery*

2.3.1. Multi objective optimization settings. We want to perform an economic, environmental and technical performance optimization. To do that we should decide what indicator to use.

- Economic: the ORC investment cost is considered as an indicator, defined by equation (7).
- Performance: The net output power, used as an objective for performance. The net power is the difference of the turbine power and the pump power for a given operating time in equation (8):

$$P_{net} = P_{turbine} - P_{pump}, \text{ kW.} \tag{8}$$

The exergetic efficiency is verified for every solution.

2.3.2. Multi objective optimization settings. We need to choose the variable of the ORC design. The system has two degrees of freedom. We choose two define two temperature to launch the energy integration model:

- The evaporation temperature of the fluid. The interval of variation is defined between 330 °K and 360 °K in order to obtain correct exergetic efficiency.
- A difference of temperature, $\Delta T_{\min}$ [5], between the exhaust gas temperature and the superheat temperature of the cycle. The interval of variation is defined between 100 °K and 400 °K. This range is for second order for the results.

We use a simulation of a population with 200 individuals with a maximum number of evaluation of 10. The computing time is enough, and we get enough design to have a precise Pareto curve of the net output power depending on the investment cost.

3. Multi objective optimization results

3.1. *Decision variables*

We need to choose the variables for the ORC design.:

- The evaporation temperature of the fluid T_{evap}. The interval of variation is defined between 330 °K and 360 °K. The criteria is to obtain good values of the exergetic efficiency.
- The superheat temperature $T_{superheating}$ is defined between 700 °K and 1050 °K.

The function to be optimized for the ORC design is defined as:

$$\min(-\dot{P}_{net}(x), Cost_{ORC}(x)), \text{ with } x \in X \text{ } (decision\_variables). \tag{9}$$

3.2. *Pareto-curves*

The multi-objective optimization converges on the solutions presented on the Pareto curve (figure 4). The trade-off between the ORC net power and the investment cost are illustrated to the Pareto curves. The best value of exegetic efficiency is reached at 25%. This is for investment cost interval between 350 and 700 Euros.

3.2.1. Method to get the ORC maps. As we want to get the ORC characteristics for each operating point, we should consider the new hot streams and cold streams, as defined in [1] and [2]. Firstly, we do the following assumption: the compressor power is the same for each operating point; we consider the flow rate to be the same. The heat cascades are defined in the energy integration method. By knowing the temperatures and the exchanged power, the energy integration module computes the power maps, the energy efficiency and the exergetic efficiency maps of the gasoline engine (figure 5). The ORC map comes as a solution of the energy integration method.

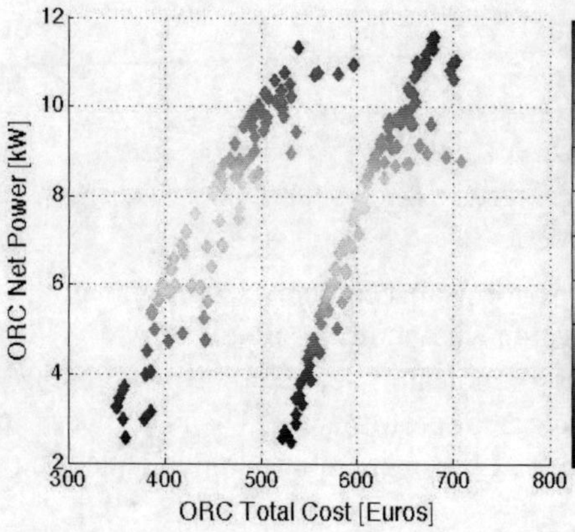

Figure 4. ORC Pareto curves: techno-economic trade-off, exergy efficiency in [-], in the colorbar [11].

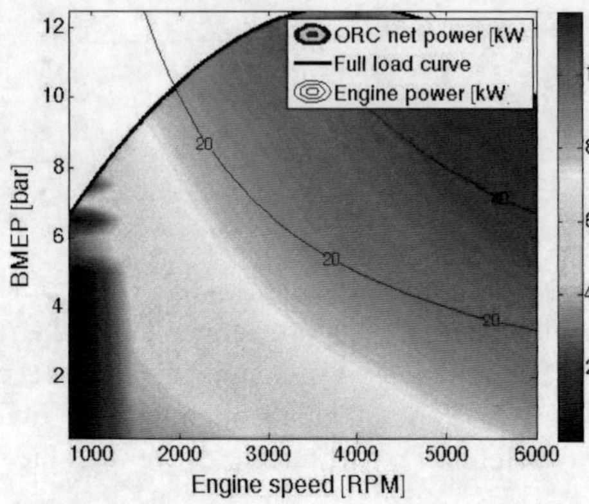

Figure 5. Map of the ORC power output, design after multi-objective optimization, ORC net power in [kW], in the colorbar [11].

The maximum net power output is reached over the engine iso-line of 30 kW. This is the area of high rotation speeds and high loads. The optimal design of the ORC design is investigated for the loaded points. The high seeds and high loads zone is also the region with the most important heat losses. The selected design is extended on the engine operating field. One considers that the pump operates with permanent parameters.

3.3. Simulation results on driving cycles

This study extends the methodology for design of vehicle energy systems on other drive cycles. The peri- urban cycle represents the home to work commuting. The holiday cycle is representative for long distances drives. The New European driving cycle (NEDC) is well known as reference and is as well considered. Table 1 summarizes their characteristics.

Table 1. Drive cycles characteristics.

Cycle [a]	Distance, km	Duration, s	Average speed, km/h	Min acceleration, m/s²	Max acceleration, m/s²
NEDC	11.023	1180	32.26		
Peri-urban	39	2440	57	–2	2
Holiday	847	28800	105	–1.9	1.9

[a] The peri-urban and holiday cycles are based on the parts of the WLTP cycle.

3.3.1. Comparison of different configurations. The gasoline engine in the hybrid electric powertrain obtains an additional flow of electrical power thanks to the heat recovery from the ORC. The studied architecture is a mild hybrid electric powertrain, as presented in figure 2. The hybrid electric powertrain has already an electric energy recovery system. This category of recovery system has an electric energy storage in the high voltage battery and is an effective solution for low speed drives, where the power delivered by the electric part of the powertrain is enough to drive the vehicles on purely electric drive. The fuel consumption is zero. The selected ORC design is optimized for the area with high rotation speeds and high loads. This is a complementary zone, where the mild hybrid electric powertrain is not acting. The simulations results are displayed in the table 2.

The fuel consumption result on NEDC for the basic configuration of the vehicle is 4.6 l/100 km.

The Mild HEV is a low degree electrical hybridization of the vehicles powertrains, characterized with low capacity of the high voltage battery – between 1.2 kWh and 6 kWh. The efficiency improvement of the MHEV on the NEDC and the Peri-urban cycles is around 10 %, because these cycles except the urban part, have also a large extra urban part, which is more energy demanding and where the MHEV mode is not active. On the holiday cycle, the MHEV is not used. In conclusion the Mild HEV is an adapted technology for the city drive and can be consider as effective technology for urban drives.

The ORC design proposed for the net power map of the figure 5 is optimal for high load and high speed engines operating points. For these points we have high temperature exhaust gases. Thus the waste heat recovery potential is important. The impact proposed ORC design is estimated on the different driving cycles. At first just the ORC impact is considered. From the results is table 2 is visible that on the combined cycles (with urban part and extra urban part) – NEDC and peri-urban the efficiency improvement is 24% and respectively 34%. The ORC is well used in the charged part of the cycles. The fuel consumption reduction potential on holiday cycle is around 40%. The fuel consumption is below 4.5 l/100 km. The selected ORC design is an efficient equipment to recover the waste heat in the high speed demanding drives. In these zones the engine power demand is high.

It is as well interesting to analyse the aggregation of the MHEV and the ORC. For the combined cycles the combination of a MHEV and ORC on a small downsized engine brings the best fuel saving potential – around 40%. The MHEV contributes to the efficiency improvement on the peri-urban drives and the ORC on the extra-urban drives. For the "holiday" long drive cycle, the ORC is the major contributor of the improvement.

The lowest consumption for is obtained for NEDC cycle, because of its particularity to combine urban and highway drives at very smooth and progressive speed changing conditions. By their construction, based on WLTP, the peri-urban and the holiday cycles combines more transients and higher accelerations. In all configurations, their consumption majors the NEDC consumption.

In conclusion the mix of a low degree electric hybridization technology such as the Mild HEV and a technology for the waste heat recovery as the organic Rankine cycle is efficient for mixed usages of the vehicle. For the C-Segment vehicle the aggregation of the two energy recovery options is efficient. The powertrain efficiency increase for NEDC is of around 40%. The vehicle consumption is 3.32 l/100 km.

Table 2. Driving cycles and fuel consumption results on C-Segment Vehicle

Driving Cycle	Vehicle type	Fuel consumption, l/100 km	Powertrain efficiency gain, %
NEDC	C-Segment 1.2	4.72	-
	C-Segment 1.2 MHEV	4.16	10.8
	C-Segment 1.2 ORC	3.71	24
	C-Segment 1.2 ORC MHEV	3.32	38.4
Peri-urban	C-Segment 1.2	5.56	-
	C-Segment 1.2 MHEV	5.05	10
	C-Segment 1.2 ORC	4.14	34.2
	C-Segment 1.2 ORC MHEV	3.98	39.7
Holidays	C-Segment 1.2	5.94	-
	C-Segment 1.2 MHEV	5.86	1.3
	C-Segment 1.2 ORC	4.27	38.9
	C-Segment 1.2 ORC MHEV	4.35	36.2

The system is interesting to be realised and to be tested. The practical difficulty is related to the ORC installation. The Mild HEV architecture is well known in the state of the art and exists in serial production. The discussed ORC system presents challenges on the direct mechanical connexion between turbine and the electric generator, because of the high the rotation speed that they have to operate. The choice and the realisation of the small size ORC components (pump and turbine) is as well challenging.

4. Conclusions

This article presents a study of the efficiency performance and the economic optimization for a middle class vehicle with a downsized gasoline engine. Two technological options are studied for the efficiency improvement of the baseline vehicle powertrain:

- Low degree electric hybridization technology – called Mild Hybrid electric powertrain;
- Waste heat recovery system for two heat sources the engine water circuit and the exhaust gases, means an external thermodynamic cycle – ORC.

The ORC is designed on an optimal way for high load and high speed engine operating conditions. The design is optimized considering techno- economic performances indicators. For this evaluation a multi-objective optimization methodology is applied, including the energy integration. For this reason, the best fuel consumption improvement is obtained for the holiday cycle, characterised by long ways drives at high speed. For the combined cycles such as NEDC and peri-urban the combination of ORC and MHEV bring the best fuel saving potential, superior to 35% in comparison with normal ICE vehicle.

The main conclusion coming from the driving cycles evaluation is the MHEV is an efficient technology for the drives with speed variations. The selected ORC design is efficient on the cycles using the charged area of the engine. For these operating points, the small gasoline engine is highly charged, due to the high power vehicle demands. The MHEV and the ORC are complementary technologies and their combination is especially efficient on the combined cycles containing urban and extra urban part.

References

[1] Dimitrova Z and Maréchal F 2015 Energy integration on multi-periods and multi-usages for hybrid electric and thermal powertrains *Energy* **83** 539–50

[2] Dimitrova Z and Maréchal F 2015 Energy integration study on a hybrid electric vehicle energy system, using process integration techniques *Applied Thermal Engineering* **91** 834–47

[3] Lümmen N, Nygård E, Koch P E and Nerheim Lars M 2018 Comparison of organic Rankine cycle concepts for recovering waste heat in a hybrid powertrain on a fast passenger ferry *Energy Conversion and Management* **163** 371–83

[4] Capata R, Toro C 2014 Feasibility analysis of a small-scale ORC energy recovery system for vehicular application *Energy Conversion and Management* **86** 1078–90

[5] Briggs I, McCullough G, Spence S and Douglas R 2014 Whole-vehicle modelling of exhaust energy recovery on a diesel-electric hybrid bus *Energy* **65** 172–81

[6] Leduc P, Smague P, Leroux A and Henry G 2017 Low temperature heat recovery in engine coolant for stationary and road transport applications *Energy Procedia* **129** 834–42

[7] Ezzat M F , Dincer I 2019 Development and exergetic assessment of a new hybrid vehicle incorporating gas turbine as powering option *Energy* **170** 112-19

[8] Glover S , Douglas R , De Rosa M , Zhang X , Glover L 2015 Simulation of a multiple heat source supercritical ORC (Organic Rankine Cycle) for vehicle waste heat recovery *Energy* **93** 1568–80

[9] Yue C and Wang P 2019 Thermal analysis on vehicle energy supplying system based on waste heat recovery ORC *Energy Procedia* **158** 5587-95

[10] Muralidhar N , Himabindu M , Ravikrishna R V 2018 Modeling of a hybrid electric heavy duty vehicle to assess energy recovery using a thermoelectric generator *Energy* **148** 1046-59

[11] Dimitrova Z, Lourdais P and Maréchal F 2015 Performance and economic optimization of an organic Rankine cycle for a gasoline hybrid pneumatic powertrain *Energy* **86** 574–88

IOP Conf. Series: Materials Science and Engineering **664** (2019) 012010 doi:10.1088/1757-899X/664/1/012010

Accumulation of metabolic carbon dioxide (CO_2) in a vehicle cabin

R A Angelova [1], D G Markov, I Simova, R Velichkova and P Stankov

Centre for Research and Design in Human Comfort, Energy and Environment, Technical University – Sofia, 8 Kliment Ohridski Blvd., 1000 Sofia, Bulgaria

[1] E-mail: joy_angels@abv.bg

Abstract. In traffic congestions, the typical reaction of the drivers is to switch-on the recirculation mode of the vehicle's air conditioning system. In the summertime, in many cities with high air humidity and temperatures, people typically drive their cars in recirculation mode as well, to ensure the best cooling. However, the recirculation mode of the air in the cabin prevents O_2 from entering and leads to an increase in CO_2. In confined spaces (e.g., a vehicle cabin), the raised CO_2 concentration due to human exhalation and metabolism is associated with increased health symptoms, decreased concentration, and impaired performance. The paper presents an experimental study on the accumulation of carbon dioxide in a vehicle cabin from occupants breathing and metabolism. The concentration of CO_2 and O_2 were measured in terms of the number of occupants (1-4) in the car. The change in the indoor environmental parameters in the car cabin was also monitored. The results obtained show the increment of the CO_2 concentration at the expense of oxygen availability, together with the apparent impact of the number of people in the vehicle.

1. Introduction

The outdoor concentration of carbon dioxide (CO_2) is considered to be around 380 parts per million (ppm) [1], though the global average CO_2 concentration level of May 2019, measured at Mauna Loa Observatory, Hawaii is 414.66 ppm (NOAA, 2019) [2].

Perisly [3] reported that CO_2 levels in the outdoor air in an urban area could reach up to 500 ppm. According to [4], the annual CO_2 concentration in Beijing for the period 2013-2016 was 374.4, 382.7, 389.0, and 392.8 ppm for the respective year, being highest in winter months and lowest in summer. Beijing is not, however, among the most congested cities in the world. The first place is occupied by Moscow, followed by Istanbul, Bogota, Mexico City, Sao Paulo, London, Rio de Janeiro, Boston, Saints Petersburg, and Rome [5].

There is an undoubted relationship between the traffic congestions in a city and the CO_2 in the cars (vehicles). In traffic congestions, the typical reaction of the drivers is to switch-on the recirculation mode of the vehicle's air conditioning system. In the summertime, in many cities with high air humidity and temperatures, people typically drive their cars in recirculation mode as well, to ensure the best cooling. However, the recirculation mode of the air in the cabin prevents O_2 from entering and leads to an increment of the CO_2. In confined spaces (e.g., a vehicle cabin), the CO_2 concentration rises fast due to human exhalation and metabolism. The high CO_2 concentration is associated with increased health symptoms, decreased concentration, and impaired performance.

Content from this work may be used under the terms of the Creative Commons Attribution 3.0 licence. Any further distribution of this work must maintain attribution to the author(s) and the title of the work, journal citation and DOI.
Published under licence by IOP Publishing Ltd

The research in [6] investigated the hypothesis that higher concentrations of CO_2 in confined spaces, without changes in ventilation rate, have a harmful influence on the decision-making performance of the occupants. The hypothesis was based on the findings in [7, 8] that concentrations of CO_2 in the range 2000-5000 ppm in a controlled indoor environment negatively affected the proofreading of text. These conclusions contradicted to previous findings that CO_2 concentration up to 5000 ppm in enclosures did not affect occupants' perception, work performance, or health [9]. The authors of [1] investigated nine primary factors and combinations that demonstrate the level of performance in a controlled indoor environment with three different levels of CO_2 concentration: 600, 1000, and 2500 ppm. They have found that seven of the factors (basic activity, task orientation, applied activity, initiative, flexibility in approach to the task, information usage, and basic strategy) were significantly reduced at 2500 ppm and the results were statistically proven. Only the information search and the focused activity were not affected by the CO_2 in the studied concentrations.

In the case of recirculation mode of the ventilation system, the CO_2 concentration in the vehicle cabin can rise fast due to the human metabolism and exhalation and at the expense of the oxygen (O_2) concentration [10]. The increment is steepest in small volume of the vehicle cabin [6]. However, there are no limits for the CO_2 levels in cars, though the ASHRAE standard 62-2001 [11] suggests a permissible level of 1200-1300 ppm for the cabin interior.

The paper presents an experimental study on the accumulation of carbon dioxide in a vehicle cabin from occupants breathing and metabolism. The concentration of CO_2 and O_2 were measured in terms of the number of occupants (1-4) in the car. The change in the indoor environmental parameters in the car cabin was also monitored. The results obtained show the increment of the CO_2 concentration at the expense of oxygen availability, together with the apparent impact of the number of people in the vehicle.

2. Experimental methodology

The experiments were performed in a vehicle (car) cabin with switched on recirculation (REC) mode: the fresh air intake was prohibited, and only cabin air was recirculated. The passengers (subjects) were sitting inside the cabin without physical activity. Thus, minimum concentration levels of CO_2 were measured, due to regular metabolic activity and respiration rate at about 6 l/min [6].

Two carbon-dioxide sensors were used for the study: one in the front, and one in the back side of the cabin (figure 1). One sensor for monitoring of the O_2 concentration, together with the air temperature, and relative humidity was placed in front of the cabin. The used software allowed direct monitoring of the changes in the measured indoor air parameters (figure 2). The accuracy of the O_2 and CO_2 concentrations were ±10% of the measured value. The measuring frequency was 0.1 Hz.

The study was performed in 4 stages: each stage involved a different number of subjects (4 at the beginning and 1 at the end of the measurements). The measurement aimed to reach CO_2 concentration of 2500 ppm, and this goal limited each stage in terms of time. The car cabin was ventilated (open doors) after each stage.

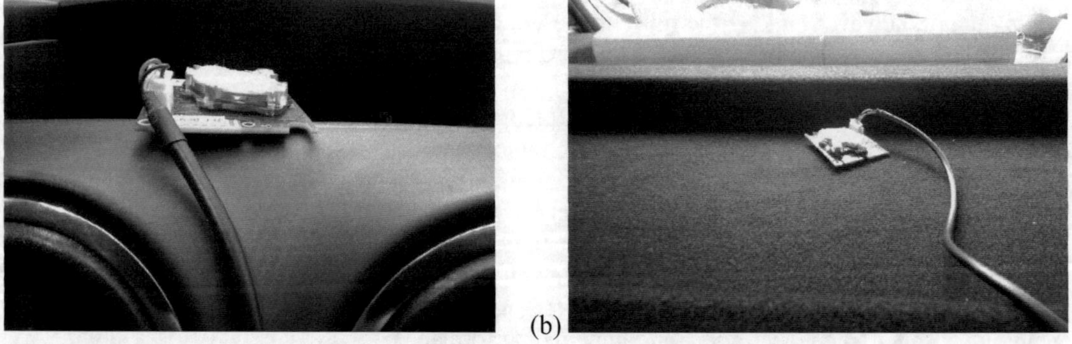

(a) (b)

Figure 1. Tests sensors: (a) front CO_2 sensor; (b) back CO_2 sensor.

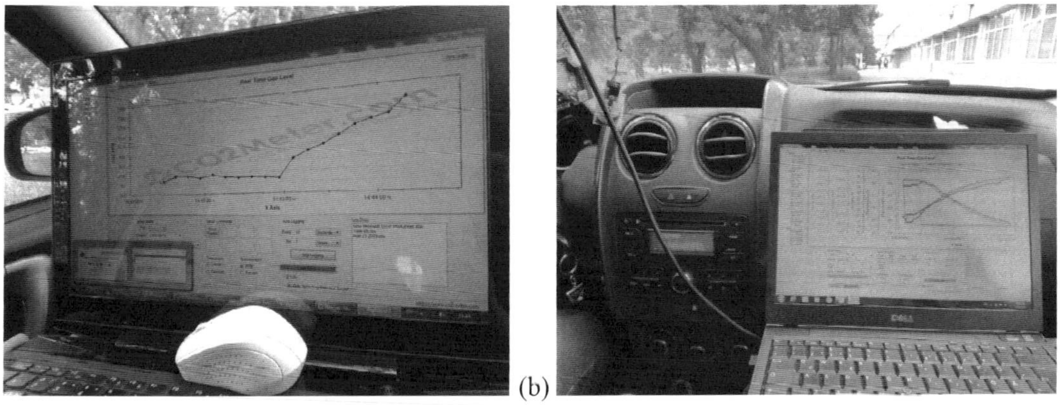

(a) (b)

Figure 2. Online monitoring of the parameters in the car cabin: (a) CO_2 concentration (front); (b) O_2 concentration and indoor environmental parameters.

3. Results and discussions

Figure 3 presents a comparison between the from and rear sensor for CO_2 concentration measurement when 3 subjects occupied the car cabin. The performed statistical analysis of the results (two-tailed t-test) showed that there was no difference between the two sets of data (confidence level 95%). Therefore, the results for the CO_2 concentration from the front sensor only are further discussed.

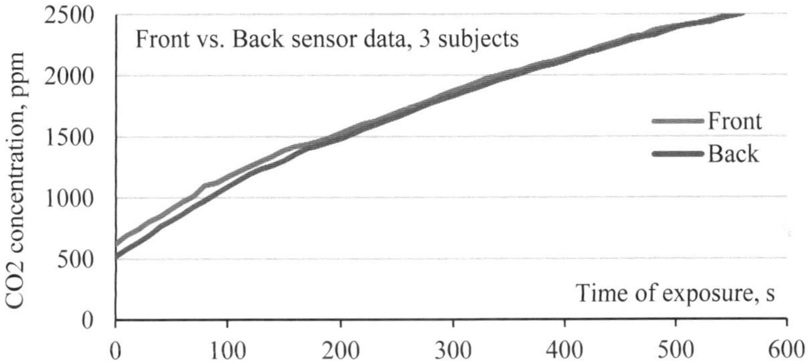

Figure 3. Comparison between front and back sensor data for CO_2 concentration (3 subjects in the car cabin).

The effect of the number of occupants on the CO_2 concentration in the cabin is summarized in figure 4. The results showed that reaching the set limit of 2500 ppm was fastest in the largest number of passengers in the car (4 subjects, REC mode). The reason is the natural exhalation process: for 340 s all passengers in the car, including the driver, were exposed to the increased carbon dioxide (2517 ppm). Therefore, for approx. 5 minutes in a congestion or recirculation mode for other reasons, a family with two children going on holiday could be subjected to the negative effect of CO_2. This is especially dangerous for the driver, who can be expected to have a reduced performance [1], being at the same time the only one, responsible for the safety of the passengers.

The increment of the CO_2 concentration in the cabin was at the expense of oxygen availability. Figure 5 presents the results for the O_2 concentration, measured for each of the stages. Though CO_2 concentration in the cabin air reached 2500 ppm in the case of 4 passengers, the oxygen decrement was lowest, due to the short time of the exposure. On the contrary, the regular breathing of a single person in the car cabin in REC mode led to 1.7% decrement in the O_2 concentration, because of the very long time of the exposure (37 min and 20 s) and the gradual oxygen depletion.

IOP Conf. Series: Materials Science and Engineering **664** (2019) 012010 doi:10.1088/1757-899X/664/1/012010

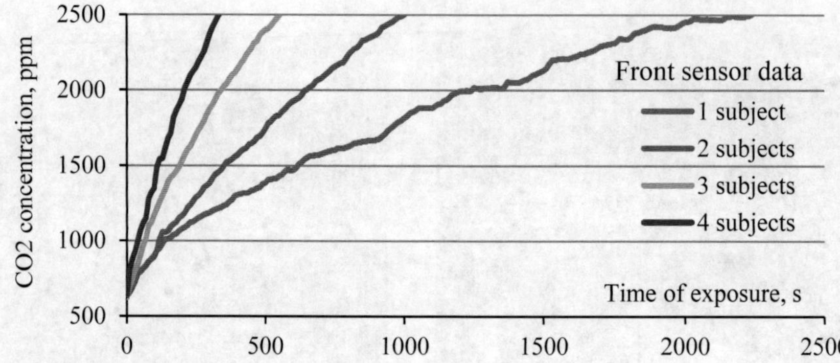

Figure 4. CO_2 concentration in the car cabin: effect of the number of passengers.

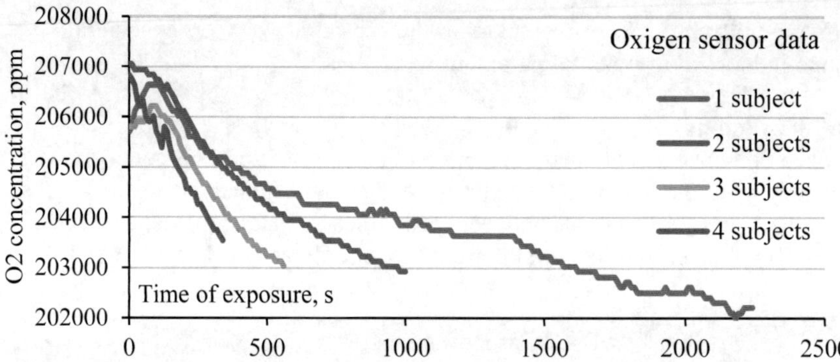

Figure 5. O_2 concentration in the car cabin: effect of the number of passengers.

Figure 6 illustrates the rate of CO_2 concentration increment for each of the tested stages. The presence of 4 passengers in the car led to an increase of the carbon dioxide in the air with 5.53 ppm/s. When only the driver was in the car, the CO_2 concentration in the cabin air (REC mode) increased with 0.84 ppm/s.

A trendline was added in figure 6 to search for a regression equation that describes the influence of the number of the occupants in the cabin and the CO_2 concentration change rate. The proposed regression model (regression coefficient $R = 0.99$) is:

$$CR_{CO2} = 0.4881e^{0.6244N}, \tag{1}$$

where CR_{CO2} is the rate of increment of the CO_2 concentration, ppm/s; N is the number of the occupants.

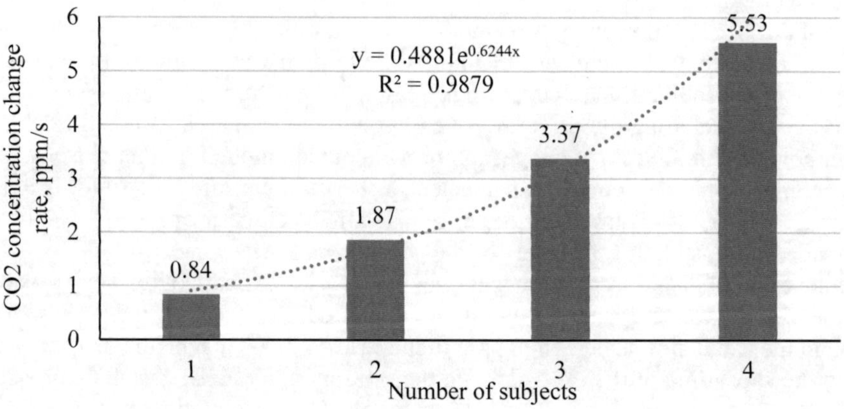

Figure 6. CO_2 concentration change rate: effect of the number of passengers.

Figure 7 summarizes the in-cabin environmental conditions during the measurements. The most significant change in the air temperature (figure 7 a) and relative humidity (figure 7 b) were for the stage with only one subject in the car. The main reason is the prolonged exposure. The thermophysiological comfort of the passengers in the cabin will be of particular interest in the next experiments within this research study.

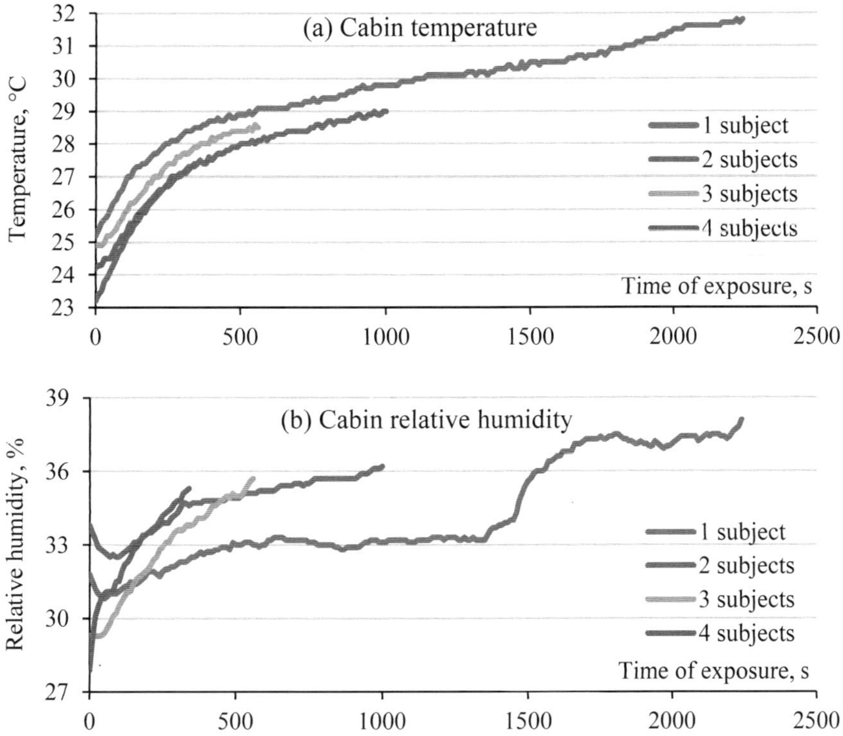

Figure 7. Cabin environmental parameters: (a) air temperature; (b) relative humidity.

4. Conclusion

The results obtained clearly show potential safety problems, related to both the increment of the CO_2 concentration and the depletion of oxygen in the vehicle cabin. The comparison between the results from the front and back sensors showed that the driver is affected by the CO_2 concentration to the same level, like the passengers in the car, due to the effective air mixing. In REC mode, when three, four or five occupants are in the car, the CO_2 concentration may reach dangerous for the driver's performance levels even for several minutes (e.g., in a city traffic congestion). The same is valid for family trips during the summer when the outside air in the passenger compartment is discontinued (REC mode) to assure the desired cooling. A solution of the problem is the regular or automatic switched-off of REC mode after several minutes of internal air recirculation.

Acknowledgments

The publication is supported by Bulgarian Science Fund of the Ministry of Education and Science ДН17 /12/12.12.2017.

References

[1] Satish U, Mendell M J, Shekhar K, Hotchi T, Sullivan D, Streufert S and Fisk W J 2012 Is CO2 an indoor pollutant? Direct effects of low-to-moderate CO2 concentrations on human decision-making performance *Environmental health perspectives* **120(12)** 1671–7

[2] National Oceanic and Athmospheric Administration (NOAA) 2019 https://www.esrl.noaa.gov/gmd/ccgg/trends/
[3] Persily A K Evaluating building 1997 IAQ and ventilation with indoor carbon dioxide. *American Society of Heating, Refrigerating and Air-Conditioning Engineers, Inc., Atlanta, GA (United States)*
[4] Cheng X L, Liu X M, Liu Y J and Hu F 2018 Characteristics of CO2 concentration and flux in the Beijing urban area *J. Geophysical Research: Atmospheres* **123(3)** 1785–801
[5] Newman K 2019 Cities with the world's worst traffic congestion *US News*
[6] Mathur G D 2018 Effect of cabin volume on build-up of cabin carbon dioxide concentrations from occupant breathing in automobiles *SAE Technical Paper*
[7] Kajtar L, Herczeg L and Lang E 2003 Examination of influence of CO2 concentration by scientific methods in the laboratory *Proc. of Healthy Buildings* 176–181.
[8] Kajtar L, Herczeg L, Lang E, Hrustinszky T and Banhidi L 2006 Influence of carbon-dioxide pollutant on human well-being and work intensity *Proc. of Healthy Buildings* 85–90
[9] Persily A K 1997 Evaluating building IAQ and ventilation with indoor carbon dioxide *American Society of Heating, Refrigerating and Air-Conditioning Engineers Inc. (Atlanta United States)*
[10] Scott J L, Kraemer D G and Keller R J 2009 Occupational hazards of carbon dioxide exposure *J. of Chemical Health and Safety* **16(2)** 18–22
[11] ASHRAE. ANSI/ASHRAE Standard 62-2001 Ventilation for Acceptable Indoor Air Quality, 2006 Supplement (American Society of Heating, Refrigerating and Air-Conditioning Engineers, Inc., Atlanta, GA.; 2001)

IOP Conf. Series: Materials Science and Engineering **664** (2019) 012011 doi:10.1088/1757-899X/664/1/012011

Optimal three-loop cascade PI-P-PI controller for electro-hydraulic power steering system

Jordan Kralev[1], Alexander Mitov[2,3], Tsonyo Slavov[1], Ilcho Angelov[2]

[1] Dept. of Systems and Control, Technical University of Sofia, Bulgaria
[2] Dept. of Hydroaerodynamics and Hydraulic Machines, Technical University of Sofia, Bulgaria

[3] E-mail: a_mitov@tu-sofia.bg

Abstract. The present paper presents an optimally tuned proportional integral differential (PID) system which calculates the driving signal for the switching valves in order to achieve steering trajectory tracking. The designed cascade control algorithm consists of three PID controllers respectively for the spool valve position, for the effective flow rate and for the steering cylinder position. The stages of the cascade control algorithm synthesis are presented. The goals of control algorithm are to achieve fast transients with minimal overshot and steady state error. Controllers' are tuned consequently by starting from the innermost loop based on an identified single-input multiple output (SIMO) model of the steering system by optimization algorithm with a quadratic cost. After one of the controllers is successfully tuned the model of the corresponding closed loop system is approximated and employed in the optimization procedure for the next control loop from the cascade system. The developed control algorithm is programmed into a specialized vehicle microcontroller and tested on a steering system test bench in the laboratory.

1. Introduction

Most of future off-road construction and agriculture machines will be fully automatized and operating without continuous human presence. Such development of autonomous and remote-controlled vehicles is dependent on implementation of precise and reliable steering servo system. The steering system is essential for machine operation since machine functional capacity cannot be effectively employed if the machine is held stationary. Moreover, the usual operational environment for such machines is off-road so the mobility of the machine is extremely important. The considerable mass of construction machines and the harsh off-road environment require powerful steering mechanism. Therefore, the hydraulic transmission is employed which works as an amplifier of the operator command [1].

In conventional (non-electronic) steering system the force amplification is achieved with a spool-sleeve valve (figure 1). It is a rotary kind of valve which is coupled to the operator's steering handle (the steering wheel). A small valve displacement initiated by the machine operator directs a portion of flow rate to the left or to the right chamber of the main steering cylinder. Since the cylinder piston is mechanically linked to the wheels the vehicle begins to turn in the commanded direction.

However, in order to steer the vehicle with an electronic signal a proportional spool valve, pilot operated with switching micro-valves [2] replaces the rotary spool-sleeve valve (figure 2). Switching micro-valves (also known as digital valves) are not the only option to achieve electronic steering. Its alternatives are direct operated directional spool valve or steer-by-wire strategy. The leading

Content from this work may be used under the terms of the Creative Commons Attribution 3.0 licence. Any further distribution of this work must maintain attribution to the author(s) and the title of the work, journal citation and DOI.
Published under licence by IOP Publishing Ltd

IOP Conf. Series: Materials Science and Engineering **664** (2019) 012011 doi:10.1088/1757-899X/664/1/012011

manufactures of hydraulic steering systems favour the switching valves due to its mechanical simplicity and low cost. Usually four micro-valves are connected in a bridge configuration where two of the valves are normally opened and the other two are normally closed. A solenoid based electric circuit drives each micro-valve between its two states. The micro-valve bridge is connected to the left and right endpoints of a translational spool valve which direct the flow from the pump to the main cylinder. In order to drive the spool valve to the left the upper and lower micro-valves from the right side of the bridge are switched.

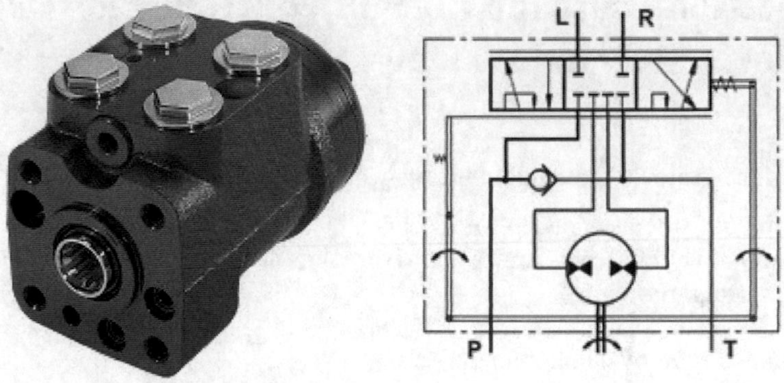

Figure 1. Conventional steering unit based on spool-sleeve rotational valve. Left is a picture of the steering unit. Right is a hydraulic schematic of the valve with no reaction (closed centre).

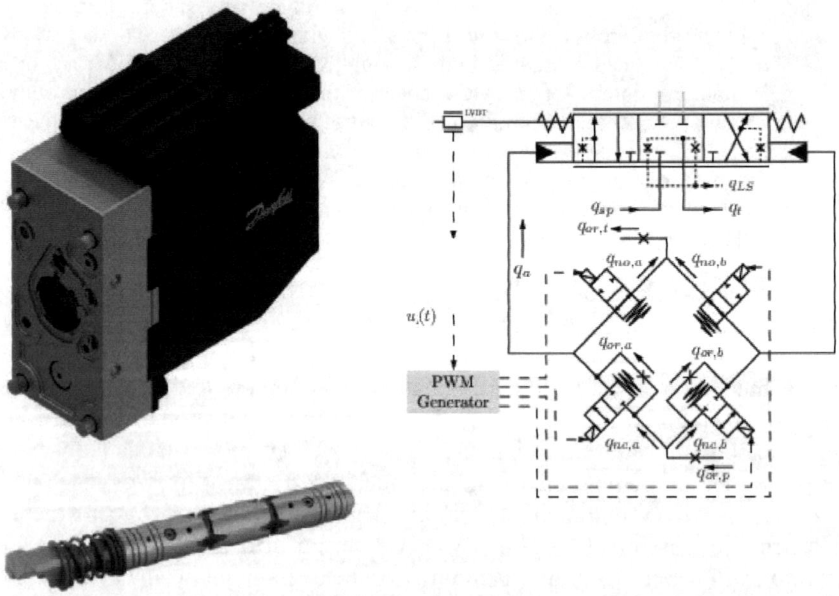

Figure 2. Proportional spool valve, pilot operated with switching micro-valves for electrohydraulic steering. Left is a picture of the spool valve and the switching valve block containing also the solenoid electronics. Right is a hydraulic schematic of the switched hydraulic system.

Various problems arise in automatic control of the discussed pilot operated steering system. First the switching of the valves is a nonlinear discrete-event process which induces pressure and flow rate oscillations through the hydraulic system with relatively high frequency (around 40÷100 Hz). The oscillations cannot be compensated with the software controller since they are inherent in the system architecture. The second problem is positive overlap in the spool valve which means that minimal amount of translation is required to open the valve in each direction. The valve overlap is mechanically designed to prevent spontaneous steering drifts which could compromise the safety. From control point of view the positive valve overlap looks like a dead band - a static nonlinearity in the loop. The third problem is the nonlinear relationship between flow rate and pressure variables which is caused by the varying geometrical shapes of the valve openings and by the continuity equation for throttle. Hence even if the spool valve position is precisely controlled the generated flow would not vary linearly with its translation. The fourth problem is unpredictable nature of the steering loads acting on the wheels and translated to the piston of the main cylinder. The control system must guarantee the commanded steering angle despite the external load.

The article proposes a complicated cascade control system which addresses these problems. It is composed of three linear controllers connected such that the output of one controller is a reference signal for the next (a cascade interconnection). Such configuration is advantageous when a single control action has to account for several output variables of the plant [3, 4]. The investigated system is of SIMO type (single input multiple output). The manipulated variable is the duty cycle of the pulse-width modulated signal determining the average openings of the switching valves. The first output signal is the spool valve position measured with a linear variable differential transducer (LVDT) sensor. The second output is the consumed flow rate of the system measured with a geared flow meter. The last output is the piston position of the steering cylinder. The goal of the system is to tracking fast enough a generated reference trajectory for the steering.

2. Mathematical model of the steering system

To investigate quantitatively the power steering system an accurate enough mathematical model of its dynamic response is necessary. The two main approaches to modelling of electrohydraulic steering systems are the analytic analysis and the identification from experimental data [5]. This article uses a black-box state-space discrete-time model obtained with identification procedure [6] presented in details in a previous authors' article in press. Here only the resultant model is summarized which has the following form

$$\begin{aligned} x(k+1) &= Fx(k) + Gu(k) + K_v e(k) \\ y(k) &= Cx(k) + Du(k) + e(k), \end{aligned} \tag{1}$$

where $x(k)$ is a state vector with components $x_1(k)$, $x_2(k)$ and $x_3(k)$ at a time instant $t = kT_S$ of the system temporal evolution ($T_S = 25ms$ is the sampling time). The state variables represent linear combinations of unknown but fixed internal system variables [7]. The $u(k)$ is the manipulated variable which in our case is the duty cycle applied to the solenoids of the switching valve bridge. The signal $u(k)$ is in the range $\pm 4V$, with the positive values activate the right upper and lower valves and the negative values active the left upper and lower valves. The output signals (spool position y_{spool}, flow rate y_{flow} and piston position y_{piston}) are collected into the vector $y(k)$. The matrices F, G, C, D and K_v determine the structure and parameterization of the linear dynamic model. Their values are estimated from experimentally collected data when a random binary input signal is generated. Their values are

$$\begin{gathered} F = \begin{pmatrix} -1.05 & 0.19 & 6.27 \\ -0.04 & 1 & 0.198 \\ -0.371 & 0.033 & 2.044 \end{pmatrix},\ G = 10^{-3}\begin{pmatrix} -0.37 \\ -0.034 \\ -0.23 \end{pmatrix}, C = \begin{pmatrix} -31 & 0.67 & -6.58 \\ -0.584 & 0.007 & -0.25 \\ 0.059 & -2.019 & 0.125 \end{pmatrix}, \\ D = 10^{-3}\begin{pmatrix} 0.117 \\ 0.014 \\ 0.0021 \end{pmatrix}, K_v = \begin{pmatrix} -0.025 & -0.023 & -0.022 \\ -0.002 & 0.0066 & -0.429 \\ -0.007 & -0.01 & 0.0033 \end{pmatrix}. \end{gathered} \tag{2}$$

The signal $e(k)$ is a vector with three components which represent the residual in the model calculated as a difference between the experimentally recorded output signal and the predicted from the model. It is described as a white noise random signal with a zero mean and covariance

IOP Conf. Series: Materials Science and Engineering **664** (2019) 012011 doi:10.1088/1757-899X/664/1/012011

$$E(ee^T) = 10^3 \begin{pmatrix} 7.92 & 0.034 & 0.011 \\ 0.034 & 0.03 & 0.001 \\ 0.011 & 0.001 & 0.003 \end{pmatrix}. \quad (3)$$

Parameters of the matrices are tuned such that the residual covariance to be minimized meaning that the model is approximating the experimental data from the plant. As can be seen from the covariance matrix the model is most accurate on its third output (the piston position), followed by the flow rate and spool position. The mathematical model can be represented also as a transfer functions in frequency domain

$$W_{spool}(z) = \frac{y_1(z)}{u(z)} = \frac{0.013z^2+0.01z-0.022}{z^3-2z^2+1.19z-0.186}; \quad (4)$$

$$W_{flow}(z) = \frac{y_2(z)}{u(z)} = 10^{-4}\frac{2.4z^2+1.7z-4.3}{z^3-2z^2+1.19z-0.186}; \quad (5)$$

$$W_{piston}(z) = \frac{y_3(z)}{u(z)} = 10^{-5}\frac{1.4z^2+5.1}{z^3-2z^2+1.19z-0.186}, \quad (6)$$

where the operator $z = e^{j\omega}$ with ω the frequency in rad/s and z^{-1} corresponding to an unit delay in the temporal domain. It can be noted that the denominators of all the transfer functions are the same. The difference is in the zeros and the static gain.

The step response of the model is presented in the figure 3. The response of the spool position is aperiodic since a mechanical spring counteracts the pressure generated form the valve bridge along the spool. The flow rate response is almost immediate with the spool translation but requires a little more time to settle. It is aperiodic too since the pump works with a constant pressure. And the third output is the steering piston stoke which is linearly increasing as expected. The figure contains also the upper and the lower bound for each signal calculated from the residual covariance. These bounds represent the uncertainty of the model and for the presented system they are small enough. However, this uncertainty indicates that a closed-loop control is required in order to achieve the desired reference tracking.

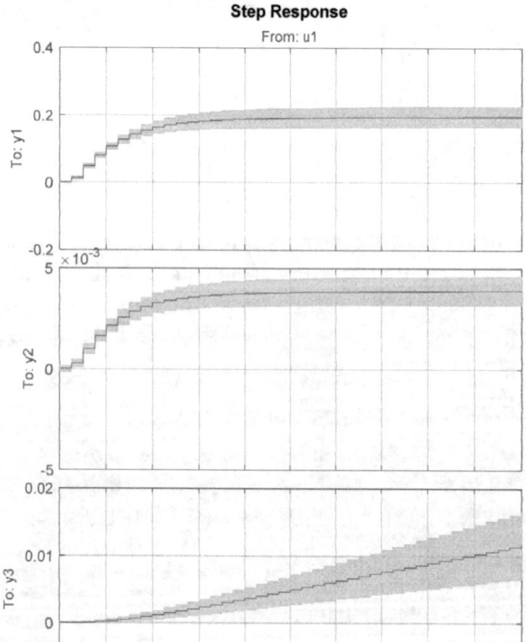

Figure 3. Output signals when unit signal is applied to the input. y_1 is the proportional spool position, y_2 is the flow rate and y_3 is the steering cylinder position.

3. Design of the control system

The presented mathematical model is used to tune the proposed control structure which is based on a three-loop cascade of PI regulator for spool position, P regulator for the flow rate and again PI regulator for the piston position. It is summarized with the diagram from figure 4. The *spool controller* is selected as PI since its response should be as fast as possible. The equation for the spool PI regulator in discrete-time domain

$$u(k) = K_{sp,p}\left(r_{sp}(k) - y_{spool}(k)\right) + K_{sp,i}x_{sp,i}(k) \tag{7}$$

calculates the control signal $u(k)$ at time instant kT_S based on the deviation from the set-point $r_{sp}(k)$ and its integrated value according to forward Euler approximation

$$x_{sp,i}(k) = x_{sp,i}(k-1) + T_S\left(r_{sp}(k) - y_{spool}(k)\right) + w_{sp}(k), \tag{8}$$

where $w_{sp}(k) = K_{w,sp}(u_{sat}(k) - u(k))$ is the anti-windup term preventing over-integration in $x_{sp,i}(k)$ which would delay the settling of the spool position transient to its set-point due to saturation $u_{sat}(k)$ of the control action $u(k)$ (i.e. *wind-up effect*). The anti-windup $K_{w,sp}$ is tuned enough up to prevent the controller from hitting the lower saturation bound immediately after the upper saturation bound is reached for positive value of the set-point. The real saturation bounds of the control signal $u(k)$ are $\pm 4V$ but if they are actually reached the switching valve electronics would raise an exception and would stop working, therefore, the practical bounds are between $\pm 3.5V$. The parameters of the PI regulator which have to be tuned by optimization are its proportional gain $K_{sp,p}$ and its integral gain $K_{sp,i}$.

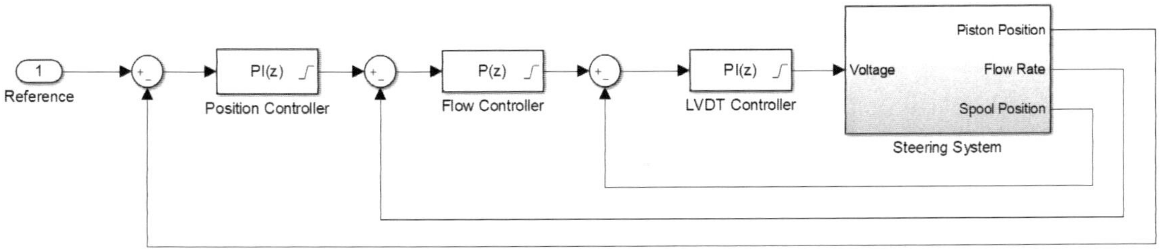

Figure 4. Three-loop cascade control system for the electrohydraulic power steering.

The spool position controller would guarantee the following of the set-point signal $r_{sp}(k)$ corresponding to setting the directional spool valve to a desired position which would cause the development of the flow rate $y_{flow}(k)$ leading to actual translation of the steering cylinder piston. However, the mathematical relationship between the spool valve opening and flow rate is not linear due to geometrical shape of the valve. Also the design of the valve is dependent on the safety requirement for minimal spool valve translation before actual opening of the valve (positive overlap). The identified linear model of the process is focused on the transient time-constants but such nonlinearities [8] cannot be accounted for with linear transfer function. So additional *flow controller* is designed as a P controller

$$r_{sp}(k) = K_{flow,p}\left(r_{flow}(k) - y_{flow}(k)\right), \tag{9}$$

where the parameter $K_{flow,p}$ is optimized up to minimize the error between required flow rate $r_{flow}(k)$ and actually measured $y_{flow}(k)$. Such controller leads to "linearization" of the relationship between valve opening and flow which makes the tuning of the piston position regulator easier [9]. The value of the $K_{flow,p}$ is limited from going to high because the controller also amplifies the measurement noise from the gear flowmeter device. The amplification of noise in detected in the optimization procedure by measuring the standard deviation of the control action and weighting it accordingly. The bounds for the flow rate controller $r_{sp}(k)$ are determined from the spool valve range of motion which is ± 2.5 mm.

The last controller from the chain is the steering piston position controller which calculated the flow rate set-point such that the desired steering angle to be achieved faster

$$r_{flow}(k) = K_{cyl,p}\big(r(k) - y_{piston}(k)\big) + K_{cyl,i}x_{cyl,i}(k), \tag{10}$$

where $r(k)$ is commanded position of the steering cylinder from the operator or remote control system and $K_{cyl,p}$ and $K_{cyl,i}$ are the controller parameters for tuning. Since there are limits for the flow rate set-point $r_{flow}(k)$ which are ± 10 l/min the integral term

$$x_{cyl,i}(k) = x_{cyl,i}(k-1) + T_S\left(r(k) - y_{piston}(k)\right) + w_{cyl}(k) \tag{11}$$

also include an *anti-windup* term.

3.1. Spool controller optimization

The main problem with the presented controller structure remains the tuning of the controller parameters which are $\theta = \left(K_{sp,p}, K_{sp,i}, K_{flow}, K_{cyl,p}, K_{cyl,i}\right)$. First the spool valve controller parameters $K_{sp,p}$ and $K_{sp,i}$ are tuned such that cost function value

$$J_{sp}(K_{sp,p}, K_{sp,i}) = \textstyle\sum_{k=0}^{100}\left(1 - y_{spool}\big(k, K_{sp,p}, K_{sp,i}\big)\right)^2 \tag{12}$$

is minimized according to *nonlinear-least squares* method. The minimization is an iterative process which requires multiple simulations of the model for different values of the optimized parameters $(K_{sp,p}, K_{sp,i})$. After each simulation the value of the cost function J_{sp} is evaluated and based on the gradient its derivative a new guess for parameters is examined which would minimize further J_{sp}. This optimization procedure is presented on figure 5. The results from the optimization are shown in table 1 for initial values of the parameters $K_{sp,p} = 0.01$ and $K_{sp,i} = 1.67$.

Table 1. Iterations for optimal tuning of the spool valve controller.

Iteration	Function evaluations	New cost value	Norm of step
0	3	7.56323	
1	6	4.8853	1.73024
2	9	3.74238	2.48575
3	12	3.37197	2.27601
–	–	–	–
8	27	3.32255	0.00878
9	30	3.32255	0.00275

The final values of the controller parameters are $K_{sp,p} = 7.7464$ and $K_{sp,i} = 1.2736$. These values are reached after nine iterations and 30 cost function evaluations. The cost function value is reducing from 7.56 to 3.32 or 56%. The optimization is stopped because the change of the cost function value between two consecutive iterations is below a predefine absolute tolerance. As can be seen the norm of the step is decreasing too which indicate the approaching of the function local minimum.
Simulation results of the spool valve controller are presented in figure 6 for the output signal (spool valve translation) and in figure 7 for the corresponding control signal. The figures compare the performance of the initial controller parameters with the controller parameters after the optimization. The optimized controller is faster in reaching the set-point. The overshoot in both controller is comparable but its area is smaller after the optimization.

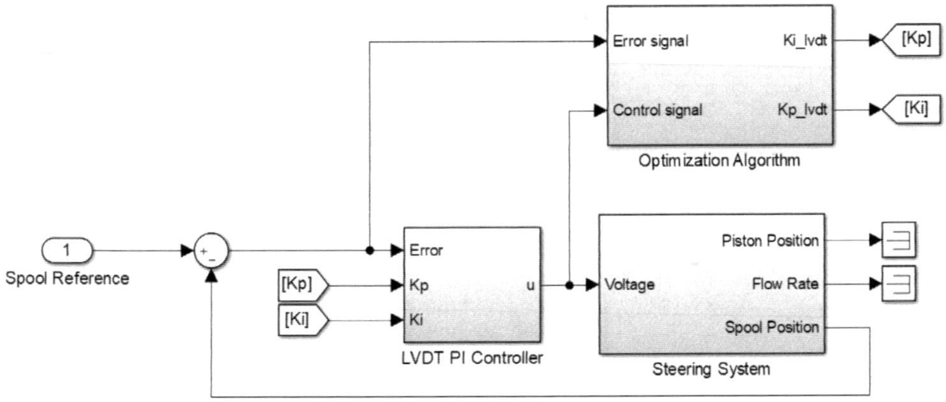

Figure 5. Optimal tuning of the spool valve controller.

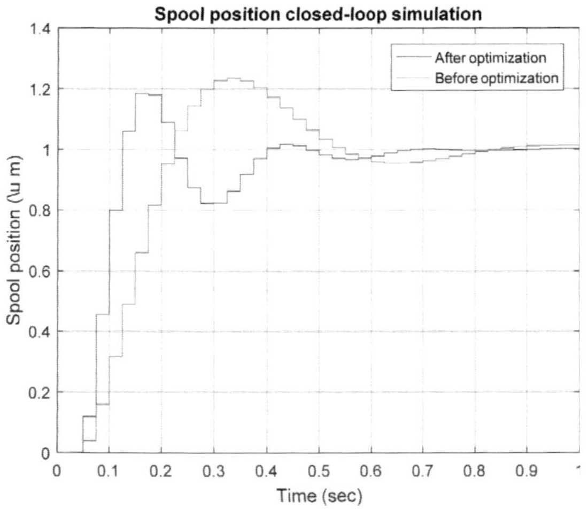

Figure 6. Spool valve closed-loop step response.

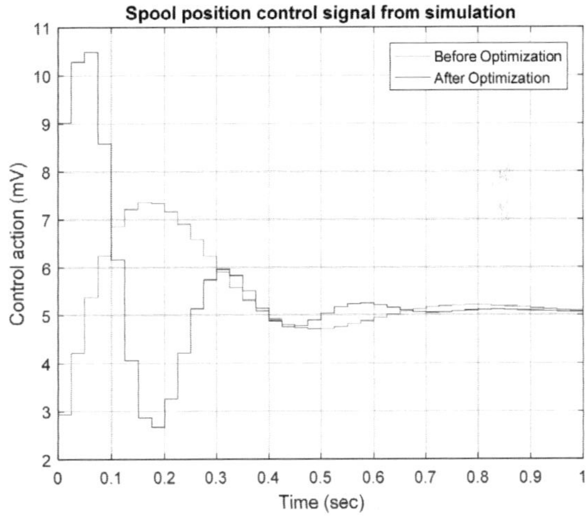

Figure 7. Spool valve closed-loop control signals.

3.2. Flow rate and piston position controllers' optimization

The second optimization procedure is executed to tune the parameter K_{flow} of the flow rate controller and the parameters $K_{cyl,p}$ and $K_{cyl,i}$ of the piston position controller (figure 8). These parameters are tuned such that the following cost function

$$J_{sp}(K_{sp,p}, K_{sp,i}) = \sum_{k=0}^{100} \left(1 - y_{spool}(k, K_{sp,p}, K_{sp,i})\right)^2 + 0.1u(k) \quad (13)$$

is minimized according to *nonlinear-least squares* method. The results from the optimization are shown in table 2 for initial values of the parameters $K_{cyl,p} = 1$, $K_{cyl,i} = 0.5$ and $K_{flow} = 7$. The final values of the controller parameters are $K_{cyl,p} = 15.8$, $K_{cyl,i} = 0.044$ and $K_{flow} = 10.7$. These values are reached after nine iterations and 120 cost function evaluations. The cost function value is reducing from 10.0482 to 9.1321 or 9%. The optimization is stopped because the change of the cost function value between two consecutive iterations is below a predefine absolute tolerance. As can be seen the norm of the step is decreasing too which indicate the approaching of the function local minimum.

Simulation results of the spool valve controller are presented in figure 9 for the output signal and in figure 10 for the corresponding control signal. The figures compare the performance of the initial controller parameters with the controller parameters after the optimization. Both controllers have similar performance but the optimized controller offer less oscillatory control action.

Table 2. Iterations for optimal tuning of flow rate and piston position controller.

Iteration	Function evaluations	New cost value	Norm of step
0	4	10.0482	
1	8	10.0122	0.2316
2	12	9.9996	0.1781
3	16	9.4736	0.2515
–	–	–	–
28	116	9.1326	0.1907
29	120	9.1321	0.1947

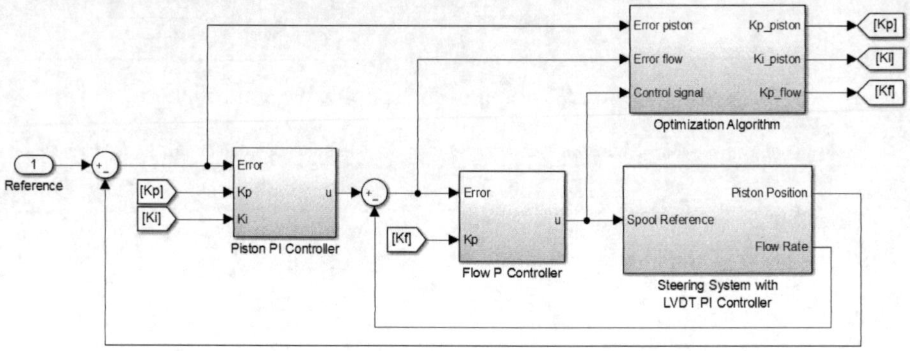

Figure 8. Optimal tuning of flow rate and piston position controllers.

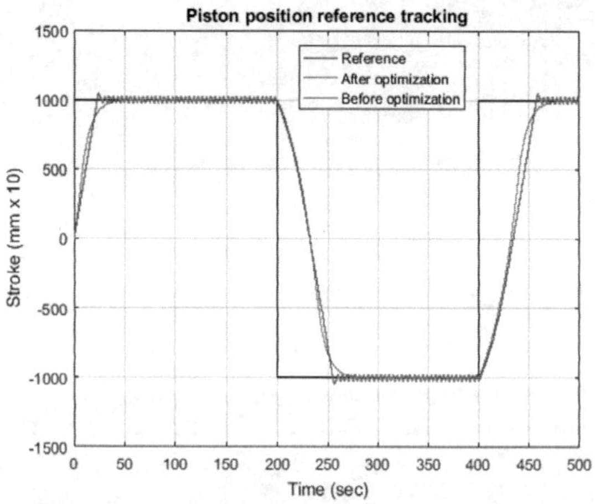

Figure 9. Piston position closed-loop step response.

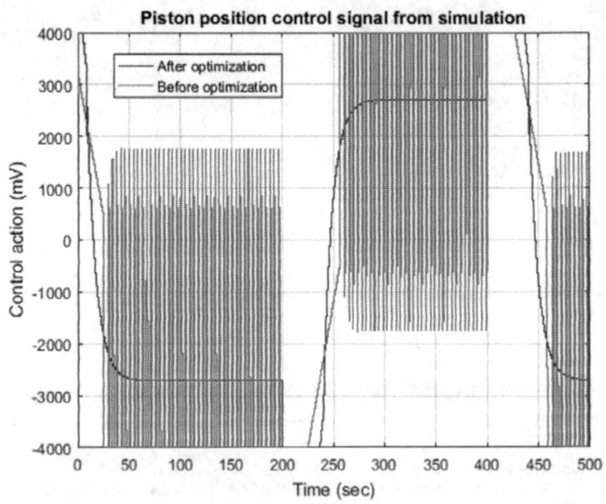

Figure 10. Piston position closed-loop control signals.

4. Experimental results

4.1. Embedded code generation

The designed cascade controller is evaluated on a laboratory system for examination of electrohydraulic systems [10]. It is composed of steering cylinder with 300 mm stroke range of motion, Danfoss OSPE electrohydraulic steering unit and Danfoss PVES switching valve block. The target microcontroller is Danfoss MC012-022 suitable for vehicle automation and diagnostic applications. The microcontroller is programmed in PLUS 1 IDE with CFC and ST languages. The cascade controller is developed as a Simulink block diagram. Then Simulink PLC coder facilitates the automatic code generation of ST function block corresponding to the Simulink diagram of the control algorithm.

4.2. Recorded signals

Presented experimental results compare the performance of the initial controller (tuned manually) and the optimized controller. The generated reference signal drives the steering piston 10 cm to the left for 20 s followed by a command for 10 cm to the right for 20 s. Such signal is kind of extremal and allows transients in the closed-loop control system to be fully investigates and clearly compared. The response of the laboratory steering system is shows on figure 11. It is interesting to note that when moving in a positive direction both controllers show similar performance. But when moving in the opposite direction the optimized controller reaches the reference signal two times faster. The parameters of the optimized controller are increased with respect to the initial controller so faster response is expected. However, when moving in positive direction the laboratory system seem irresponsible to the rising of the parameter values. Figure 12 present the corresponding control actions. The amplitudes of the control signal of the optimized controller are slightly increased.

Figures 13 and 14 compares the flow rate reference tracking for the initial and for the optimized controller. Actually these figures show the performance of the internal flow rate P controller. Both controllers achieve good reference tracking [11]. The measured flow rate signal is disturbed by a sensor noise so the value of K_{flow} should be limited. Figures 15 and 16 compare the spool valve position reference tracking. Both controllers achieve good tracking when the reference is below 0.6mm and present some steady error for higher levels. This effect is related to the reactivity of the passing flow

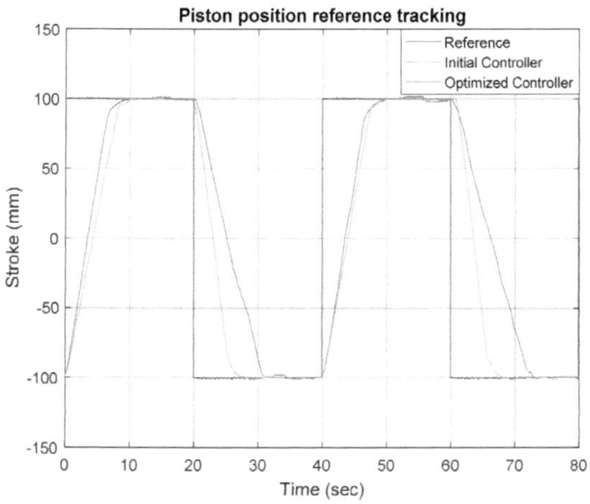

Figure 11. Experimental step response of the closed-loop system.

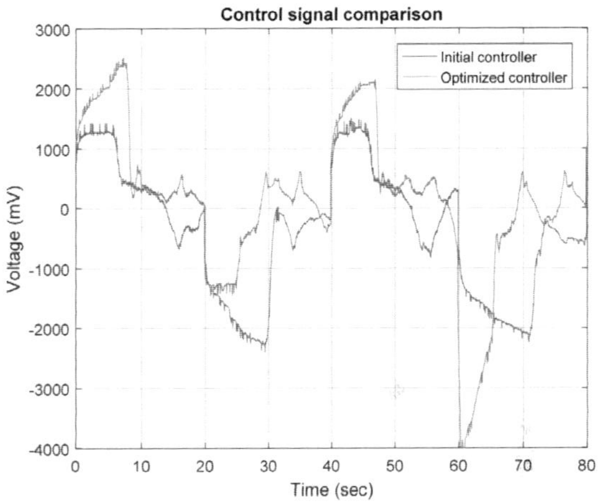

Figure 12. Experimental control action of the closed-loop system.

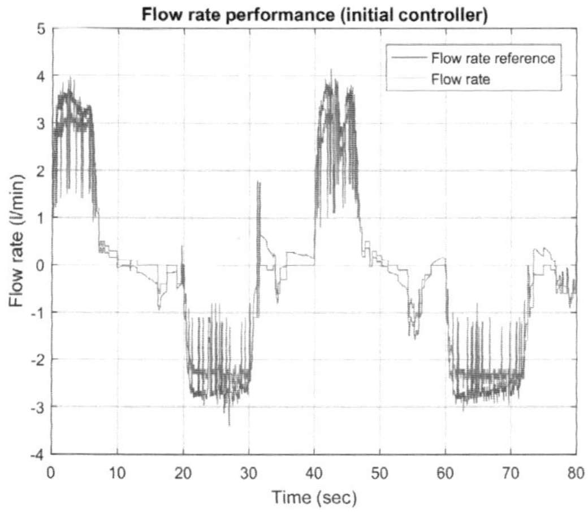

Figure 13. Flow rate – initial controller.

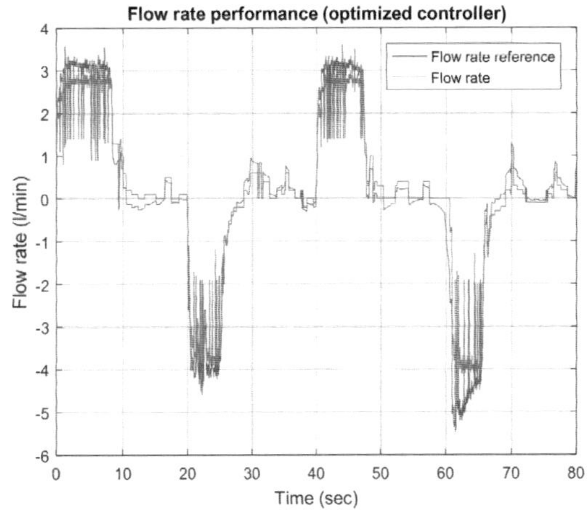

Figure 14. Flow rate – optimized controller.

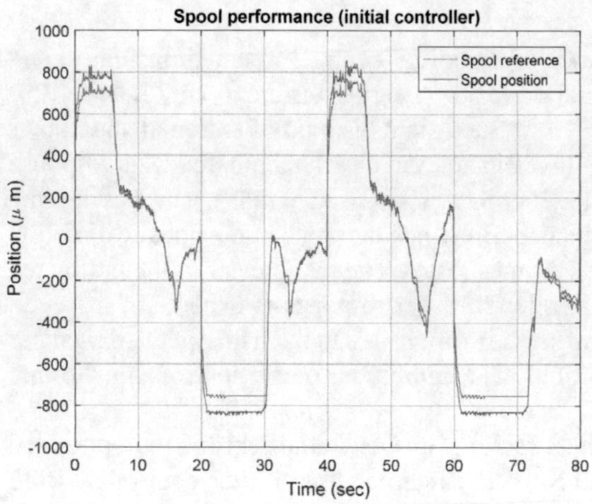

Figure 15. Proportional spool valve position – initial controller.

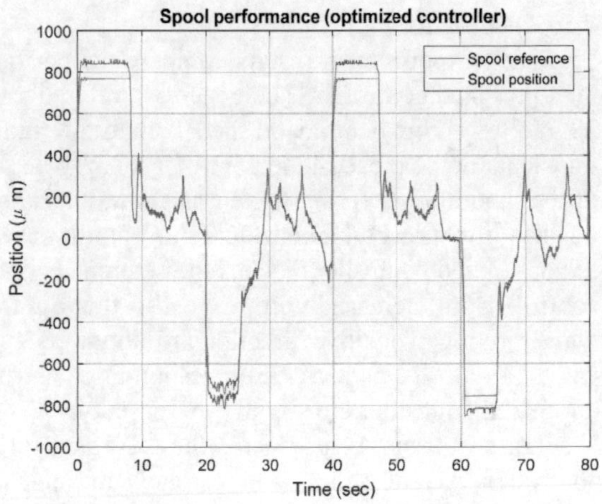

Figure 16. Proportional spool valve position – optimized controller.

5. Conclusion

The article presented a PI-P-PI cascade control system for electrohydraulic steering of heavy duty machines. The three controllers are cascaded such that the output of one controller is a reference signal for the next. Several output variables of the plant are utilized in calculation of the effective control action which is the duty cycle of PWM signal [12] driving a switching valve bridge. The first output signal is the spool valve position measured with a linear variable differential transducer (LVDT) sensor. The second output is the consumed flow rate of the system measured with a geared flowmeter. The last output is the piston position of the steering cylinder. After manual initial tuning the performance of the closed loop system is increased by a nonlinear least squares optimization procedure. The comparative analysis of the performance of the system tuned manually and optimized control system is done. Results shows that the parameters of the optimized controller are increased with respect to the initial controller so faster response is expected. This statement is experimentally confirmed not only by the control signal records but also by the cylinder piston tracking records when the system working with the both different controllers.

Acknowledgments

This work was supported by the Bulgarian Ministry of Education and Science under the National Research Programme "Low carbon energy for transport and live", approved by DCM # 577 / 17.08.2018.

Reference

[1] Pfeffer P and Harrer M 2011 *Lenkungshandbuch. Lenksysteme, Lenkgefühl, Fahrdynamik von Kraftfahrzeugen* (Vieweg + Teubner Verlag)

[2] Danfoss 2016 *OSPE Steering Valve* (Technical Information 11068682)

[3] Goodwin G C, Graebe S F and Salgado M E 2001 *Control System Design* (Prentice-Hall Inc.)

[4] Åström K J and Hägglund T 2001 The future of PID control *Control engineering practice* **9(11)** 1163–75

[5] Schulze T and Weber J 2017 Model based system identification for hydraulic deep drawing presses *Proc. 15th Scandinavian Int. Conf. on Fluid Power SICFP17 (Linköping: Sweden)* 69–79

[6] Zhao P, Wang S, Li X and Zhang B 2011 A novel method of parameter identification of nonlinear electrohydraulic servo systems *Proc. 2011 Int. Conf. Fluid Power and Mechatronics* 547–51

[7] Proca A and Keyhani A 1998 Identification of power steering system dynamic models *Mechatronics* **8** 255–70

[8] Kemmetmuller W, Muller S and Kugi A 2007 Mathematical modeling and nonlinear controller design for a novel electrohydraulic power-steering system *IEEE/ASME Transactions on mechatronics* **12(1)** 85–97
[9] Zhang Q, Wu D, Reid J F and Benson E R 2002 Model recognition and validation for an off-road vehicle electrohydraulic steering controller *Mechatronics* **12(6)** 845–58
[10] Angelov I and Mitov A 2016 Test bench for experimental research and identification of electrohydraulic steering units *10th Int. Fluid Power Conf. (Dresden)* **3** 225–236
[11] Daher N and Ivantysynova M 2014 A virtual yaw rate sensor for articulated vehicles featuring novel electro-hydraulic steer-by-wire technology *Control Engineering Practice* **30** 45–54
[12] Graovac D, Koppl B, Scheffer M, Kiep A and Purschel M 2008 Optimal PWM method for electric and electro-hydraulic power steering applications *2008 IEEE Power Electronics Specialists Conf.* 3213–9

Study of the stresses in the front suspension components of a car passing over speed breakers

Z A Georgiev [1] and L P Kunchev

Technical University – Sofia, Department of Combustion Engines, Automobile Engineering and Transport, 8 Kliment Ohridski Blvd., 1000 Sofia, Bulgaria

[1] E-mail: zlaty.georgiev@gmail.com

Abstract. To reduce the speed of vehicles before pedestrian crossings or as a warning for lowering the speed, vertical deflection devices or road profiles are placed transversely to the direction of travel. The type of profiles is different – speed bumps and humps with different height and length. Some of the profiles have the task of alerting the driver of changing conditions or potential danger - such are the rumble strips, while others aim to force the driving speed below 30 km/h. The amplitudes of the second profiles are considerably larger than those of the first and can cause significant stresses in the front suspension elements and its connections. The purpose of the study is to obtain information on the load on the front suspension components when passing over speed breakers.

1. Introduction

Speed breakers are divided into several types depending on their dimensions: speed humps, speed bumps and speed tables. They are designed to limit the speed of traffic depending on urban conditions: residential streets, school areas or public areas [1]. The speed at which vehicles approach the breakers decreases with increasing height of the calming devices [2]. This applies for drivers familiar with road specifics or in the case of good road markings [3]. In the absence of these conditions, the driver may approach the calming device with inappropriately high speed, which is perceived as a jolt to the vehicle and its passengers [4]. Also some studies have shown that over time drivers get used to speed breakers and speeds rise again [5]. In this case, the suspension elements are subjected to high dynamic forces which result in increased stresses. In residential areas with high concentration of speed breakers, increased air pollution [6] and noise [7] are observed. Due to the abovementioned disadvantages of the calming devices, the optimal design of speed breakers has been an object of research [8]. Good design slows down the traffic to safety levels without causing any impact on the suspension elements [9] and excessive discomfort in the passengers [10].

This study examines the available speed breakers in an urban road network. Depending on their dimensions and the approaching speed, different vertical and longitudinal accelerations in the hub carrier (wheel axle) are recorded. These accelerations are used as input parameters in a finite element model of a MacPherson strut suspension. The study presents the stresses in the suspension elements obtained when passing over speed breakers and the impact of their dimensions on the dynamic forces in the wheel axle.

2. Experimental determination of wheel axle accelerations

To determine the accelerations in the wheel axle, two accelerometers are mounted on the hub carrier. They are uniaxial – one measuring the accelerations in the vertical and the other in the longitudinal

Content from this work may be used under the terms of the Creative Commons Attribution 3.0 licence. Any further distribution of this work must maintain attribution to the author(s) and the title of the work, journal citation and DOI.
Published under licence by IOP Publishing Ltd

direction. The accelerometers are piezoelectric and fitted with the suitable "Brüel & Kjær" charge amplifiers. They are connected to the HBM DAQ device 401 and via it the oscillograms are recorded in a computer (figure 1).

The suspension is complete with a 205/55 R16 radial tyre on a steel wheel. The air pressure in the tyre is 2.1 bar. The measurements are conducted on dry pavements at ambient temperature of 25 °C. The speeds of passing over the calming devices are 20, 30 and 40 km/h.

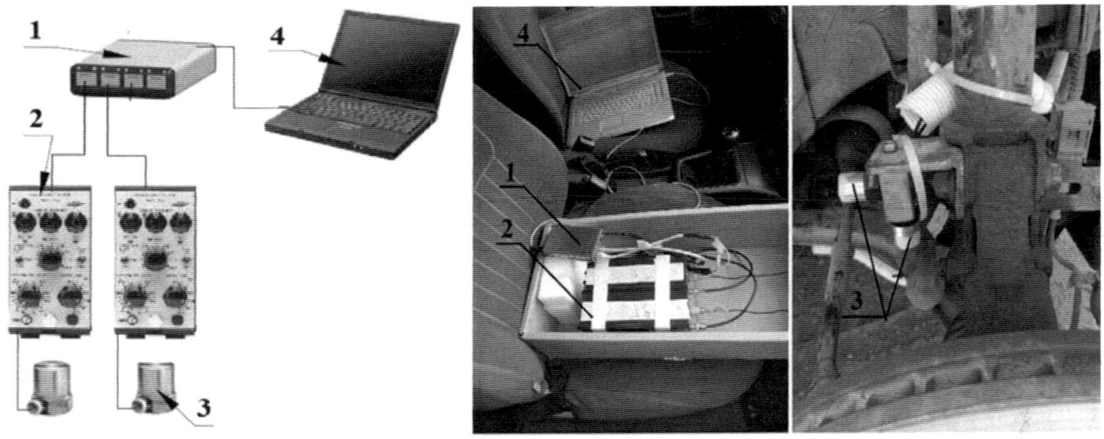

Figure 1. Layout of the measuring system: 1 – DAQ DC module, 2 – charge amplifiers, 3 – piezoelectric accelerometers, 4 – PC.

3. Measuring the profile of the speed breakers

For measuring the profile of the speed breakers, a metric levelling rod and self-levelling laser level shown in figure 2 are used. The levelling rod is placed at the beginning of the speed breaker and it is moved to points, spaced at equal distance of 50 mm between them. The finish point is at the other end of the calming device.

A total of eight different size speed breakers are measured. These include temporary rubber and permanent concrete ones. Some of them are shown in figure 3.

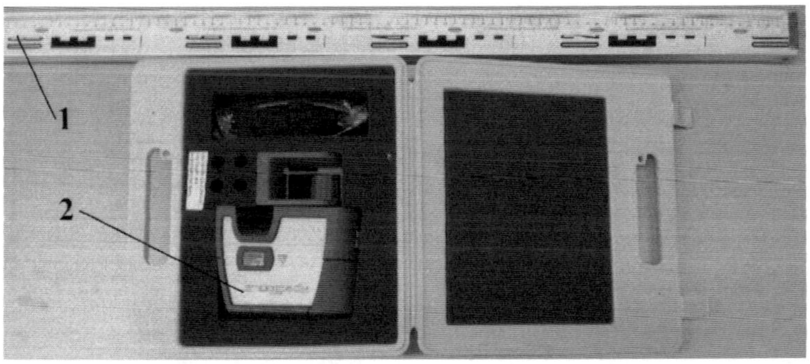

Figure 2. Equipment for measuring the profile of speed breakers:
1 – metric levelling rod;
2 – self-levelling laser level RoboToolz RT-7610-5.

Figure 3. Sample of the measured speed breakers.

4. Processing the experimentally obtained results

To determine the "angle of attack" of a calming device, a drawing consisting of the speed breaker profile and the vehicle pneumatic tyre is made. A tangent line to the speed breaker curve is drawn. This line passes through the common point between the tyre and the speed breaker curve. The angle between this tangent and the pavement surface is the angle of attack – angle α in figure 4. For the specific pneumatic tyre, its static deflection of 16 mm, under the weight of the vehicle, is taken into account.

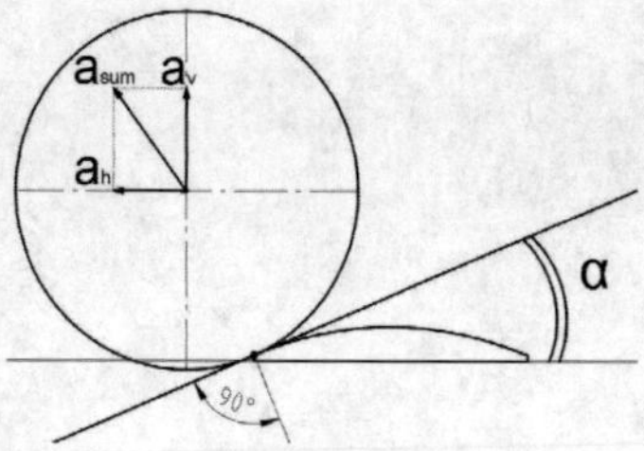

Figure 4. The angle of attack of a speed breaker and the resultant acceleration which it causes in the axle of the wheel.

From the measured maximum vertical and horizontal accelerations (figure 4), the total (resultant) acceleration a_{sum} is determined:

$$a_{sum} = \sqrt{a_v^2 + a_h^2}, \tag{1}$$

where a_v and a_h are respectively the maximum measured vertical and horizontal accelerations of the hub carrier.

The resulting total accelerations are compared for different passing speed and angles of attack.

5. Suspension modelling

A structural finite element model of the studied MacPherson strut suspension is made. The model is implemented in SolidWorks Simulation software. The coil spring is assumed to be made of steel C60 with elastic modulus $E = 2.1\times10^{11}$ N/m^2, Poisson coefficient $\mu = 0.28$ and density $\rho = 7800$ kg/m^3. The hub carrier is assumed to be made of grey cast iron with elastic modulus $E = 6.6\times10^{10}$ N/m^2, Poisson coefficient $\mu = 0.27$ and density $\rho = 7200$ kg/m^3. All other metal elements are assumed to be made of mild steel with elastic modulus $E = 2\times10^{11}$ N/m^2, Poisson coefficient $\mu = 0.29$ and density ρ=7900 kg/m^3. The rubber mounts of the suspension are presented as distributed springs without their own mass. The rubber mounts of the lower control arm have a radial stiffness of 1.2×10^6 N/m (position 1 in figure 5). The top mount of the shock absorber has a normal stiffness of 1×10^6 N/m (position 3 in figure 5). The shock absorber is modelled as a rod system made up of two solid elements: a rigidly connected piston and a piston rod as well as a twin-tube solid element. A spring-damper element connects the piston and the bottom of the inner tube (position 2 in figure 5). The spring-damper connector limits the movement of the piston and piston rod in radial direction by its high tangential stiffness. It also has a damping coefficient of 2000 N.s/m in axial direction.

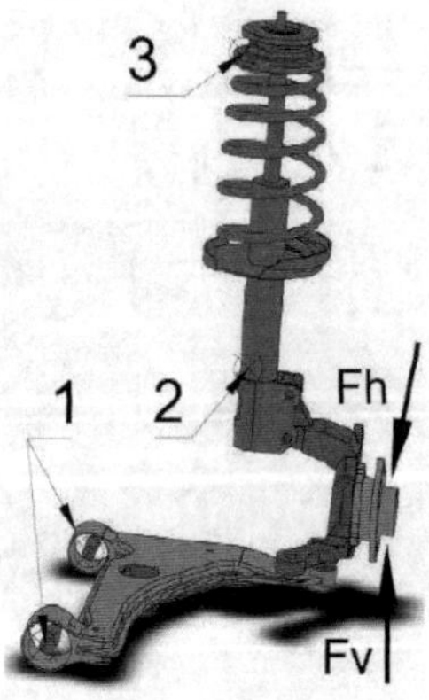

Figure 5. Model of the studied MacPherson strut suspension: 1 – rubber mounts of the lower control arm; 2 – spring-damper connector between the piston and the inner tube; 3 – top mount of the shock-absorber.

Input parameters are the vertical and horizontal dynamic forces in the wheel hub (axle) when passing over a speed breaker. The dynamic forces are the product of the measured maximum accelerations with the unsprung mass. The unsprung mass is 45 kg. It includes the mass of the tyre, the steel wheel, the bearing hub, the hub carrier, the brake caliper and disc, the lower arm, and two-thirds of the mass of the coil spring and the shock absorber. Also a static vertical force of 4000 N is applied to the wheel hub.

"Modal Time History Study" available in the SolidWorks Simulation software is used to solve the numerical dynamic study. It allows for the disturbing force to be set as a function of time. The aim of the model is to find the elements of the suspension with the highest stresses caused by passing over a calming device. The resulting stresses estimate the necessary speed of going over a speed breaker to reach the yield strength of material for the endangered suspension element.

6. Results

Table 1 shows the resulting total acceleration and the attack angle for the different speed breakers. The values of the accelerations are measured for three different speeds of motion.

Table 1. Resulting accelerations for different speed of passing over a speed breaker.

Speed breaker	Angle of attack (degrees)	Resulting acceleration (g)		
		Speed 20 km/h	Speed 30 km/h	Speed 40 km/h
№1	5.0	1.16	1.41	2.23
№2	8.1	1.17	2.05	2.91
№3	8.9	1.19	2.12	2.83
№4	10.4	1.89	3.18	4.11
№5	10.8	1.66	3.05	3.61
№6	13.7	2.50	3.81	4.95
№7	15.2	2.24	3.37	4.87
№8	22.6	2.25	3.81	5.39

From the results it can be seen that speed breakers with a larger angle of attack cause even higher accelerations in the wheel axle. Figure 6 shows the resulting total accelerations as a function of the angle of attack and the speed of passing over the speed breakers. The angles of attack correspond to calming devices of different heights. Figure 7 presents the dimensions of calming devices for different angles of attack and a height of 60 mm. The angle of attack decreases proportionally to an increase in length.

The model study is conducted for the highest recorded accelerations at the road measurements. They are registered for speed breaker №8 (Table 1) with the maximum angle of attack. The results show that the suspension elements with highest stresses are the control arm and the piston rod (figure 8).

Stresses in the suspension elements increase proportionally to the disturbing forces at the wheel axle. The disturbing forces depend on the speed of passing over the speed breakers. Figure 9 shows the maximum stresses in the control arm and the piston rod as a function of the speed of passing over a speed breaker with 22.6° angle of attack. There is linear dependence between them. This allows us to calculate the estimated speed for which the stresses in one element of the suspension would reach its yield strength. In this case, the estimated speed for the material of the piston rod to reach its yield strength of 350 MPa is 65 km/h.

The mesh is constructed by linear tetrahedral solid elements with a total size of 10 mm and tolerance of 0.5 mm. A numerical solution with smaller size mesh elements (total size of 5 mm and tolerance of 0.25 mm) is performed. The stresses in the endangered elements for the second solution have a deviation less than 5%.

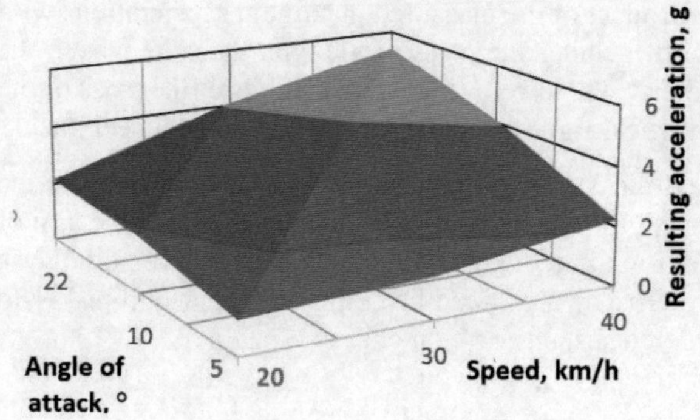

Figure 6. The resulting acceleration in the wheel axle as a function of the angle of attack of a speed breaker and the speed.

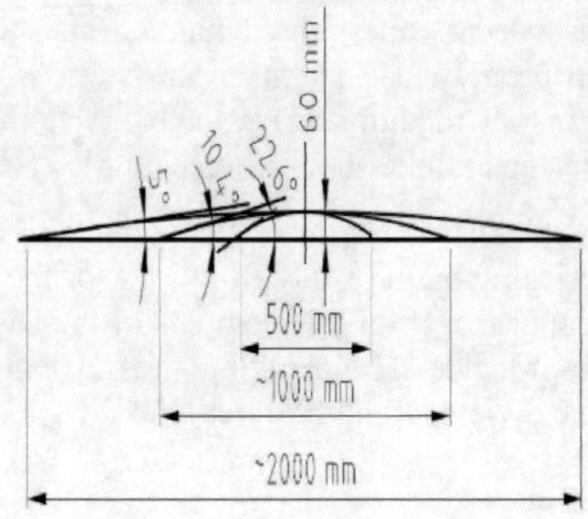

Figure 7. Speed breakers with equal height and different angles of attack.

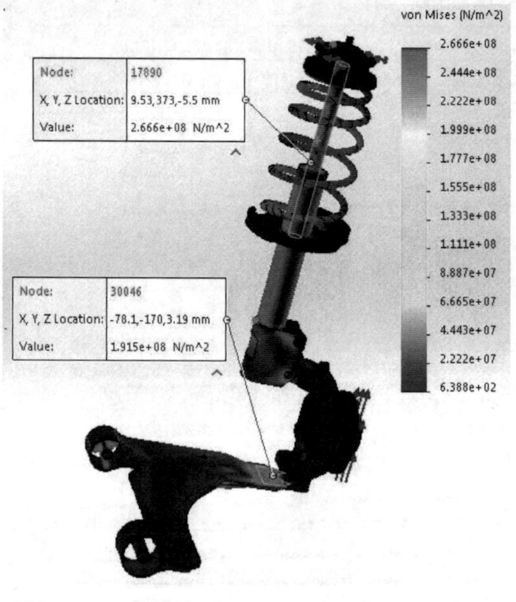

Figure 8. Stresses von Misses in the suspension elements caused by passing over a calming device with an angle of attack of 22.6° at a speed of 40 km/h.

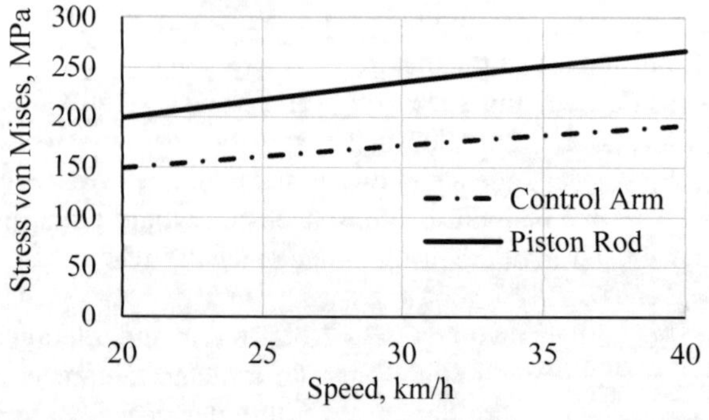

Figure 9. Maximum stresses in the control arm and the piston rod as a function of the speed of passing over a speed breaker with 22.6° angle of attack.

7. Conclusion

The experimental study shows that the angle of attack determines the force of dynamic impact on the wheel axle. The angle of attack depends on the geometric dimensions of the speed breakers such as the height at the midpoint and the length of the base of the cross section.

The dynamic forces are proportional to the speed of passing over the calming devices.

The theoretical FEM study shows that the highest stresses are recorded in the piston rod of the shock absorber and the control arm. At an estimated speed of 65 km/h, possible for urban conditions, the resulting stresses can reach the yield strength of the material. For the purposes of the study the assumed material is mild steel with yield strength of 350 MPa. Stresses higher than this could lead to plastic deformations of the elements and disturbance of the suspension geometry.

From the numerical study, by comparing the stresses caused by the static and the dynamic load, a dynamic coefficient can be calculated. In this case it is $k = 1.82$.

Speed breakers with an angle of attack higher than 20° are not suitable for speeds above 30 km/h. They can only be used in residential areas.

References

[1] Gupta A 2014 Study on speed profile across speed bumps (A Project Report Department of Civil Engineering National Institute of Technology Rourkela)
[2] Antic B, Pesic D, Vujanic M and Lipovac K 2013 The influence of speed bumps heights to the decrease of the vehicle speed – Belgrade experience *Safety Science* **57** 303–12
[3] Smith D, Hallmark S, Knapp K and Thomas G 2002 Temporary speed hump impact evaluation *Iowa Department of Transportation*
[4] Aslan S, Karcioglu O, Katirci Y, Kandis H, Ezirmik N and Bilir O 2005 Speed bump–induced spinal column injury *American Journal of Emergency Medicine* **23** 563–4
[5] Ripley D and Klingaman M 1998 City of Iowa City Memorandum: Teg drive traffic calming evaluation (Iowa City Iowa June 4)
[6] Baltrenas H P, Janusevicius T and Chlebnikovas A 2017 Research into the impact of speed bumps on particulate matter air pollution *Measurement* **100** 62–7
[7] Kokowski P and Makarewicz R 2006 Predicted effects of a speed bump on light vehicle noise *Applied Acoustics* **67** 570–9
[8] Hakan Lav A, Bilgin E and Hilmi Lav A 2018 A fundamental experimental approach for optimal design of speed bumps *Accident Analysis and Prevention* **116** 53–68
[9] Yuanlonga W, Wanzhonga Z, Guana Z, Qiangb G and Chunyana W 2018 Suspension mechanical performance and vehicle ride comfort applying a novel jounce bumper based on negative Poisson's ratio structure *Advances in Engineering Software* **122** 1–12
[10] Gedik A, Bilgin E, Hilmi Lav A and Artan R An investigation into the effect of parabolic speed hump profiles on ride comfort and driving safety under variable vehicle speeds: A campus experience *Sustainable Cities and Society* **45** 413–21

IOP Conf. Series: Materials Science and Engineering **664** (2019) 012013 doi:10.1088/1757-899X/664/1/012013

Neural model of projecting flexural strength of cement concrete intended for airfield pavements

Małgorzata Linek

Kielce University of Technology, Faculty of Civil Engineering and Architecture, Department of Transportation Engineering, Aleja Tysiąclecia Państwa Polskiego Street 7, 25-314 Kielce, POLAND

E-mail: linekm@tu.kielce.pl

Abstract. This work present to the mathematical model in the form of Artificial Neural Network (ANN), intended for projecting concrete flexural strength. Input data was classified according to the type of component material and its content in concrete mix (cement contents, coarse aggregate, fine aggregate, water and admixtures). In order to determine mathematical model, a multilayer, one-way perceptron network was used, recursion network with sigmoidal neurons. The model assumes that neurons are gathered in some layers (one input layer, hidden layers and one output layer). The conducted cross-section of the impact of variable parameters values (learning constant α and momentum values η) on the accuracy of representation of flexural strength was analyses. Assessment criterion was assumed as consideration the lowest mistake level and 100% compliance. According to the obtained results ANN was assumed to be the best representing network for constant value of momentum 0.3, learning constant of 0.04 and 9 neurons in a hidden layer. Very good coincidence of component models with experiment results was achieved. At testing stage, the coincidence was achieved at the level of 99.25%, in case of the assumed network structure. During model verification by means of experimental results, the average coincidence was 99.68%.

1. Introduction

ANN were proved, among others, in the following works [1-3]. This research paper concerns the issue of projecting the flexural strength of concrete intended for airfield pavements as far as structure durability is concerned. One of the basic parameters of hardened concrete, which has the direct impact on the durability thereof in case of airfield pavement, is flexural strength [4, 5]. Using own collected database containing strength, laboratory and field test results, the mathematical model in the form of ANN, intended for projecting concrete strength, was presented.

2. Experimental studies

The testing base used for the purposes of ANN modelling included cement concretes intended for airfield pavements. Aggregate compositions of mixes in case of the analysed concretes were diversified in regard to the content of individual components of concrete mix. In case of each out of 6500 mixes, coarse aggregate was the base element. Fine aggregate in the form of washed sand was also included as aggregate composition of mixes. Due to the diversification of individual mix components, aggregate content was taken into consideration with the change every 1kg/m^3. The designed aggregate compositions each time complied with PN-EN 206-1 [6], NO 17A-204 [7]. Mix composition included clean-clinker Portland cement in various amounts CEM I 42.5 (which fulfilled the requirements of PN-

Content from this work may be used under the terms of the Creative Commons Attribution 3.0 licence. Any further distribution of this work must maintain attribution to the author(s) and the title of the work, journal citation and DOI.
Published under licence by IOP Publishing Ltd

EN 197-1 [8]) and water (complied with PN-EN 1008 [9]), designing diversified water-content ratio with reference to cement amount (w/c), according to the requirements of PN-EN 206-1 [6]. Water-cement ratio significantly impacts concrete durability [7]. Due to the fact that limiting the value of w/c coefficient impacts the reduction of hardened concrete permeability, at the same time increasing durability thereof, in case of pavement concretes, especially concretes intended for airfield pavement application, the value of w/c ratio is reduced to maximum level of 0.4 [5, 7]. Limit content of individual mix components have been presented in the table 1.

Table 1. Limit component contents of the analysed concrete mixes.

Elements / Contemt	Cement kg/m^3	Coarse aggregate kg/m^3	Fine aggregate kg/m^3	Water kg/m^3	Admixtures kg/m^3
Minimum	335	1149	358	120	0
Maximum	380	1401	604	150	6.4

Concretes curing in standard conditions, (according to the requirements of PN-EN 12390-2[10]) were considered for the purposes of this research paper. On the 28th day since concreting, each sample, prepared and cured in accordance with the requirements of [10], was subject to destructive tests by means of hydraulic press in compliance with PN-EN 12390-4:2009 [11]. Maximum force destructing the sample, obtained according to the measurements, taking into consideration the assumed constant loading speed of 0.05MPa/s, was the basis for defining flexural strength of individual samples. Samples of bar shape and dimensions of 150×150×700 mm were used in the course of the research. Dimensions of samples and load diagram were in compliance with the requirements of [7] (figure 1).

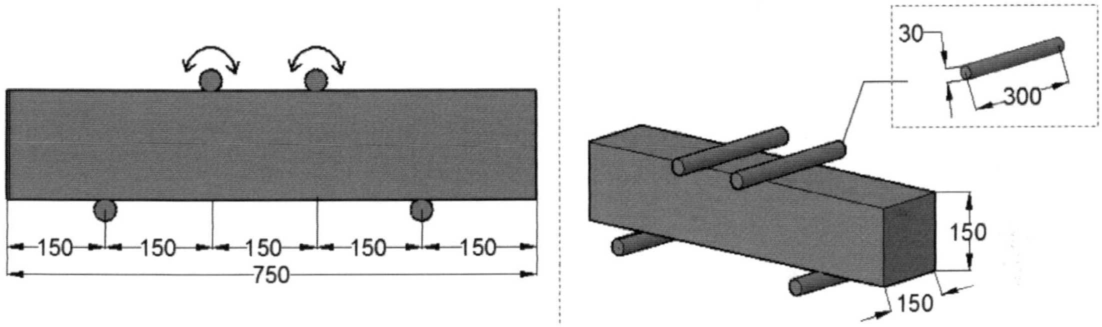

Figure 1. Dimensions of samples and load diagram.

In case of each sample, in compliance with the requirements of [12], the flexural strength f_{cf} was determined, according to the formula (1), where F refers to maximum load registered in the course of sample destruction, l – spacing between supports (l = 150 mm), b – width of the sample (b = 150 mm), h – height of the sample (h = 150 mm).

$$f_{cf} = \frac{F\,l}{b\,h^2} \qquad (1)$$

The strength values obtained in this way were the input base to assess flexural strength of cement concrete intended for airfield pavements using ANN. All obtained test results concerning individual research series corresponded the criterion of minimum strength according to the [7] standard and they were statistically essential. Therefore, they were the reliable source of input data for ANN identification, reflecting concrete strength in case of various compositions thereof.

3. Results

3.1. Network learning

The essence of ANN is the selection of network structure (the number of layers and the number of neurons in particular layers) and defining the parameters thereof (constant values and weights). According to the analyses using the artificial neural networks were created regressive model, where it was expected to provide a specific numerical value being the solution to the problem. In case of the analysed case this value was identified with the assumed flexural strength value. In order to conduct cross-section analysis of the influence of variables parameters values (α and η) on the accuracy of representation of flexural strength, analyses including matrix of 90×90 elements. Values of learning constant α ranged from $\alpha_{min} = 0.01$ to $\alpha_{max} = 0.09$ every 0.01. Momentum values η were assumed from $\eta_{min} = 0.01$ to $\eta_{max} = 0.09$ every 0.01. In the case of each combination α and η the constant value was assumed to be $\varepsilon = 0.01$.

According to the obtained results it was proved that the selection of α and η has significant impact on the extent of the obtained representation. Tables 2, 3 and 4 present exemplary summary of a given internal layers structure, selected in the course of experiment. Table 2 presents the impact of learning constant values, ranging from 0.03 to 0.07, on the accuracy of flexural strength representation for the previously assumed constant values $\eta = 0.3$ and $\varepsilon = 0.01$. Table 3 presents the impact of momentum value, ranging from 0.3 to 0.7, on the accuracy of flexural strength representation for the previously assumed constant values $\alpha = 0.04$ and $\varepsilon = 0.01$.

Analysing the data presented in table 2, the initial tolerance of learning constant was assumed at 0.04 for further analyses. With such assumed tolerance, the least value of the generated mistake (Δy_{sr}) and the highest compliance of results was obtained. The assessment of representation accuracy of values from the experiment ($y_{rz\,i}$) and coefficients generated by neural network ($y_{ANN\,i}$) was defined on the basis the sum of squares of differences which was determined according to the relationship (2).

$$\Delta\, y_{sr} = \frac{\sum_{i=1}^{N}(y_{ANN\,i} - y_{rz\,i})^2}{N} \tag{2}$$

Table 2. Impact of learning constant value α on the accuracy of flexural strength representation (assuming constant value of momentum $\eta = 0.4$ and $\varepsilon = 0.01$).

No ANN	α	η	Number of inputs	Number of neurons in the hidden layer	Number of outputs	Percentage of positive events	Δy_{sr}
Z/1-1	0.03	0.40	5	3	1	100	0.0051724
Z/1-2	0.04	0.40	5	3	1	100	0.0050352
Z/1-3	0.05	0.40	5	3	1	100	0.0055646
Z/1-4	0.06	0.40	5	3	1	100	0.0066406
Z/1-5	0.07	0.40	5	3	1	100	0.0084439

Table 3. Impact of momentum value η (from 0.3 to 0.7) on the accuracy of flexural strength representation (assuming value of learning constant $\alpha = 0.04$ and $\varepsilon = 0.01$).

No ANN	α	η	Number of inputs	Number of neurons in the hidden layer	Number of outputs	Percentage of positive events	Δy_{sr}
Z/2-1	0.04	0.30	5	4	1	100	0.0036493
Z/2-2	0.04	0.40	5	4	1	100	0.0036632
Z/2-3	0.04	0.50	5	4	1	100	0.0037701
Z/2-4	0.04	0.60	5	4	1	100	0.0041493
Z/2-5	0.04	0.70	5	4	1	100	0.0057739

Analysing the data presented in table 3 the initial tolerance of momentum was assumed at 0.3 for further analyses. In case of such assumed tolerance, the least value of the generated mistake (Δy_{sr}) and the highest compliance of results was obtained.

Analysing the data presented in table 4 the initial number of neurons equalling to 9 was assumed in a hidden layer. In case of such tolerance, the least value of the generated mistake and the highest compliance of results was obtained. Table 5 presents exemplary summary of the selected network structures in case of the diversified number of hidden layers, selected experimentally.

Table 4. Impact of neurons number (3-14) in a hidden layer on the accuracy of representation of flexural strength at constant value of momentum 0.3, learning constant 0.04 and $\varepsilon = 0.01$.

The number of neurons in the hidden layer	3	4	5	6	7	8
Percentage of positive events	100	100	100	100	100	100
Δy_{sr}	0.0051505	0.0036632	0.0052850	0.0039173	0.0052694	0.0063905
The number of neurons in the hidden layer	9	10	11	12	13	14
Percentage of positive events	100	70,8	70,8	70,8	100	100
Δy_{sr}	0.0033323	0.0709030	0.0718920	0.0694870	0.0071414	0.0063639

Table 5. Impact of neurons number (1-10) in a hidden layer on the accuracy of representation of flexural strength at constant value of momentum 0.3, learning constant 0.04 and $\varepsilon = 0.01$.

No. ANN	α	η	number of inputs	Number of neurons in the hidden layer I	II	number of outputs	percentage of positive events	Δy_{sr}
Z/3-1	0.04	0.30	5	9	1	1	99.2	0.0163880
Z/3-2	0.04	0.30	5	9	2	1	58.4	0.1357900
Z/3-3	0.04	0.30	5	9	3	1	58.4	0.1352300
Z/3-4	0.04	0.30	5	9	4	1	100	0.0053105
Z/3-5	0.04	0.30	5	9	5	1	58.4	0.1318200
Z/3-6	0.04	0.30	5	9	6	1	58.4	0.1319000
Z/3-7	0.04	0.30	5	9	7	1	58.4	0.1217300
Z/3-8	0.04	0.30	5	9	8	1	58.4	0.1259200
Z/3-9	0.04	0.30	5	9	9	1	58.4	0.1187800
Z/3-10	0.04	0.30	5	9	10	1	58.4	0.1254600
Z/3-11	0.04	0.30	5	1	9	1	99.2	0.0134200
Z/3-12	0.04	0.30	5	2	9	1	100	0.0096353
Z/3-13	0.04	0.30	5	3	9	1	58.4	0.1242800
Z/3-14	0.04	0.30	5	4	9	1	58.4	0.1254000
Z/3-15	0.04	0.30	5	5	9	1	58.4	0.1235000
Z/3-16	0.04	0.30	5	6	9	1	58.4	0.1239000
Z/3-17	0.04	0.30	5	7	9	1	58.4	0.1257200
Z/3-18	0.04	0.30	5	8	9	1	58.4	0.1255400
Z/3-19	0.04	0.30	5	10	9	1	58.4	0.1243800

According to the obtained analysis results ANN was assumed the best representing network for constant value of momentum 0.3, learning constant of 0.04, 9 neurons in a hidden layer and assumed $\varepsilon = 0.01$. In case of network of 5-9-1 structure, the first digit (5) refers to the number of inputs, next (9) the number of neurons in a hidden layer, and the last one (1) refers to one output identified with the assumed compressive strength value. This structure was adopted as a result of accuracy tests of representing real flexural strength values by the network. As the assessment criterion was assumed the lowest mistake level (Δy_{sr}) and 100% compliance.

Figure 2 presents the courses of the learning in case of network with neurons in a hidden layer. It can be observed that minimum values Δy_{sr} amount to 0.008, and maximum values amount to 0.250 in the case of the assumed number of iterations.

According to the obtained results of ANN analysis it was proved that the determined network can be recognized reliable for 170000 iterations.

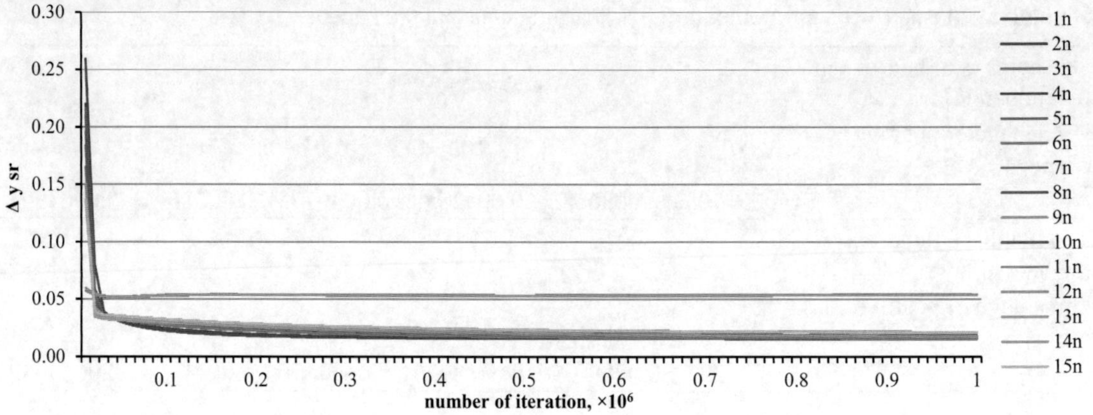

Figure 2. The course of learning process by artificial neural networks of diversified neurons number in a hidden layer.

3.2. Network testing

Out of the collection of 6500 diversified concrete mix compositions intended for airfield pavements, 20 % was selected randomly and intended for the educated network testing. Table 6 presents approximate contents of individual mix components intended for network testing.

For the purposes of this research paper, the analysis included concretes curing in standard conditions and examined on 28th day since concreting. The research process, aimed at defining flexural strength, was in compliance with the one presented in section 2. Figure 3 presents graphic summary of data determined at testing stage by the educated network.

Table 6. Limit component contents of the analysed concrete mixes.

Elements / Contemt	Cement kg/m^3	Coarse aggregate kg/m^3	Fine aggregate kg/m^3	Water kg/m^3	Admixtures kg/m^3
Minimum	300	1200	400	110	0.1
Maximum	490	1700	840	190	3.6

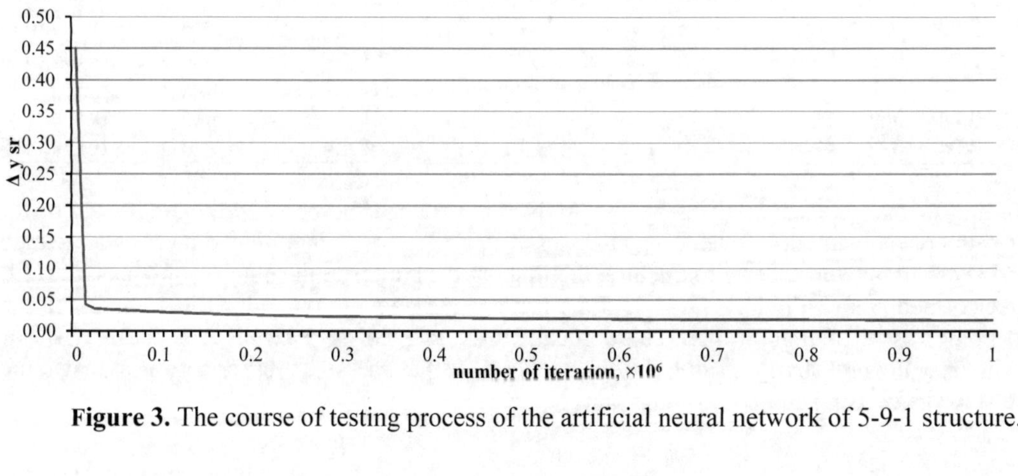

Figure 3. The course of testing process of the artificial neural network of 5-9-1 structure.

According to the obtained characteristics (table 7), it was proved that testing the artificial neural network model reflects the values determined by the network. The obtained average coincidence value from 1300 events is very high. Average value of coincidence is 99.26 %.

Table 7. Testing of ANN model.

	Input data					Output data		
L.p.	C [a] kg/m^3	Kg [a] kg/m^3	Kd [a] kg/m^3	W [a] kg/m^3	D [a] kg/m^3	from the network MPa	during testing MPa	accuracy %
1.	370	1510	630	144	0.1	5.9	6.0	98.33
2.	320	1580	590	127	1.7	7.1	7.1	100.00
3.	390	1150	840	150	1.5	6.3	6.3	100.00
4.	380	1290	760	152	2.2	5.2	5.2	100.00
5.	310	1650	430	120	0.6	6.8	6.8	100.00
6.	360	1430	610	140	2.0	5.4	5.5	98.18
7.	280	1370	790	115	2.9	6.9	6.9	100.00
8.	410	1620	580	158	2.3	4.5	4.6	97.83
9.	290	1430	820	110	2.5	6.1	6.1	100.00
10.	490	1350	810	195	1.5	5.8	5.8	100.00

[a] C – Cement; Kg – Coarse aggregate; Kd – Fine aggregate; W – Water; D – Admixtures

3.3. Network verification

The educated and tested ANN network of 5-9-1 structure was intended for experimental verification. The collected data, out of 20% of all laboratory tests, were subject to verification - this data were not intended for network learning or testing. Tables 8 and 9 includes the obtained analyses results. Table 8 presents approximate contents of individual mix components intended for network testing.

Table 8. Limit component contents of the analysed concrete mixes.

Elements / Contemt	Cement kg/m^3	Coarse aggregate kg/m^3	Fine aggregate kg/m^3	Water kg/m^3	Admixtures kg/m^3
Minimum	260	1100	380	100	0.0
Maximum	430	1670	690	165	4.4

Table 9. ANN model verification.

	Input data					Output data		
L.p.	C [a] kg/m^3	Kg [a] kg/m^3	Kd [a] kg/m^3	W [a] kg/m^3	D [a] kg/m^3	from the network MPa	during testing MPa	accuracy %
1.	268	1310	650	110	0.9	6.4	6.5	98.46
2.	295	1590	410	125	1.2	7.3	7.3	100.00
3.	315	1280	720	120	2.1	6.5	6.5	100.00
4.	390	1420	480	150	0.8	5.9	5.8	98.31
5.	410	1360	620	156	0.0	5.2	5.2	100.00
6.	365	1250	680	148	0.8	6.8	6.8	100.00
7.	403	1530	450	153	3.4	7.3	7.3	100.00
8.	340	1610	400	130	2.1	6	6.1	98.36
9.	355	1330	420	135	3.5	7.1	7.1	100.00
10.	280	1470	530	114	4.0	6.9	6.9	100.00

[a] C – Cement; Kg – Coarse aggregate; Kd – Fine aggregate; W – Water; D – Admixtures

According to the obtained characteristics (table 9), it was proved that verification the artificial neural network model reflects the values determined by the network. The obtained average coincidence value from 1300 events is very high. Average value of coincidence is 99.68 %.

4. Conclusions

It should be concluded that the assumptions concerning the construction of ANN model and the course of analysis can anticipate with credibility the parameters of composite material – concrete. The results of the conducted analyses prove the purpose of application of the presented method in modelling the composition of concrete mix and projecting the flexural strength of concrete after 28 days of curing in standard conditions.

The analysed structure 5-9-1 make it possible to draw conclusions regarding material parameters even in the case of limited number of elements. The network analysing five neurons in an input layer, diversified in terms of the selection of hidden neurons amount, complies with model assumptions. It was proven that a 5-9-1 network, with learning constant of 0.04, momentum 0.3 and $\varepsilon = 0.01$, is the most favourable structure for the assumed data. Very good coincidence of component models with experiment results was achieved. At testing stage, the coincidence achieved was at the level of 99.25%, in case of the assumed network structure. During model verification by means of experimental results, the average coincidence was 99.68%.

References

[1] Albuthbahak O M and Alkhudery H H 2018 Artificial neural network model for flexural design of concrete hydraulic structures *Int. J. Civil Engineering and Technology* **9(2)** 265–74

[2] Köroglu M A, Ceylan M, Arslan M H and Ilki A 2012 Estimation of flexural capacity of quadrilateral FRP-confined RC columns using combined artificial neural network *Engineering Structures* **42** 23–32

[3] Lepadatu D, Barbuta M, Rujanu M, Judele L andMitroi R 2018 Fly ash concrete with fibers: comparison of tensile strength using neural network and design of experiments methods *Environmental Engineering and Management Journal* **17** 1321–8

[4] Glinicki M A 2011 Durability of concrete in road surfaces. Effect of microstructure, materials design, diagnostics (in Polish) *IBDiM (Warsaw)* **66**

[5] Nita P 2005 *Concrete Surface Airport. Theory and Structural Dimensioning* (Air Force Institute of Technology, Warsaw)

[6] PN-EN 206-1: 2003 Concrete-Part 1: Specification, performance, production and conformity - Polish standard.

[7] NO 17-A204:2015 Airfield concrete pavements - Requirements and test methods for cement concrete pavements - Polish standard.

[8] PN-EN 197-1:2012 Cement, part 1: Composition, specyfications and conformity criteria for common cements - Polish standard.

[9] PN-EN 1008:2004 Mixing water for concrete - Specification for sampling, testing and assessing the suitability of water, including water recovered from processes in the concrete industry, as mixing water for concrete - Polish standard.

[10] PN-EN 12390-2:2011 Testing hardened concrete - Part 2: Making and curing speciments for strength tests - Polish standard.

[11] PN-EN 12390-4:2011 Testing hardened concrete - Part 4: Compressive strength - Specification for testing machines - Polish standard.

[12] PN-EN 12390-5:2011 Testing hardened concrete - Part 5: Flexural strength of test specimens - Polish standard.

Rationalization of cross-sections of the composite reinforced concrete span structure of bridges with a monolithic reinforced concrete roadway slab

G L Vatulia[1,3], O V Lobiak[1], S V Deryzemlia[1], M A Verevicheva[1] and Ye F Orel[2]

[1] Structural Mechanics and Hydraulics Department, Ukrainian State University of Railway Transport, Feuerbach sq. 7, 61050 Kharkiv, Ukraine
[2] Railway Survey and Design Department, Ukrainian State University of Railway Transport, Feuerbach sq. 7, 61050 Kharkiv, Ukraine

[3] Email: glebvatulya@gmail.com

Abstract. The article presents the developed rationalization technique of composite steel reinforced concrete sections with steel open section beams based on the criterion of equal strength of the section elements that are extremely distant from the neutral line. Algorithms for search for geometric parameters of a composite section limited to a certain range of values are implemented to achieve the equal strength condition. The dimensions of the individual elements which are parts of the cross-section are obtained from the condition of the constant ratio of the distances from the neutral axis to the extreme concrete and steel fibers. The numerical methods were used for calculation of continuous three-span composite reinforce concrete bridge. The technique implements the steps of bridge construction, taking into account the contact yield of the composite section, the redistribution of forces between the elements, and the effect of elastic-plastic and rheological properties of materials. The generalized kinetic curve was utilized for evaluation of concrete creep together with the phenomenological equations for the development of deformations based on a colloid-chemical representation of the mechanism for long-term concrete deformation. The proposed methodology is implemented in the LIRA-SAPR software package based on the Building Information Model Technology (BIM) and the Finite Element Method (FEM).

1. Introduction

The efficiency of composite reinforced concrete span structures with the use of a monolithic reinforced concrete roadway slab is verified by calculation at the design stage, and will primarily depend on the adequacy of the simulation of the task. However, the applicable Ukrainian building codes [1] do not contain a technique of such calculations, which complicates the application of this type of structures. A rational choice of preliminary design parameters of the composite cross-section (such as beam height, slab thickness, upper and lower chord width, wall thickness, etc.) is also problematic. The article describes the practical experience of finite-element simulation of the work of a span structure using the example of a continuous composite reinforced concrete road bridge, taking into account the stages of operation of the construction, from the construction completion to the late service life. The proposed methodology implements the technique of rationalization and evolutionary transformations of the model (calculation based on the phases of construction) taking into account the influence of the elastic-plastic and rheological properties of the materials.

Content from this work may be used under the terms of the Creative Commons Attribution 3.0 licence. Any further distribution of this work must maintain attribution to the author(s) and the title of the work, journal citation and DOI.
Published under licence by IOP Publishing Ltd

2. The rationalization technique for a composite cross-section

2.1. Problem statement

Figure 1 shows a composite beam cross-section taking bending moment. Based on the analysis of[2-5] with proposals on reinforcement of steel bars with concrete, the article [6] presented the rationale that the location of the slab in the compressed area of I-section is the effective reinforcement in case of perception of deformation during bending is (figure 1). This shape of the cross-section is acceptable, in particular, during the reconstruction of the structure for reinforcement purpose.

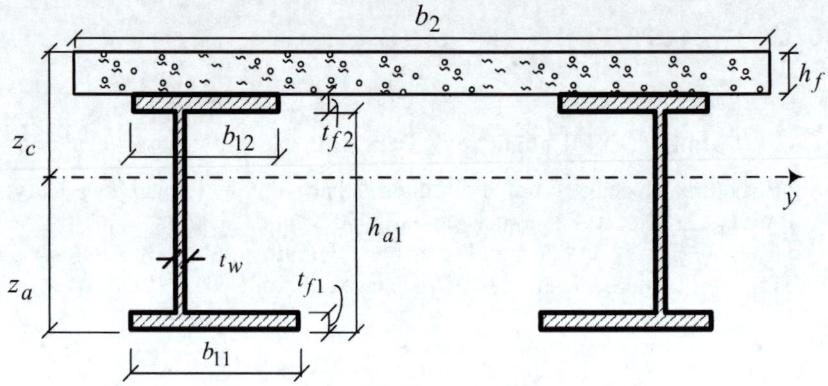

Figure 1. Transverse cross-section.

The section will be rational provided that one of the requirements for optimal design – equal strength – is taken into account. Equal strength for a section means that the stresses in the most distant from the neutral axis of the fibers reach their limiting values simultaneously. For concrete, this is $\sigma_c = f_{cd}$ (estimated value of compressive strength of concrete) in compressed fibers, in steel $\sigma_a = f_{yd}$ (estimated resistance at the yield limit) in stretched fibers.

2.2. Rationalization algorithm

Let us determine the position of the neutral axis using the transformed section method described in [7]. Cumulative axial force acting in the transverse section should be zero, i.e.

$$\int \sigma_a \mathrm{d}A + \int \sigma_c \mathrm{d}A = 0, \tag{1}$$

where dA – cross-section area.

In bending:

$$\sigma_a = \frac{E_a z}{\rho}; \quad \sigma_c = \frac{E_c z}{\rho}, \tag{2}$$

where E_a, E_c – elastic modulus of steel and concrete; z – distance from the described point to the neutral axis; ρ –curvature radius of neutral layer.

When (2) is substituted in (1), we get:

$$\int_a z\,\mathrm{d}A + n\int_c z\,\mathrm{d}A = 0, \tag{3}$$

where $n = E_c / E_a$.

Determination of the maximum bending moment (figure 1). According to the axis deflection equation, maximum estimated bending moment M_{Ed} is determined by expression:

$$M_{Ed} = \frac{\gamma_{M0} f_{yd} J_{tr}}{z_a}, \tag{4}$$

where γ_{M0} – reliability factor of the steel; J_{tr} – inertia moment of the transformed section, determined by formula:

$$J_{tr} = J_a + nJ_c, \tag{5}$$

where J_a, J_c – inertia moments of the steel and concrete parts of the section relative to the neutral axis of the section.

Substituting in condition (3) the integrands in accordance with figure 1, we obtain an equation for z, the solution of which is the position of the neutral axis z_a:

$$z_a = \frac{b_{11}t_{f1}^2 + t_w\left(h_{a1}^2 - t_{f1}^2\right) + b_{12}\,t_{f2}\left(2h_{a1} + t_{f2}\right) + nb_2h_f\left(h_{a1} + t_{f2} + \frac{h_f}{2}\right)}{2\left[b_{11}t_{f1} + t_w\left(h_{a1} - t_{f1}\right) + b_{12}t_{f2}\right] + nb_2h_f}. \tag{6}$$

The calculated cross-section corresponds to the maximum bending moment. Here, the greatest stresses occur in the most distant from the neutral axis fibers, which, according to the strength condition, should not exceed the estimated values:

$$\sigma_{a\max} = \frac{M_{Ed}z_a}{J_{\text{tr}}} \le \gamma_{M0} f_{yd}; \tag{7}$$

$$\sigma_{c\max} = n\sigma_{tr\max} = n\frac{M_{Ed}z_c}{J_{tr}} \le f_{cd}, \tag{8}$$

where $\sigma_{tr\max}$ – stress of transformed cross-section.

To achieve the equal strength of the cross-section and, thus, the optimal design, the strength conditions of concrete and steel at the same bending moment M_{Ed} should be met in the form of equalities. Hence:

$$\frac{M_{Ed}}{J_{tr}} = \frac{f_{cd}}{n\,z_c} = \frac{\gamma_{M0} f_{yd}}{z_a}, \tag{9}$$

and the dimensions of individual elements constituting the cross-section should be found from the condition of constant ratio of the distances to the extreme fibers of concrete and steel from the neutral axis:

$$\frac{z_c}{z_a} = \frac{f_{cd}}{n\gamma_{M0} f_{yd}}. \tag{10}$$

Figure 1 shows that

$$z_c = h_{a1} + h_f + t_{f2} - z_a. \tag{11}$$

From (10) and (11), we find

$$z_a = \frac{\left(h_{a1} + h_f + t_{f2}\right)n\gamma_{M0} f_{yd}}{f_{cd} + n\gamma_{M0} f_{yd}}. \tag{12}$$

The equation for determining the optimal dimensions of the composite section is obtained by equating the right member of the condition of equal strength (12) and expression (6) for the neutral axis:

IOP Conf. Series: Materials Science and Engineering **664** (2019) 012014 doi:10.1088/1757-899X/664/1/012014

$$\frac{b_{11}t_{f1}^2 + t_w\left(h_{a1}^2 - t_{f1}^2\right) + b_{12}t_{f2}\left(2h_{a1} + t_{f2}\right) + nb_2h_f\left(h_{a1} + t_{f2} + \dfrac{h_f}{2}\right)}{2\left[b_{11}t_{f1} + t_w\left(h_{a1} - t_{f1}\right) + b_{12}t_{f2}\right] + nb_2h_f} = \frac{\left(h_{a1} + h_f + t_{f2}\right)n\gamma_{M0}f_{yd}}{f_{cd} + n\gamma_{M0}f_{yd}}. \quad (13)$$

Fulfillment of condition (13) is achieved numerically by exhaustive search through all optimized cross-section dimensions.

Since equation (13) is non-linear, various combinations of dimensions of equal-strength sections can be obtained. Of these, the one must be chosen for which $M_{max} = M_{Ed}$, where M_{max} is maximum bending moment occurring in a beam from a given external load.

Value M_{max} is chosen among the maximum values of moment on spans 1 and 2 (M_{max1} and M_{max2}), as well as the values of moment at support $|M_{sup}|$. Since the beam section is piecewise constant within each span, the method of forces is used to determine bending moments.

To determine the maximum design moment M_{Ed} (4), moments of inertia of the steel and concrete components of the section are expressed (figure 1):

$$J_a = 2\left[\frac{b_{11}t_{f1}^3}{12} + b_{11}t_{f1}\left(z_a - \frac{t_{f1}}{2}\right)^2\right] + 2\left[\frac{t_w\left(h_{a1} - t_{f1}\right)^3}{12} + t_w\left(h_{a1} - t_{f1}\right)\left(z_a - \frac{h_{a1}}{2} - \frac{t_{f1}}{2}\right)^2\right] +$$

$$+ 2\left[\frac{b_{12}t_{f2}^3}{12} + b_{12}t_{f2}\left(h_{a1} + \frac{t_{f2}}{2} - z_a\right)^2\right]; \quad (14)$$

$$J_c = \frac{b_2h_f^3}{12} + b_2h_f\left(h_{a1} + t_{f2} - z_a + \frac{h_f}{2} - z_a\right)^2. \quad (15)$$

When inertia moments (14), (15) are known, we can find inertia moment of the transformed section J_{tr} (5) and the maximum bending moment (4).

The per-meter cost of the beam is determined by formula:

$$C = V_a\gamma_a c_a + V_c c_c = \left(A_a \cdot 1\right)\gamma_a c_a + \left(A_c \cdot 1\right)c_c, \quad (16)$$

where c_a – cost of the steel (UAH/ton); c_c – cost of the concrete (UAH/m^3); V_c and A_c – volume of lineal meter (m^3) and area of the concrete part of the cross-section (m^2); V_a and A_a – volume of lineal meter (m^3) and area of the steel part of the cross-section (m^2); γ_a – volume-weight of the steel.

3. Concrete creep

Inelastic deformations are known to occur in concrete structures perceiving continuous action of loads; such deformations can exceed the initial, conditionally elastic deformations several times. The most critical problem of creep is in composite reinforced concrete beams [8]. The presence of a rigid steel part in the steel and reinforced concrete section, in contrast to reinforced concrete beams, determines a different mechanism of concrete creep influence, which is the redistribution of forces between concrete and steel. As a result, compressive forces in concrete are reduced and bending moment increases in a steel beam.

The Building Norms of Ukraine harmonized with Eurocode 2 [9] are based on the simplest approach establishing dependencies between deformations and time with the creep coefficient $\varphi(t, t_0)$.

According to this method, ultimate deformations of creep depend on concrete strength, age and relative moisture, and creep development in time t depends on the relative moisture and section size:

$$\varepsilon_{c,t} = \left[1 + \varphi_0\beta\left(t, t_0\right)\right]\sigma_c / E_c, \quad (17)$$

where $\beta(t, t_0)$ is coefficient describing the development of creep over time; t_0 – age of the concrete at the time of the first load; φ_0 – theoretical creep coefficient.

This technique implements 90% of the creep deformations of concrete in 20-25 years of operation of the structure.

Due to the acceptable integrity and good implementation using the state-of-the-art software systems, we have adopted this method as the basis for further calculations of composite reinforced concrete spans.

At the same time, an alternative method for assessing the long-term deformation of concrete, based on the colloid-chemical presentation of the creep mechanism, has been acknowledged earlier in studies of CFST columns [10] and reinforced concrete spans structures [11].

4. Implementation of the technique

4.1. Design solution

The presented rationalization technique was taken as the basis for designing a continuous composite reinforced concrete span bridge with a monolithic reinforced concrete roadway slab (figure 2). The bridge diagram is 34.0+45.0+34.0 m, the total length is 117.30 m. The bridge dimension is G 0.75(pedestrian part)+1(right-of-way)+11.7(traffic way)×2+1(right-of-way)+1.5(pedestrian part) m. The custom-design intermediate poles are made of monolithic reinforced concrete on the foundation of cast in situ piles.

Design load: heavy single load – NK-100 (according to State Building Norms in Production DBN V.1.2-15:2009), motor vehicles load – six A-15 bands (the load from a convoy of two-axle 30-ton trucks), pedestrian load – 2.0 kPa.

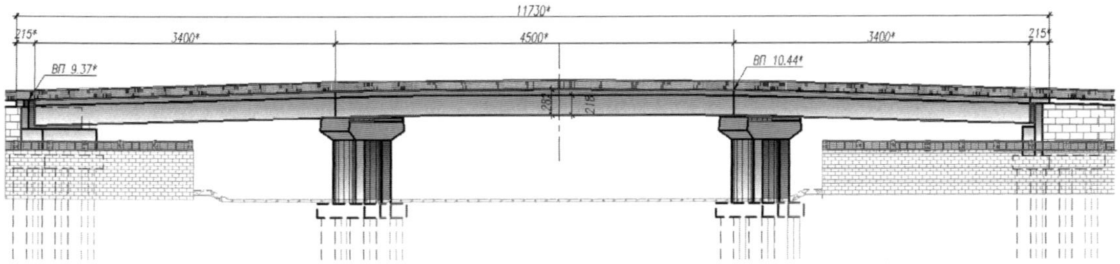

Figure 2. Bridge diagram.

The span structure in the cross-section is taken as six box-shaped main beams with chords of variable cross-section and dimensions established in the first approximation according to the results of the proposed rationalization technique. The distance between the beams is 4.4 m. Plate diaphragms with apertures are installed with a pitch of 1.5 m in a checkerboard pattern. Transverse beams are installed with a pitch of 3 m.

The reinforced concrete slab of the span structure with a thickness of 220 mm is combined with the main beams using shear bolts KKB-22/150 according to DIN32000-3. The compressive strength of concrete is C28/35. The transverse beams are made of welded I-beam and installed with a pitch of 3 m. The diaphragms with a pitch of 1.5 m are installed to achieve stability of the main beam walls.

4.2. Design scheme

Modern design of transport structures, as a rule, involves the use of industrial software systems as the main tool for computer-aided modeling [12-15]. The final calculation of the span can be fully conducted by means of LIRA-SAPR 2018 Pro software complex [16].

Computer-aided modeling is carried out to verify and finalized the obtained results taking into account the multistage work of the span (structural non-linearity), physical non-linearity and concrete creep, redistribution of forces due to the compliance of a flexible connection over the contact between concrete and steel.

The following hypotheses were taken for designing a finite element model of the span.

1. Calculation involves plotting a plate and rod design scheme using universal rod FE for simulation of a girder network and physically non-linear FE of the shell for simulation of a roadway slab.

2. The model is built within the entire span structure. The elements of the section are combined into joint work with the help of absolutely rigid bodies, as well as binodal finite elements of elastic coupling with a given stiffness to simulate the operation of shear bolts (figure 3).

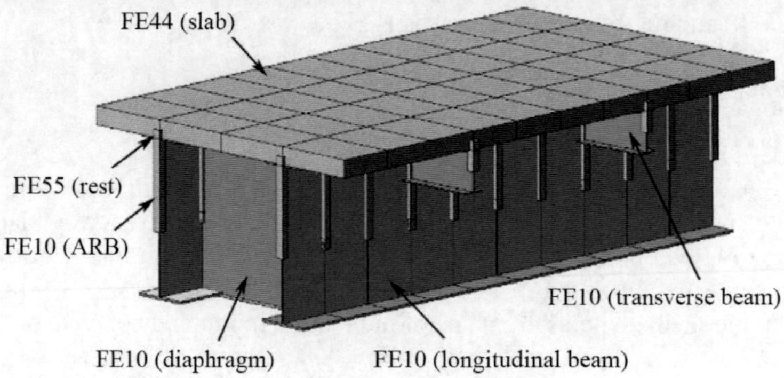

Figure 3. Types of conventional finite elements.

3. The strength and deformability of concrete is characterized by parabolic and linear deformation diagrams $\sigma_c - \varepsilon_c$ depending on its class, as well as long-term deformation diagrams $\varphi(t) - t$. The creep calculation considered the relative humidity RH = 74% and the age of concrete of the mounted slab to be 28 days. Diagram $\sigma_a - \varepsilon_a$ without a yield plateau is used to account for reinforcing steel.

4. The structural non-linearity of the system was revealed with the design method "Mounting" which considered the evolutionary transformations. Besides, each building stage had the finite elements specified for it. And the load corresponding to this stage was calculated. For finite elements included for several stages all the efforts were summed up. In this case, three loading (assembly) stages are considered. The loads perceived by the beams at the first stage (1st part of constant loads) include the deadweight of the beams, diaphragms, as well as the load from the weight of the reinforced concrete roadway slab. The loads perceived by the combined cross-section in the second stage (the 2nd part of constant loads) include the deadweight of asphalt concrete, railings and barriers. The loads perceived in the third stage include one of the imposed loads.

5. The model of the span structure is made in accordance with the statically indeterminate scheme and hinged support along the axes of the supports.

4.3. Calculation results

At the stage of analysis of the calculation results, the problem was solved to finalize the accepted dimensions of the combined cross section at a constant thickness and the concrete class of the roadway slab, for which reinforcement was selected.

As a result, two sizes were selected with the solution of the search problem. The first one is for the 18 m long sections of the span symmetrically above the supports, and the second one is for the 25 m and 27 m long sections located in the spans.

The calculation results in this article are presented in the form of graphs of forces plotted for cross sections in the central span of the bridge and on the support (figure 4), as well as graphs of shear forces over the contact of concrete and steel (figure 5).

The results of calculations are summarized in table 1.

The estimated cross-sections obtained by combining the rationalization algorithm and the finite-element simulation method are shown in figure 6 and table 2.

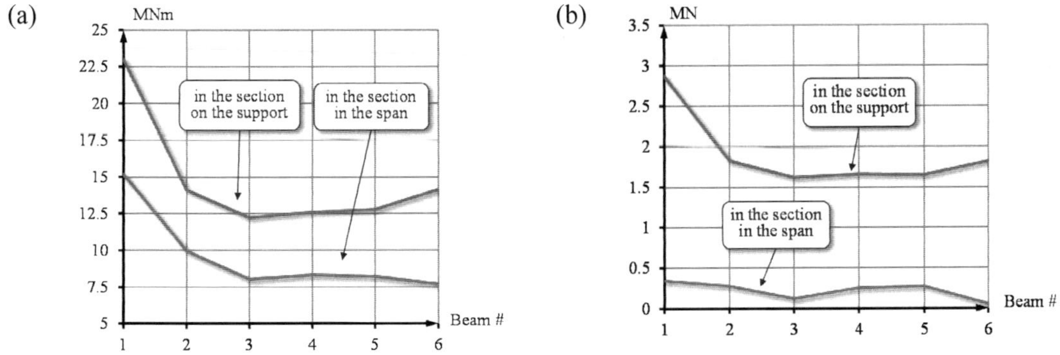

Figure 4. Diagrams of forces (a – moment and b – lateral forces) in transverse cross-sections of the bridge span structure.

Table 1. The results of calculations of the span structure.

Force	Value	Units
Bending moment in the main beam on the support	22.60	MNm
Lateral force in the main beam on the support	2.82	MN
Bending moment in the main beam in the span	14.90	MNm
Lateral force in the longitudinal beam in the span	0.34	MN
Strength index of the combined cross-section	1.16	–
Shear force along the span structure over the contact between the main beam and the slab	650.00	kN
Shear force across span structure and the slab	269.00	kN
Bending moment in the roadway slab	27.40	(kNm)/m
The required number of shear bolts along the span structure	24.00	pcs/m

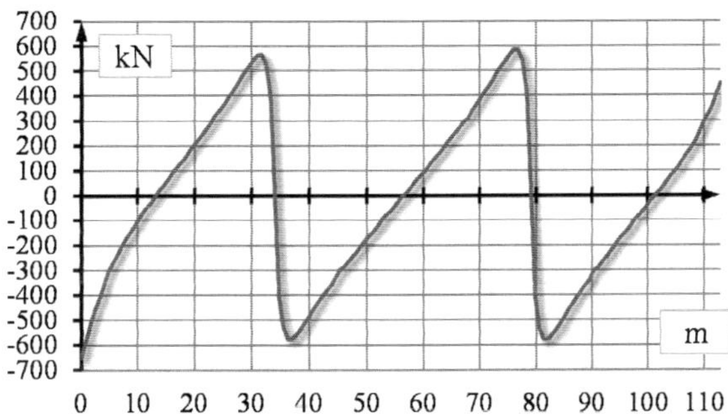

Figure 5. Diagram of shear forces over the concrete-steel contact.

Table 2. Results of section sizes calculation.

Rationalization parameter	Combined section obtained by rationalization algorithm	Combined section obtained by LIRA SAPR	
		in the bridge span	above the bridge support
Slab thickness, cm	15.5	22.5	22.5
Wall height, cm	206	213	214
Upper chord width, cm	53	60	60
Lower chord width, cm	63	70	90
Upper chord thickness, cm	2	2	2
Lower chord thickness, cm	3	3	4
Stiffness moment, cm^3	107388	131703	158400

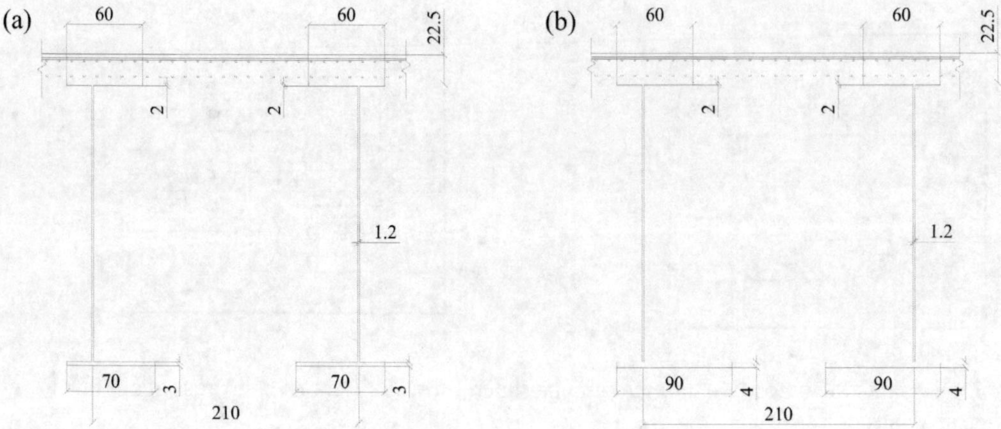

Figure 6. Size (cm) of transversal cross-sections of a span structure: a – in the span; b – on the support.

5. Conclusion

The calculations show that when the loading sequence of the span model and the contact yield of concrete and steel are taken into account, it leads to a decrease in the influence of the roadway slab on the load-bearing capacity of the structure. As a result the cross-section obtained by the rationalization algorithm was corrected with finite-element modeling by stiffness moment increase for combined section in the bridge span on 18.5% and on 32% – above the bridge support. It has also been established that the plastic properties of materials and long processes that occur in the combined cross-section affect the stress-strain state, increasing bending moment and deformation in steel beams. The efficiency of the composite reinforced concrete will not decrease with time, and will depend on the rational composition of the concrete to minimize creep deformations and the optimal ratios of the design parameters. Using the proposed simulation technique, the actual load-bearing capacity of the bridge span structure was established and a rational design of the composite cross-section was assigned to meet the modern loads of NK-100 and A-15.

References

[1] DBN V.2.3-14:2006 Bridges and pipes. Design rule (Kyiv: Min. bud-va, arkhit. ta zhytlovo-kom. hosp)

[2] Nie J, Xiao Y and Chen L 2004 Experimental studies on shear strength of steel-concrete composite beams *J. of Str. Engineering* **130(8)** 1206–13

[3] Nie J. and Zhao J. 2009 Flexural Behavior of Steel Plate-Concrete Composite Beams, *Key Engineering Materials* **400-402** 37–42

[4] Vasdravellis G and Uy B 2014 Shear strength and moment-shear interaction in steel-concrete composite beams *J. of Str. Engineering* **140(11)** 04014084

[5] Liang Q Q, Uy B and Ronagh H R 2005 Strength analysis of steel-concrete composite beams in combined bending and shear *J of Str. Engineering* **131** 1593–99

[6] Kitov Yu P, Verevicheva M A and Kravtsiv L B 2011 Expediency of I-beams strengthening by their inter flange hollows concreting *Nauk.-tekhn probl. such. zalizob.* **74** 318–25

[7] Kitov Yu P, Verevicheva M A, Vatulia G L, Orel Ye F and Deryzemlia S V 2017 Design solutions of the structures with the optimal internal stress distribution *MATEC Web of Conf.* **133** 03001

[8] Ranzia G, Leoni G and Zandoninic R 2013 State of the art on the time-dependent behaviour of composite steel–concrete structures *J of Constr. Steel Research* **80** 252–63

[9] CEN 2004 EN 1992-1-1: Eurocode 2: Design of concrete structures - Part 1-1: General rules and rules for buildings (Brussels: BSI)

[10] Lobiak O V, Vatulia G L and Orel Ye, 2017 Simulation of performance of circular CFST columns under short-time and long-time load *MATEC Web of Conf.* **116** 02036

[11] Lobiak O, Plugin A, Kravtsiv L and Kovalova O, 2018 Modelling of motorway bridge spans under modernization with consideration of rheological properties of the materials *MATEC Web of Conf.* **234** 04004

[12] da Silva A R and Sousa J B M 2009 A family of interface elements for the analysis of composite beams with interlayer slip *Finite Elements in Analysis and Design* **45(5)** 305–14

[13] Zhao G and Li A Numerical study of a bonded steel and concrete composite beam *Computers & Structures* **86(19-20)** 1830–8

[14] Zona A and Ranzi G 2011 Finite element models for nonlinear analysis of steel–concrete composite beams with partial interaction in combined bending and shear *Finite Elements in Analysis and Design* **47** 98–118

[15] Chung W and Sotelino E D 2006 Three-dimensional finite element modeling of composite girder bridges *Engineering Structures* **28(1)** 63–71

[16] Gorodetsky A and Evzerov I 2007 *Computer models of structures* (Kyiv: Fact)

Web-based virtual reality for simulation of a heavy manipulator with a freely suspended payload

R P Mitrev[1,4], B N Tudjarov[2] and T S Todorov[3]

[1] Department of Logistics Engineering, Technical University of Sofia, Sofia 1000, Bulgaria
[2] Department of Computer Systems, Technical University of Sofia, Sofia 1000, Bulgaria
[3] Department of Theory of Mechanisms and Machines, Technical University of Sofia, Sofia 1000, Bulgaria

[4] E-mail: rosenm@tu-sofia.bg

Abstract. The paper presents the development of a Web-based Virtual Reality (VR) environment for simulation and visualization of the motion of a hydraulically driven heavy manipulator with a freely suspended payload. Two main problems are resolved in the paper. First, a combined mathematical model of the manipulator, considered as a 2 DOF discrete mechanical model is derived. It represents the dynamics of the two key functional subsystems of the manipulator – mechanical subsystem, consisting of a link and freely suspended payload and hydraulic subsystem, consisting of a hydraulic cylinder. Second, by the use of the contemporary Web technologies a VR environment for simulation, visualization and animation of the obtained results is developed. The produced VR environment is easy reused and shared with other users and thus considerably facilitates the design, investigation, and e-learning of such type of system.

1. Introduction

During the last decades, the computers became an indispensable part of the engineering design process of the mobile and stationary heavy lifting manipulators and similar hydraulically driven manipulators. To effectively deal with the growing complexity of such type of systems and to introduce the new designs to market faster, a large variety of CAE products which facilitate the design, analysis or E-learning of machines and mechanisms are used by the designers. Such type of systems include modules for preprocessing, numerical analysis and results post-processing with possibilities to resolve a large variety of tasks – from motion simulation (providing information for the position, velocity, acceleration, joint reactions, inertial forces and power requirements of all the components of a moving mechanism) [1] to Finite Element Analysis (providing information about stresses, deformations and vibration modes) [2]. Despite the widespread distribution of the CAE systems due to their undeniable advantages, they cannot be considered as an “ultimate and universal design tool” for the practicing engineer. One of the reasons for that is the evolving during the last years need to consider the technical systems, independently of their scale, as consisting of interacting subsystems from different physical domains and having its own dynamics. An established approach is to use dedicated software tools for modeling of the different subsystems and by co-simulation to investigate the overall system behavior [3]. Another widely adopted approach for the multi-domain system model development is the use of the open source [4] or commercial [5, 6] software tools for modeling of integrated and complex mechatronic systems.

Content from this work may be used under the terms of the Creative Commons Attribution 3.0 licence. Any further distribution of this work must maintain attribution to the author(s) and the title of the work, journal citation and DOI.
Published under licence by IOP Publishing Ltd

Software systems where active involvement of the user in the system model development is required are becoming increasingly popular [7, 8]. Web-based platforms and client-server systems [9-11], exploiting the advantages of the network computing provide valuable opportunities to use standard libraries for computation, visualization, animation and results plotting.

However, despite the vast number of approaches and available software systems supporting the design process, the model development is a time consuming and tedious process. The issue of the effective design and investigation of the heavy manipulator considered as a multi-domain physical system and results visualization and animation is still open from the point of view of the effectiveness, easy reuse and sharing with other users. The general aim of the presented paper is by using the advantages of the contemporary Web technologies to develop a prototype of a Web-based Virtual Reality (VR) environment for simulation, visualization, and animation of the motion of a hydraulically driven heavy manipulator, considered as an integrated multi-domain physical system. The produced VR environment will considerably facilitate the design, investigation, and E-learning of such type of system. A key issue in the software system development is the derivation of a mathematical model, properly representing the jointly dynamic behavior of the mechanical and hydraulic functional subsystems of the manipulator.

2. Development of a mathematical model of the system

A discrete 3D model of a heavy lifting manipulator with a freely suspended payload is shown in figure 1. It consists of a fixed base 1 and three rotating links: a column 2, rotating about the vertical axis; an arm 3 and forearm 4, both rotating about horizontal axes to determine the payload 5 motion. The manipulator hydraulic drive system consists of a rack and pinion mechanism 6, intended for column rotation, and hydraulic cylinders 7 and 8, intended for arm and forearm rotation respectively. All crane elements are considered as rigid.

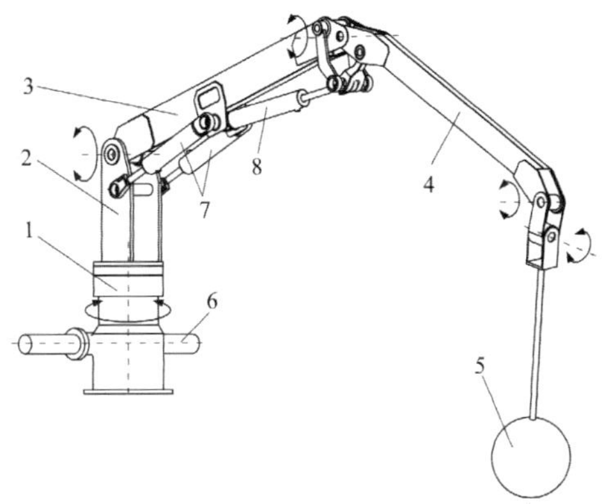

Figure 1. Discrete 3D model of the hydraulic manipulator.

For the current study, the manipulator is regarded as a planar two-link kinematic chain with two revolute joints. The column 2 and arm 3 are fixed, the forearm 4 and payload 5 are driven in the vertical plane by the hydraulic cylinder 8 through a four-bar linkage. In order to derive a dynamic model suitable for the study of the link motions, in the current paper we make the following assumptions: 1) the payload is considered as a point mass; 2) the stiffness of the links is neglected and they are considered as rigid bodies; 3) the inertia properties of the hydraulic cylinder and linkage are included into the inertial parameters of the forearm; 4) the backlashes and the friction forces in the joints are neglected.

2.1. Mathematical model of the mechanical subsystem

Figure 2 shows a layout of the manipulator with the consideration of the payload swinging. The mechanical system is modeled as an open kinematic chain undergoing planar motion. The links are interconnected by rotational joints and are subjected to forces generated by the hydraulic actuator and gravity. Each link is characterized by its geometrical and inertia parameters. Investigation of the dynamic behavior requires the solution of the forward dynamics problem for a mechanism with two rotational degrees of freedom – generalized coordinate θ_1 describing the rotation of the forearm and θ_2, describing the rotation of the rope together with the payload.

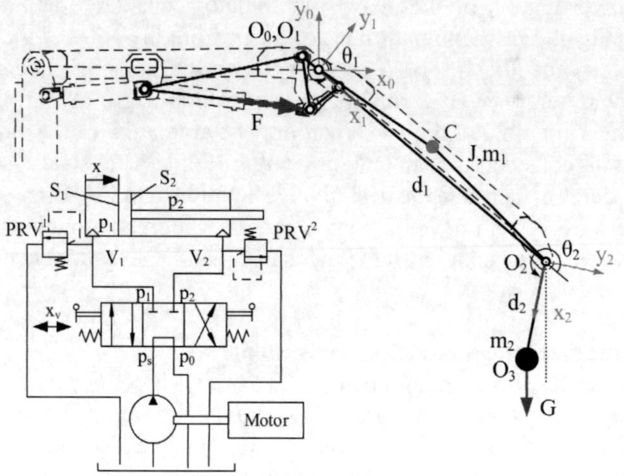

Figure 2. Schematic of the mechanical and hydraulic subsystems of the manipulator.

The dynamic equations of motion of the manipulator are derived in a systematic way using the Lagrange formalism [12]:

$$\frac{d}{dt}\frac{\partial L_a}{\partial \dot{q}_i} - \frac{\partial L_a}{\partial q_i} = Q_i, (i = 1,2), \tag{1}$$

where the Lagrangian L_a represents the difference between the kinetic K and potential P energies of the system studied; Q_i are the generalized forces associated with the generalized coordinates.

The kinetic and potential energies of the system are positive definite quadratic forms of the respectively generalized velocities $\dot{\mathbf{q}} = \begin{bmatrix} \dot{\theta}_1 & \dot{\theta}_2 \end{bmatrix}^T$ and generalized coordinates $\mathbf{q} = \begin{bmatrix} \theta_1 & \theta_2 \end{bmatrix}^T$.

2.1.1. Kinetic energy of the mechanical system. As is shown in figure 2, a local coordinate system (c.s.) $\{x_i y_i\}$ is connected to every link of the kinematic chain. The global c.s. $\{x_0 y_0\}$ is connected to the center of the forearm rotational joint O_1, the origin of c.s. $\{x_1 y_1\}$ coincides with the origin of the global c.s and $\{x_2 y_2\}$ is situated at the center of the rotational joint O_2. The transition between the local coordinate systems is performed by the use of transformation matrices in the following form:

$$\mathbf{T}_i^j(\alpha, a_x, a_y) = \begin{bmatrix} \cos\alpha & -\sin\alpha & a_x \\ \sin\alpha & \cos\alpha & a_y \\ 0 & 0 & 1 \end{bmatrix}, \tag{2}$$

where i and j denote the numbers of the adjacent coordinate systems; α is the angle between the coordinate systems; $[a_x\ a_y]^T$ represents the vector from the origin of c.s. i to the origin of c.s. j expressed in the c.s. i. The kinetic energy of the system depends on the system geometrical configuration and the velocities of the links. Taking into account the used in figure 2 notations, the transformation matrices are: $\mathbf{T}_0^1(\theta_1, 0, 0)$ and $\mathbf{T}_1^2(\theta_2, d_1, 0)$

The total kinetic energy of the system comprises the kinetic energies of the forearm K_1 and swinging payload K_2. Using the notations in figure 2 and according to König's theorem, the total kinetic energy of the mechanical system is obtained as follows:

$$K = \sum_{i=1}^{2} K_i \,, \tag{3}$$

where: $2K_1 = J\dot{\theta}_1^2 + m_1\left(\dot{x}_C^2 + \dot{y}_C^2\right), 2K_2 = m_2\left(\dot{x}_{O_3}^2 + \dot{y}_{O_3}^2\right)$.

In (3) the following notations are used: $\dot{x}_C$ and $\dot{y}_C$ denote the absolute velocities of the mass centers of the forearm. Similarly, $\dot{x}_{O_3}$ and $\dot{y}_{O_3}$ denote the absolute velocities of the payload. All velocities are obtained by a time differentiation of the corresponding position vectors according to the global coordinate system (c.s.). Position vectors are obtained by the use of the already defined transformation matrices:

$${}^0\mathbf{r}_C = \mathbf{T}_0^1\left({}^1\mathbf{r}_C\right); \tag{4}$$

$${}^0\mathbf{r}_{O_3} = \mathbf{T}_0^1\mathbf{T}_1^2\left({}^2\mathbf{r}_{O_3}\right), \tag{5}$$

where by ${}^0\mathbf{r}_C = \begin{bmatrix} x_C & y_C & 1 \end{bmatrix}^T$ and ${}^0\mathbf{r}_{O_3} = \begin{bmatrix} x_{O_3} & y_{O_3} & 1 \end{bmatrix}^T$ are denoted the vectors of the points C and O_3 correspondingly in the global coordinate system; by ${}^1\mathbf{r}_C = \begin{bmatrix} L_C^x & L_C^y & 1 \end{bmatrix}^T$ and ${}^2\mathbf{r}_{O_3} = \begin{bmatrix} d_2 & 0 & 1 \end{bmatrix}^T$ are denoted the coordinates of the same points in the corresponding local coordinate system.

Using Eqs. (3÷5), the kinetic energy of the system takes the following form:

$$\begin{aligned} 2K = J + m_1\left(\left(L_C^x\right)^2 + \left(L_C^y\right)^2\right) + m_2\left(d_1^2 + d_2^2\right) + 2m_2 d_1 d_2 \cos\theta_2 \dot{\theta}_1^2 + \\ 2d_2 m_2\left(d_2 + d_1\cos\theta_2\right)\dot{\theta}_1\dot{\theta}_2 + m_2 d_2^2 \dot{\theta}_2^2 \end{aligned} \tag{6}$$

2.1.2. Potential energy of the mechanical system. The potential energy of the system is computed as a sum of the potential energies of the forearm P_1 and the payload P_2:

$$P = \sum_{i=1}^{2} P_i \,. \tag{7}$$

The potential energies of the forearm and payload are computed as the work, required to raise its gravity center according to the x0-axis, i.e. they are a function of the system geometrical configuration:

$$P = g\left(m_1 L_C^y \cos\theta_1 + \left(m_1 L_C^x + d_1 m_2\right)\sin\theta_1 + d_2 m_2 \sin\left(\theta_1 + \theta_2\right)\right). \tag{8}$$

2.1.3. Differential equations of the mechanical system. Using the derived equations for the kinetic and potential energies and performing the mathematical operations in Eq. (1), the nonlinear dynamic equations of the motion are obtained in the form:

$$\mathbf{M}(\mathbf{q})\ddot{\mathbf{q}} + \mathbf{H}(\mathbf{q},\dot{\mathbf{q}}) + \mathbf{G}(\mathbf{q}) = \mathbf{Q}\,. \tag{9}$$

The notations used are as follows: Symmetric and positive definite inertia matrix $\mathbf{M}(\mathbf{q})_{2\times2}$, consisting of inertia terms; Vector $\mathbf{H}(\mathbf{q},\dot{\mathbf{q}})_{2\times1}$, consisting of centrifugal and Coriolis terms; Vector $\mathbf{G}(\mathbf{q})_{2\times1}$, consisting of terms, proportional to the weight of the links; Vector $\mathbf{Q}_{2\times1}$, consisting of generalized forces and moments, associated with the respective generalized coordinates: $Q_1 = \tau, Q_2 = 0$ where τ denotes the driving torque applied to the first joint. The elements of the matrices are as follows:

$$M_{11} = J + m_1\left(\left(L_C^x\right)^2 + \left(L_C^y\right)^2\right) + m_2\left(d_1^2 + d_2^2\right) + 2m_2 d_1 d_2 \cos\theta_2,\; M_{21} = M_{12} = d_2 m_2\left(d_2 + d_1\cos\theta_2\right);$$

$$M_{22} = m_2 d_2^2\;; H_{11} = -d_1 d_2 m_2 \sin\theta_2\left(2\dot{\theta}_1\dot{\theta}_2 + \dot{\theta}_2^2\right), H_{21} = d_1 d_2 m_2 \sin\theta_2 \dot{\theta}_1^2\;;$$

$$G_{11} = g\left(\left(L_C^x m_1 + d_1 m_2\right)\cos\theta_1 + m_2 d_2 \cos\left(\theta_1 + \theta_2\right) - m_1 L_C^y \sin\theta_1\right), G_{21} = m_2 g d_2 \cos\left(\theta_1 + \theta_2\right).$$

2.1.4. Determination of the driving torque. The mapping $\tau(F)$ of the hydraulic cylinder force F to the joint torque τ is established utilizing the principle of virtual work [12]. It allows the formulation of the required equations without having to consider the forces generated in the linkage joints. The layout of the linkage is shown in figure 3 and using the adopted notations one can write:

$$\tau = -F\frac{L_5\sin\left(\varphi_1 - \varphi_2\right)\sin\left(\varphi_4 - \theta_1\right)}{\sin\left(\varphi_2 - \varphi_4\right)}, \tag{10}$$

where all angles are measured from the arm axis *n-n*.

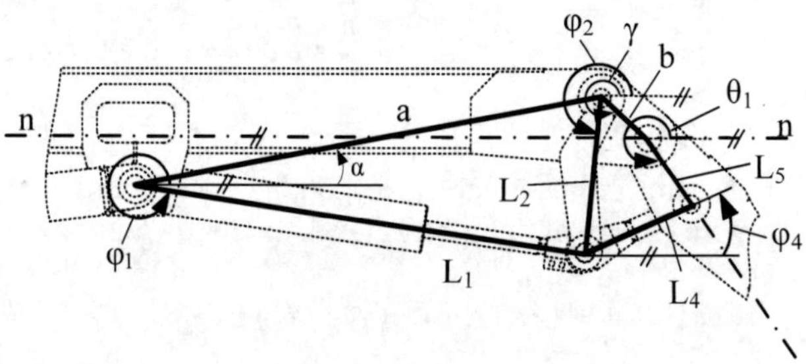

Figure 3. Layout of the forearm driving mechanism.

Using the notations in figure 3 and performing position analysis of the considered linkage one can compute the needed angles by the sequential application of the following equations:

$$A = -2L_2\left(b\cos\gamma + L_5\cos\theta_1\right); \tag{11}$$

$$B = -2L_2\left(b\sin\gamma + L_5\sin\theta_1\right); \tag{12}$$

$$D = \sqrt{A^2 + B^2}\;; \tag{13}$$

$$\sigma = \operatorname{atan2}\left(B/D, A/D\right); \tag{14}$$

$$C = b^2 + L_3^2 + L_5^2 - L_4^2 + 2bL_5\cos\left(\gamma - \theta_1\right); \tag{15}$$

$$\varphi_2 = \sigma + \arccos\left(-C/D\right); \tag{16}$$

$$\cos\varphi_1 = \left(a\cos\alpha + L_2\cos\varphi_2\right)/L_1\;; \tag{17}$$

$$\sin\varphi_1 = \left(a\sin\alpha + L_2\sin\varphi_2\right)/L_1\;; \tag{18}$$

$$\varphi_1 = \operatorname{atan2}\left(\sin\varphi_1, \cos\varphi_1\right); \tag{19}$$

$$\cos\varphi_4 = \left(b\cos\gamma + L_5\cos\theta_1 - L_3\cos\varphi_2\right)/L_4; \tag{20}$$

$$\sin\varphi_4 = \left(b\sin\gamma + L_5\sin\theta_1 - L_3\sin\varphi_2\right)/L_4; \tag{21}$$

$$\varphi_4 = \operatorname{atan2}\left(\sin\varphi_4, \cos\varphi_4\right). \tag{22}$$

The velocity of the piston can be determined by the following equation:

$$\dot{x} = \dot{L}_1 = -\dot{\theta}_1 \frac{L_5\sin(\varphi_1-\varphi_2)\sin(\varphi_4-\theta_1)}{\sin(\varphi_2-\varphi_4)}. \tag{23}$$

2.2. Mathematical model of the hydraulic subsystem

In figure 2 is shown a fragment of the manipulator hydraulic subsystem. The flow to the cylinder is controlled by the spool direction control valve. Two double cross-over relief valves PRV1 and PRV2 are incorporated in the hydraulic circuit in order to relieve the excessive pressure caused by the sudden closure of the direction control valve or occurred inertial loads during the motion of the links. The force in the hydraulic cylinder is calculated as a function of the chambers pressures p_1 and p_2:

$$F = p_1S_1 - p_2S_2 - b\dot{x}, \tag{24}$$

where by b is denoted the viscous friction coefficient; by $\dot{x}$ is denoted the velocity of the piston; by S_1 and S_2 are denoted the piston and piston minus rod areas correspondingly.

A simplified dynamical model of the hydraulic system is presented under the following assumptions: constant supply pressure, compressible oil, and absence of oil leakage in the hydraulic system. Based on the continuity equation [13] for the compressible oil and referring to figure 2, one can write the equations for the pressures in the hydraulic cylinder chambers:

$$\dot{p}_1 = \left(\frac{\beta}{V_1 + S_1x}\right)\left(Q_1 - S_1\dot{x} - Q_v^1\right); \tag{25}$$

$$\dot{p}_2 = \left(\frac{\beta}{V_2 + S_2(h-x)}\right)\left(S_2\dot{x} - Q_2 - Q_v^2\right), \tag{26}$$

where h is the hydraulic cylinder stroke, β is the fluid bulk modulus, V_1 and V_2 are the constant oil volumes, subjected to compression. The flow rates Q_1 and Q_2 through the direction control valve are determined by the following equations:

$$Q_1 = \begin{cases} if\ x_v \geq 0\ then\quad c_d w x_v(t)\sqrt{\dfrac{2|p_s - p_1|}{\rho}}sign(p_s - p_1) \\ else\ c_d w x_v(t)\sqrt{\dfrac{2|p_1 - p_0|}{\rho}}sign(p_1 - p_0) \end{cases} \tag{27}$$

$$Q_2 = \begin{cases} if\ x_v \geq 0\ then\quad c_d w x_v(t)\sqrt{\dfrac{2|p_2 - p_0|}{\rho}}sign(p_2 - p_0) \\ else\ c_d w x_v(t)\sqrt{\dfrac{2|p_s - p_2|}{\rho}}sign(p_s - p_2) \end{cases} \tag{28}$$

where c_d is the orifice discharge coefficient, w is the direction control valve area gradient, $x_v(t)$ is the direction control valve opening, represented as a function of time, ρ is the oil density, p_s is the supply pressure, p_0 is the drain pressure.

If the dynamics of the relief valves is neglected and turbulent flow is assumed, then the flows Q_v^i (i=1,2) through the valves PRV1 and PRV2 correspondingly is governed by the equations:

$$Q_v^i = c_{dv} A_v\left(\Delta p^i\right)\sqrt{\frac{2}{\rho}\left|\Delta p^i\right|}\,sign\left(\Delta p^i\right),\quad i=1,2, \tag{29}$$

where the pressure-dependent orifice passage area $A_v(\Delta p_i)$ is described by the following piecewise linear function:

$$A_v\left(\Delta p^i\right)=\begin{cases}0 & if\ \Delta p^i \le p_{set}\\ \dfrac{A_{\max}}{p_{reg}}\left(\Delta p^i - p_{set}\right) & if\ p_{set} < \Delta p^i < p_{\max}\\ A_{\max} & if\ \Delta p^i \ge p_{\max}\end{cases} \tag{30}$$

where Δp_i is the relief valve pressure differential for valve i, A_{max} is the fully open relief valve passage area, p_{set} is the relief valve preset pressure, p_{reg} is the relief valve regulation range, p_{max} is the valve pressure at the maximum opening, and c_{dv} is the orifice discharge coefficient.

3. Experimental setup and validation of the mathematical model

The experimental setup with a kinematic structure, similar to the considered manipulator (figure 1) is used to obtain experimental data for the force in the forearm hydraulic cylinder (figure 1, pos.8). The experimental setup consists of (figure 4): a rotating column 1, an arm 2, a forearm 3 and a hydraulic power unit 4. The mechanical structure is equipped with sensors for measuring the pressures at the cap end and the rod end of the forearm 6 and arm 5 driving cylinders; 2) the displacement of the arm and the forearm driving cylinders 7 and 8. Analog outputs from the sensors are transmitted to the computer measurement system 9 by the use of an analog-to-digital converter. For the simulation of the manipulator working cycle has used the law of motion of the spool of the hydraulic directional control valve, shown in figure 5. This pattern corresponds to the following sequence of the forearm motion – forearm raising (from initial angle θ_l= –15° to angle 24.5°), dwell in the upper position and lowering to the initial position.

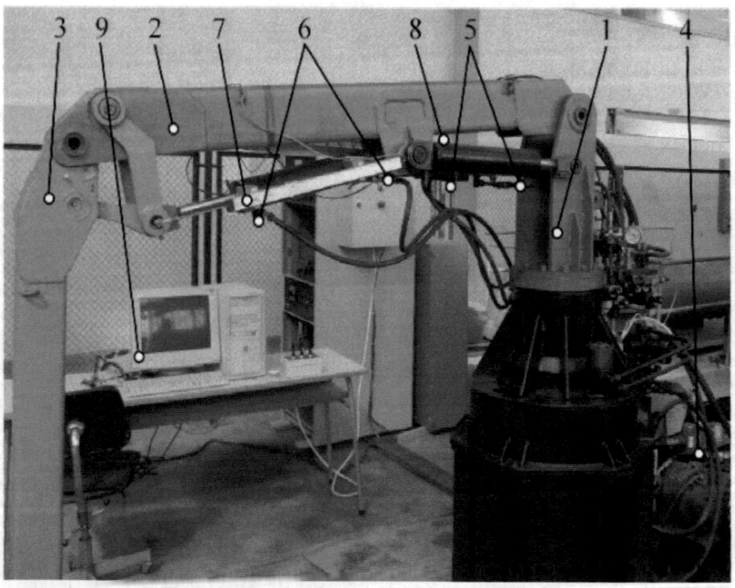

Figure 4. View of the small scale manipulator and sensor placement.

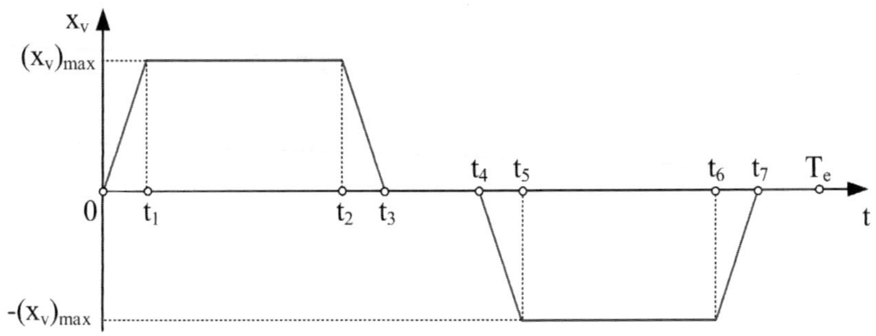

Figure 5. Law of motion of the valve spool.

The following numerical values of the system parameters are used: $(x_v)_{max} = 0.0005$ m, $t_0 = 0$ s, $t_1 = 0.05$ s, $t_2 = 10.15$ s, $t_3 = 10.2$ s, $t_4 = 14.4$ s, $t_5 = 14.45$ s, $t_6 = 19.5$ s, $t_7 = 19.55$ s, $T_e = 24$ s, $L_2 = 0.292$ m, $L_3 = 0.292$ m, $L_4 = 0.2$ m, $L_5 = 0.147$ m, $a = 0.908$ m, $b = 0.122$ m, $\alpha = 100$, $\gamma = 320.13°$, $S_1 = 1.96\times10^{-3}$ m^2, $S_2 = S_1/2.08$, $V_1 = V_2 = 35.7\times10^{-5}$ m^3, $h = 0.5$ m, $w = 0.003$, $c_d = 0.62$, $\beta = 1\times10^9$ Pa, $p_s = 20$ MPa, $p_0 = 0$, $\rho = 855$ kg/m^3, $b = 85000$ Ns/m, $d_1 = 2$ m, $d_2 = 1.1$ m, $J = 80$ kg.m^2, $m_1 = 100$ kg, $m_2 = 400$ kg, $g = 9.81$ m/s^2, $L_C^x = 1\text{m}$, $L_C^y = 0.1\text{m}$.

The experimental data for the driving hydraulic cylinder (figure 4, pos.8) force, computed by the use of the measured chamber pressures, is shown simultaneously with the obtained by the presented mathematical model hydraulic cylinder force. Although some differences are observed during the duty cycle, the good correlation between the experimental data and simulation results allows us to conclude that the adopted mathematical model may confidently be used to simulate the dynamic problems present. The differences between the curves are mainly due to the simplified mathematical model and inaccuracies in determining the numerical values of the system parameters.

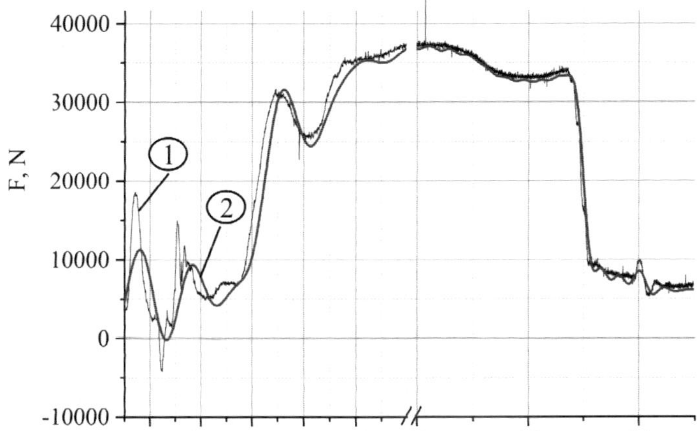

Figure 6. Experimentally and theoretically obtained graphs of the forearm hydraulic cylinder force, pos.1 – experimental data; pos.2 – theoretical data.

4. Development and testing of a Web-based Virtual Reality environment

The system of equations (9-30) describing the system dynamical behavior is programmed in Matlab environment and an available routine for the solution of stiff differential equations is used. In addition, functionality for performing input-output operations is added. The Matlab file is compiled and is used for the development of a Web-based software system for simulation of the manipulator motion. The block scheme of the software system, represented in figure 7 is described briefly as follow:

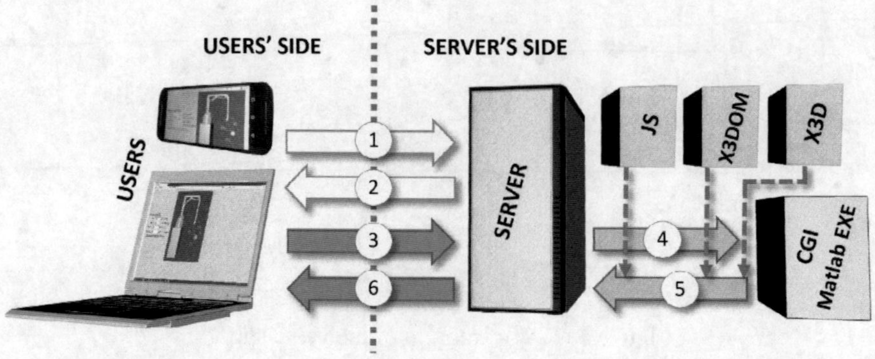

Figure 7. Framework of the developed experimental Web-based system.

1) The users request a Web page/resource by entering the Web page address or clicking on a link to it;
2) The server [14] sends back the requested HTML page with an initial information "FORM" tag on it (the URL of the CGI application is specified as a part of the "FORM" by using the attribute "ACTION")
3) The form's fields are filled out and the data are sent to the Server;
4) The Common Gateway Interface (CGI) is a part of the Web's Hypertext Transfer Protocol (HTTP) and by using CGI Web server passes user's request to the Matlab executable file.
5) The server receives data back from the "exe" file as a HTML5 [15] page containing links to:

- X3DOM (JavaScript framework for the creation of declarative 3D scenes in Web pages, without need to use any plugin for to display X3D scenes by using WebGL enabled browsers);
- X3D (eXtensible 3D graphics) [16] model of the manipulator;
- JavaScript file (JS) - assures main functionality in the system: a) adds parts to the X3D model according to the request of the user and finishes building of the model of manipulator; b) controls the simulation process by getting calculation data from the Matlab CGI executable application and by the use of JavaScript methods "getElementById" (access the objects in the scene marked by the values of the attributes "id") and "setAttribute" (change the values of the attributes "translation" and "rotation") the movements of the objects are realized; c) by capturing keyboard events additional functionality is added - for the payload trajectory plotting and for changing some parameters of the scene visualization;

6) Generated, as it is described above, HTML5 Web page is forwarded back to the user.

The manipulator is depicted in figure 8 (a) and its 13 modeled elements are marked by numbers. By using points (marked by "P" and number) are shown the connection places between elements (including the axes of relative rotation). The model of the manipulator is developed by the use of the X3D language – XML (eXtensible Markup Language) based language for 3D modeling. By the XML language models are presented in a tree structure (hierarchical) in compliance with the syntax, defined in the World Wide Web Consortium standard.

Because the model of the manipulator is not a pure tree structure and there are three cycles shown by dashed lines (see figure 8), the X3D model option for the inheritance [17] of the movements is used. The dashed connections are removed in order to transform the structure of the model into a tree (corresponding to the X3D language) and in addition to calculate at every simulation step the geometrical configuration of the triangles *P3P4P5*, *P5P6P7* and *P6P7P8* (figure 8) according to the specific geometric constraints.

In figure 9 are shown work screens of the developed 3D simulation scene and payload trajectories for different initial values of the generalized coordinates. As additional information, different output characteristics of the model can be plotted - for example, in figure 10 (a) and (b) are shown, correspondingly, the simulated curves of the hydraulic cylinder force and the hydraulic cylinder pressures.

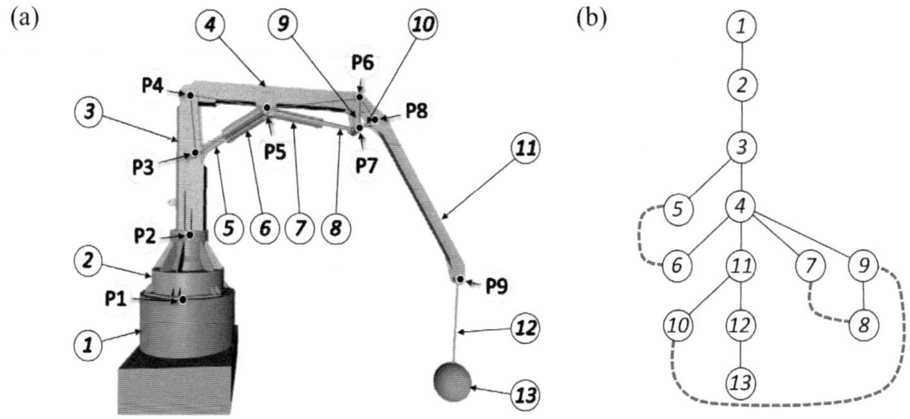

Figure 8. Structure of the model of manipulator.

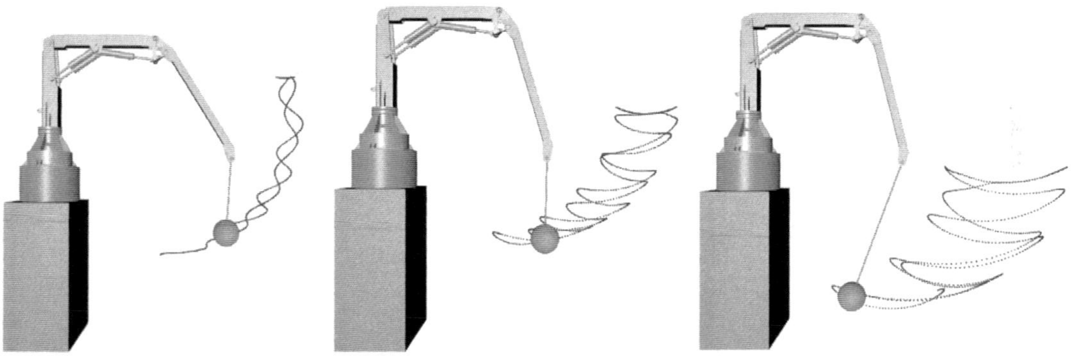

Figure 9. 3D simulation scene and the payload trajectory for different user's inputs.

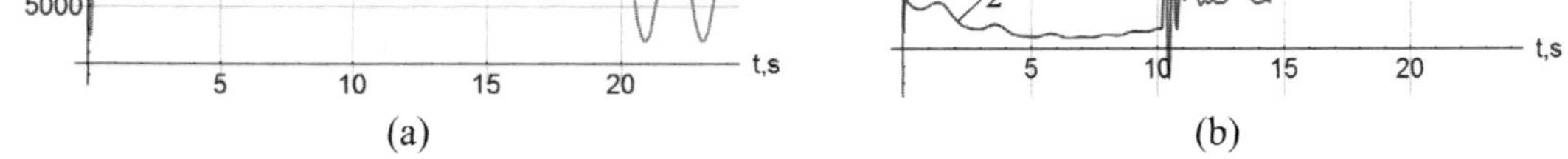

Figure 10. Output characteristics: (a) Time evolution of the hydraulic cylinder force; (b) Time evolution of the pressures in the hydraulic cylinder chambers: 1 – piston end chamber; 2 - rod end chamber.

5. Conclusions

This paper introduces the development of an integrated Web-based Virtual Reality environment for simulation and visualization of the motion of a hydraulically driven heavy manipulator with a freely suspended payload. The iterative nature of the design process requires multiple repetitions of the same actions leading to the design of a competitive manipulator, satisfying predefined requirements. The developed Web-based software system can significantly shorten the design time and reduce the cost of the design work. The animation capability allows the user to thoroughly understand the machine behavior during performing various motions. Presentations of the experiments are done through realistic Web visualization, capable to show to the learner processes or relationships which are not usually visible. The high-quality 3D models contribute to the intuitive perception of information and therefore provide significant educational potential. As a result of the practical tests performed, it can be concluded that the developed system is of great value and it can serve as a basis for the development of a system with greater design capabilities. As a future work we can point out the development of a fully-featured software system with the addition of a database of 3D models of machines and computational modules.

Acknowledgments

The authors would like to thank the Research and Development Sector at the Technical University of Sofia for financial support.

References

[1] Hroncová D, Delyová I and Frankovský P 2014 Kinematic analysis of mechanisms using MSC Adams *Applied Mechanics and Materials* **611** 83–9
[2] Du J and Wand Z 2009 Applying the ANSYS simulation to study the cushion performance of paper-pulp on the structural unit *Int. Workshop on Intelligent Systems and Applications* 1–3
[3] Sosa-Méndez D, Lugo-González E, Arias-Montiel M and García-García R 2017 ADAMS-MATLAB co-simulation for kinematics, dynamics, and control of the Stewart–Gough platform *International Journal of Advanced Robotic Systems* **14** 1–10
[4] Kinnander A, Sjölund M and Pop 2016 A Industrial evaluation of integrated performance analysis and equation model debugging for equation-based models *Modeling, Identification and Control,* **37** 225–36
[5] Villa-Villaseñor N and Galindo-Orozco R 2015 20-Sim implementation of bond graph models of flexible photovoltaic arrays based on data sheets *20th International Conference on Methods and Models in Automation and Robotics (MMAR)* 260–5
[6] Chu Y, Æsøy V, Ehlers S and Zhang H 2015 Integrated multi-domain system modelling and simulation for offshore crane operations *Ship Technology Research* **62** 36–46
[7] Campbell M and Cheng H 2007 Teaching computer-aided mechanism design and analysis using a high-level mechanism toolkit *Comput. Appl. Eng. Educ.* **15** 277–88
[8] Nguyen V, Zehn M and Marinković D 2016 An efficient co-rotational fem formulation using a projector matrix *Facta Universitatis Series: Mechanical Engineering* **14** 227–40
[9] Katwyk K and Cheng H 1997 XLINKAGE: A Web-based analysis and simulation tool for planar mechanical systems *Proc. DETC'97 ASME Design Engineering Technical Conferences*
[10] Mitrev R, Tudjarov B and Kubota N 2013 Web based solutions for mechanical engineering *Proc. Conf. IWACIII '2013 (Shanghai China)* GS2-9.1-GS2-9.6
[11] Song L, Wu X and Yang Z 2008 Research on web-based optimization for path generation synthesis of planar four-bar linkage *IEEE Int. Conf. Mechatronics and Automation* 1085–8
[12] Baruh H 2006 *Analytical Dynamics* (Boston:McGraw–Hill) p 744
[13] Watton J 2009 *Fundamentals of Fluid Power Control* (Cambridge University Press)
[14] Coar K and Bowen R 2008 *Apache Cookbook: Solutions and Examples for Apache Administrators (* O'Reilly Media Inc) p. 308
[15] Meyer J 2018 *HTML5 and JavaScript Projects* (Apress) p 417
[16] Brutzman D and Daly L 2007 *X3D: Extensible 3D Graphics for Web Authors* (Elsevier) p 417
[17] Tudjarov B and Hristov V 2017 Concerning the possibilities of ensuring the correctness of the movements in Web-based modelling and simulation of mechanical assemblies, *CAx Technologies* **5** 49–54

Machine for accelerated cyclic corrosion tests through alternate immersion in salt solution

Nikolay Nikolov[1], Veselin Tsonev, Kamen Penkov, Nikola Kuzmanov and Boris Borisov

Technical University – Sofia, "Strength of materials" Department, 8 Kliment Ohridski Blvd., 1000 Sofia, Bulgaria

[1] E-mail: nyky@tu-sofia.bg

Abstract. Metal structures corrosion is a problem of global economic importance. Typically, the corrosion process runs slowly, which requires accelerated methods for corrosion behaviour examination. The conditions when accelerated methods are used differ from the real ones, which affects the results obtained. One of the methods that relatively successfully reproduces corrosion in a natural atmosphere is included in the standard EN ISO 11130:2018 – Alternate immersion test in salt solution. Unlike other widespread methods, uneven corrosion occurs here, including spots and pits, which greatly affect the mechanical behaviour of the material. The commercial devices for such tests are quite expensive. This justifies the development and production of one's own equipment as the one presented in this paper. The design, construction and production of a low-cost device are hereby described. The first results of the tests performed with this machine are also given. The corrosion rate of rods, 6 mm in diameter, from widely used steel grades – S235JR and S235JRC – was determined. The duration of the test was 42 days. The results were compared with those from another corrosion test method with the same duration – immersion in a solution of 3.5% NaCl.

1. Introduction

Cyclic Corrosion Testing (CCT) is state-of-the-art accelerated method for corrosion research in conditions close to those in the real world. Specialized chambers are usually used where a corrosive solution is sprayed on the test specimens. A variety of factors, which affect the corrosion process, are controlled – temperature and humidity, solution parameters, UV radiation. As a result, the devices used for corrosion testing have high complexity and high cost.

A simplified alternative for CCT is an alternate immersion test (AI-test) in salt solution – this is a standard method described in [1]. Alternating immersion and withdrawal of the test piece is realized relatively easy by mechanical movement of the test piece [2] or by evacuating the corrosive solution [3]. Besides being simple to implement this method has the advantages of the more complex CCT – it implements the high corrosion rate and the duration of the test is reduced; irregular corrosion is achieved which is important when the mechanical properties of the corroded test pieces are estimated.

The accelerated cyclic corrosion test by alternate immersion in salt solution has a variety of applications: testing of different types of coating [4]; study of the effect of surface roughness and surface machining on corrosion behaviour [5]; study and comparison of the corrosion behaviour of different types of metals and alloys [6]; the effect of the carrion on precious metals in the jewellery industry [7]; study of the effect of corrosion on the properties of specific part features i.e. railway wheels [8], bridge

Content from this work may be used under the terms of the Creative Commons Attribution 3.0 licence. Any further distribution of this work must maintain attribution to the author(s) and the title of the work, journal citation and DOI.
Published under licence by IOP Publishing Ltd

cables [9] or bolt joints [10]. In the majority of articles the results are emphasized without giving detailed descriptions of method implementation. An exception is [3], where two containers and a simple system of two pumps and two timers are used to transfer the solution to and from the test piece, as well as [2], where a system with stepper motor is used to move the test pieces in and out of the solution. As an appendix to the standard [1] an example machine is given which is based on the principle of moving the test piece.

The AI-tests is not always automated. In [11] is implemented a cycle with immersion for one hour in 3.5 % NaCl solution, followed by oven-drying at 60°C for 11 hours. In the same study the effects of implementing the CCT method on steel wires is compared with the results from galvanostatic method – another widely used method for accelerated corrosion tests. It is demonstrated that corrosion caused by galvanostatic method is evenly distributed, while corrosion caused by alternate immersion in salt solution is highly unevenly distributed which is characteristic of natural atmospheric corrosion. This affects the mechanical properties and more specifically the stress-strain curve obtained from tensile test. Uneven corrosion leads to brittle behaviour of the material even at small weight loss and the tensile strength rapidly decreases. Evenly distributed corrosion leads to significantly slower reduction of the yield strength with weight loss and the test piece retails the elastic behaviour even with significant weight loss.

The goal of this work is to create a test machine for performing AI-tests. The machine must meet the following requirements:

- to assure a high corrosion speed;
- to create unevenly distributed corrosion similar to the corrosion occurring in natural atmosphere, because later the test pieces will be subjected to mechanical tests;
- the standard testing conditions must be ensured automatically and for unlimited amount of time;
- to allow the setting the time intervals by the user;
- to have a low cost.

As a result of a survey, a decision was made for the machine to meet the guidelines given in standard [1] – Alternate immersion test in salt solution.

Using the newly developed system tensile tests are going to be made to determine the mechanical behaviour of corroded S235JR and S235JRC steel rods with 6 mm in diameter.

2. Particularities of EN ISO 11130:2018 [1]

2.1. Corrosive solution

The composition of the solution is selected according to the operating conditions which are going to be simulated. In the standard [1] and its annexes, a total of four compositions are described. A neutral salt solution is used for simulating the corrosion effect of sea environment, which is prepared by dissolving sodium chloride in distilled or deionized water with conduction no greater than 2 mS/m at 25 °C ± 2 °C. The concentration of the solution is 35 g/l ± 1 g/l. This solution is going to be used in future studies with the developed machine.

The standard has three additional guidelines with regard to the solution:

- maximum amount of allowed contaminants;
- pH level should be in the range 6-7 at 25 °C ± 2 °C. A correction of the pH level can be made by adding distilled hydrochloric acid or sodium bicarbonate solution;
- it is recommended that the volume of the solution is no less than 3 l/dm^2 with respect to the surface area of the tested pieces.

To meet these guidelines the necessary agents and testing equipment for measuring weight, pH and electrical conductivity are purchased. The container volume is selected to be 45 l, considering the needed test pieces number and size.

2.2. Equipment

The equipment must include:

- a system for automatic execution of the immersion and withdrawal cycles. The time for full immersion or withdrawal should be no greater than 2 min;
- a containers for the solution which should be made of glass or plastic polymer.

Only one type of metal, alloy or coating should be immersed in each container. The materials in contact with the solution should not be affected by it and should not change its properties. The implementation of inert plastics or glass is recommended.

The test piece holders should meet the following requirements:

- to ensure free contact of the test pieces with the solution;
- should not be limit the airflow after withdrawal;
- should not be retain solution after withdrawal;
- the solution which drains from any of the pieces after withdrawal should not get in direct contact with another test piece.

Moderated air circulation is recommended which would ensure air-drying of the test piece within 40 min. Additional guidelines in regard to the air circulation:

- the air should not be fed using overpressure;
- stale air should be avoided;
- recommended air temperature 27 °C ± 1 °C;
- recommended relative humidity 45 % ± 6 %.

To meet these guidelines during the development of the machine the following is done:

- a control unit for managing the immersion and withdrawal is created, based on programmable microcontroller;
- all parts which are in contact with the solution or with the test pieces are made of inert transparent polymers;
- a proper, simplified fixture is devised to hold the test pieces. This fixture is described below;
- a chamber with temperature and humidity control assuring medium air circulation is created. The chamber is described below.

2.3. Procedure

Before the test, the test piece should be carefully cleaned using an appropriate method. The following procedure has been adopted – degreasing in alkaline solution (pH 12) at 65 °C for 15 min; rinsing in distilled water; submerging in acidic solution (500 ml hydrochloric acid, 3.5 ml hexamethylenetetramine and 500 ml distilled water [12]), for 10 min; rinsing in distilled water; drying with ethanol to remove remaining water. After cleaning the test pieces initial weight m_1 are weighed with precision scales (with accuracy of 0.001 g).

As the object of a further study will be mainly 6 mm in diameter cylindrical rods the fixture has been made to accommodate such test pieces.

The recommended cycle includes 10 min of immersion in the solution followed by 50 min of air-drying. The recommended temperature of the solution is 25 °C ± 1 °C. These parameters were taken into consideration when the machine was designed.

The guideline, which states that in immersed position the test pieces must be fully submerged and at least 10 mm below the surface of the solution, is also met.

The level of the solution in the container should be kept by adding distilled water to compensate for the losses caused by evaporation. It is recommended that the solution is changed every 168 h or when the pH level changes with more than 0.3 units with respect to the initial pH. The design of the construction ensures user-friendly process for changing of the solution.

After the test is completed, the test pieces have to be extracted from the fixture, cleaned, rinsed and dried. To do this the following procedure is adopted: cleaning with a soft brush; immersion in the previously described acidic solution for 10 min; rinsing in distilled water; drying with ethanol to remove remaining water. After the cleaning is done the test pieces are weighed again and their final weight m_2 is recorded.

IOP Conf. Series: Materials Science and Engineering **664** (2019) 012016 doi:10.1088/1757-899X/664/1/012016

3. Design of the machine

The general view of the newly developed machine is shown in figure 1. The machine comprises four subassemblies:

- chamber – figure 2;
- moving mechanism – figure 3;
- test piece fixture and solution container – figure 4;
- electronic control unit (ECU) – figure 5.

The chamber shown in figure 2, consists of frame 1, made by powder coated steel beams. Four clear PPMA plates 2 and MDF top 4 are attached to the frame.

Ventilation holes are cut on two opposing plates. One set of holes is situated in the bottom part of the plate and vents 6 are mounted on them. Fresh air enter the chamber using these vents. The other set of holes is situated in the upper part of the plate and fans 3 are mounted on them. The air leaves the chamber through these holes. The fans switch on and off automatically if the temperature or the humidity in the chamber exceed the given values. Additionally, four incandescent light bulbs 5 are mounted on the frame. They are turned on and off automatically when it is necessary to increase the air temperature or the solution temperature.

(a) (b)

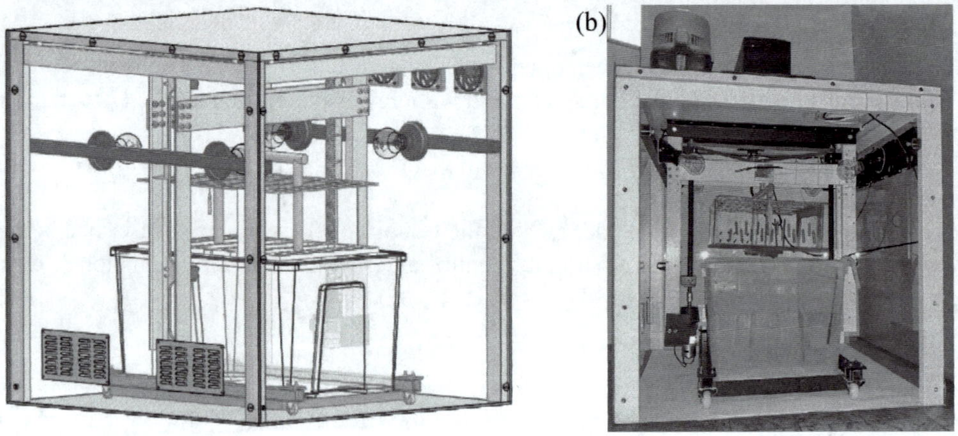

Figure 1. General view of the built machine: (a) 3D model; (b) Photo.

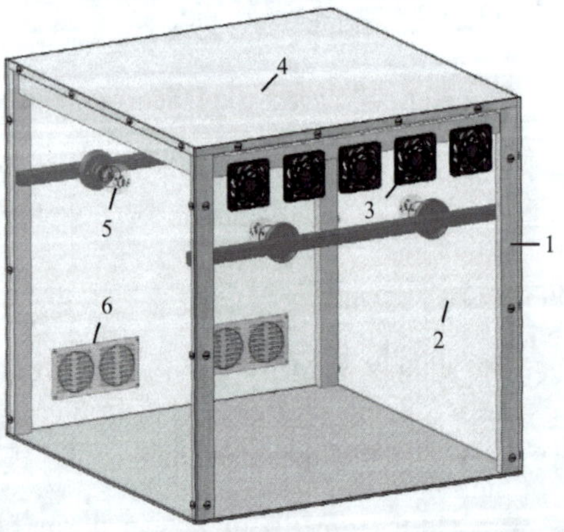

Figure 2. Chamber: 1 – frame; 2 – plates; 3 – fans; 4 – top; 5 – heating lamps; 6 – vents and ventilation holes.

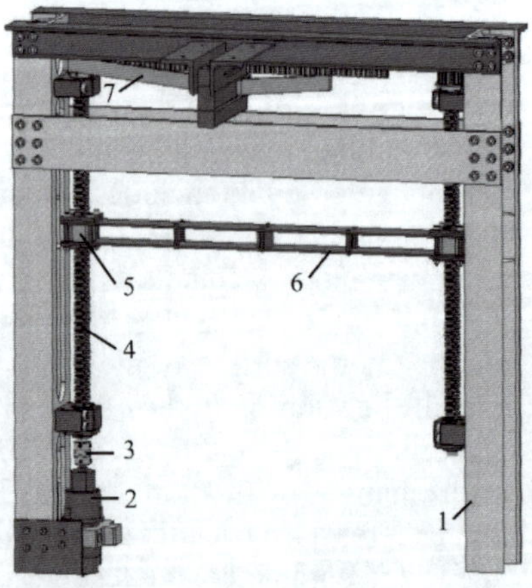

Figure 3. Moving mechanism: 1 – frame; 2 – DC motor with reduction gear; 3 – flexible coupling; 4, 5 – leadscrew; 6 – traverse; 7 – timing belt.

The moving mechanism (figure 3) consists of frame 1, to which a DC motor with reduction gear 2 is attached. It rotates screw 4 with the help of flexible coupling 3. Screw 4 in combination with nut 5 creates leadscrew transmission which converts the rotary movement of the motor into linear movement of traverse 6. The test piece fixture (figure 4) is mounted on the traverse. The movement of the second lead screw is done using toothed belt 7.

The test piece fixture and the solution container are shown in figure 4. The fixture is made up of two horizontal plates 1 and 2 which are forming frame with the vertical elements 3. In lower plate 1, 107 blind holes with a diameter of 7 mm are drilled. In upper plate 2, 107 through holes with the same diameter are drilled. The through holes are coaxial to the blind holes from plate 1. The cylindrical test pieces 5 with a diameter of 6 mm and a length of 160 mm are placed in these holes. The test pieces are designed to be subjected to tensile and torsion tests. In the upper plate additional 36 through holes with a diameter of 3 mm are drilled in which test pieces 6 can be tied. These test pieces are from the same material but with smaller dimensions. Test pieces 6 are not appropriate for mechanical tests but are suitable to determine the amount and the rate of corrosion. Holder 4 is mounted on the fixture and it is connected to the traverse of the moving mechanism (figure 3).

The material used for the stand is PMMA. During assembly a high-strength water and UV resistant adhesive was used. All elements of the fixture are designed in such a manner that they do not block the contact of the test piece with the solution during immersion, nor to prevent the draining of the solution and the airflow during drying.

Figure 4 also shows container 7 with a capacity of 45 l which holds the NaCl solution. The container is placed on wheels 8 which makes it easier to extract it from the chamber, for example to change the solution.

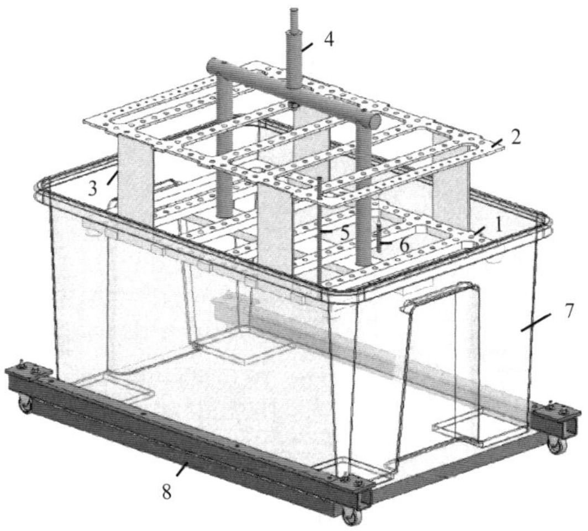

Figure 4. Test piece fixture and water solution container: 1 – lower plate; 2 – upper plate; 3 – connecting elements; 4 – holder; 5 – test piece for mechanical testing; 6 – test piece for determining the amount and rate of the corrosion; 7 – container; 8 – wheel.

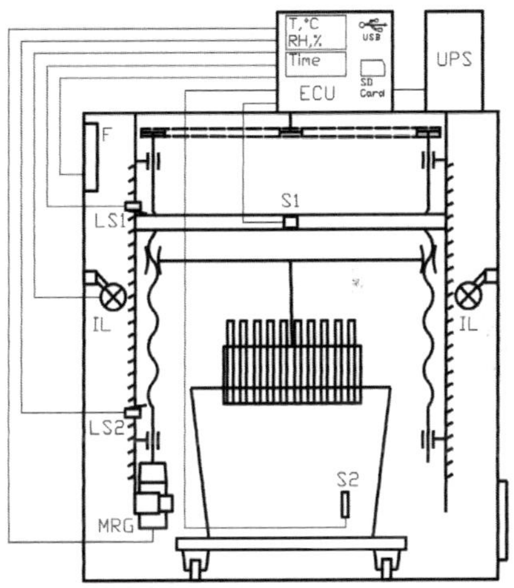

Figure 5. Control system scheme: ECU – electronic control unit; UPS – uninterruptible power supply; F – fans; LS1 and LS2 – limit switches; S1 – sensor for the humidity and temperature of the air; S2 – sensor for the NaCl solution temperature; IL – incandescent light bulb; MRG – motor with reduction gear.

Control system scheme is shown in Figure 5. It consists of electronic control unit (ECU), a temperature and humidity sensor (S1), a solution temperature measurement sensor (S2), terminal switches (LS1 and LS2) and uninterruptible power supply (UPS). The electronic control unit is equipped with a microcontroller, USB port, SD-card slot and two LCD displays. It can be connected to PC via USB interface for firmware update. The software used in the ECU is specially written and allows:

- programming the time intervals for immersed and withdrawn state of the fixture;
- programming the speed of immersion and withdrawal;
- setting the desired air humidity, air temperature and solution temperature.

The LCD displays shows the values given by the sensors, the air temperature and air humidity thresholds and the real time. The displayed information is also recorded in non-volatile memory (SD-card) for later use. Using the sensors data and the parameters of the program, the ECU controls the following subsystems:

- the DC motor (MRG in figure 5), which drives the fixture with the test pieces;
- incandescent lamps (IL), which are used to increase the temperature in the chamber;
- fans (F), which are used to reduce the temperature and the humidity of the air in the chamber.

For ease of maintaining the microclimate in the chamber, in the room where the machine is situated the temperature of the air is maintained 22-25 °C and the humidity of the air is maintained 40-45 %. The room is periodically ventilated.

The machine is equipped with UPS, which allows its independent operation for up to 24 h if the main power supply fails. Thus the uninterruptedness of the test is ensured as standard [1] requires.

4. Alternate immersion testing

The goals of this study are:

- to verify the normal operation of the machine and its compliance with standard [1];
- to investigate the corrosion behaviour of two varieties of one of the most commonly-used structural steel – S235. The steel is in the shape of rods. The first type of steel (S235JR) is represented by hot rolled rods with a measured diameter of 5.9 mm. The other type of steel (S235JRC) is represented by cold drawn rods with a measured diameter of 5.98 mm. The chemical composition of the studied steels is shown in table 1;
- to carry out a comparison of the results of the corrosion behaviour of the same steels at continuous immersion in 3.5 % water solution of NaCl (continuous immersion, CI-test).

Table 1. Chemical composition of the studied rods.

Element	C	Si	Mn	P	S	Cr	Cu	N
S235JR, %	0.170	–	1.400	0.035	0.035	–	0.500	0.012
S235JRC, %	0.080	0.200	0.080	0.027	0.027	0.100	–	–

The corrosion AI-testing are made using the standard procedure [1], described in section 2, using the machine described in section 3. Total 6 durations of the AI-test are used – 7, 14, 21, 28, 35 and 42 days. A set 3 test pieces of each material are tested at each duration. The total number of test pieces for this study is 36.

A photograph of the experiment is shown in figure 1 (b). The shape and the size of the test piece are shown in figure 6 (a). Figure 6 (b) shows a photograph of a test piece from S235JRC calibrated steel rod before the immersion. Figure 6 (c) shows the photograph of a test piece from the same material after AI-test with a duration of 35 days. There is a thick layer of corrosion products in the form of flakes with little mechanical strength, which disintegrate when touched. Figure 6 (d) shows a photograph of the same test piece as figure 6 (c) but cleaned using previously described procedure. Figure 6 (e) shows a photograph of a test piece from S235JR hot-rolled steel rod, after AI-test with a duration of 35 days.

The corrosion study results processing starts with visual inspection of the corrosive behaviour. The photographs in Figure 6 (d) and (e) show the non-uniform corrosion and the complex texture of the surface. The damage caused by corrosion as type and extent are comparable. In hot-rolled steel the pittings are predominant, but there are also large cavities, as well as longitudinal grooves. In cold-drawn steel the cavities prevail, but there are also pittings and the longitudinal grooves are more and deeper as compared to the other grade of steel. These differences can be explained by the different type of microstructure of the material due to the different type of manufacturing.

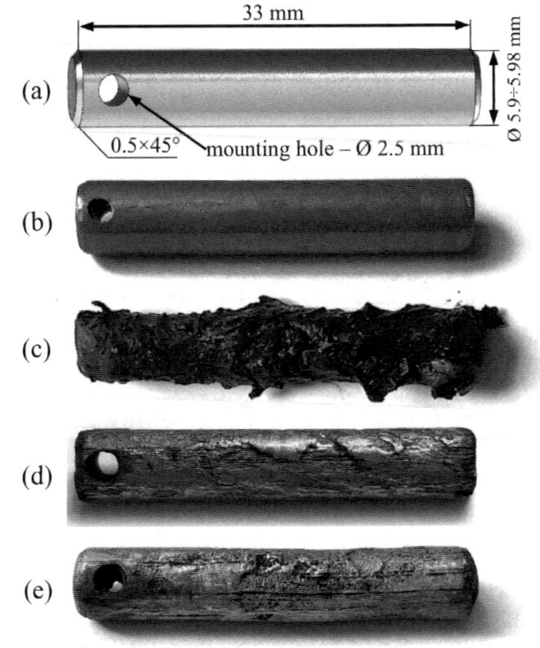

Figure 6. Test pieces before and after AI-test with a duration of 35 days. *a*) 3D model with dimensions; *b*) test piece of S235JRC steel before testing; *c*) uncleaned test piece after AI-test; *d*) cleaned test piece of S235JRC steel; *e*) cleaned test piece of S235JR steel.

Using measured masses before and after the experiment (m_1 and m_2), the loss of mass Δm_i, g, the percentage loss of mass $\Delta m_{p,i}$, % and the specific loss of mass $\Delta m_{s,i}$, g/m^2 for each test piece is calculated:

$$\Delta m_i = m_1 - m_2, \text{g}; \quad \Delta m_{p,i} = \frac{m_1 - m_2}{m_1}100, \% \quad \Delta m_{s,i} = \frac{m_1 - m_2}{A_i}, \frac{\text{g}}{\text{m}^2}, \tag{1}$$

where A_i, m^2 is surface area of the test piece.

The corrosion rate k_i for each test piece is

$$k_i = \frac{\Delta m_i}{A_i t}, \frac{\text{g}}{\text{m}^2\text{day}}. \tag{2}$$

where t, days is duration of the AI-test.

The average the loss of mass Δm, the average percentage loss of mass Δm_p, the average specific loss of mass Δm_s and the average corrosion rate k for each set of three test pieces are

$$\Delta m = \frac{1}{3}\sum_{i=1}^{3}\Delta m_i, \text{g}; \quad \Delta m_p = \frac{1}{3}\sum_{i=1}^{3}\Delta m_{p,i}, \%; \quad \Delta m_s = \frac{1}{3}\sum_{i=1}^{3}\Delta m_{s,i}, \frac{\text{g}}{\text{m}^2}; \quad k = \frac{1}{3}\sum_{i=1}^{3}k_i, \frac{\text{g}}{\text{m}^2\text{day}}. \tag{3}$$

The experimentally obtained results for both tested steels are represented numerically in table 2 and graphically in figures 7 and 8.

From table 2 and figure 7 it can be seen that during the first month of testing, the corrosion speed of the two steels with common chemical composition is different. The hot-rolled steel S235JR has higher corrosion rate. This steel has rougher surface texture which is more susceptible to corrosion. When the duration of the test is increased the difference between the corrosion rates of the two steels decreases. After AI-test with duration of one moth the surface of the two types of test pieces acquire common texture and after the 35-th day the corrosion rate becomes practically the same.

IOP Conf. Series: Materials Science and Engineering **664** (2019) 012016 doi:10.1088/1757-899X/664/1/012016

Table 2. Loss of mass and corrosion rate of S235JR and S235JRC steels in AI-tests performed in 7-42 days.

t days	Δm g	Δm_p %	Δm_s g/m^2	k g/m^2.day
S235JR steel				
7	0.118	1.75	167.98	24.00
14	0.308	4.60	440.18	31.44
21	0.433	6.49	618.63	29.46
28	0.570	8.47	813.73	29.06
35	0.709	10.52	1012.17	28.92
42	0.848	12.66	1211.08	28.84
S235JRC steel				
7	0.109	1.54	152.85	21.84
14	0.263	3.78	369.94	26.42
21	0.447	6.39	628.75	29.94
28	0.615	8.78	865.06	30.90
35	0.733	10.47	1031.51	29.47
42	0.877	12.55	1233.13	29.36

Figure 7 shows the change of the corrosion rate at continuous immersion of test pieces from the same two types of steel in 3.5% of sodium-chloride solution (CI-test) [13]. It can be seen that the corrosion rate is approximately 20 times smaller as compared to the AI-test. Despite the fact that in some cases the CI-test is closer to the natural working environment, the corrosion rate of the CI-test is so small that the duration of the experiment must be many months and even years. This makes it unsuitable for accelerated corrosion testing. On the other hand, the previously described AI-test gives rapid results, but additional research is needed to determine the relation between corrosion behaviour in an AI-test and corrosion behaviour in a real working environment.

Figure 8 shows the specific loss of mass in time for the two steels, determined using the AI-test. These correlations can be very useful for determination of the required time for performing the accelerated AI-test to achieve a corrosion effect close to corrosion in a real working environment. To achieve this goal the specific loss of mass and the exposure time of the studied object in its working environment, should be known.

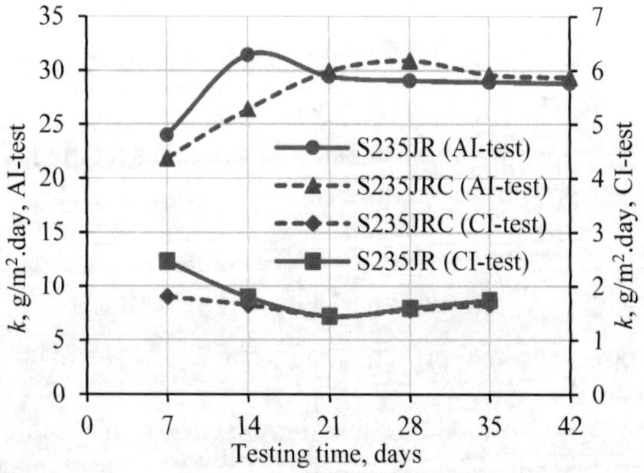

Figure 7. Corrosion rate of S235JR and S235JRC steels at Alternate immersion (AI) test and at Continuous immersion (CI) test in 3.5% water solution of NaCl.

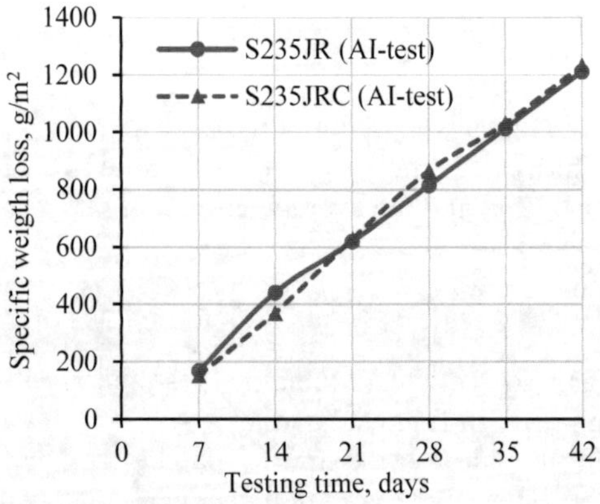

Figure 8. Specific loss of mass, AI-test.

5. Conclusions

The machine described in this paper stands apart for its relatively simple and reliable design as well as for its low price. The conducted tests show that it can be used successfully to perform accelerated cyclic corrosion tests by alternate immersion in salt solution, which comply with standard [1]. The tests performed with structural steels S235JR and S235JRC can be used to study their corrosion behaviour. The comparison between the AI and CI-tests can be useful when choosing a method for corrosion testing. The experimentally obtained dependence of the specific loss of mass as a function of the test duration is needed to determine the duration of the AI-test, which will ensure the comparability of the results with corrosion due to exposure to real working environment conditions.

Acknowledgements
This work is a part of the research project in support of PhD students № 182ПД0022-04 funded by Research Centre at Technical University – Sofia.

References

[1] EN ISO 11130:2018 Corrosion of metals and alloys – Alternate immersion test in salt solution
[2] Hassel A W, Bonk S, Tsuri S and Stratmann M 2008 A universal alternating immersion simulator for accelerated cyclic corrosion tests *Materials and corrosion* **59**(2) 175–180
[3] Paolo Z, Curtarello A, Maiorana E, Pellegrino C, De Rossi N, Savio G and Concheri G, 2017 Influence of corrosion morphology on the fatigue strength of bolted joints *Procedia Structural Integrity* **5** 409–415
[4] Zhou Q, Sun G X, Hu B and Zhao P 2012 Initial exploitation of corrosion resistance and adhesion of 10 kinds of silane on A3 steel *Advanced Materials Research* **460** 94–7
[5] Chiavari C Bernardi E Martini C Passarini F Motori A and Bignozzi M C 2012 Atmospheric corrosion of Cor-Ten steel with different surface finish: Accelerated ageing and metal release *Materials Chemistry and Physics* **136**(2–3) 477–86
[6] Su G and Gao X 2017 Comparison of medium manganese steel and Q345 steel on corrosion behavior in a 3.5 wt% NaCl solution *Materials* **10**(8) 938
[7] Surnam B Y R Audit B S and Soopee A 2017 Use of titanium and tungsten as an alternative to gold in jewellery manufacturing *Iranian J. Mat. Sc. and Engineering* **14**(3) 11–20
[8] Ren X Wu F Xiao F and Jiang B 2015 Corrosion induced fatigue failure of railway wheels *Engineering Failure Analysis* **55** 300–16
[9] Perier V Dieng L Gaillet L Tessier C and Fouvry S 2009 Fretting-fatigue behaviour of bridge engineering cables in a solution of sodium chloride *Wear* **267**(1–4) 308–14
[10] Paolo Z Curtarello A Maiorana E and Pellegrino C 2017 Numerical analyses of corroded bolted connections *Procedia Structural Integrity* **5** 592–9
[11] Chen G Hadi M N Gao D and Zhao L 2015 Experimental study on the properties of corroded steel fibres *Construction and Building Materials* **79** 165–72
[12] EN ISO 8407:2014 Corrosion of metals and alloys – Removal of corrosion products from corrosion test specimens
[13] Tsonev V Nikolov N Marcheva Y 2018 Influence of plastic deformation of S235JR steel rods on their mechanical properties and corrosion behavior in NaCl solution *MATEC Web of Conferences* **234** 04001

IOP Conf. Series: Materials Science and Engineering **664** (2019) 012017 doi:10.1088/1757-899X/664/1/012017

Hydroacoustic stand for evaluating underwater sound systems in a measurement pool

Iliyan Iliev[1] and Nikolai Kolev

Electronic Department, Nikola Vaptsarov Naval Academy, 73 Vassil Drumev Str., 9002 Varna, Bulgaria

[1] E-mail: i.iliev@naval-acad.bg

Abstract. Measurement of sonar parameters in laboratory conditions is a powerful tool for development, control and diagnostics. The realization of such a tool requires the construction of a measuring pool and its equipment with digital channels for receiving and generating sonar signals. An essential requirement for the measuring equipment is to quantify time and amplitude signals with the necessary accuracy and speed. The paper presents a new stand for evaluating the parameters of active and passive sonar systems. The stand is based on the PXI console of National Instruments and the underwater receiving channel of Bruel & Kjaer. Software solutions for the synthesis, transmission, reception and analysis of acoustic signals with different pulse lengths and intrapulse modulations have been developed through the LabVIEW software platform. VI instruments were developed for digital processing, and visualization of signals, and the ability to measure Sonar Transducer parameters in a test tank.

1. Analysis of the basic requirements for a stand for testing of underwater sound systems

Testing of sonar equipment on a stand makes it considerably easier to evaluate its parameters than to perform similar measurement in real conditions. In order for the results to be reliable and accurate, the measurement conditions must correspond to the actual limitations imposed, the parameters of the measuring apparatus shall allow the generation of test signals with characteristics corresponding to the sonar under test as well as the reception of signals emitted by it. Sonar systems in both versions, passive and active, have been widely used to solve tasks of a different nature both in the water and in the air. Underwater active sonars are the main means of detecting and classifying underwater objects as well as conducting hydrographic measurements. The range of these measurements includes bottom mapping, bottom bathymetry, determination of sound wave velocity, salinity, temperature and velocity of water layers. In the air, active sonars are used for ultrasonic distance measurements, car parktronic, ultrasonic echography, defectoscopy, and in other applications. Passive sonars have found a broad application for detecting underwater objects through their noise and measuring its parameters (intensity, bandwidth, spectral lines, spectral level, and correlation time). In these systems, the measurement of object parameters, distance, bearing, radial velocity, and own noise is performed by evaluating the parameters of the received signal. The range of these parameters includes amplitude, latency, Doppler Frequency, Strength, Spectrum, and Spectral Lines in the Spectrum. To measure these parameters, testing systems must support narrowband processing techniques for active systems and broadband processing for passive systems. To limit the impact of external disturbances, it is required that measurement is performed in a test tank for underwater sonar systems or in anechoic chambers for air sound systems.

Content from this work may be used under the terms of the Creative Commons Attribution 3.0 licence. Any further distribution of this work must maintain attribution to the author(s) and the title of the work, journal citation and DOI.
Published under licence by IOP Publishing Ltd

The main modules of the presented stand include a test tank, a microprocessor unit, electroacoustic transducers, a calibrated input amplifier and an output power amplifier. For non-distortion measurements in a test tank a far field condition between hydrophones must be met [1, 2].

$$d \geq \frac{a^2}{\lambda}, \tag{1}$$

where a is the biggest dimension of the hydrophone and λ is the wave length; d is the separation between hydrophones.

Also the pulse length τ must be long enough for a steady-state condition of oscillation to be met [3] and the spectrum of signal must be enough to eliminate the impact of reflections from the water boundaries [4, 5].

The microprocessor block is used to perform the following functions:

- Control the stand operating modes;
- Receive and convert the analog signal into a digital form;
- Analysis of an active sonar signal or underwater object noise in the received input signal mixture;
- Synthesis of signal with duration and frequency band corresponding to the signal of the tested system.

The microprocessor system uses the Direct Digital Synthesis (DDS) technique to generate test signals. This technique requires the rapid processing of the microprocessor system to meet the condition:

$$f_s \geq f_0 N, \tag{2}$$

where f_s is frequency of the microprocessor system; f_0 – signal frequency; N – number of samples to generate signal.

For periodic signal synthesis, DDS generators use a table with digital signal values for one period (LookUp Table – LUT). To generate the sinusoidal signal values with the LUT table, the following formula is used:

$$U(n) = sin\left(n\frac{360}{N}K\right) + U_0, \tag{3}$$

where n is a sample number from 0 to N; K – number of DAC quantum levels; U_0 – the DC voltage generated at the output of the DAC.

The number of samples in the LUT table N is determined by the expression:

$$N = \frac{f_s}{f_0}. \tag{4}$$

2. Functional description of a hydroacoustic stand for tests of underwater sound systems

The input amplifier provides the necessary sensitivity of the receiving channel. The basic requirements for it are to be calibrated and matched with the receiving hydrophone, to have bandwidth equal to the bandwidth of the test signal, to provide the required amplification factor, and not to cause nonlinear signal distortion. The power amplifier provides the power for transmission of test signals in the measuring pool. Its basic requirements are to be impedance matched with the acoustic transducer, its frequency band to permit the transmission of the test signals and to have calibrated power output.

The described stand is designed to study the operation of an underwater sound sonar simulator and measure the parameters of electro-acoustic transducers. The stand is built on the hardware of Bruel & Kjaer and National Instruments and NI LabVIEW software. The stand uses the 8100 and 8104 hydrophones with a bandwidth of up to 100 kHz, the NEXUS type 2692 Bruel & Kjaer's wideband two-channel pre-amplifier for calibrated amplification of the received signals. For power amplification of the emitted signals, a type 2713 Bruel & Kjaer amplifier is used with a maximum power of 100 W. The computer analysis, synthesis and display block is based on National Instruments digital units including PXI console, PXI high-speed digitizer type NI 5922 and PXI signal generator type NI 5402. They allow for the input of 6 MHz bandwidth signals and synthesis of signals with a bandwidth of up to 20 MHz. The functional diagram of the stand is shown in figure 1.

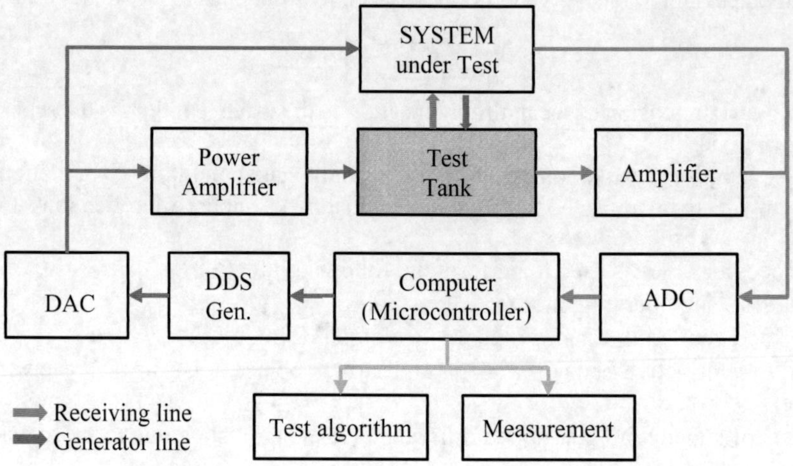

Figure 1. Functional scheme of hydroacoustic stand.

2.1. Functional description of the hydroacoustic stand

Functionally, a generator, receiving and control channels are implemented in the stand. The generator channel produces and emits the test signals (continuous or pulsed, with or without modulation) for analysis of the input stages of the test apparatus or for evaluation of the antenna electroacoustic parameters. The receiving channel handles both broadband and narrowband signals. The use of high-speed ADC NI 5922 and digital signal processing ensures high stability of the spectral estimation [6] and accuracy of power spectral measurements [7]. The stand also allows for measurement of the sensitivity and level of emission of an electroacoustic transducer.

2.2. Measurement methodology for transmitting sensitivity of an electroacoustic transducer.

Functional diagram of the transducer transmitting sensitivity measurement stand is presented in figure 2. The green lines show electrical input to industrial PXI system from power amplifier and preamplifier while the red lines show transducer stimulus signal to the power amplifier.

Important parameters of an electroacoustic transducer are his receiving and transmitting sensitivities. The receiving sensitivity is [8-10]:

$$S_H = \frac{e}{p}, \tag{5}$$

where e is the effective electrical voltage at its output, V; p – effective acoustic pressure in the place of the hydrophone, Pa. The required dimension of the receiving sensitivity is logarithmic – dB re 1µPa/V.

Transmitting sensitivity is:

$$S_V = \frac{p}{E}, \tag{6}$$

where E is the electrical voltage at the input of the transducer, V; p – the acoustic pressure at distance 1 m from the transducer, Pa.

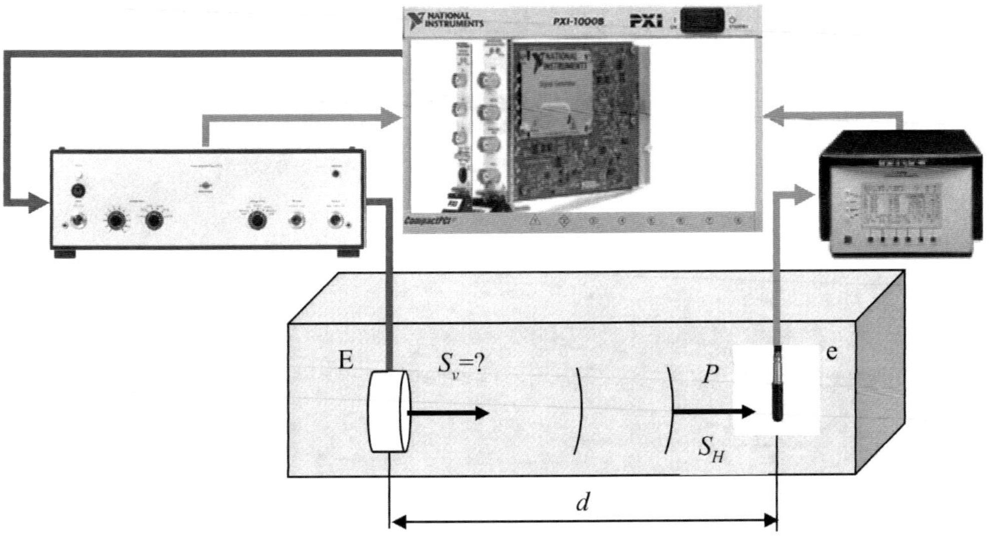

Figure 2. Test tank experimental setup for measurement of transmitting sensitivity.

The required dimension of the transmitting sensitivity is dB re V/μPa at a distance of 1 m from the transducer. The experimental study of the transducer transmitting sensitivity allows for its performance to be estimated as a function of frequency.

The unknown transmitting sensitivity of an electroacoustic transducer is measured with relative calibration by the comparison method [11, 12]. The pressure at distance d from the transducer is measured with previously calibrated hydrophone. If an effective voltage E is applied at the transducer with transmitting sensitivity S_V, the sound pressure at distance d will be:

$$p = \frac{ES_v}{d}. \tag{7}$$

In expression (6) it is assumed that a spherical wave is transmitted from the transducer and its effective pressure p is measured at a relatively far distance d. It is known that for radiated sphere wave front the pressure at distance d will be d times smaller (inverse proportion) than the pressure at da stance of 1m from the transducer which is S_V. From (5) and (7) it follows that the unknown transmitting sensitivity is:

$$S_v = \frac{dp}{E} = \frac{de}{ES_H}. \tag{8}$$

In logarithmic units expression (8) is:

$$20\log_{10} S_V = -20\log_{10} E - 20\log_{10} S_H + 20\log_{10} d + 20\log_{10} e. \tag{9}$$

Usually we measure effective voltage e at the output of the calibrated hydrophone with sensitivity S_H through calibrated charge preamplifier BK2692 connected to the hydrophone.

A calibrated power amplifier BK2713 and preamplifier BK2692 are used to measure precisely voltages E at the input of the transducer and voltage e at the output of the preamplifier. Distance d is controlled through LabVIEW motion interface and is precisely measured. The control of the emitted pulses is done through the LabVIEW virtual instrument, figure 3. In order to estimate the quality of the assumption that an impulsive sound field may have sphere wave front according to (7) a measurement was made of the SPL of transmitting hydrophone BK8104 with varying from 1.32 m to 3.72 m distance to the receiver for frequency 20 KHz.

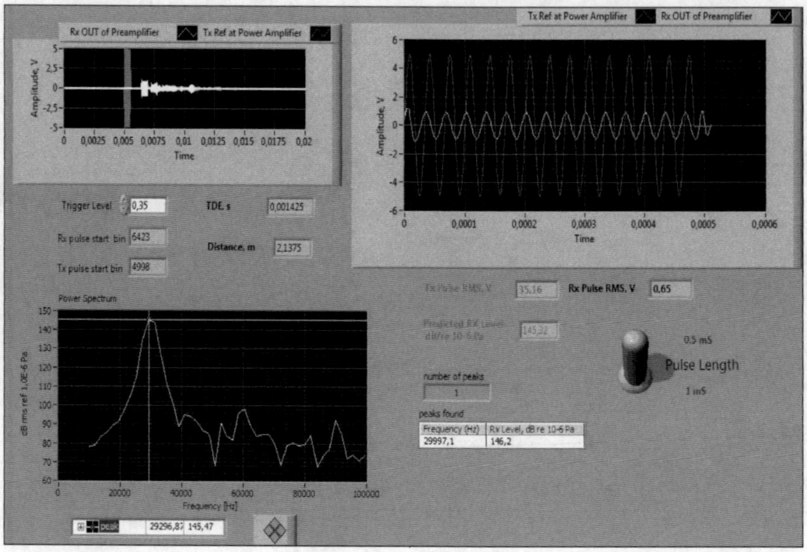

Figure 3. Front panel of virtual instrument for estimation of test tank pulse signals SPL.

Transmitted pulses have a pulse width of 0.5 ms. The depth of the transmitting hydrophone is 1.7 m, while the depth of the receiving hydrophone BK8100 is 1 m. The mean error between the theoretical SPL and measured SPL is -0.07 dB, while the RMSE error is 0.23 dB, figure 4.

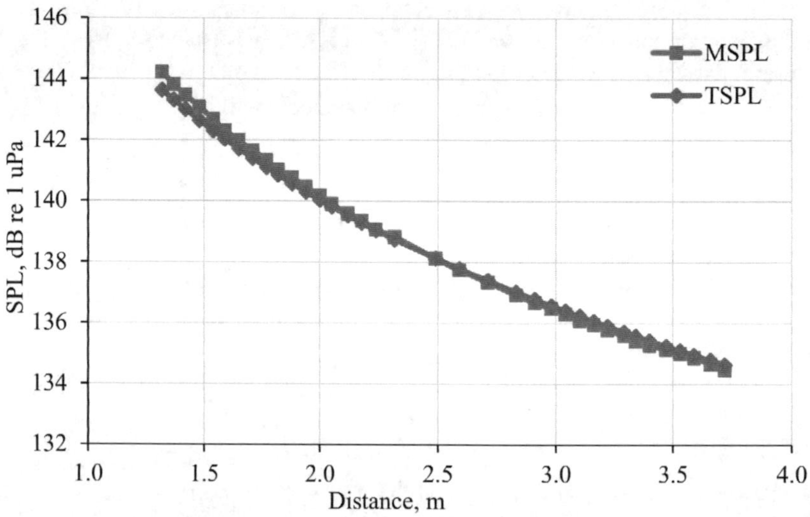

Figure 4. Comparison between measured SPL (MSPL) and theory SPL (TSPL) for spherical wave front at different distances from the transducer BK8104.

2.3. Experimental results.

The paper presents the results of two experiments. The first is a microprocessor system test built with the "Arduino Due" microcontroller, which is designed to detect an active sonar pulse and to synthesize an echo signal with a duration and frequency corresponding to the sounding pulse. For this purpose, the stand is programmed to work with a test algorithm in the presented sequence in figure 5. The test pulse signal is synthesized by a DDS generator operating in arbitrary mode. A time selective window is used to eliminate reflections and spectral leakage. The time domain and spectral diagrams of the test (sonar) and echo signals are shown in figures 6 and 7.

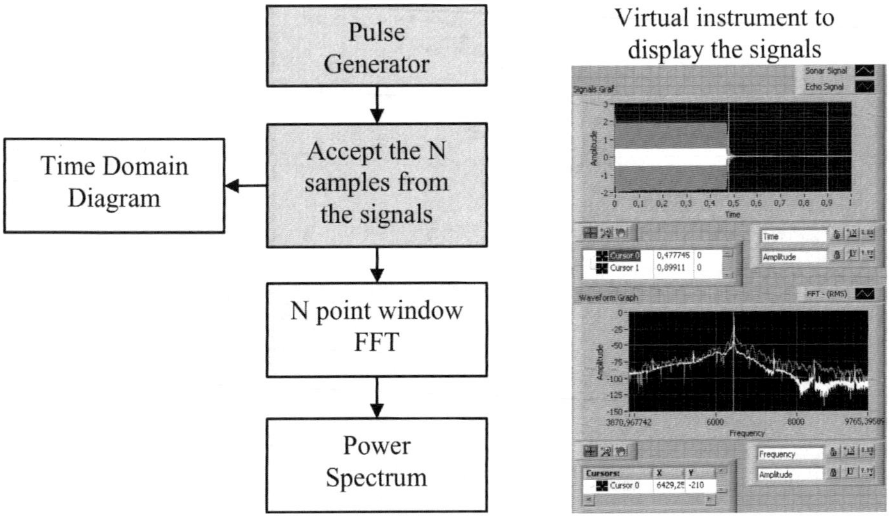

Figure 5. Algorithm and virtual instruments use for testing a Hydroacoustic signal imitators.

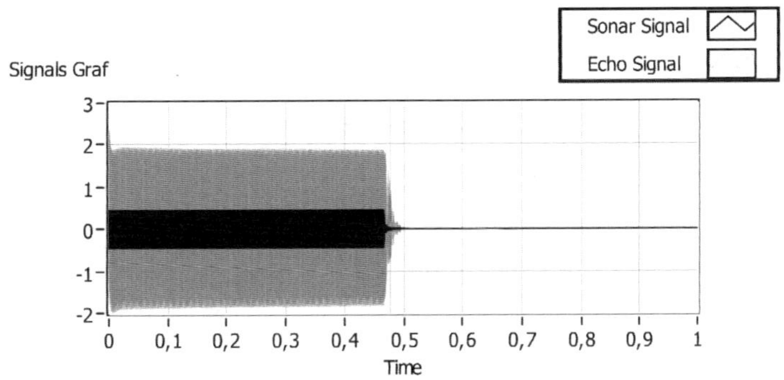

Figure 6. Pulse signals with frequency 6.4 kHz and duration t = 470 ms.

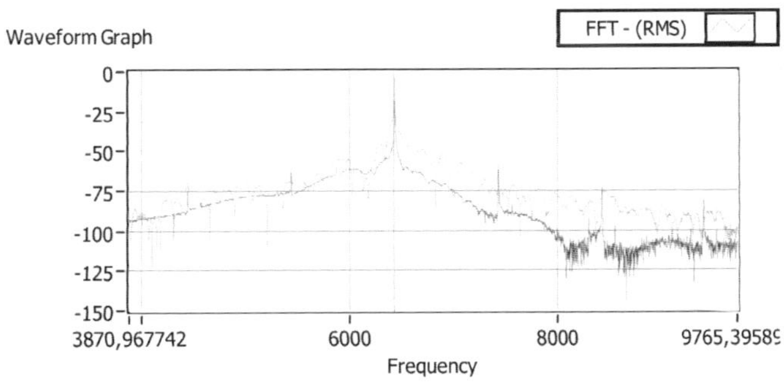

Figure 7. Spectrum of received pulse signals (Sonar and Echo) at a frequency of 6.4 kHz and duration of $t = 470$ ms.

The second experiment demonstrates the possibilities of hydrophone sensitivity measurement at the stand. Comparison of measured transmitting sensitivity of BK8100 hydrophone compared with data from BK nomograme is given in figure 8.

The experimental results for the transmitting sensitivity of the broadband transducer are presented in figures 9 and 10.

IOP Conf. Series: Materials Science and Engineering **664** (2019) 012017 doi:10.1088/1757-899X/664/1/012017

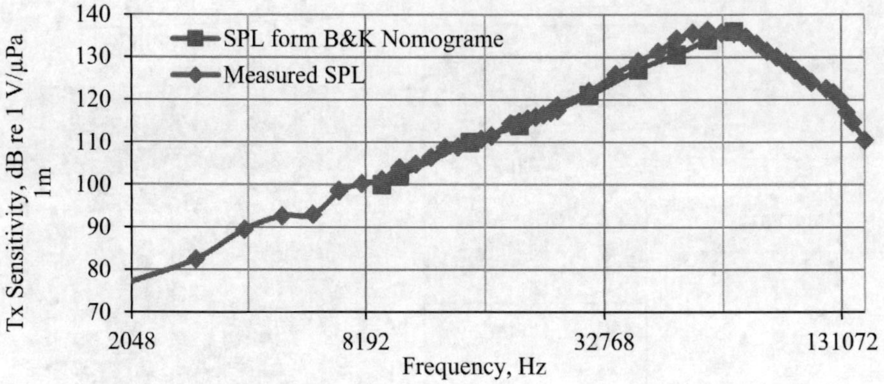

Figure 8. Measured transmitting sensitivity of BK 8100 hydrophone.

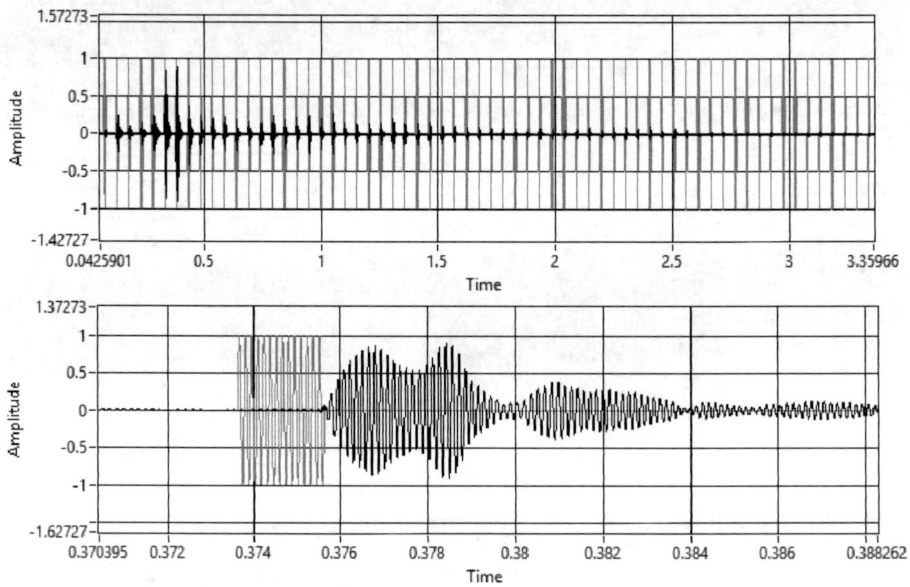

Figure 9. Pulse train of 2 ms transmitted from a broadband transducer and received with the calibrated hydrophone channel with incremental frequency from pulse to pulse.

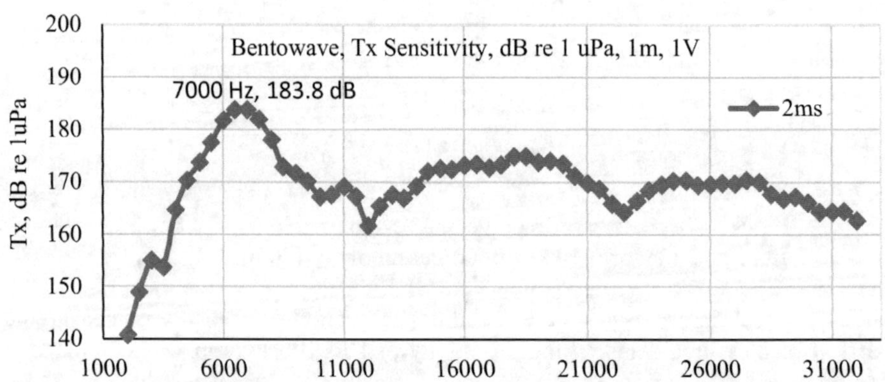

Figure 10. Measured transmitting sensitivity of a broadband transducer "BII-8080".

3. Conclusion

The demonstrated Stand is built up through LabVIEW sequenced development software modules that allow for the hardware to be configured in order to perform various measurements. Thanks to the built-in libraries for DDS functional generator and ADC, the transducer generates sound signals with and without modulation, and allows for the development of digital processing algorithms in the interest of highly accurate measurements. A methodology for test stand sonar transducer transmitting sensitivity measurement is presented based on software gating procedure. Broadband sonar imitator transducer transmitting sensitivity is measured with the stand in the test tank.

Acknowledgements

This paper has been produced under project BG05M2OP001-2.009-0037-C01 "Support for the development of PhD students, post-doctoral students, young scientists and lecturers at the Nikola Vaptsarov Naval Academy", with Nikola Vaptsarov Naval Academy, Varna as beneficiary, funded by the Intelligent Growth Science and Education Operational Program, co-financed by the European Union through the European structural and investment funds.

References

[1] Bobber R J 1970 *Underwater Electroacoustic Measurements* (Naval research laboratory, Washington)
[2] Foote K G 2014 Discriminating between the nearfield and the farfield of acoustic transducers *J. Acoustical Society of America* **136(4)** 1511
[3] Iliev I and Kolev N 2018 Pulse system for evaluation of parameters of electro-acoustic transducers in a hydroacoustic tank *IOP Conf. Ser.: Earth Environ. Sci.* **172** 012021
[4] Isaev E 2014 A quality criterion for obtaining free-field conditions when calibrating a hydroacoustic receiver in a water tank with reflecting sides *Measurement Techniques* **57(5)** 549–56
[5] Barrera-Figueroa S, Rasmussen K and Jacobsen F 2003 A time-selective technique for free-field reciprocity calibration of condenser microphones *J. Acoust. Soc. Am.* **114**(3) 1467–76
[6] Mochizuki K, Uchino M and Morikawa T 2007 Frequency-stability measurement system using high-speed ADCs and digital signal processing *IEEE Transactions on Instrumentation and Measurement* **56(5)** 1887–93
[7] Sidhu T S 1999 Accurate measurement of power system frequency using a digital signal processing techniqu *IEEE Transactions on Instrumentation and Measurement* **48(1)** 75–81
[8] Lurton X 2010 *Introduction to Underwater Acoustics: Principles and Applications*. (Springer-Verlag Berlin)
[9] Camp L 1970 *Underwater Acoustics* (Wiley Interscience, New York)
[10] Robinson S P 2018 Signal-modelling methods applied to the free-field calibration of hydrophones and projectors in laboratory test tanks *Meas. Sci. Technol.* **29** 085001
[11] Van Buren L 1991 Procedure for the in situ calibration of sonar transducers *J. Acoustical Society of America* **90(48)**
[12] Levin P A 1973 Calibration of hydrophones *Brüel&Kjær Technical Review* **1** 3–17

IOP Conf. Series: Materials Science and Engineering **664** (2019) 012018 doi:10.1088/1757-899X/664/1/012018

Modelling of a semi-active vehicle suspension implemented with magneto-rheological dampers

Julian Genov

Technical University – Sofia, 8 Kliment Ohridski Blvd., 1000 Sofia, Bulgaria

E-mail: j_genov@mail.bg

Abstract. This paper describes a procedure for a simulation design aimed to achieve improved performance of the vehicle suspension, using magneto-rheological (MR) semi-active dampers. It is used a spatial so-called "Full car"-model of the vehicle. The model of the kinematic excitations from the unevenness of the road pavement is described in detail. The function of the MR damper is represented by linear and non-linear analytical models. Based on them, is obtained and its inverse neural-network model needed for the suspension control purposes.

1. Introduction

When the vehicle's suspension is synthesized, the simultaneous provision of ride comfort and stability of the vehicle on the road surface is required. The ride comfort assesses the degree of the impact of the road roughness on the driver's health and the functional abilities as well as the degree of vibrational and shock protection of the transported cargo and transport equipment. Evaluatively, the ride comfort is related to the reduction of the vibrational displacement and especially of the vibrational acceleration of the sprung masses (chassis, coupe, and driver). The stability of the vehicle has several indicators, and in this study is considered in the context of this one, linked to the presence of sufficiently large forces of interaction between the tires and the pavement, providing the necessary level of adhesion. In this regard, stability requires a reduction in the vibrational characteristics of the unsprung masses (mainly the car's tires and some suspension elements) and, above all, the reduction of the dynamic component of the normal reaction between the tire and the road pavement. Additionally, this component takes an effect and on the operational condition of the road. These two criteria are contradictory, as in principle the first one requires a softer suspension and the other one requires a stiffer. The compromise between them in the passive suspension is limited. Furthermore, in the different frequency ranges of the excitations, these requirements having changes quantitatively. The relationship between the elasticity of the front and the rear suspension with respect to both criteria is also ambiguous and also changes in the different frequency ranges [1-11]. In this regard, the use of controllable dampers allows for a much higher quality of the compromise between them. Semi-active dampers, although more limited in function than the active dampers, do not require extra energy (except this for the control) and are more reliable, making them more widely applicable.

2. The full car model

The model is shown in figure 1. The CY and CZ axles are central inertia axes, and CX is the principal inertia axis of the model's sprung mass. It is with seven degrees of freedom: the displacements of the unsprung masses in a vertical direction – z_1, z_2, z_3, z_4 and the bounce z_c, the pitch angle θ_x and the roll angle θ_y of the vehicle's sprung mass, denoted from the situation of the static equilibrium:

Content from this work may be used under the terms of the Creative Commons Attribution 3.0 licence. Any further distribution of this work must maintain attribution to the author(s) and the title of the work, journal citation and DOI.
Published under licence by IOP Publishing Ltd

$$q = \begin{bmatrix} q^{us} \\ q^{s} \end{bmatrix}, \quad q^{s} = \left[z_c, \theta_x, \theta_y\right]^T, q^{us} = \left[z_1, z_2, z_3, z_4\right]^T \tag{1}$$

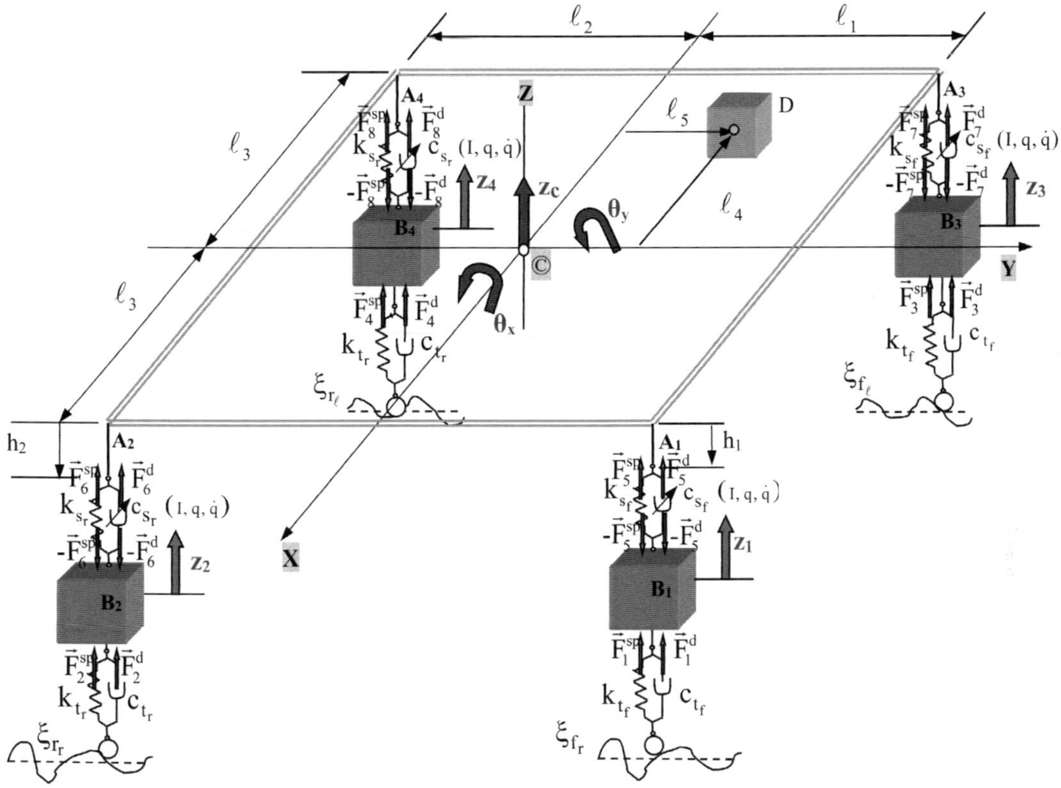

Figure1. Full car model of the vehicle's suspension.

2.1. The parameters of the model

m_{usf}, m_{usr}, m_s – masses of the unsprung part of the front and rear suspension and of the sprung parts;

I_{cx}, I_{cy} – mass inertia moments towards the axis CX and CY;

k_{t_f}, k_{t_r}, c_{t_f}, c_{t_r}, – coefficients of elasticity and viscous damping of the front and rear tires;

k_{s_f}, k_{s_r} – coefficients of elasticity of the front and rear suspension;

F_i^d, $i = 5, .. 8$ – forces in the semi-active dampers;

ξ_{fr}, ξ_{rr}, ξ_{fl}, ξ_{rl} – kinematic excitations from the road roughness;

l_1, l_2 – distances between the centre of gravity (*CG*) of the sprung part and the axes of the front and rear suspension respectively;

$2l_3$ – length of the wheel track (it is accepted that *Cyz* is a plane of a mass-geometrical symmetry);

l_4, l_5 – coordinates of the driver's seat position from *CG*, $q_D = [-l_4, l_5]$.

2.2. Mechano-mathematical model

The displacements of the suspension points $A_i(x_i, y_i, z_i)$, $i = 1,..4$ of the sprung mass are:

$$\vec{r}_{A_i} = \vec{r}_C + \vec{\theta} \times \vec{\rho}_{A_i}, \; i = 1,..4, \tag{2}$$

where: $\begin{matrix}\vec{r}_c=[0,0,z_c][\vec{i},\vec{j},\vec{k}]^T\\ \vec{\theta}=[\theta_x,\theta_y,0][\vec{i},\vec{j},\vec{k}]^T\end{matrix}$, $\vec{\rho}_A=\begin{bmatrix}\vec{\rho}_{A_1}\\ \vec{\rho}_{A_2}\\ \vec{\rho}_{A_3}\\ \vec{\rho}_{A_4}\end{bmatrix}=\Bigg[\underbrace{\begin{bmatrix}l_3\\ l_3\\ -l_3\\ -l_3\end{bmatrix}}_{\rho_{Ax}}\underbrace{\begin{bmatrix}l_1\\ -l_2\\ l_1\\ -l_2\end{bmatrix}}_{\rho_{Ay}}\underbrace{\begin{bmatrix}-h_1\\ -h_2\\ -h_1\\ -h_2\end{bmatrix}}_{\rho_{Az}}\Bigg]\begin{bmatrix}\vec{i}\\ \vec{j}\\ \vec{k}\end{bmatrix}=\rho_{A_x}\vec{i}+\rho_{A_y}\vec{j}+\rho_{A_z}\vec{k}$,

and in the vertical direction: $z_A=\left[z_{A_1},z_{A_2},z_{A_3},z_{A_4}\right]^T=\underbrace{\left[1_{4x1},\ \rho_{A_y},\ -\rho_{A_x}\right]}_{\alpha_{As}}q_s=\underbrace{\left[0_{4x4},\alpha_{A_S}\right]}_{\alpha_A}q$. (3)

The dynamic deformations of sprung elements are:

$$\delta_i^s=z_i-z_{A_i},\ \text{i=1,..4}\Rightarrow\delta^s=\left[\delta_1^s,\delta_2^s,\delta_3^s,\delta_4^s\right]=-\underbrace{\left[-E_4,\alpha_{A_S}\right]}_{\alpha_{\delta^s}}q. \quad (4)$$

The dynamic components of the forces acting on the sprung mass are:

$$F^s=\begin{bmatrix}F_5\\ F_6\\ F_7\\ F_8\end{bmatrix}=\begin{bmatrix}F_5^{sp\ dyn}+F_5^d\\ F_6^{sp\ dyn}+F_6^d\\ F_7^{sp\ dyn}+F_7^d\\ F_8^{sp\ dyn}+F_8^d\end{bmatrix}=\text{diag}\Big\{\underbrace{[k_{s_f},\ k_{s_r},k_{s_f},\ k_{s_r}]}_{k^s}\Big\}\delta^s+\begin{bmatrix}F_5^d=f_{fr}(I,q,\dot{q})\\ F_6^d=f_{rr}(I,q,\dot{q})\\ F_7^d=f_{fl}(I,q,\dot{q})\\ F_8^d=f_{rl}(I,q,\dot{q})\end{bmatrix}=-\text{diag}(k^s)\alpha_{\delta^s}q+\begin{bmatrix}f_{fr}(I,q,\dot{q})\\ f_{rr}(I,q,\dot{q})\\ f_{fl}(I,q,\dot{q})\\ f_{rl}(I,q,\dot{q})\end{bmatrix}, \quad (5)$$

where: $f_{fr}(I,q,\dot{q}), f_{rr}(I,q,\dot{q}), f_{fl}(I,q,\dot{q}), f_{rl}(I,q,\dot{q})$ are the forces in controllable dampers, functionally depend from $q,\dot{q}$ and from the controlling signal (current in this case) I; diag(.) – function that denotes the diagonal matrix with the argument matrix at the main diagonal.

Their sum is

$$F_\Sigma^s=\sum_{i=5}^{8}F_i=-[k^s\alpha_{\delta^s},0_{1x7}]\begin{bmatrix}q\\ \dot{q}\end{bmatrix}+1_{1x4}U,\ U=\left[f_{fr}(I,q,\dot{q}),f_{rr}(I,q,\dot{q}),f_{fl}(I,q,\dot{q}),f_{rl}(I,q,\dot{q})\right]^T. \quad (6)$$

The moments towards the CG and their sum towards the axes are:

$$\vec{M}_{c\ i}^s=\vec{\rho}_{A_i}\times(F_{4+i}^{sp\ dyn}+F_{4+i}^d)\vec{k},\ i=1,..4,\ M_c^s=\begin{bmatrix}M_{cx}^s\\ M_{cy}^s\end{bmatrix}=\underbrace{\begin{bmatrix}\rho_{A_y}{}^T\\ -\rho_{A_x}{}^T\end{bmatrix}}_{l^s}F^s=-\left[l^s\text{diag}(k^s)\alpha_{\delta^s},0_{2x7}\right]\begin{bmatrix}q\\ \dot{q}\end{bmatrix}+l^sU. \quad (7)$$

The tire's deformations are

$$\delta_{us}=\left[\delta_1^{us},\delta_2^{us},\delta_3^{us},\delta_4^{us}\right]^T=E_4\xi-\underbrace{\left[E_4,0_{4x3}\right]}_{\alpha_{\delta^{us}}}q,\ \left(\begin{matrix}\delta_i^{us}=\xi_i-z_i,\ \text{i=1,..4}\\ \xi=[\xi_1,\xi_2,\xi_3,\xi_4]^T=\left[\xi_{fr},\xi_{rr},\xi_{fl},\xi_{rl}\right]^T\end{matrix}\right). \quad (8)$$

For linear elastic and damping forces in the tires it is obtained:

$$F^{tyre}=\begin{bmatrix}F_1\\ F_2\\ F_3\\ F_4\end{bmatrix}=\begin{bmatrix}F_1^{sp\ dyn}+F_1^d\\ F_2^{sp\ dyn}+F_2^d\\ F_3^{sp\ dyn}+F_3^d\\ F_4^{sp\ dyn}+F_4^d\end{bmatrix}=\Bigg[\text{diag}\Big\{\underbrace{[k_{t_f},\ k_{t_r},k_{t_f},\ k_{t_r}]}_{k^{us}}\Big\}\delta^{us}+\text{diag}\Big\{\underbrace{[c_{t_f},\ c_{t_r},c_{t_f},\ c_{t_r}]}_{c^{us}}\Big\}\dot{\delta}^{us}\Bigg]=$$
$$-\left[\text{diag}(k^{us})\alpha_{\delta^{us}},\text{diag}(c^{us})\alpha_{\delta^{us}}\right]\begin{bmatrix}q\\ \dot{q}\end{bmatrix}+\left[\text{diag}(k^{us}),\text{diag}(c^{us})\right]\begin{bmatrix}\xi\\ \dot{\xi}\end{bmatrix}. \quad (9)$$

and the acting on the unsprung masses forces are:

$$F^{us}=\left[F_1^{us},F_2^{us},F_2^{us},F_2^{us}\right]^T=F^{tyre}-F^s\Rightarrow$$
$$F^{us}=-\left[\mathrm{diag}(k^{us})\alpha_{\delta^{us}}-\mathrm{diag}(k^s)\alpha_{\delta^s},\mathrm{diag}(c^{us})\alpha_{\delta^{us}}\right]\begin{bmatrix}q\\ \dot{q}\end{bmatrix}+\left[\mathrm{diag}(k^{us}),\mathrm{diag}(c^{us})\right]\begin{bmatrix}\xi\\ \dot{\xi}\end{bmatrix}-U, \tag{10}$$

The differential equations of movement in a matrix presentation are:

$$m\ddot{q}=\begin{bmatrix}F^{us}\\ F_\Sigma^s\\ M_c^s\end{bmatrix}=-[k,c]\underbrace{\begin{bmatrix}q\\ \dot{q}\end{bmatrix}}_{X}+d_\Lambda\underbrace{\begin{bmatrix}\xi\\ \dot{\xi}\end{bmatrix}}_{\Lambda}+d_U U\Rightarrow m\ddot{q}=-[k,c]X+d_\Lambda\Lambda+d_U U, \tag{11}$$

where: $m = \mathrm{diag}([m_{usf}, m_{usr}, m_{usf}, m_{usr}, m_s, I_{cx}, I_{cy}])$,

$$k=\begin{bmatrix}\mathrm{diag}(k^{us})\alpha_{\delta^{us}}-\mathrm{diag}(k^s)\alpha_{\delta^s}\\ k^s\alpha_{\delta^s}\\ l^s\mathrm{diag}(k^s)\alpha_{\delta^s}\end{bmatrix};\ c=\begin{bmatrix}\mathrm{diag}(c^{us})\alpha_{\delta^{us}}\\ 0_{1x7}\\ 0_{2x7}\end{bmatrix};\ d_\Lambda=\begin{bmatrix}\mathrm{diag}(k^{us}),\ \mathrm{diag}(c^{us})\\ 0_{3x8}\end{bmatrix};\ d_U=\begin{bmatrix}-E_4\\ 1_{1x4}\\ l^s\end{bmatrix}=\begin{bmatrix}-E_4\\ 1_{1x4}\\ \rho_{A_y}^{\ T}\\ -\rho_{A_x}^{\ T}\end{bmatrix}.$$

With introducing the state and the outputs variables vectors:

$$X=[q,\ \dot{q}]^T,\ Y=\left[q,\ddot{z}_d,-F^{tyre}\right]^T\in R^{12x1}, \tag{12}$$

it is obtained the state-space presentation:

$$\begin{vmatrix}\dot{X}=AX+B_\Lambda\Lambda+B_U U\\ Y=CX+D_\Lambda\Lambda+D_U U\end{vmatrix}, \tag{13}$$

where:

$$A=\begin{bmatrix}0_{7x7} & E_7\\ -m^{-1}k & -m^{-1}c\end{bmatrix}\in R^{14x14},\ B_\Lambda=\begin{bmatrix}0_{7x8}\\ m^{-1}d_\Lambda\end{bmatrix}\in R^{14x8},\ B_U=\begin{bmatrix}0_{7x4}\\ m^{-1}d_U\end{bmatrix}\in R^{14x4},$$

$$C=\begin{bmatrix}I_7,0_{7x7}\\ A_{(12,:)}+\ell_5A_{(13,:)}+\ell_4A_{(14,:)}\\ [\mathrm{diag}(k^{us}),0_{4x3},\mathrm{diag}(c^{us}),0_{4x3}]\end{bmatrix}\in R^{12x14},\ D_\Lambda=\begin{bmatrix}0_{7x8}\\ B_{\Lambda(12,:)}+\ell_5B_{\Lambda(13,:)}+\ell_4B_{\Lambda(14,:)}\\ -[\mathrm{diag}(k^{us}),\mathrm{diag}(c^{us})]\end{bmatrix}\in R^{12x8},$$

$$D_U=\begin{bmatrix}0_{3x4}\\ B_U(12,:)+\ell_5B_U(13,:)+\ell_4B_U(14,:)\\ 0_{4x4}\end{bmatrix}\in R^{12x14}.$$

2.3. Modelling of the kinematic excitations

In order to establish a correlation between disturbances in the right and the left track, is proposed an approach similar to this in [12, 13], and based on the summing of two uncorrelated processes, obtained by white noise filtration.

The algorithm is next:

- It is formed a base signal ξ_b with spectral density S_{ξ_b} with three components S_{ξ_1}, S_{ξ_2} and S_{ξ_3} and with the next requirements for them:

$$S_{\xi_1}(\Omega)=const=S_e;\ S_{\dot{\xi}_2}(\Omega)=const=S_s,\ S_{\ddot{\xi}_3}(\Omega)=const=S_a\Rightarrow$$
$$S_{\xi_b}(\Omega)=S_e+\Omega^{-2}S_s+\Omega^{-4}S_a, \tag{14}$$

where Ω is a spatial frequency.

Table 1. Values of the spectral density for different pavement types [14].

Pavement's type	$S_e.10^8$, m^3	$S_s.10^6$, m	$S_a.10^6$, m^{-1}
smooth asphalt	0	0.955	0
asphalt	0	1.910	0.027
rough asphalt	0.04775	3.183	0.032
smooth concrete	0	0.159	0
concrete	1.59	3.183	0.040
rough concrete	1.59	5.570	0.048

- The obtained signal is filtered by low- and high-pass first-order filters:

$$W_{lp}(s)=\Omega_c(\Omega_c+s)^{-1},\ W_{hp}(s)=s(\Omega_c+s)^{-1},\tag{15}$$

where the cut-off frequency is $\Omega_c=\sqrt{2}\pi/L$ (for length of the wheel track $2l_3$ = 1.5 m, L = 5 m [15]).

- By the spectral densities of the filtered signals:

$$S^{lp}(\Omega)=\Omega_s^{\ 2}(\Omega_s+\Omega^2)^{-1}S_{\xi_b}(\Omega)\text{ and }S^{hp}(\Omega)=\Omega^2(\Omega_s+\Omega^2)^{-1}S_{\xi_b}(\Omega)\tag{16}$$

are obtained these for the forming signals:

$$S_{\xi_z}(\Omega)=S_{\xi_b}(\Omega)+S^{lp}(\Omega),\ S_{\xi_{\theta_y}}(\Omega)=S^{hp}(\Omega),\tag{17}$$

and themselves signals ξ_z and ξ_{θ_y} by applying an inverse Fourier transform:

$$\xi_z(\text{s})=\text{F}^{-1}\{S_{\xi_z}(\Omega)\},\ \xi_{\theta_y}(s)=\text{F}^{-1}\{\xi_{\theta_y}(\Omega)\},\tag{18}$$

where with $s(t)$ is denoted the vehicle's position on the road.

- Finally the kinematic excitations are obtained as:

$$\left|\begin{array}{l}\xi_l(s)=\xi_z(s)+\xi_{\theta_y}(s)\\ \xi_r(s)=\xi_z(s)+\xi_{\theta_y}(s)\end{array}\right.\Rightarrow\left|\begin{array}{l}\xi_{fr}(t)=\xi_r(s(t)+l_1+l_2),\ \xi_{f\ell}(t)=\xi_\ell(s(t)+l_1+l_2)\\ \xi_{rr}(t)=\xi_r(s(t)),\ \xi_{r\ell}(t)=\xi_\ell(s(t))\end{array}\right..\tag{19}$$

On the figure 2 are shown the excitations under the left and right vehicle's track.

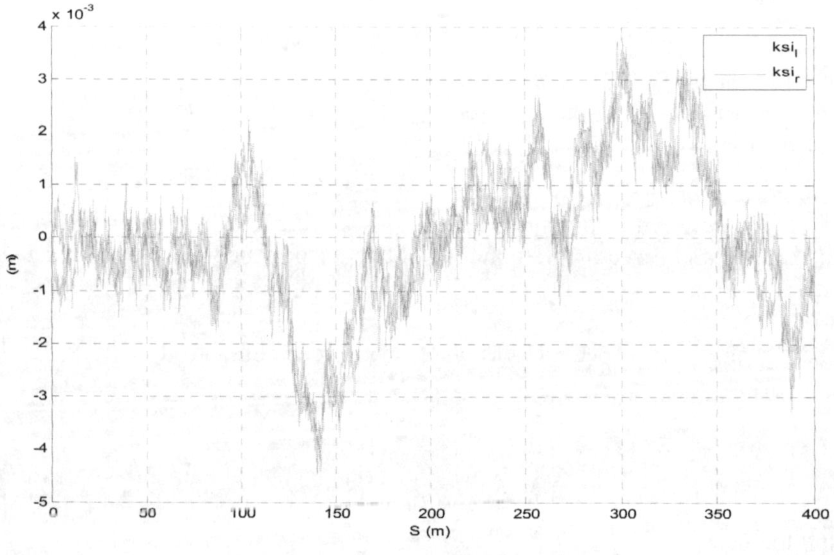

Figure 2. Excitations $\xi_l(s)$ and $\xi_r(s)$ under the left and right vehicle's track.

3. Model of the magneto-rheological fluid damper

The semi-active MR dampers have controlled rheology that is due to the presence in their basic fluid of microscopic polarizable particles. Under the impact of the magnetic field, arisen by the controlling current signal, those particles form chain-domains, which leads to a significant increase of the damping coefficient.

The controlling current I_{contr} in MR dampers varying in the diapason $[0, I_{max}]$, and usually $I_{max} = 1$A.

3.1. Linear model

The linearized model of the damper has the next presentation:

$$F^d = \left[c^l + \left(c^u - c^l\right) I_{contr} I_{max}^{-1}\right]\left(\dot{z}_B - \dot{z}_A\right), \quad (20)$$

where c^l and c^u are the minimal and maximal values of the damping coefficient.

3.2. Phenomenological model

The experimental characteristics of the MR damper are shown on figure 3. They have such nonlinearities as hysteresis, saturation and others. A nonlinear mechanical secondary model proposed in [16] based on the hysteresis presentation in [17, 18] is shown on figure 4.

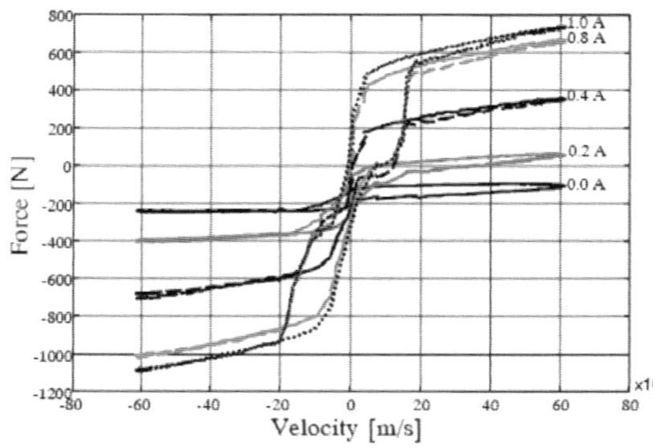

Figure 3. Characteristics of MR damper for different values of the current.

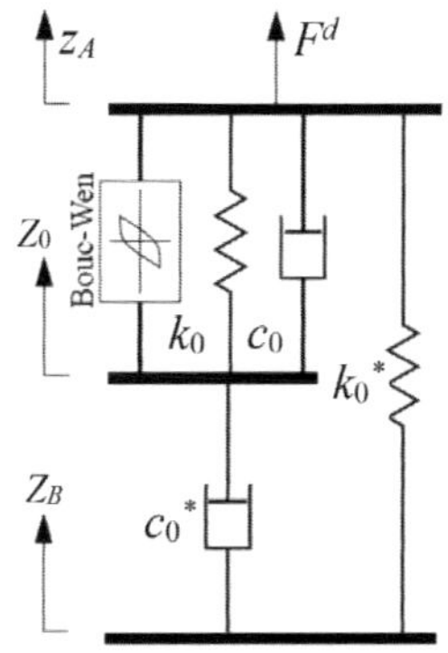

Figure 4. Nonlinear model of the MR damper [16].

The damper's force is

$$F^d = k_0\left(z_0 - z_A\right) + c_0\left(\dot{z}_0 - \dot{z}_A\right) - k_h\zeta + k_0^*(z_B - z_A + \delta_0) = k_0^*(z_B - z_A + \delta_0) + c_0^*\left(\dot{z}_B - \dot{z}_0\right), \quad (21)$$

where δ_0 accounting the presence of a hydraulic accumulator.

From the balance of the forces on the z_0 is obtained:

$$\dot{x}_0 = \left[c_0^*\dot{z}_B + c_0\dot{z}_A - k_0\left(z_0 - z_A\right) + k_h\zeta\right] / \left(c_0 + c_0^*\right). \quad (22)$$

The evolutionary variable ζ is determined by the dependence:

$$\dot{\zeta} = -\gamma\left|\dot{z}_0 - \dot{z}_A\right|\zeta\left|\zeta\right|^{\chi-1} + \beta\left(\dot{z}_0 - \dot{z}_A\right)\left|\zeta\right|^{\chi} - \nu\left(\dot{z}_0 - \dot{z}_A\right). \quad (23)$$

The variables k_h, c_0 и c_0^*, in first approximation, are presented by the linear dependences:

$$k_h = k_h^c + k_h^v I', \ c_0 = c_0^c + c_0^v I', \ c_0^* = c_0^{*c} + c_0^{*v} I', \ \dot{I}' = (I - I')/\tau_r. \quad (24)$$

The obtained values of the model's parameters for MR damper RD-1005-3 of "Lord Corporation" Ltd USA [19] are given in table 2. These values are averaged from those identified from a set of experiments at different parameters of the excitation varying in the ranges: frequency [0,1÷15] Hz, control current [0÷1] A, and amplitude of [0,5÷4] mm, by processing of the experimental data. The identification is based on an optimization procedure.

Table 2. Parameter's values of the MR damper model RD1005-3 of Lord Corporation Ltd.

k_h^c N/m	k_h^v N/mA	k_0 N/m	c_0^c Ns/m	c_0^v Ns/mA	k_0^* N/m	c_0^{*c} Ns/m	c_0^{*v} Ns/mA	χ	τ_r s	γ m^{-2}	β m^{-2}	ν
14500	71500	1400	780	2230	540	28000	500	2.5	$5.3.10^{-3}$	$0.5.10^{6}$	$0.8.10^{6}$	170

The vector of the identification parameters is

$$\Omega=[k_h^c,k_h^v,k_0,c_0^c,c_0^v,k_0^*,c_0^{*c},c_0^{*v},\gamma,\beta,\nu]. \tag{25}$$

If $F^d(t_i),\delta(t_i),\ \dot{\delta}(t_i)$, are the discrete in time values respectively of the damper force, deformation and its velocity, the identification of Ω is reduced to the least-squares optimization problem:

$$\min_{\Omega\in\Theta}\left\{J(\Omega)=\sum_{i=1}^{n}\left[F^d(t_i)-\hat{F}^d(t_i)\right]^2\right\}, \tag{26}$$

where t_i is a discrete time and $\hat{F}^d(t_i)$ is the damper's force calculated by the relation (21).

The set $\Theta\in R^{11}$, in which the solution is sought, sets the minimum and maximum values of the vector elements Ω:

$$\Theta_L\le\Theta\le\Theta_H, \tag{27}$$

where: Θ_L = [3500, 17380, 700, 525, 522, 125, 7075, 73, $0,5.10^6$, $0.5.10^6$, 75];
Θ_H = [28000, 139000, 2000, 4200, 4200, 2200, 120000, 600, $7.2.10^6$, $7.2.10^6$, 602].

Some of the parameters are with fixed values:

- $\delta_0 = F_{offset}/k_0^*$, $F_{offset} = -175$ N – origins shifting of the damper characteristic, due to accumulator availability;
- $\chi = 2.5$ and $\tau_r = 1/190$ s.

The solution to the task is difficult due to the large size of the vector Ω. Therefore, a Monte Carlo optimization method is used, based on the generation of pseudo-random solutions (points) evenly distributed in the hypercube Θ based on the Sobolev's Lp Tau-sequences [20]. The best results obtained are for the values shown in table 2.

3.3. Neural networks (NN) model

From the experimental investigation becomes clear that the upper values are not constant and varying with the amplitude, frequency of the vibrations and with the level of the control signal. For this reason, the use of a model based on neural networks for the approximation of the experimental data is very successfully. It also may be used and as an approximation of the analytical model of the damper.

The structure of the used neural model is shown on figure 5.

The Input layer is formed by the values: of the damper's deformation and its first and second derivatives $d(i), \dot{d}(i), \ddot{d}(i)$, for the current discrete-value of the time, the control current $I_{contr}(i)$ and the damper's force $F^d(i-1)$ in the previous step.

The hidden layer consist of 14 neurons with non-linear logical sigmoid activation functions $g^{(1)}(x) = 2(1 + e^{-2x})^{-1} - 1$.

The Output layer defines the damper force $F^d(i)$ for the current step. The index i is related to the discrete time moment $t_i = i\Delta t$, $i = 0, 1, \ldots$, where Δt is the sampling period. This layer is with a linear activation functions $g^{(2)}(x) = x$.

For the training of the artificial neural network the "Back-propagation" algorithm is used. For the better results is used the data normalization with regard to their minimal and maximal values.

The training is going through in 300 epochs. From the training dataset 60% are used for the training, and 40% for the verification of the trained artificial neural network.The obtained results are given on figure 6.

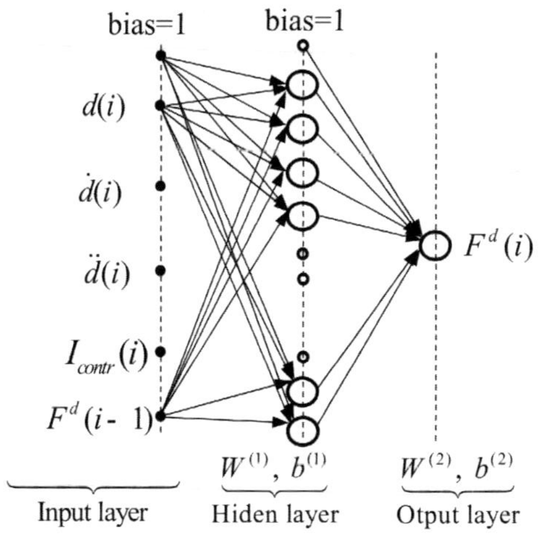

Figure 5. Neural model of MR Damper.

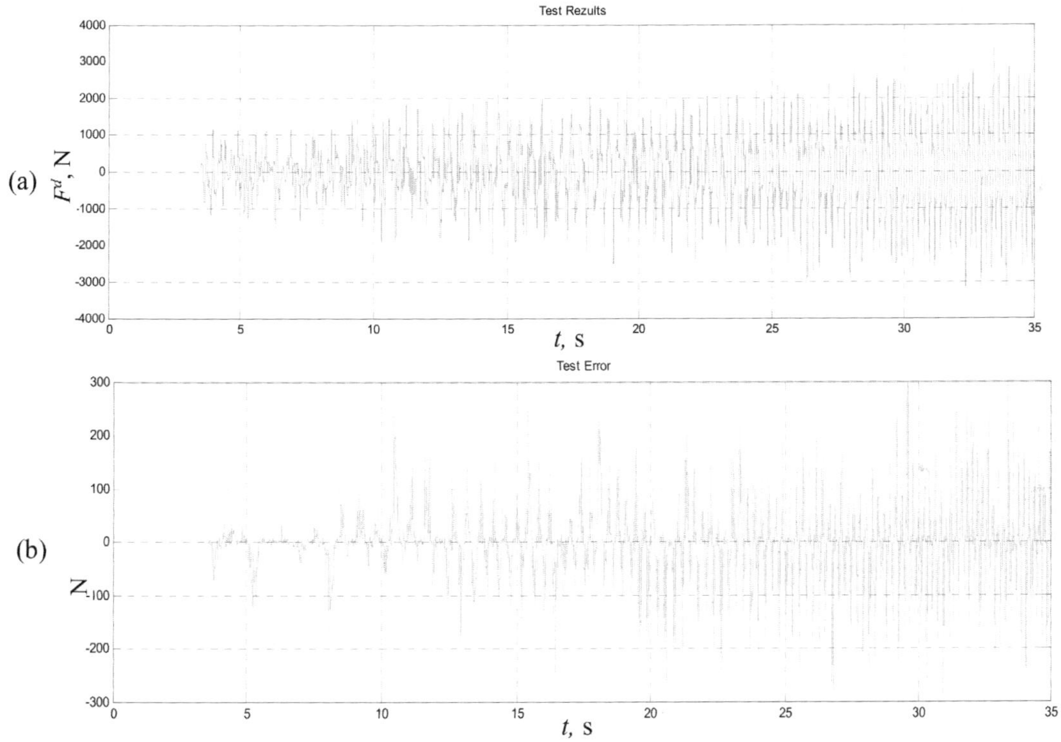

Figure 6. (a) Results the obtained NN model of MR damper; (b) The error in comparison with the experimental results.

3.4. Inverse model of MR damper.

For the purposes of the control, it is needed an inverse damper's model.

3.4.1. Linear model. Using the presentation (20) of the linearized model, the inverse model has the form:

$$\tilde{I} = \frac{I_{\max}}{(\mathrm{c}^{u} - \mathrm{c}^{l})}\left[\frac{1}{(\dot{z}_B - \dot{z}_A)}F^d - \mathrm{c}^l\right], \quad I_{contr} = \begin{cases} \tilde{I}, & \text{за } \tilde{I} \in [0, I_{\max}] \\ 0\ , & \text{за } \tilde{I} < 0 \\ I_{\max}, & \text{за } \tilde{I} > I_{\max} \end{cases}. \tag{28}$$

IOP Conf. Series: Materials Science and Engineering **664** (2019) 012018 doi:10.1088/1757-899X/664/1/012018

3.4.2. Nonlinear model. The inversion of the phenomenological model is rather complicated and in most cases is in a contradiction with the condition for real-time control. For this reason, it is more felicitous the using on inverse NN model. Its structure is similar to the forward model - figure 7. The experimental control voltage and this calculated from the obtained inverse model are shown on figure 8. It can be seen from figure 8 that a very good approximation of the experimental data has been obtained. Some of the training data are shown in figure 9.

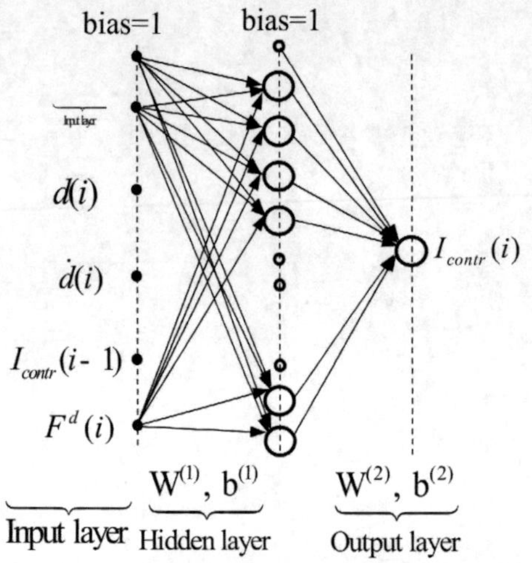

Figure 7. Inverse NN model of MR Damper.

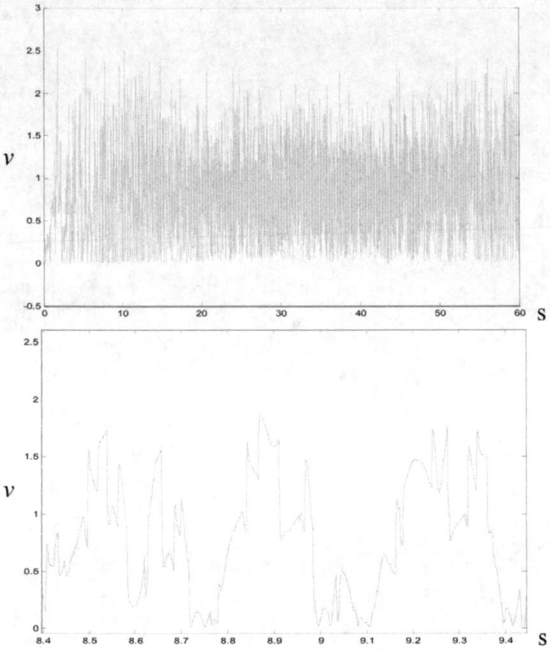

Figure 8. The control voltage and this calculated from the obtained inverse model.

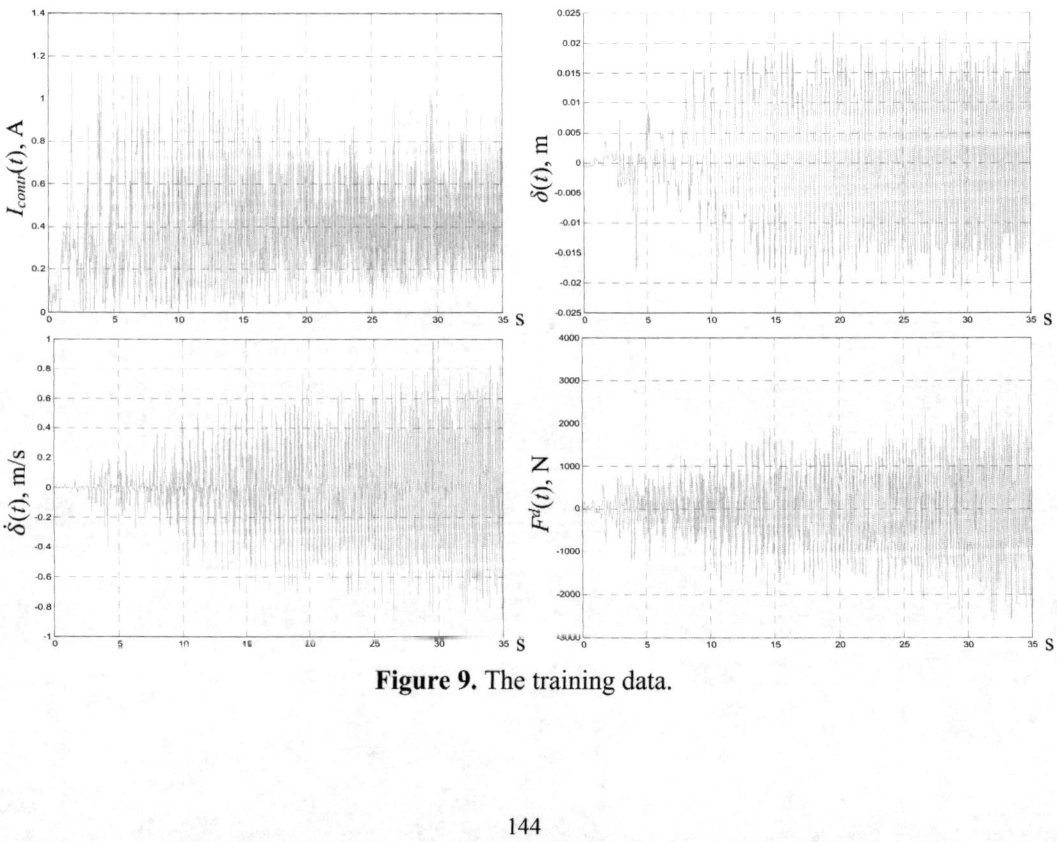

Figure 9. The training data.

4. Conclusions

The paper studies a spatial model of the vehicle's suspension. The model of the kinematic excitations from the unevenness of the road pavement is described in detail. Correlation between the disturbances under the left and right wheel-trucks, as well as the time delay between the front and rear axles of the vehicle, are taking into account. The linearized, phenomenological and neural network (NN) forward and inverse models of the magneto-rheological damper used in the vehicle's semi-active suspension are discussed. Appropriately selected neural structure and the learning variables allow the obtained artificial neuronal model to approximate the behaviour of the MP damper to a very good extent. For the purpose of the control is obtained and its inverse NN model.

Acknowledgements
This work was supported by the Bulgarian Ministry of Education and Science under the National Research Program "Low carbon energy for transport and live", approved by DCM №577/17.08.2018.

References

[1] Campbell C 1981 *Automobile Suspensions* (Chapman and Hall Ltd. London) p 213
[2] Best A 1984 Vehicle ride–stages in comprehension *Physics in Technology* **15** 205–10
[3] Gillespie Th 1992 *Fundamentals of Vehicle Dynamics* (SAE Engineers Inc.) p 472
[4] Sharp R and Pilbeam C 1993 Achievability and value of passive suspension designs for minimum pitch response, *Int. Conf. "Vehicle ride and handling 1993 - Birmingham"* **9** 243–59
[5] Crolla D and King R 2000 Olley's "Flat Ride" revisited *Dynamics of vehicles on roads and on tracks, Proc. of the 16th IAVSD Symp. Pretoria SA* 762–74
[6] Wang J 2001 *Theory of Ground Vehicles* (John Wiley & Sons, Inc.) p 528
[7] Milliken W and Milliken D 2002 *Chassis Design, Principles and Analysis* (SAE Int.) p 676
[8] Sharp R 2002 Wheelbase filtering and automobile suspension tuning for minimizing motions in pitch *SAGE J. – Proc. Inst. Mech. Engineers Part D: J. Automobile Engineering* **216(12)** 933–46
[9] Cao D, Rakheja S and Su Ch-Yi 2008 Heavy vehicle pitch dynamics and suspension tuning. Part I: unconnected suspension *Vehicle System Dynamics* **46(10)** 931–52
[10] Jazar R 2017 *Vehicle Dynamics Theory and Application, Third Edit.* (Springer International Publishing AG) p 985
[11] Marzbani H, Lazar D and Fard M 2017 Vehicle Flat Ride Dynamics (*in Intelligent Interactive Multimedia Systems and Services 2017*), eds Giuseppe De Pietro et al. (Springer Int. Publ. AG) pp 278–88
[12] *Parkhilovskii* I 1968 Investigations of the probability characteristics of the surfaces of distributed types of roads *Autumn Promotion* **8** 18–22
[13] Rill G 1983 The influence of correlated random road excitation processes on vehicle vibration *Vehicle System Dynamics* **12(1–3)** 50–4
[14] Karamihas S 2005 *Critical Profiler Accuracy Requirements, Rept. UMTRI-2005-24* (The Univ. of Michigan Transportation Research Inst.) 125
[15] Sayers M 1988 *Dynamic Terrain Inputs to Predict Structural Integrity of Ground Vehicles - Report for Contract F3615-87-M-3021* (The Univ. of Michigan) p 120
[16] Spencer Jr. B, Dyke S, Sain M and Carlson J 1997 Phenomenological model of a magnetorheological damper journal of engineering mechanics *ASCE* **123 (3)** 230–8
[17] Bouc R 1967 Forced vibrations of a mechanical system with hysteresis, *Proc. 4th Conf. Non-linear Oscillations (Prague, 5-9 September 1967)* p 315
[18] Wen Y 1976 Method for random vibration of hysteretic systems *J. Engineering Mechanics Division* **102(2)** 249–63
[19] Lord Corporation, 2003, RD-1005-3 Product Bulletin
[20] Sobol I, Asotsky D, Kreinin A and Kucherenko S 2011 Construction and comparison of high-dimensional Sobol' generators *Wilmott* **56** 64–79

Multicriterial synthesis of the control law for a semi-active spatial model of the vehicle suspension

Julian Genov

Technical University – Sofia, 8 Kliment Ohridski Blvd., 1000 Sofia, Bulgaria

E-mail: j_genov@mail.bg

Abstract. This article explores the task for the multicriterial synthesis of the control-law from the type Linear Quadratic Regulator (LQR), for a semi-active vehicle suspension. It is consists of obtaining the optimal compromise between the requirements for sufficient ride comfort and vehicle stability on the road pavement. The relevant object-functions are defined. It is proposed an algorithm for a multicriterial hierarchical approach for identifying in the Pareto's set of solutions, this solution that realizes the optimal compromise in the sense of John Nash's sustained strategy from the antagonistic games theory. Based on this approach, a new algorithm for the synthesis of LQR is proposed, combined with a compensator, in which the weight matrices being the subject of optimization. This synthesis was applied for a spatial model of a semi-active suspension of a vehicle.

1. Introduction

The quality criteria of the vehicle vibration insulation are contradictory. The main problem in the suspension synthesis is to be obtained such compromise that satisfied all of them in maximal extent. In general, ensuring enough ride comfort requires one softer suspension, while better vehicle stability and better traction of the tire with the pavement require a rigid one. Obviously, synthesis requires the solution of a multi-criterion task [1-12], allowing the reaching to the optimal compromise, satisfying to the maximum extent of both requirements. The using of active and semi-active suspensions gives significantly bigger possibilities in this direction. The semi-active suspension is more widely spread because of its reliability, low price and exploitation period and not at last position due to the absence of a necessity of a source of external energy.

The most commonly used approach to achieving the required compromise control is through the linear quadratic regulator. A synthesis of this type of control is described in [13, 14]. Its applications for control of the semi-active vehicle suspensions are discussed in [15-30]. In these researches are defined characteristics related to the comfort and stability which, together with the control (the force in the semi-active damper), by means of weight matrices, form the minimized function. These matrices are set by various means, most often intuitively or on the basis of normalizing of the characteristics with respect to their mean, maximum or acceptable values. However, in this way it is obtained different laws of the control. Depending on the choice of the weights they combine the quality criteria with varying degrees of importance. Thus in the best case, are obtained solutions that belong in the Pareto-set of non-dominated solutions [31], but for which ones no ground to be claimed, that they represent namely the optimal compromise. Another important moment is that due to the presence of saturation and hysteresis in the characteristic of the controllable semi-active damper [32], the generated solution, based on the theory of linear regulator, is quasi-optimal.

Content from this work may be used under the terms of the Creative Commons Attribution 3.0 licence. Any further distribution of this work must maintain attribution to the author(s) and the title of the work, journal citation and DOI.
Published under licence by IOP Publishing Ltd

2. Synthesis of LQR

The state space representation of the spatial model of the vehicle suspension is:

$$\begin{vmatrix} \dot{X} = AX + B_\Lambda \Lambda + B_U U \\ Y = CX + D_\Lambda \Lambda + D_U U \end{vmatrix}, \tag{1}$$

where:

- the general coordinates q and state vector X are:

$$q = \begin{bmatrix} q^{us} \\ q^{s} \end{bmatrix}, \quad \begin{matrix} q^s = \left[z_c, \theta_x, \theta_y\right]^T - \text{ bounce, ride, roll of the sprung mass} \\ q^{us} = \left[z_1, z_2, z_3, z_4\right]^T - \text{vertical displacements of unsprung masses} \end{matrix}, \quad X = \begin{bmatrix} q \\ \dot{q} \end{bmatrix};$$

- the vector of the input excitations Λ, which are due to the roughness $\xi(t)$ of the road pavement, and acting under the four vehicle wheels is:

$$\xi(t) = [\xi_{fr}(t),\ \xi_{rr}(t),\ \xi_{fl}(t),\ \xi_{rl}(t)]^T, \quad \begin{matrix} fr \text{ is denotes front-rigth,} & fl \text{ is denotes front-left} \\ rr \text{ is denotes rear-rigth,} & rl \text{ is denotes frear-left} \end{matrix}; \quad \Lambda = \begin{bmatrix} \xi(t) \\ \dot{\xi}(t) \end{bmatrix};$$

- the vector of the control, consisting of the forces in the semi-active dampers, acting on the sprung mass, is:

$$U = \left[f_{fr}(I,q,\dot{q}), f_{rr}(I,q,\dot{q}), f_{fl}(I,q,\dot{q}), f_{rl}(I,q,\dot{q}) \right]^T, \ I \text{ is the damper's controlling current};$$

- the vector of the outputs variables is:

$$Y = \left[q^T, \ddot{z}_d^{(w)}, -F^{tire}\right]^T, \quad \begin{vmatrix} \ddot{z}_d^{(w)} - \text{ the filtered acceleration in the place of the driver seat [33]} \\ F^{tire} - \text{ the dynamical parts of the forces between the tires and the road pavemets} \end{vmatrix};$$

- A, B_Λ, B_U – state matrices, C, D_Λ, D_U – matrices of the output variables.

The synthesis of the linear quadratic regulator is related to minimizing the functional:

$$J(U) = X^T(T_r)FX(T_r) + \int_0^{T_r} \left(Y^T QY + U^T RU\right)dt \tag{2}$$

where: $F \in \mathbf{R}^{14x14}$ and $Q \in \mathbf{R}^{12x12}$ are positive semi-definite, and $R \in \mathbf{R}^{4x4}$ $R \in ?^{4x4}$ is positive definite weighting matrices (in this consideration $F = 0_{14\times14}$ and Q, R are diagonal);

T_r is time-interval in which the regulating process is realizes.

Taking into account (1) in (2) it is obtained:

$$J = \int_0^{T_r} \left[\left(X^T Q_X X + 2X^T N_{XU} U + U^T R_U U \right) + \left(2X^T N_{X\Lambda} \Lambda + 2U^T N_{U\Lambda} \Lambda + \Lambda^T R_\Lambda \Lambda \right) \right] dt \tag{3}$$

where:

$$Q_X = C^T QC, \ \in \mathbf{R}^{14x14}, \ N_{X\Lambda} = C^T QD_\Lambda \in \mathbf{R}^{14\,x\,8}, \ N_{U\Lambda} = D_U{}^T QD_\Lambda \in \mathbf{R}^{4\,x\,8},$$
$$R_\Lambda = D_\Lambda{}^T QD_\Lambda \in \mathbf{R}^{8\,x\,8}, \ R_U = D_U{}^T QD_U + R \in \mathbf{R}^{4x4}$$

Hamiltonian in normalized form $\bar{H}$ and canonical equation about conjugate momentum P are:

$$\begin{vmatrix} \bar{H} = \left(X^T Q_X X + 2X^T N_{XU} U + U^T R_U U \right) + \left(2X^T N_{X\Lambda} \Lambda + 2U^T N_{U\Lambda} \Lambda + \Lambda^T R_\Lambda \Lambda \right) + P^T \left(AX + B_U U + B_\Lambda \Lambda \right) \\ \dot{P} = -\dfrac{\partial \bar{H}}{\partial X} = -\left(2Q_X X + 2N_{XU} U + A^T P + 2N_{X\Lambda} \Lambda \right), \ P(T_r) = FX(T_r) = 0_{2n\,x\,1} \end{vmatrix} \tag{4}$$

The kinematic excitation may be directly measured or identified in real time. In this case, control efficiency can be increased by including additional compensation that depends on the value of the input excitation. Based on this idea, the solution from equation (4) for conjugate momentum is proposed to be in the form:

$$P(t) = 2\left(K(t)X(t) - \Psi\left(t, \Lambda(t)\right) \right), \tag{5}$$

where $\Psi(t, \Lambda(t))$ is this additional part represented the compensation by the input excitation in the controller.

Substituting the time-derivative of the conjugate momentum:

$$\dot{P} = 2\left(\dot{K}X + KAX + KB_U U + KB_\Lambda \Lambda - \dot{\Psi}\right),$$

in equation (4) it is obtained:

$$\left[\left(\dot{K} + KA + A^T K + Q_X\right)X + \left(KB_U + N_{XU}\right)U\right] + \left[-\dot{\Psi} - A^T\Psi + \left(KB_\Lambda + N_{X\Lambda}\right)\Lambda\right] = 0 \qquad (6)$$

According the Pontryagin's minimum principle the optimal control is:

$$\left.\frac{\partial \overline{H}}{\partial U}\right|_{U=U^{extr}} = 0 \Rightarrow U^{extr} = -R_U^{-1}\left(N_{XU}{}^T + B_U{}^T K\right)X - R_U^{-1}\left(N_{U\Lambda}\Lambda - B_U{}^T\Psi\right)$$
$$\frac{\partial^2 \overline{H}}{\partial U_F{}^2} = 2R_U > 0 \Rightarrow U^{extr} \equiv U^{\min} \qquad (7)$$

After substitution $U^{\min}$ in (6) it is obtained a system of two matrix Riccati differential equations:

$$\dot{K} + KA + A^T K + Q_X - \left(KB_U + N_{XU}\right)R_U^{-1}\left(N_{XU}{}^T + B_U{}^T K\right) = 0$$
$$\dot{\Psi} + \left[A^T - \left(KB_U + N_{XU}\right)R_U^{-1}B_U{}^T\right]\Psi + \left[\left(KB_U + N_{XU}\right)R_U^{-1}N_{U\Lambda} - \left(KB_\Lambda + N_{X\Lambda}\right)\right]\Lambda = 0 \qquad (8)$$

For $T_r \Rightarrow \infty$ and if the matrixes A, B, C, D, Q, R are constant, the boundary conditions are [34]:

$$\left|\begin{matrix}\lim\limits_{T_r\to\infty} K(t) \Rightarrow \overline{K} = const \\ \lim\limits_{T_r\to\infty} \Psi(t,\Lambda) \Rightarrow \overline{\Psi}(\Lambda)\end{matrix}\right. \Rightarrow \left|\begin{matrix}K = const = \overline{K} \\ \Psi = \overline{\Psi}(\Lambda)\end{matrix}\right.$$

by which the upper system is transformed into an algebraic:

$$\overline{K}\,\overline{A} + \overline{A}^T\overline{K} - \overline{K}B_U R_U{}^{-1}B_U{}^T\overline{K} + \overline{Q}_x = 0$$
$$\overline{\Psi}(\Lambda) = \left[A^T - \left(\overline{K}B_U + N_{XU}\right)R_U{}^{-1}B_U{}^T\right]^{-1}\left[\overline{K}B_\Lambda + N_{X\Lambda} - \left(\overline{K}B_U + N_{XU}\right)R_U{}^{-1}N_{U\Lambda}\right]\Lambda \qquad (9)$$

where: $\overline{A} = A - B_U R_U{}^{-1}N_{XU}{}^T$, $\overline{Q}_x = Q_X - N_{XU}R_U{}^{-1}N_{XU}{}^T$.

The solution $\overline{K}$ of the first equation is obtained by numerical determination (using Q-R algorithm [35]) of the 2n-of number stabile eigenvalues of corresponding Hamiltonian matrix [36]:

$$H = \begin{bmatrix} \overline{A} & B_U R_U{}^{-1}B_U{}^T \\ -\overline{Q}_X & -\overline{A}^T \end{bmatrix}. \qquad (10)$$

The solution of the second equation is obtained directly by substitution in it the solution $\overline{K}$ of the first equation.

Finally the control formed from LQR is:

$$U^{LQR} = -G_X X - G_\Lambda \Lambda, \qquad (11)$$

where:

$$G_X = R_U^{-1}\left(N_{XU}{}^T + B_U{}^T\overline{K}\right),\ G_\Lambda = R_U^{-1}\left\{N_{U\Lambda} - B_U{}^T\left[A^T - G_X{}^T B_U{}^T\right]^{-1}\left[\overline{K}B_\Lambda + N_{X\Lambda} - G_X{}^T N_{U\Lambda}\right]\right\}.$$

The structural scheme of a vehicle's suspension implemented with magneto-rheological semi-active dampers and their control realized by the synthesized LQR is shown in figure 1. The inverse model of the MR damper is realized by artificial neural network.

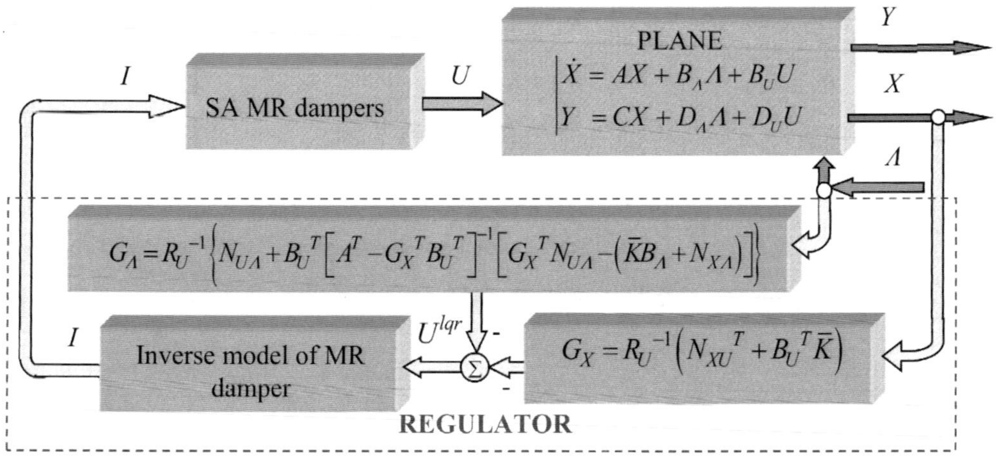

Figure 1. Structural scheme of the LQR-control of semi-active vehicle suspension.

3. Multi-objective approach

The control obtained in the synthesis of the linear quadratic regulator minimizes the integral weighted function, formed by the target characteristics and by the control itself. It depends on the concrete values of the weighting matrices and essentially represents some solution from the Pareto-set. In order to obtain the optimal compromise solution, it is proposed determination of the weighting matrices based on a two-level hierarchical optimization approach, whereby the optimal compromise is considered in the sense of John Nash's sustained strategy from the antagonistic games theory [37].

3.1. The definition of the optimization task

The target functions are the mean square values (RMS) of: the unsprung co-ordinates, the sprung coordinates, the acceleration at the driver's seat attachment and the dynamic components of forces between the tires and road pavement. In general, the first and fourth criteria relate to sustainability and the second and third to ride comfort but do not overlap entirely with the requirements. In this way, the including of the first two criteria leads to a more effective reduction of the level of the vibration field. The vector criterion is:

$$J(\Theta) = \left[J_1^t(\Theta),\ J_2^t(\Theta),\ J_3^t(\Theta),\ J_4^t(\Theta)\right], \qquad (12)$$

$$\left|\begin{array}{l} J_1(\Theta) = 0.25\sum_{i=1}^{4}\left(\sqrt{\frac{1}{T}\int_0^T \left\|z_i(t,\Theta)\right\|^2 dt}\right),\ J_3(\Theta) = \sqrt{\frac{1}{T}\int_0^T \left\|\ddot{z}_d^{(w)}(t,\Theta)\right\|^2 dt},\ J_4(\Theta) = 0.25\sum_{i=1}^{4}\left(\sqrt{\frac{1}{T}\int_0^T \left\|F_i^d(t,\Theta)\right\|^2 dt}\right) \\ J_2(\Theta) = 0.33\left(\mu_{z_C}\sqrt{\frac{1}{T}\int_0^T \left\|z_c(t,\Theta)\right\|^2 dt} + \mu_{\theta_{Cx}}\sqrt{\frac{1}{T}\int_0^T \left\|\theta_{Cx}(t,\Theta)\right\|^2 dt} + \mu_{\theta_{Cy}}\sqrt{\frac{1}{T}\int_0^T \left\|\theta_{Cy}(t,\Theta)\right\|^2 dt}\right) \end{array}\right.$$

For the purpose of reducing the number on the optimized parameters, the weighted matrices are presented in the form:

$$Q = \mu_Q \bar{Q},\ \ R = \mu_R \bar{R}, \qquad (13)$$

where the normalized matrices $\bar{Q}$ and $\bar{R}$ are constructed as:

$$\left|\begin{array}{l} \bar{Q} = diag\left(\left[Q_{us}, Q_{us}, Q_{us}, Q_{us}, Q_s, Q_s, Q_s, Q_{\ddot{z}_d}, Q_F, Q_F, Q_F, Q_F\right]\right) \\ \bar{Q}_{z_1} = \bar{Q}_{z_2} = \bar{Q}_{z_3} = \bar{Q}_{z_4} = Q_{us},\ \bar{Q}_{z_c} = \bar{Q}_{\theta_{Cx}} = \bar{Q}_{\theta_{Cy}} = Q_s,\ \bar{Q}_{F_1} = \bar{Q}_{F_2} = \bar{Q}_{F_3} = \bar{Q}_{F_4} = Q_F \end{array}\right. \quad \left|\begin{array}{l} \bar{R} = diag([R, R, R, R]) \\ \bar{R}_1 = \bar{R}_2 = \bar{R}_3 = \bar{R}_4 = R \end{array}\right.$$

and the scaling coefficients for the parameters' values reconciling μ_Q and μ_R are:

$$\mu_Q = diag\left(\left[\mu_{us},\ \mu_{us},\ \mu_{us},\ \mu_{us},\ \mu_s,\ 50\mu_s,\ 8.13\mu_s,\ \mu_{\ddot{z}_d},\ \mu_F,\ \mu_F,\ \mu_F,\ \mu_F\right]\right) \quad \mu_R = diag\left(\left[\mu_U, \mu_U, \mu_U, \mu_U\right]\right)$$

$$\left|\begin{array}{l} \mu_{z_i} = 0.25 / RMS^2\left(z_i(t)\right) = \mu_{us},\ i = 1,..4,\ \mu_{z_c} = 0.33 / RMS^2\left(z_c(t)\right) = \mu_s, \\ \mu_{\theta_{Cx}} = 0.33 / RMS^2\left(\theta_{Cx}(t)\right) \approx 50\mu_s,\ \mu_{\theta_{Cy}} = 0.33 / RMS^2\left(\theta_{Cy}(t)\right) \approx 8.13\mu_s, \\ \mu_{\ddot{z}_d} = 1 / RMS^2\left(\ddot{z}_d(t)\right),\ \mu_{F_i} = 0.25 / RMS^2\left(F_i^d(t)\right) = \mu_F,\ i = 1,..4; \end{array}\right. \quad \left|\begin{array}{l} \mu_{U_i} = 1 / RMS^2\left(U_i(t)\right) = \mu_U, \\ i = 1,..4 \end{array}\right.$$

in which the root mean square (RMS) values are obtained from implementation with passive suspension.

The vector of the optimized parameters is formed from the elements forming the weighted matrices and from the coefficients of elasticity of the front and rear suspensions $\tilde{\Theta} = \left[k_{s_f},\ k_{s_r}\right]$:

$$\Theta = \left[Q_{us},\ Q_s,\ Q_{\ddot{z}_d},\ Q_F,\ R,\ \tilde{\Theta}\right] \in D, \tag{14}$$

where the definition area is of the form "lower-upper boundaries"

$$D = \left[\Theta^{\ell},\ \Theta^{u}\right]. \tag{15}$$

3.2. *Algorithm for multiobjective synthesis*

The algorithm consists of the next steps:

- It is performed optimization for each of the criteria (12) in regard to Θ and are defined:

$$\Theta_i^{opt} \Rightarrow J_i\left(\Theta_i\right) = \min_{\Theta \in D},\ i \in n_J = \{1,..4\}; \tag{16}$$

- A functional space is formed for a defining the strategy in the context of the Games theory:

$$J(\Theta) \in D_{game} = \left\{J(\Theta) \in \mathbf{R}_{\geq}^{n_J} \left| J^{\min} \leq J(\Theta) \leq J^{\max}\right.\right\}, \tag{17}$$

where:

$$\left|\begin{array}{l} J^{\min} = [J_1\left(\Theta_1^{opt}\right),..J_4\left(\Theta_4^{opt}\right)]^T, \\ J^{\max} = [J_1^{\max},..J_1^{\max}]^T;\ J_k^{\max} = \max\limits_{\Theta_i^{opt}\ |i \in (1,..4) \wedge i \neq k} J_k\left(\Theta_i^{opt}\right),\ k = 1,..4 \end{array}\right.$$

and all object functions are transforming in sense of the cooperative game's strategy:

$$J_k^*(\Theta) = \left(J_k(\Theta) - J_k^{\min}\right)\left(J_k^{\max} - J_k^{\min}\right)^{-1} \in [0,1],\ k = 1,..4, \tag{18}$$

through which it goes to the task of minimizing the vector criterion:

$$J^*(\Theta) = \left[J_1^*(\Theta),..J_k^*(\Theta),..J_{n_J}^*(\Theta)\right]^T; \tag{19}$$

- A generalized criterion is formed as:

$$S(\Theta, \Psi) = \Psi J^*(\Theta), \tag{20}$$

with a weighted vector $\Psi = [\lambda_1,..\lambda_4] \in \Upsilon = \left\{\Psi \in \mathbf{R}_{\geq}^4 \left| \sum_{j=1}^{4} \lambda_j = 1\right.\right\}$,

and whose minima $\Theta_p(\Psi)$ form the Pareto-set for the multi-criteria task:

$$D_p = \left\{\Theta^p(\Psi) \in D \left| \left(S(\Theta^p, \Psi) = \min_{\Theta}\right) \wedge \left(J(\Theta^p) \in D_{game}\right)\right.\right\}; \tag{21}$$

- The stationary solution according to the Nash theory is sought as the solution from the Pareto set, which minimizes L^P-norm in the space of the transformed object-functions:

$$\min_{\Psi \in \Upsilon} L^p\left(J^*\left(\Theta_p(\Psi)\right)\right) = \min_{\Psi \in \Upsilon} \left[\sum_{k=1}^{n_J}\left(J_k^*\left(\Theta_p(\Psi)\right)\right)^p\right]^{1/p} ; \qquad (22)$$

- Thus the optimization process takes place on two levels. For each search-attempt for minimization of (22), generates a value of the weighted vector Ψ and is performed the minimization of (20), whereby is defines the Pareto-solution corresponding to Ψ. The search direction, setting the value of Ψ for the next step, is successful if the corresponding Pareto-solution realizes a minimization of L^p-norm (22) in comparison with the norm's value at the current search-step.

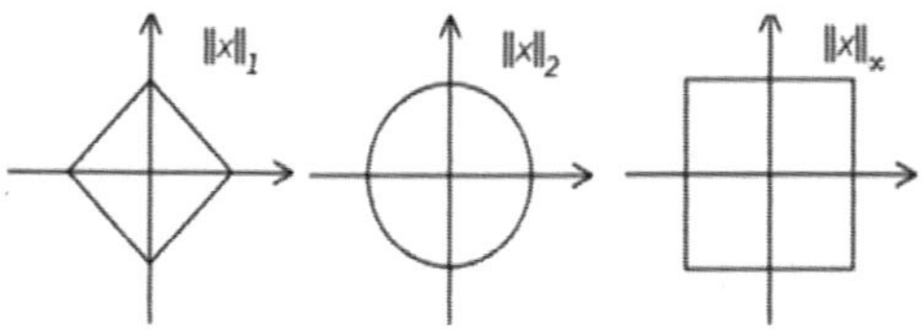

$$L^p\left(J^*\left(\Theta_p\right)\right) \underset{p \Rightarrow \infty}{\Rightarrow} \max_{k=1,..n_J} J_k^*\left(\Theta_p\right)$$

Figure 2. Representation of the L^p-norm for $p = 1, 2, \infty$.

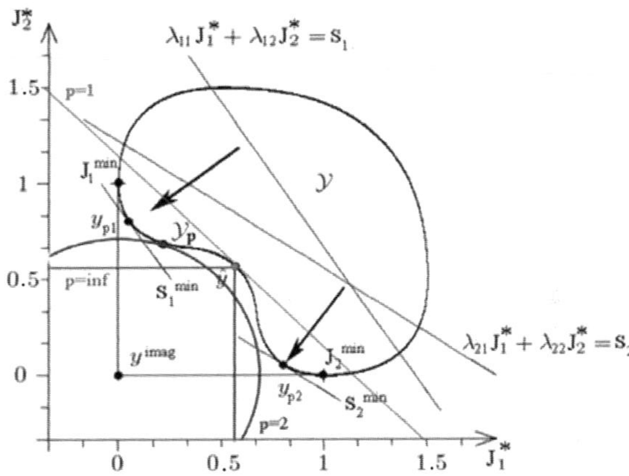

Figure 3. Sustained multi-criteria solutions obtained for $p = 1, 2, \infty$.

In figure 2 are shown the forms of the L^p-norm for different values of the power p. The norm, with the growth of the power p, tends towards the function of the maximum and respectively the solution of the optimization task, to the mini-max von Neumann's solution [38], i.e. it will have a signification of a John Nash's sustained strategy and respectively of the optimal compromise solution [37]. On figure 3 is shown the Pareto's set for two object-functions (the part of the line between the points corresponded to $J_1^{\min}$ and $J_2^{\min}$) and the different optimal compromise's solutions obtained for the values of the norm's power p respectively 1, 2 and ∞.

4. Illustration of the proposed algorithm

4.1. *Parameters of the Full car model:*

- front and rear unsprung masses, sprung mass, moments of inertia about principal transverse and longitudinal axes:

$$m_{us_f} = 40 \text{ kg},\ m_{us_r} = 35 \text{ kg},\ m_s = 580 \text{ kg},\ I_{Cx} = 1580 \text{ kgm}^2, I_{Cy} = 430 \text{ kgm}^2;$$

- coefficients of elasticity and damping in the tires:

$$k_{t_f} = k_{t_r} = 175.5 \text{ kN/m},\ c_{t_f} = c_{t_r} = 80 \text{ Ns/m};$$

- distances from the centre of gravity to the front and rear axes, half-track, driver seat's coordinates about longitudinal and transverse axes:

$$\ell_1 = 1.2 \text{ m}, \ell_2 = 1.4 \text{ m}, \ell_3 = 0.75 \text{ m}, \ell_4 = 0.32 \text{ m}, \ell_5 = -0.4 \text{ m};$$

- boundaries of variation of the optimized parameters:

$$\begin{matrix} & & Q_{us} & Q_s & Q_{\ddot{z}_d} & Q_F & R & k_{s_f} & k_{s_r} \\ \begin{bmatrix} X^\ell \\ X^u \end{bmatrix} & = & \Big[\begin{matrix} 0 \\ 1.5 \end{matrix} & \begin{matrix} 0 \\ 1.5 \end{matrix} & \begin{matrix} 0 \\ 1.5 \end{matrix} & \begin{matrix} 0 \\ 1.5 \end{matrix} & \begin{matrix} 0.05 \\ 1.5 \end{matrix} & \begin{matrix} 12000 \\ 28000 \end{matrix} & \begin{matrix} 12000 \\ 28000 \end{matrix} \Big] \end{matrix}$$

- RMS and scaling coefficients for the parameters' values reconciling are given at a table 1 (they are obtained from the optimal compromise solution for a passive vehicle suspension)

Table 1. RMS and scaling coefficients for the parameters' values reconciling.

RMS	z_{us}, m	z_s, m	$\ddot{z}_d$, ms^{-2}	F^d, N	U, N
	0.0074	0.0074	0.45	390	160
μ	μ_{us}, m^{-2}	μ_s, m^{-2}	$\mu_{\ddot{z}_d}$	μ_F	μ_U
	4565	6026	4.94	1.64×10^{-6}	39.06×10^{-6}

4.2. Results

On table 2 are given the results from optimization by each of the fourth object-functions. On the table 3 is given the game strategy defined by the main object function J_3 and J_4. On table 4 is given the obtained optimal compromise solution for this strategy for the power of L^p-norm $p = 6$, which converges to this for the infinity. It is seen that both functions are equally far from their optimal values and their normalized values are approximately equal: $J_3^* = 0.418$ and $J_4^* = 0.425$, as what is the goal of the optimal compromise. This result is shown and in figure 4 for $p = 6$ (≈ such as for infinity), and also is shown and the result for $p = 2$. The values of the matrixes of the LQR - controller for the realization of this optimal compromise are given in table 5.

Table 2. Optimization by the each of the object functions.

	Values of the optimized parameters							Values of the object functions			
	Q_{us}	Q_s	$Q_{\ddot{z}_d}$	Q_F	R_U	k_{sf} N/m	k_{sr} N/m	J_1 m	J_2 m	J_3 m/s^2	J_4 N
J_1	0.811	0.202	0.788	1.192	0.053	21531	23906	0.00830	0.0089	0.321	352
J_2	1.392	1.286	0.161	0.015	0.364	15844	13218	0.00850	0.0067	0.264	439
J_3	0.010	0.022	0.005	0.013	0.050	12000	12000	0.00847	0.0077	0.246	450
J_4	0.812	0.202	0.788	1.182	0.053	21531	23906	0.00850	0.0089	0.321	352

Table 3. Game strategy by J_3 and J_4.

	J_1, m	J_2, m	J_3, m/s^2	J_4, N
J_{opt}	0.00847	0.0077	0.246	352
$J_{\max}$	0.00850	0.0089	0.321	450

Table 4. Optimal compromise solution by J_3 and J_4 ($p = 6$ approximating infinity).

Values of the optimized parameters							Values of the object functions			
Q_{us}	Q_s	$Q_{\ddot{z}_d}$	Q_F	R_U	k_{sf} N/m	k_{sr} N/m	J_1 m	J_2 m	J_3 m/s^2	J_4 N
0.054	0.68	0.633	0.445	0.2086	13750	25750	8.51×10^{-3}	8.02×10^{-3}	0.278	388
								normalized values:	0.418	0.425

Table 5. Values of the LQR matrixes.

$G_X.10^{-3}$													
–13.96	–1.513	2.779	7.775	2.563	47.55	–29.054	-632.6	0.012	0.066	–0.02	2.211	5.89	–3.3
–1.201	–13.74	4.019	–1.916	–5.531	–62.1	–23.258	8.856	–0.567	–0,041	0.025	1.724	–7.025	–2.94
2.433	2.678	–2.78	9.780	–99.61	40.9	6.122	72.277	–0.044	–503	0.095	2.294	5.408	2.07
6.040	–2.318	4.486	–8.504	–7.748	–68.2	10.07	–5.178	0.025	0.112	–0,517	1.843	–7.446	2.22

G_Λ							
x10^3	x10^3	x10^3	x10^3	x10^{-3}	x10^{-3}	x10^{-3}	x10^{-3}
20.043	1.258	0,208	–14.568	9.134	0.573	0.094	–6.640
2.613	22.120	–6.733	3.979	1.191	10.080	–3.068	1.813
0.007	–10.685	10.077	–7.530	0.003	–4.870	4.592	–3.430
–9.000	3.860	–2.268	18.456	–4.100	1.762	–1.033	8.410

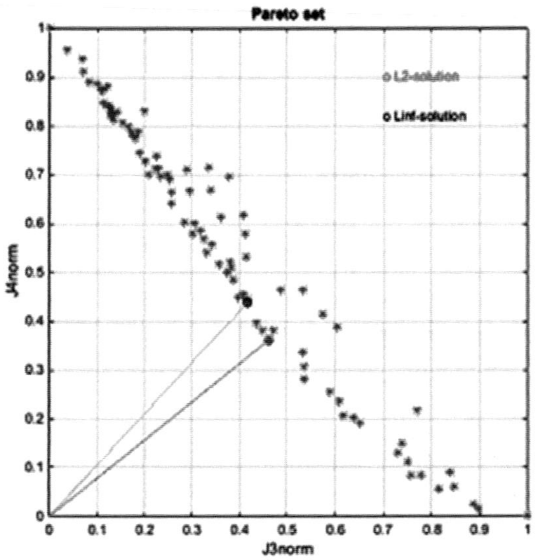

Figure 4. Pareto's set and the optimal compromise solution for synthesis of LQR for p=2 and p = 6.

5. Conclusion

A multicriterial approach for a vehicle suspension synthesis has been proposed, combining the theoretical formulations for V. Pareto's non-dominated solutions and the Theory of antagonistic games by J. Nash. The solved task is related to the determination of the optimal compromise between ride comfort and on-road stability. Appropriate target functions being proposed for quantitative evaluations. The approach considered is applied for the multicriterial synthesis of a semi-active suspension controlled by a linear quadratic regulator. The LQR is combined with a compensator by the input excitations. Another news in the approach is including the elements of the weighted matrixes in the optimized parameters. It is given a concrete illustration of the proposed algorithm for a multicriterial synthesis of the control law for a semi-active spatial model of the vehicle suspension.

Acknowledgements

This work was supported by the Bulgarian Ministry of Education and Science under the National Research Program "Low carbon energy for transport and live", approved by DCM №577/17.08.2018.

References

[1] Pisarev A, Belnikolovski B 1974 Optimal dynamical synthesis of three mass aggregate with elastic cardan shafts, *STU of ME*, (in Bulgarian)

[2] Abuid B, Jordanov I, Belnikolovski B 1995 Optimal dynamic synthesis for the design of a mechanism with single stage spur gears *J. Theoretical and Applied Mechanics – Sofia* **3**

[3] Belnikolovski B, Jordanov I, Nikolov N 1995 Computer aided optimal design of elastic suspended presses, *J. Theoretical and Applied Mechanics – Sofia* **25(3)** 52-61
[4] Cheshankov B, Belnikolovski B, Jordanov I 1996 Multicriteria optimization of a machine aggregate with single stage spur gears *Proc. of Sixth conference on nonlinear Vibrations, Stability and Dynamic of Structures, Virginia Polytechnic Institute and State University* 1–4
[5] Cheshankov B, Belnikolovski B, Jordanov 1997 I Multicriteria optimization of a machine aggregate with single stage spur gears *Proceedings of the 1st Int. Conf. on Control of Oscillations and Chaos*, St. Petersburg, Sec. A 267-70
[6] Vroemen B and de Jager B Multiobjective Control: An Overview 1997 *Proc. 36th IEEE Conf. Decision and Control (CDC'97), San Diego, Califorпnia*, pp. 440–5
[7] Belnikolovski B 1998 *Dynamics of gears with straight teeth* (DSc Dissert., TU-Sofia, Bulgarian)
[8] Belnikolovski B 1998 Optimal dynamic synthesis for the design of a mechanism with two stage spur gears *Proc. XXV National seminar of Dynamic of mechanical systems – Varna*
[9] Miettinen K 1999 *Nonlinear Multiobjective Optimization* (Kluwer Academic Publisher) p161
[10] Liu G,. Yang J, Whidborne J, 2003 *Multiobjective Optimization and Control* (Research Studies Press LTD, GB) p 330
[11] Jahn J 2011 *Vector Optimization Theory, Applications, and Extension - Second Edition*, (Springer-Verlag Berlin Heidelberg) p 497
[12] Statnikov R 1999 *Multicriteria Design: Optimization and Identification* (Kluwer Academic Dordrecht) p190
[13] Anderson O and Moore J 1989 *Optimal Control Linear Quadratic Methods* (Prentice-Hall International Canberra, Australia) p 394
[14] Stengel R 1994 *Optimal Control and Estimation* (Dover Publ. Inc. N.Y.) p 337
[15] Taghirad H, Esmailzadeh E 1997 Automobile passenger comfort assured through Lqg/Lqr active suspension *J. Vibration and Control* **4(5)** 603–18
[16] Naidu D 2003 *Optimal Control Systems* (CRC Press LLC, USA) p 460
[17] Shono T, Nagasaka N, Nishi M, Takagi Y 2003 Multy-mode control for a active isolation system with an elastic table *JSME Int. Jour. Ser. C* **46(3)** 835–40
[18] Sonh H-Ch, Hong K-T, Hong K-Sh and Yoo W-S 2004 An adaptive LQG control for semi-active suspension system *Int. J. Vehicle Design* **34(4)** 309–26
[19] Nguyen T-A 2006 *Application of Optimization Methods to Controller Design for Active Suspensions* Doktor-Ing. genehmigte Diss. (Techn. Univ. Cottbus) p 152
[20] Imsland L, Johansen T, Grip H 2007 On nonlinear unknown input observers applied to lateral vehicle velocity estimation on banked roads *Int. J. Control* **80** 1741–50
[21] Shirahatti A, Prasad P, Panzade P, Kulkarn M 2008 Optimal design of passenger car suspension for ride and road holding *J. Braz. Soc. of Mech. Sci. & Eng.* **30** 66–76
[22] Pakshin P and Peaucelle D 2009 LQR parametrization of static output feedback gains for linear systems with Markovian switching and related robust stabilization *48th IEEE Conf. on Decision and Control (Shanghai, P. R. China)* 1157–62
[23] Huang Hs 2009 *Controller Design for Stability and Rollover Prevention of Multi-Body Ground Vehicles with Uncertain Dynamics and Faults* (PhD Thesis, The Ohio State Univ.) p 125
[24] Creed B, Kahawatte N and Varnhagen S 2010 Design of an LQR control strategy for implementation on a vehicular active suspension system *MAE* **272** 29
[25] Németh B and Gáspár P 2011 Model-based LQ control design of integrated vehicle tracking systems *Transportation Engineering* **39(2)** 77–81
[26] Jahromi A, Xie W and Bhat R 2012 Ride control of passenger cars with semiactive suspension system using a linear quadratic regulator and hybrid optimization algorithm *Int. J. Mechanical and Mechatronics Engineering* **6(7)** 1280–6
[27] Rao K and Kumar S 2014 Modeling and simulation of quarter car semi active suspension system using LQR controller *Proc. 3rd Int. Conf. Frontiers of Intelligent Computing: Theory and Applications* 441–8

[28] Rao K and Pavani K 2015 Modelling and vibration control of suspension system for automobiles using LQR and PID controllers *IJLTEMAS* 64–8
[29] Nguyen Th, Vu V, Dao M, Gaspar P, András M and Ha Tr 2017 Using the LQR control method on the active suspension system of automobiles *Proc. 10th Nat. Conf. on Mechanics and the 8th Nat. Cong. of the VAM (Ha Noi, Viet Nam)* 9
[30] Maurya V and Bhangal N 2018 Optimal control of vehicle active suspension system *J. Automation and Control Engineering* **6(1)** 22–6
[31] Pareto V 1896 *Cours D'économie Politique Tome Premier* (F. Rouge, Éditeur) p 438
[32] Spencer Jr. B, Dyke S, Sain M and Carlson J 1997 Phenomenological model of a magnetorheological damper *J. Engineering Mechanics (ASCE)* **123(3)** 230–38
[33] International Standard ISO 2631-1 2010 Mechanical vibration and shock - Evaluation of human exposure to whole-body vibration - Part 1: General requirements, Amendment 1 (ISO Central Secretariat, Switzerland) p 40
[34] Schwarz H 1976 *Optimale Regelung Linearer Systeme Theor. und Exp. Methoden der Regelungstechnik Bd.12* (Verlag: Mannheim, Wien, Zurich: Bibl. Inst.) p 242
[35] Ralston A and Rabinowitz Ph 2001 *A First Course in Numerical Analysis* (Dover Books on Mathematics) p 624
[36] Stengel R 1994 *Optimal Control and Estimation* (Dover Publ. Inc. N.Y.) p 669
[37] Nash J 1950 *Non-cooperative Games* (PhD Thesis Princeton University) p 32
[38] von Neumann J and Morgenstern O 1953, *Theory of games and economic behaviour - Third edit.* (London: Geoffrey Cumberlege Oxford Univ. Press) p 641

IOP Conf. Series: Materials Science and Engineering **664** (2019) 012020 doi:10.1088/1757-899X/664/1/012020

Adjustment parameters of an internal combustion engine working with methane

R Dimitrov[1,4], K Bogdanov[1], R Wrobel[2], L Serrano[3] and V Mihaylov[1]

[1] Technical University of Varna, Department "Transport Engineering and Technologies", 9010 Varna, Bulgaria
[2] Wroclaw University of Science and technology, Department "Automotive Engineering", Wroclaw, Poland
[3] Polytechnic of Leiria, Department "Automotive Engineering", Leiria, Portugal

[4] E-mail: r_dimitrov@tu-varna.bg

Abstract: The paper presents a study of the variation of ignition advance when the engine works with methane and a comparison with values when the engine runs on gasoline. It is known that methane has a lower burning velocity than gasoline, and therefore, to obtain a maximal efficiency from the working process (the maximum value of cylinder pressure p_z should be 7-15 degrees after TDC), ignition advance should be increased, as for different operating regimes this angle should be increased with different values. A gasoline engine has been studied and were measured and analysed the ignition advance at work with gasoline and methane, with optimal air-fuel ratio (α) for both fuels. Three-dimensional graphics of ignition advance variation across the rpm and load range were made. In conclusion, on the basis of the experiments, a table is produced, which presents the values of coefficients for change of base parameters of ignition advance for all engine operating modes, in dependence on rpm and load of the engine.

1. Introduction

As it is known, methane has a lower burning velocity than gasoline [1, 2]. If a methane fuel system is added to a spark-ignition (SI) engine and to make it possible to use the maximum engine efficiency, it is necessary to change the angle of the electric spark in the direction of increase. Under standard conditions (1 bar of Absolute Pressure, temperature of 15 °C and air-fuel ratio of α = 0.8 and 1) the burning velocity according to the technical data is respectively Vg = 54.8 cm/s and Vg = 72.0 cm/s while with methane it is Vm = 49.4 cm/s and Vm = 65.8 cm/s [3]. Due to this characteristic of methane fuel, it is necessary to increase the angle of the electric spark. To obtain maximum efficiency from the combustion process (figure 1), the maximum pressure p_z is required to be 7÷15° after TDC (Top Dead Centre) [4-6]. Changing this angle through the action of a knock sensor is inapplicable because methane has an octane number of about RON 130 and is resistant to detonation burning. Most aftermarket advance processors that are used to change the ignition advance, can change the advance with a single fixed value for all engine operating modes with values typically of 3°, 6°, 9°, 12°, 15°, which under real operating conditions is a compromise option that improves performance of the engine but only in certain modes of operation [7-11]. However, there also exist advance processors that allow for changing the advance of ignition spark typically ±25° depending on the crankshaft speed. The purpose of the present study is to determine the necessary increase in the ignition advance of a SI engine retrofitted for methane operation at all speed and load regimes.

Content from this work may be used under the terms of the Creative Commons Attribution 3.0 licence. Any further distribution of this work must maintain attribution to the author(s) and the title of the work, journal citation and DOI.
Published under licence by IOP Publishing Ltd

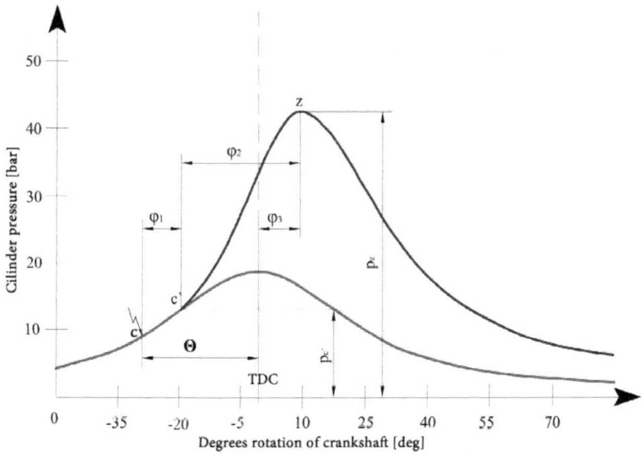

Figure 1. Base points of indicator diagram.

Figure 1 shows the main characteristics of the indicator diagram related to the influence on the combustion process. Point *c* is the moment where the electric spark is supplied. The period φ_1 (from point *c* to point *c'*) is the time for the formation of the pre-flame reaction into the cylinder of the engine. From point *c'* starts the combustion process in the cylinder. Point *z* is the top of the indicator chart corresponding to the maximum cylinder pressure. According to the theory of the internal combustion engine, there is maximum efficiency of the combustion process when p_z is obtained till 15° after Top Dead Centre (TDC) [5, 12, 13].

2. Experimental methodology

Experiments are made on SI engine with a volume of 1.275 litters, compression ratio ε = 9.75 and nominal power 46 kW at 5500 min^{-1}.

The Methane fuel system is conventional with a reducer-evaporator, where it is easy to change the air-fuel ratio of the mixture. The engine is mounted on an electric DC current dynamometer. Many regulated characteristics have been made, from where the optimal parameters of air-fuel ratio and ignition advance, required for maximum efficiency of the combustion process are determined. Also, for each measured point, an indicator diagram was also recorded using a piezo-quartz sensor of the Kistler brand. A schematic diagram of the laboratory unit is shown in figure 2.

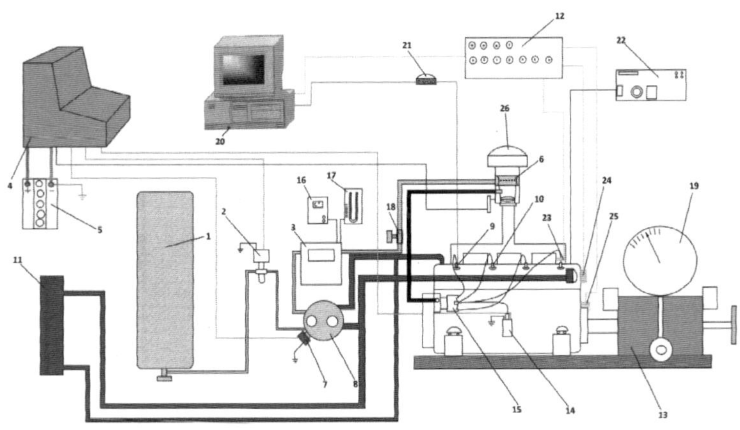

Figure 2. Experimental apparatus: 1 – methane bottle; 2 – magnetic valve; 3 – flow meter; 4 – control panel; 5 – battery; 6 – mixer; 7 – stop magnetic valve; 8 – reducer-evaporator; 9 – pressure sensor; 10 – spark plugs; 11 – cooling engine system; 12 – BNC adapter; 13 – dynamometer; 14 – ignition coil; 15 – distributor; 16 – thermometer; 17 – pressure gauge; 18 – gas regulator; 19 – brake force scale; 20 – PC system; 21 – amplifier; 22 – gas analyzer; 23 – ignition system sensor; 24 – hall effect sensor; 25 – hall effect sensor; 26 – air filter.

IOP Conf. Series: Materials Science and Engineering **664** (2019) 012020 doi:10.1088/1757-899X/664/1/012020

3. Experimental data and results

After processing and analysis of the collected experimental data, the best performance metrics for the methane engine are obtained with air-fuel ratio α = 1.00÷1.1, which is due to the type of fuel. The following figures show part of the depicted flowcharts when running the methane engine at different crankshaft speeds. Characteristic points used in the processing of the experimental tests are noted on the diagrams. The first channel of the oscillograms shows the variation of the cylinder pressure, the second channel shows the position of the TDC, the third and the fourth ones indicate spark ignition time information. Figure 3 shows an indicator diagram at 3540 min^{-1}, air-fuel ratio α = 1.00, fully open throttle, whereby the maximum pressure value is obtained at about 5° after TDC. Figure 4 shows an indicator diagram at 4530 min^{-1}, air-fuel ratio α = 1.10, a fully open throttle, wherein the angle of supply of the electric spark is about 50° before the TDC.

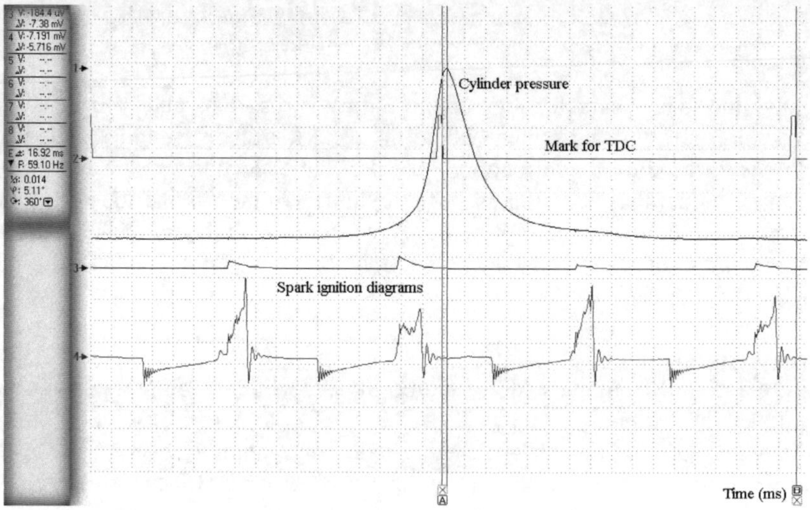

Figure 3. Indicator diagram at 3540 rpm.

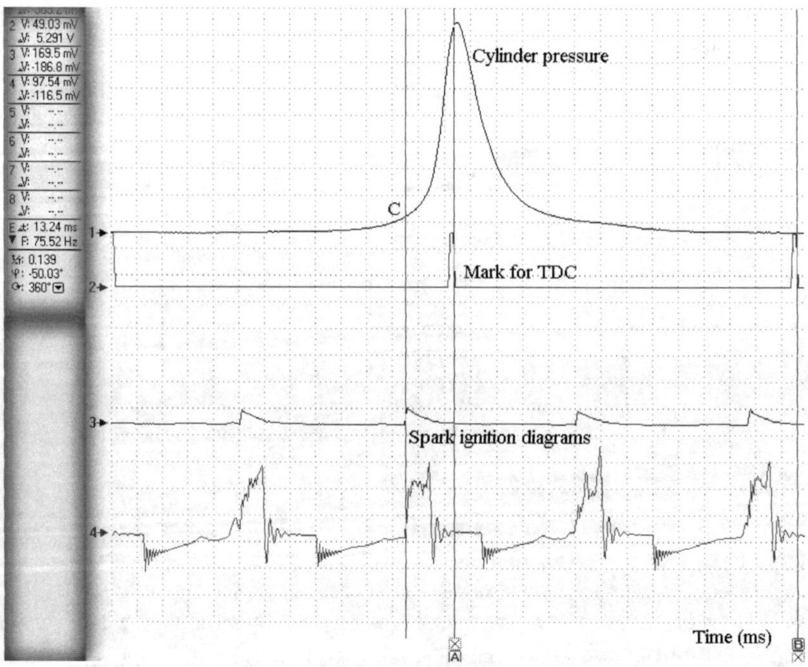

Figure 4. Indicator diagram at 4530 rpm.

After processing of the experimental results, values have been obtained for the necessary angle change of the electric spark over the entire operating speed and load range of the engine. Figures 5 and 6 show 3-dimensional diagrams for the ignition advance depending on the load and the crankshaft speed of the engine with petrol and methane respectively.

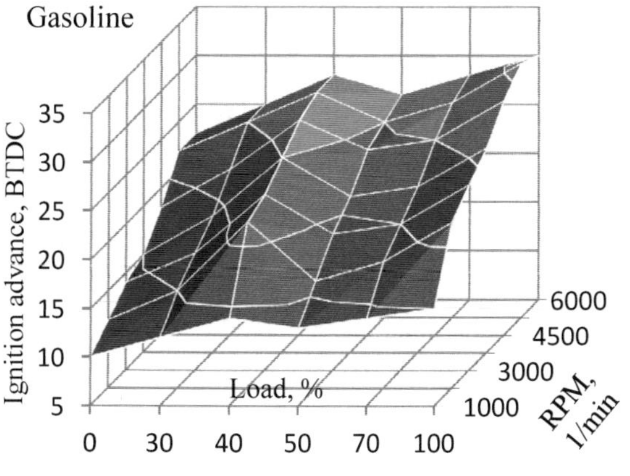

Figure 5. 3D diagram when the engine works with gasoline.

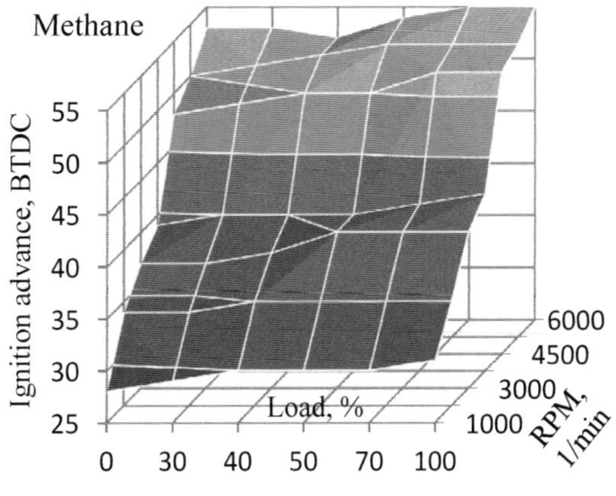

Figure 6. 3D diagram when engine works with methane.

When the engine works with gasoline, at 1000 rpm, the ignition advance changes from 10 to 15° before TDC, respectively, at a load from 0 to 100%. At 3500 rpm (maximum effective torque), the ignition advance changes from 15 to 26 degrees before TDC, respectively, at a load of 0 to 100%. At high rpm above 5000 min^{-1}, the ignition advance changes from 22 to 31° before TDC, respectively, at a load of 0 to 100%.

When the engine works with methane, at 1000 rpm, the ignition advance changes from 25 to 30° before TDC, respectively, at a load from 0 to 100%. At 3500 rpm (maximum effective torque), the ignition advance changes from 39 to 45° before TDC, respectively, at a load of 0 to 100%. At high rpm above 5000 min^{-1}, the ignition advance changes from 48 to 55° before TDC, respectively, at a load of 0 to 100%.

After analysing and processing the resulting 3D diagrams, a table (matrix 11×7) is produced giving the numerical values of coefficients with which it is necessary to increase the ignition advance of the electric spark during retrofitting and operation of a SI engine on methane. The values obtained are shown in table 1, where, depending on the crankshaft speed and the engine load, the coefficients for increasing the engine preload are given.

Table 1. Increase of ignition advance in degrees.

RPM/Load	0	10	20	30	40	50	60	70	80	90	100
1000	15	15	15	15	15	15	15	15	15	15	15
2000	20	20	20	20	20	20	20	20	15	15	15
3000	20	20	20	20	20	20	20	20	15	15	15
3500	25	25	25	25	25	20	20	20	20	20	20
4000	25	25	25	25	25	25	25	25	25	25	25
5000	30	30	30	30	30	30	30	30	25	25	25
6000	30	30	30	30	30	30	30	30	25	25	25

With a crankshaft rotation speed of 1000 min^{-1}, the required advance extension is 15° for the entire load range. With a crankshaft rotation speed of between 2000 to 4000 rpm, the required extension in advance is between 20÷25°, with the increase in engine load the coefficients decrease by 5°, due to better mixing conditions and improved thermo-dynamic engine parameters. At high crankshaft speeds, the coefficients take maximum values of up to 30° due to the reduced time for one duty cycle, and with a load of 80÷100% the ignition advance again has lower values.

4. Conclusion

The resulting 11×7 matrix allows us to use the engine with maximum efficiency over its entire speed and load range when the engine works with methane fuel.

The data obtained represent important and necessary information for adjusting and tuning the methane system to obtain maximum efficiency when converting a gasoline engine to work with methane.

The theory of maximizing engine efficiency makes it possible to obtain the values for the ignition advance coefficients at 5° without affecting the efficient engine performance.

Acknowledgments

This work received funding from the Technical University of Varna under project NP12 with a project leader Assoc. Prof. PhD Eng. R. Dimitrov.

References

[1] Petrakides S, Chen R, Gao D and Wei H 2016 Experimental study on stoichiometric laminar flame velocities and markstein lengths of methane and prf 95 dual fuels *Fuel* **182** 721–31

[2] Mannaa O, Mansour M S, Roberts W L and Chung S H 2015 Laminar burning velocities at elevated pressures for gasoline and gasoline surrogates associated with ron *Combustion and Flame* **162** 2311–21

[3] Petrakides S, Butcher D, Pezouvanis A and Chen R 2018 On the combustion of premixed gasoline-natural gas dual fuel blends in an optical SI engine *J. Power Technologies* **98** 387–95

[4] Beeckmann J, Röhl O and Peters N 2009 Numerical and experimental investigation of laminar burning velocities of iso-octane, ethanol and n-butanol *SAE Technical Paper*

[5] Heywood J 1988 *Internal Combustion Engine Fundamentals* (McGraw Hill) p 372-375

[6] Wattanavichien K and Azetsu A 2004 Studies of visualized diesohol combustion phenomena in IDI engine (CNG and Alternative Fuels, Oxygenated Fuels) *Proc. Int. Symp. diagnostics and modeling of combustion in internal combustion engines* **2004.6** 423–30

[7] Gou M, Detuncq B, Guernier C and St-Germain P 1990 Performance of a single cylinder engine fuelled by a mixture of natural gas and gasoline *SAE Technical Paper*

[8] Di Iorio S, Sementa P and Vaglieco B 2013 Experimental investigation of a methane-gasoline dual-fuel combustion in a small displacement optical engine *SAE Technical Paper*

[9] Di Iorio S, Sementa P, Vaglieco B and Catapano F 2014 An experimental investigation on combustion and engine performance and emissions of a methane-gasoline dual-fuel optical engine *SAE Technical Paper*

[10] Catapano F, Di Iorio S, Sementa P and Vaglieco B 2015 Experimental analysis of a gasoline pfi-methane di dual fuel and an air assisted combustion of a transparent small displacement SI engine *SAE Technical Paper*

[11] Iliev S 2018 *Comparison of Ethanol and Methanol Blending with Gasoline Using Engine Simulation* (In Biodiesel and Biofuels, IntechOpen)

[12] Jerzembeck S, Peters N, Pepiot-Desjardins P and Pitsch H 2009 Laminar burning velocities at high pressure for primary reference fuels and gasoline: Experimental and numerical investigation *Combustion and Flame* **156** 292–301

[13] Iliev S 2017 Investigation of n-butanol blending with gasoline using a 1-D engine model *3rd Int. Conf. Vehicle Technology and Intelligent Transport Systems* **1** 385–91

Flow characteristics of gas injectors

Zdravko Ivanov[1], Stoyan Stoyanov[1], Veselin Mihaylov[1,3] and Helder Santos[2]

[1]Technical University of Varna, Department "Transport Engineering and Technologies", 9010 Varna, Bulgaria
[2]Polytechnic of Leiria, Department "Automotive Engineering", Leiria, Portugal

[3] E-mail: v_mihaylov@tu-varna.bg

Abstract. The operation of Internal Combustion Engine (ICE) with gaseous fuels is characterized by the specificity of the fuelling process. Nowadays the most common system is the one in which after lowering its pressure the gas fuel is delivered to the engine intake manifold by means of gas injectors. This leads to a difference in the laws of fuel supply relative to the original gasoline injection system that dispenses fuel in liquid phase. Theoretical research has been done and gas leakage has been found to be critical under engine normal operating conditions, that is, it depends on fuel pressure and not on the environment in which the fuel leaks (intake manifold). An experimental test stand was set-up and multiparametric characteristics of a common construction bottom feed gas injectors were determined experimentally under conditions very close to the real working ones.

1. Introduction

The fuelling process for LPG and CNG engines has features related to the delivery of fuel in gaseous phase. The injection is carried out by electromagnetic injectors – figure 1. The movement of the needle of the injector is seen as a dynamic system involving gas, inertial, elastic and electromagnetic forces – their balance in time determines its position [1, 2]. As the pressure at the outlet of the injector is variable (the pressure in the intake manifold), to ensure the normal operation of the injector, especially constant opening time, it is necessary to guarantee a constant differential pressure upstream and downstream of the injector.

The main features of gas fuel as compared to liquid fuel are, on the one hand, gas compressibility and, on the other hand, gas leakage is critical, that is, at constant upstream pressure it depends only on that pressure and not on the pressure in the environment in which it flows.

The accessible literature lacks information on the characteristics of gas injectors under engine working conditions. In most cases the characteristics are given as volume flow rate with air, in nl/h, at upstream pressure of 2 bar(a) and downstream pressure of 1 bar(a), that is differential pressure of 1bar (in the paper "bar(a)" stands for unit of absolute pressure). The changing pressure at the intake manifold, which is around 300÷1000 mbar(a), during engine operation and the constant differential pressure, ensured by the reducer, leads to change of the upstream pressure at the injector and thus fuel mass flow changes eventually with change in the manifold pressure.

2. Theoretical background

2.1. Leakage of gas through an orifice

The process of ideal adiabatic gas leakage through a tapering orifice (such as the orifice used in the gas fuel injectors), coupled to a reservoir, depends on the ratio $\beta = P_2 / P_1$. The initial parameters – pressure,

Content from this work may be used under the terms of the Creative Commons Attribution 3.0 licence. Any further distribution of this work must maintain attribution to the author(s) and the title of the work, journal citation and DOI.
Published under licence by IOP Publishing Ltd

specific volume and temperature, of the upstream gas are P_1, v_1, T_1, and the downstream pressure is P_2. There is a value of this ratio β_c where the velocity of fluid flow is equal to the local sound velocity, and the pressure in the nozzle opening reaches P_c value, below which it cannot fall [3, 4]. This pressure is called critical pressure, and the leakage of gas through an orifice can either be subcritical, at $\beta > \beta_c$ or at most critical (choked) at $\beta \leq \beta_c$ (figure 2). This means that in a critical leak, the mass of the fluid passing through the orifice will only depend on pressure P_1 and will not depend on pressure P_2 of the environment in which it flows.

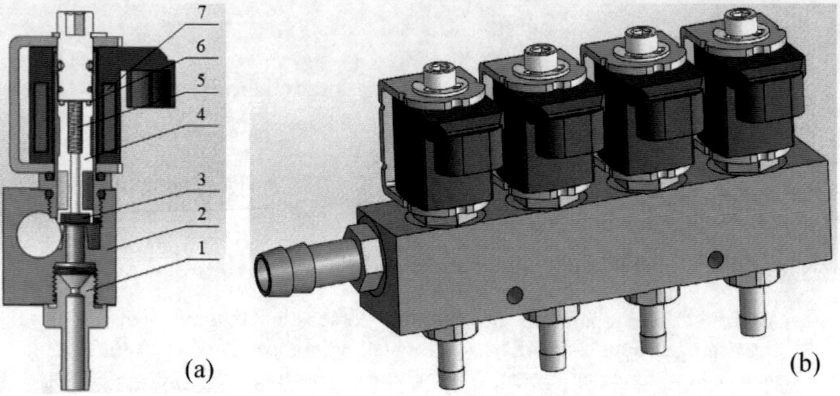

Figure 1. Section view of an injector (a) and injector assembly for 4 cylinder engine (b): 1 – Nipple with calibrated orifice; 2 – Fuel rail; 3 – Rubber insert; 4 – Needle; 5 – Spring; 6 – O-ring; 7 – Coil.

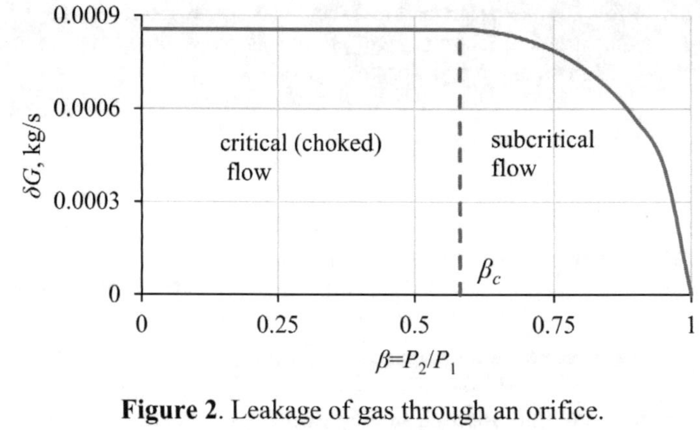

Figure 2. Leakage of gas through an orifice.

$$P_c = P_1\,\beta_c; \tag{1}$$

$$\beta_c = \left(\frac{2}{\kappa+1}\right)^{\frac{\kappa}{\kappa-1}}. \tag{2}$$

The ratio of critical pressure at the orifice exit $P_2 = P_c$ to the pressure in front of the nozzle P_1 is a constant value, depending only on the adiabatic process parameter κ, that is from the nature of the working fluid. For a mixture of 50% propane and 50% butane at 25 °C, $\kappa = 1.1136$, $\beta_c = 0.5818$.

For engine gaseous fuel delivery systems where the absolute pressure in the intake manifold is in the range of 0.3 ÷ 1.0 bar(a) and the difference with the fuel pressure is constant (about 1 bar), the ratio β is less than β_c and leakage through electromagnetic injector is always critical – figure 3.

In this case, the theoretical mass flow rate ∂G, kg/s, is determined by the following relation:

$$\partial G = A\,[\kappa\,\rho_1\,P_1\,\left(\frac{2}{\kappa+1}\right)^{\frac{\kappa+1}{\kappa-1}}]^{1/2}, \tag{3}$$

where ∂G is theoretical mass flow rate, kg/s; A – orifice cross-sectional area, m^2; P_1 – absolute upstream total pressure of the gas, Pa; ρ_1 – density of the gas at pressure P_1 and temperature T_1 , kg/m^3; κ – adiabatic index, $\kappa = \frac{c_p}{c_v}$, where c_p is specific heat of the gas at constant pressure and c_v is specific heat of the gas at constant volume.

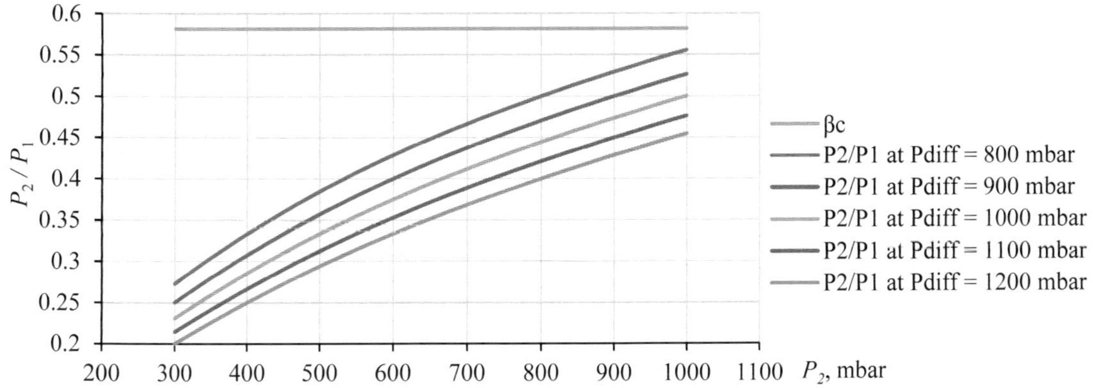

Figure 3. Ratio of P_2/P_1 at different settings of the differential pressure at the reducer.

Formula (3) refers to a theoretical leakage and does not take into account the influence of the valve construction. For this purpose, a discharge coefficient C_d is introduced which indicates the ratio of the actual mass fuel flow rate to the theoretically determined one. Thus the formula to estimate the flow rate becomes (4), where C_d – discharge coefficient:

$$\partial G = C_d\, A\, [k\, \rho_1\, P_1 \left(\frac{2}{k+1}\right)^{\frac{k+1}{k-1}}]^{1/2}. \tag{4}$$

The exact values of the discharge coefficient C_d can be determined experimentally.

2.2. *Flow simulations*

Computational fluid dynamics (CFD) is a branch of fluid mechanics that uses numerical analysis and data structures to analyse and solve problems that involve fluid flows. It can be used in various applications to predict the thermal behaviour, aerodynamic performance, the gas flow through or around a structure [5-7].

Simulations were conducted to study the effect of aerodynamic resistance of the construction on the gas flow. The results show that structural elements affect the leakage of the fuel. The pressure drop of the flowing fuel through the injector is determined by its internal aerodynamic drag and the presence of a calibrated orifice at the outlet nipple.

For the given bottom feed fuel injector design with stroke h = 0.45 mm, calibrated orifice diameter d = 2 mm, parametric simulations were performed at downstream pressure of 300 mbar to 1000 mbar abs., at differential pressure p_{diff} = 1 bar, figure 4 (the coil is not shown). The internal resistance of the injector is characterized by the transition area between points 1-2, figure 5. This drop in pressure is less than the drop in the pressure at transition area 2-3 at the exit of the injector.

The simulations performed show that for this design the main resistance of the injector is at the orifice at the injector outlet nipple.

3. Experimental set-up

To determine the actual characteristics, a non-engine test stand is used. It simulates the actual fuelling conditions in the injector. The stand makes it possible to explore the fuel delivery and to study the operation of the gas injector at different pressure values and at different pressures of the environment in which the fuel leak occurs [1, 8-12]. The stand is shown in figure 6.

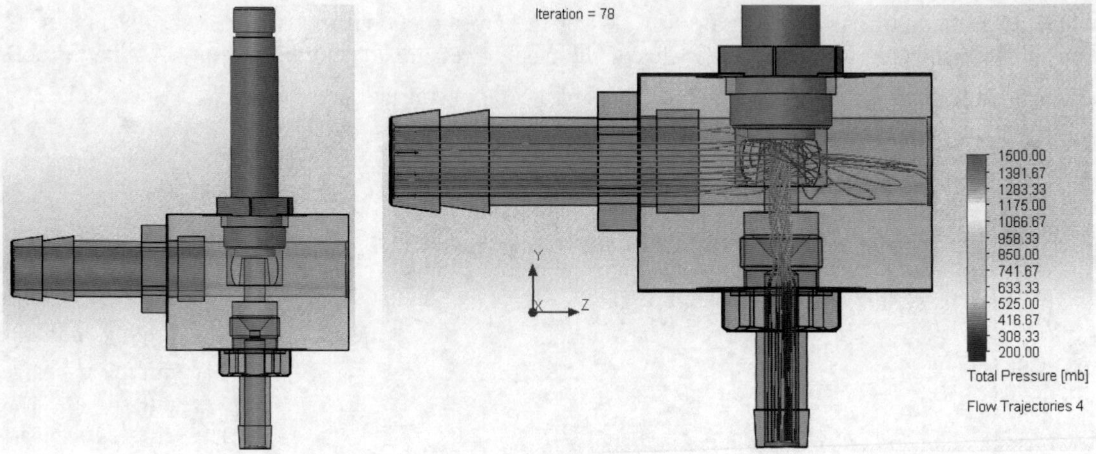

Figure 4. Flow trajectories and pressure distribution at boundary conditions – upstream pressure 1300 mbar and downstream pressure 300 mbar (absolute).

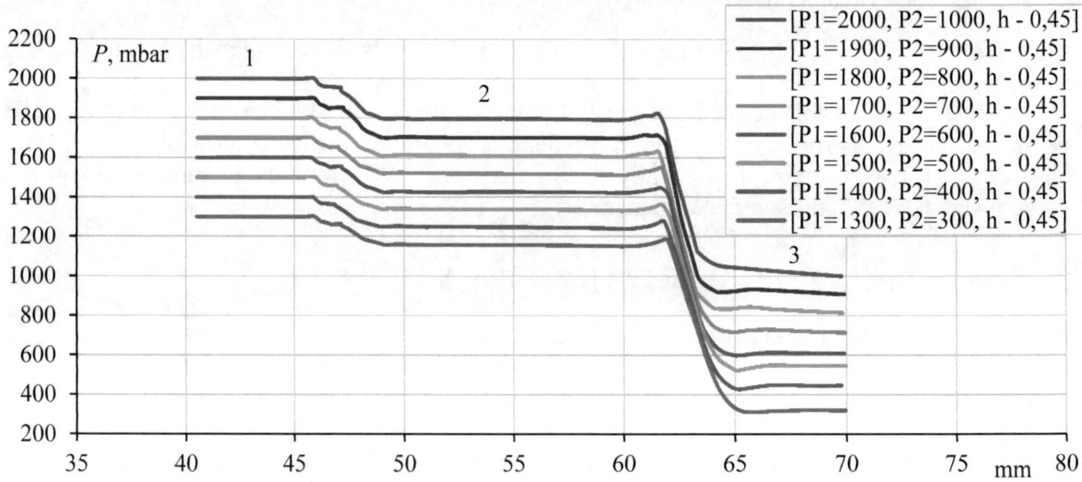

Figure 5. Pressure distribution along the fluid trajectory.

4. Results and discussions

The change of mass of injected fuel per cycle of electromagnetic injector of the discussed type is examined. The diameter of the calibrated orifice at the exit of the injector, which is defining for the fuel leakage, has a diameter $d = 2$ mm. At the inlet and outlet of the injector a constant differential pressure $p_{diff} = 1$ bar is maintained, provided by the reducer balancing system. The fuel temperature passing through the injector is maintained within the operating range of the engine.

The characteristics of the injected fuel per cycle q_c, mg, at different pressures in the engine intake manifold and for different duration of the control electric impulse are given in figure 7. In this case, the experimental results outlined in the graph are processed with first degree polynomials.

The results in figure 7 show the dependence of the injected fuel per cycle on the pressure at the intake manifold at constant duration of the control pulse. With a pulse duration of 10 ms, the injected fuel per cycle q_c increases from 9.4 mg at 300 mbar to 14.6 mg at 1000 mbar, representing an increase in the cyclic portion value of about 55%. Accordingly, for a duration of 30 ms, the increase is from 29.5 mg at 360 mbar to 46 mg at 1000 mbar, or 56% increase of fuel delivery for the same duration of the control signal. The change for intermediate values of the duration of the control pulse are shown in the graph.

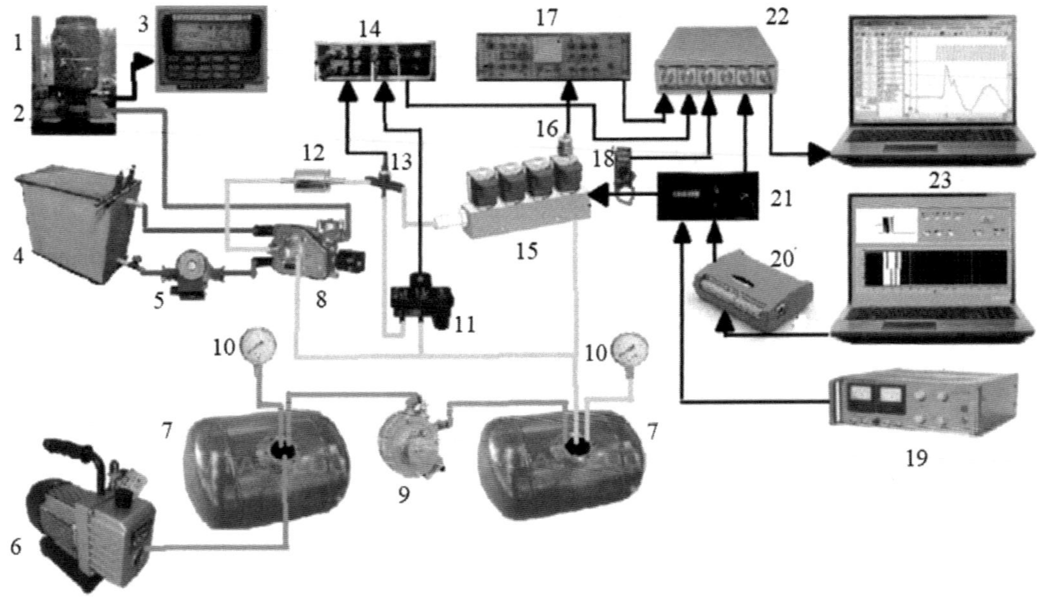

Figure 6. Experimental test stand: 1 – fuel tank; 2 – scale; 3 – scale controller with display; 4 – heated water tank; 5 – water pump; 6 – vacuum pump; 7 – tanks; 8 – gas reducer; 9 – modified reducer; 10 – pressure gauges; 11 – 2-Channel MAP pressure sensor; 12 – fuel filter; 13 – 3-Way nipple with temp. sensor; 14 – module for temperature and MAP sensors; 15 – injectors; 16 – accelerometer; 17 – chassis with charge amplifier module; 18 – current clamp; 19 – power supply; 20 – DAC for control signal generation; 21 – amplifier with counter; 22 – DAQ device; 23 – laptops.

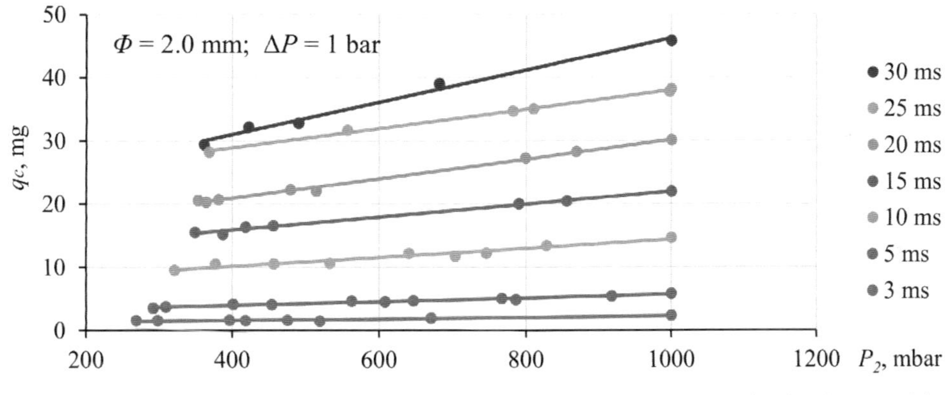

Figure 7. Mass of injected fuel per cycle vs downstream pressure (absolute), p_{diff}=1 bar, at different duration of electrical control signal.

The previous graph shows the full map of the engine operation depending on the duration of the control pulse and the pressure in the environment in which the gas is injected. In real conditions, the injector works in part of this map. Low engine loads are characterized by small values of manifold absolute pressure (MAP) and a small fuel portion per cycle determined by the corresponding duration of the control pulse. At high loads and at high mean effective pressure, ICE operates with an open throttle valve, which increases the absolute pressure in the intake manifold, and the fuel portion per cycle approximates the maximum values for the specific engine. Maximum loads are characterized by high peak values and an extended period of control pulse. For areas of the injector operation map, characterized by low pressures and long pulse duration and high pressure and short pulse values, the magnitude of the fuel portion per cycle for these modes is irrelevant or the gas injectors are excluded from the engine management for environmental considerations. The zone of possible operation of the gas injectors in real conditions is shown in figure 8 and is indicated by a blue colour.

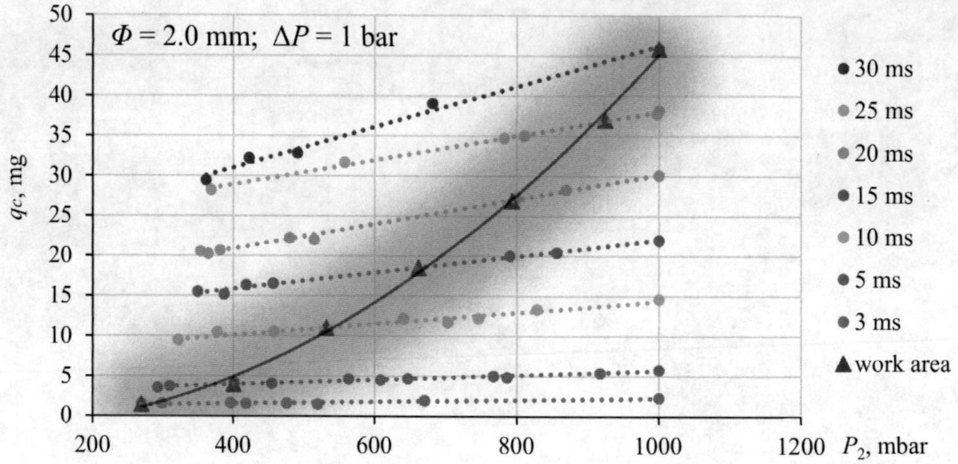

Figure 8. Working area of injector during normal engine operation.

Although the delivery of fuel with gas injectors is a non-continuous pulsed process, if we divide the injected fuel per cycle by the duration of the control signal, we could receive the mass air flow. Figure 9 shows the results for injection duration of 15 ms. It can be seen, that there is good correspondence between the experimentally determined points with the results from simulation.

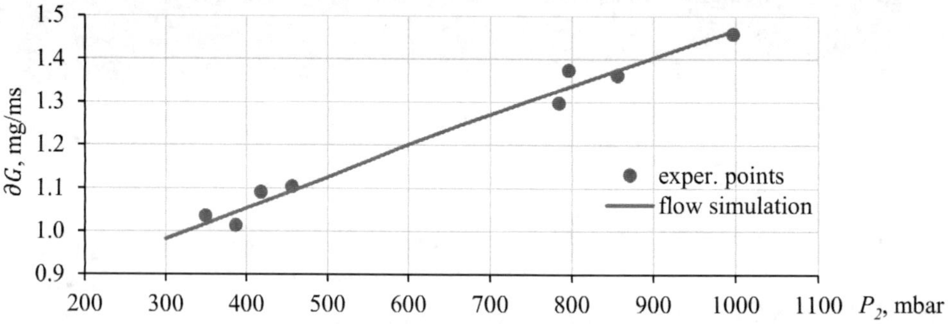

Figure 9. Mass flow rate.

5. Conclusion

Theoretical research has been done and gas leakage has been found to be critical under engine normal operating conditions.

Multi-parameter characteristics of a bottom feed type gas injectors are experimentally determined, that show the dependence of the duration of the electrical control signal on the injected fuel per cycle at varying pressure at the intake manifold and constant differential pressure.

It can be concluded that at increasing load (increasing intake manifold pressure), the proportional coefficient in the software of the gas control unit, that is used to relate the fuel flow of the gasoline and gas injectors should be corrected in negative direction.

References

[1] Ivanov Z, Mihaylov V and Mersinkov H 2014 Optical method for measurement needle lift of electromagnetic valves for LPG/CNG injection *XX Proc. Tech. Sci. Conf. ECO Varna 2014* 415-22

[2] Mitukiewicz G, Dychto R and Leyko J 2015 Relationship between LPG fuel and gasoline injection duration for gasoline direct injection engines *Fuel* **153** 526-34
[3] Potter M, Wiggert D and Ramadan B 2010 *Mechanics of Fluids* 5th ed. (Cengage) chapter 9
[4] Miller R 1996 *Flow Measurement Engineering Handbook* 3rd ed (McGraw-Hill)
[5] Matsson J 2018 *An Introduction to SolidWorks Flow Simulation* (SDC Publications) chapter 6, 9
[6] Szpica D 2018 Investigations of the flow characteristics of low-pressure gas-phase injectors under the conditions of numerical simulation *Proc. 23rd Int. Sci. Conf. Mechanika 2018* 164-68
[7] Zlateva P, Yordanov K and Dimitrov D 2014 Numerical simulation of temperature separation phenomena inside vortex tube *Acta Technica Corvininesis-Bulletin of Engineering* **7(4)** 23-26
[8] Ivanov Z and Mihaylov V 2012 Vibrational state of gas injection valves for spark-ignited (SI) engines *XVIII Proc. Tech. Sci. Conf. ECO Varna 2012* 502-10
[9] Dimitrov R, Magdziak – Tokłowicz M, Sitnik L, Reksa M, Tkaczyk M and Wróbel R 2014 A comparison of fuel consumption between the new European driving cycle test and the natural operation of a vehicle *22nd Int. Sci. Tech. Conf. Trans&Motauto* 14–16
[10] Belchev S 2019 Method for determining the parameters of injection of common rail injectors *XXVI Proc. Tech. Sci. Conf. ECO Varna 2019*
[11] Szpica D 2016 The influence of selected adjustment parameters on the operation of LPG vapor phase pulse injectors *J. Natural Gas Science and Engineering* **34** 1127–36
[12] Majerczyk A and Radzimirski S 2012 Effect of LPG gas fuel injectors on the properties of low emission vehicles *J. KONES Powertrain and Transport* **19** 401–10

IOP Conf. Series: Materials Science and Engineering **664** (2019) 012022 doi:10.1088/1757-899X/664/1/012022

Economic aspects of using hydrogen in compression ignition engine operating on gas-diesel cycle

E Dimitrov, S Pantchev, Ph Michaylov and M Peychev[1]

Department of Combustion Engines, Automobile Engineering and Transport, Technical University – Sofia, 8 Kliment Ohridski Blvd., 1756 Sofia, Bulgaria

[1] E-mail: m_peychev@tu-sofia.bg

Abstract. The article gives an overview of the economic aspects of a single-cylinder internal combustion engine dual fuel operation with diesel fuel and hydrogen. The engine operating costs are calculated. Two separate cases are investigated. The first case is at constant engine speed – 1500 rpm, constant engine power (1.5 kW) and variable hydrogen mass fraction in the total fuel mass (from 0 to 50%) and the second case is at constant engine speed – (1500 and 2000 rpm), constant hydrogen mass fraction in the total fuel mass (10%) and variable engine power (1 to 4 kW). The following parameters (as a function of the corresponding variable) are determined for both cases: expenses for hydrogen and diesel fuel (in €/hour), price of the energy obtained at the engine crankshaft (in €/kWh), expenses for the CO_2 emissions emitted by the engine (in €/hour). The weighted average of the fuel mixture price (in €/kg) is also determined for the first case. The calculations are made using average fuel prices in Germany for April 2019. A predominant growth in the operating costs is observed as a function of the growing hydrogen mass fraction and increasing engine power. A significant reduction in the price of CO_2 emissions at constant engine power and growing hydrogen mass fraction is observed due to the lower exhaust gas toxicity.

1. Introduction

Hydrogen is among the purest fuels mankind knows. During hydrogen oxidation, only water is produced. Therefore, the harmful emissions of an internal combustion (IC) engine should tend to zero if hydrogen is used as a fuel. In practice, this is not the case because the higher in-cylinder temperatures caused by the hydrogen combustion create conditions for nitrogen oxide formation. There is a lot of research dedicated to this problem proposing a wide range of solutions such as urea injection in the engine exhaust, water injection in the engine intake or innovative methods for exhaust gas recirculation [1-3]. However, the use of hydrogen as a fuel has a number of benefits, such as reduction of the exhaust gas emissions, reduction of the fuel consumption, efficiency increase and thus environment protection. If used as an auxiliary fuel in compression ignition engines, hydrogen has a major influence on exhaust gas toxicity – greatly reduces the opacity and carbon oxide content [4-5]. The main obstacle to the mass use of hydrogen as a fuel for ICE is its high price. In this relation, the current study represents an analysis of the operating costs of a diesel engine working on a gas-diesel cycle with hydrogen.

The term "gas-diesel cycle" is used to describe the operation of a compression ignition engine with two fuels simultaneously – diesel fuel and gaseous fuel, hydrogen in the current case. A small amount of diesel fuel, which self-ignites and then ignites the gaseous fuel, is injected into the cylinder. The majority of heat during gas-diesel cycle operation is normally produced by the gaseous fuel. It may be injected

Content from this work may be used under the terms of the Creative Commons Attribution 3.0 licence. Any further distribution of this work must maintain attribution to the author(s) and the title of the work, journal citation and DOI.
Published under licence by IOP Publishing Ltd

directly into the cylinder or in the intake manifold and then enters in the cylinder during the intake process. For the aim of the current study the gas-diesel cycle is accomplished via external mixture formation of the gaseous fuel (hydrogen injected in the intake manifold). In order to determine the quantity of hydrogen in fuel, a coefficient K_{H2} is defined. It represents the ratio between the hydrogen fuel mass and the total fuel mass (hydrogen and diesel fuel). The operating costs calculations are performed for variable values of this coefficient and steady operation modes – constant engine speed and load. Calculations are also made for a constant K_{H2} coefficient, constant engine speed and variable engine loads. Such results can be useful for future economic assessments of the implementation of hydrogen as a fuel for mass use in road vehicles and industrial capacities. An economic assessment is of significant importance for the industry and the sector of services in the context of constant reduction of the available resources and rise of prices if keeping the same demand level.

The high price of hydrogen presumes a negative economic assessment for the operational costs of a compression ignition engine working on diesel-hydrogen cycle. Nevertheless, an approximate calculation of the operational costs of a compression ignition engine running on gas-diesel cycle deserves attention due to the abovementioned benefits of using hydrogen as fuel. Such calculation would be of use to engine, road vehicles and hydrogen manufacturers.

2. Purpose of the study

After taking into account the previous analysis, we can summarize that the purpose of the study is to define the monetary value of the operational costs of a compression ignition engine running on hydrogen-diesel cycle.

3. Methodology of the research

Some of the data used in the current article is gathered for the purpose of another research [6] conducted by the same authors and concerning the influence of hydrogen on a compression ignition engine performance. The experiments are conducted on a test bench which consists of a compression ignition engine and a direct-current dynamometer coupled via a common shaft. The direct-current dynamometer is used to regulate the engine load. Engine parameters are shown in table 1.

Table 1. Parameters of a compression ignition engine DV 550.

Parameter	Value	Dimension
Displacement volume	550	cm^3
Bore	91.5	mm
Stroke	85	mm
Number of cylinders	1	-
Number of valves per cylinder	2	-
Compression ratio	17.5	-

Both fuels – diesel fuel and hydrogen, are injected directly into the cylinder and in the intake manifold, respectively, and the hydrogen-diesel ratio is varied during the experiment. Fuel consumption for both fuels, exhaust gas toxicity and in-cylinder pressure are measured. The experimental setup and additional information for the referring experiment can be found in the corresponding article.

In order to fulfil the purpose of the study it is necessary to make a calculation of the engine operating expenses for gas-diesel cycle: fuel expenses (diesel fuel and hydrogen), expenses for engine maintenance – oil, filters etc. and expenses for the toxic components of the exhaust gases (if applicable). These expenses should be calculated in two separate cases. In the first case – as a function of the hydrogen mass share in the total fuel amount with engine running at constant speed and load. In the second case – as a function of the engine load with engine running at constant speed and constant hydrogen mass share in the total fuel amount. A 10% mass share of hydrogen is chosen because this is the maximum achievable share on all engine modes without engine design modifications. This is due to reaching dangerously high values of the in-cylinder temperature and pressure rise rate with the increase of hydrogen content in the fuel, especially with engine running under high loads [7].

The expenses for engine maintenance should not change as a function of the hydrogen mass share in the total fuel amount. That is because under a constant load and engine speed, the oil, filters – oil filter, air filter, fuel filter, etc., do not wear with different intensity as a function of the hydrogen quantity in the total fuel. It is possible for oil to keep its finest properties longer due to the lowered opacity in blown by gases and for the diesel fuel filter to have a longer life due the lower amount of diesel fuel passing through. Such study would require extended engine tests and therefore the maintenance costs are not taken into account for the calculation of engine operating costs.

The diesel fuel expenses P_{DF} can be calculated using the following equation:

$$P_{DF} = p_{DF} B_{DF}, \text{€/h}, \quad (1)$$

where p_{DF} is diesel fuel price, €/kg; B_{DF} – diesel fuel consumption, kg/h.

Similarly, for the hydrogen fuel expense P_{H2}, we have:

$$P_{H2} = p_{H2} B_{H2}, \text{€/h}, \quad (2)$$

where p_{H2} is hydrogen fuel price, €/kg; B_{H2} – hydrogen fuel consumption, kg/h.

The total fuel expenses P_F can be calculated using (1) and (2) with the equation:

$$P_F = P_{DF} + P_{H2}, \text{€/h}. \quad (3)$$

The expenses thus calculated vary as a function of the hydrogen-diesel fuel mass ratio exact numerical value information for which we can get from the K_{H2} coefficient given with the equation:

$$K_{H2} = \frac{B_{H2}}{B_F} 100, \%, \quad (4)$$

where B_F is total engine fuel consumption:

$$B_F = B_{DF} + B_{H2}, \text{kg/h}. \quad (5)$$

In this way, we obtain the function: $P_F = f\,(K_{H2})$.

If we divide the total fuel expenses by the total fuel mass which has entered in the engine we obtain the weighted average price P_{FA} per kilogram of fuel mixture:

$$P_{FA} = \frac{P_F}{B_F} = \frac{P_{DF}+P_{H2}}{B_{DF}+B_{H2}} = \frac{p_{H2}B_{H2}+p_{DF}B_{DF}}{B_{DF}+B_{H2}}, \text{€/kg}. \quad (6)$$

The thus calculated price for a kilogram of fuel mixture gives a possibility to determine the expenses for the mechanical energy produced by the engine. The price of electricity is usually measured in €/kWh. Therefore in order to calculate the price of energy produced by the engine we may use an indicator determined from the multiplication of the weighted average price per kilogram of fuel mixture P_{FA} and the engine specific fuel consumption:

$$P_{kW} = P_{FA}\, g_e, \text{€/kWh}, \quad (7)$$

where g_e is brake specific fuel consumption, kg/kWh.

However, if calculated this way, the price accounts for the entire energy acquired from both fuels (total energy price). In order to measure the share of energy expenses for each fuel, the P_{kW} parameter (total energy price) should be weighted by the ratio of power output produced from the corresponding fuel to the total engine power output produced with the whole fuel mixture. Two indicators are derived – for the price of 1kWh of energy, acquired from hydrogen combustion (or hydrogen energy price) P_{kWH2} and for the price of 1kWh of energy acquired from the diesel fuel combustion (or diesel fuel energy price) P_{kWDF}. In order to do that, it is necessary to calculate the output power for both fuels. This is a complex task because engine power is usually calculated for a single type of fuel.

Nevertheless, with the use of some assumptions and simplifications the following formula for engine power [8] can be adapted:

$$Ne = \frac{i\, V_h\, H_{LF}\, \rho_0}{30\, \tau\, l_0}\, n\, \frac{1}{\alpha}\, \eta_v\, \eta_i\, \eta_m \text{ , kW,} \quad (8)$$

where i is number of cylinders; V_h – displacement volume per one cylinder, m^3; H_{LF} – lower heating value of fuel, kJ/kg; ρ_0 – air-fuel mixture density, kg/m^3; τ – parameter for the engine type (2-stroke or 4-stroke); l_0 – theoretical amount of air necessary for the complete combustion of 1 kg of fuel, kg/kg.; n – engine speed, min^{-1}; α– air-fuel equivalence ratio; η_v – volumetric efficiency; η_i – indicated efficiency; η_m – mechanical efficiency.

The values of parameters i, V_h, τ, ρ_0, n, η_v, η_i и η_m can be assumed constant. The number of cylinders, engine displacement volume and the τ parameter are physical parameters which cannot be changed. Engine speed is constant for every engine mode investigated in the current study. The volumetric, indicated and mechanical efficiencies can also be assumed constant for a specified engine mode (constant engine power, speed and constant hydrogen-diesel fuel ratio). With enough precision, the air-fuel mixture density can be assumed constant for the specified engine mode and the corresponding hydrogen-diesel ratio. The formula for brake engine power could hence be simplified to:

$$N_e = C\frac{H_{LF}}{l_0}\frac{1}{\alpha}, \quad (9)$$

where $C = \frac{i\, V_h\, \rho_0}{30\, \tau}\, n\, \eta_v\, \eta_i\, \eta_m = \text{const.}$

If we determine the air equivalence ratio – α using the equation:

$$\alpha = \frac{B_a}{B_F\, l_0}, \quad (10)$$

where B_a is engine air consumption, kg/h, for N_e, we have:

$$N_e = C\frac{H_{LF}}{l_0}\frac{B_F\, l_0}{B_a} = CH_{LF}\frac{B_F}{B_a}. \quad (11)$$

For a single operating mode, the air consumption is constant and the brake power can be simplified to:

$$N_e = C_1 H_{LF} B_F, \quad (12)$$

where: $C_1 = \frac{C}{B_a}$.

So the engine power is proportional to the heat produced by the fuel:

$$N_e = C_1 H_{LF} B_F = C_1 Q_F, \quad (13)$$

where Q_F is heat produced by fuel, kJ/h.

Then for the power output produced by the hydrogen combustion:

$$N_{eH2} = C_1 H_{LH2} B_{H2} = C_1 Q_{H2}, \quad (14)$$

where H_{LH2} is lower heating value of hydrogen.

Similarly, for the power output produced by the diesel fuel combustion and for the total power output:

$$N_{eDF} = C_1 H_{LDF} B_{DF} = C_1 Q_{DF}; \quad (15)$$

$$N_{e\Sigma} = N_e = C_1 H_{LF} B_F = C_1 Q_F, \quad (16)$$

where: H_{LDF} – lower heating value of diesel fuel.

The lower heating value of the hydrogen-diesel fuel mixture H_{LF} is calculated as a weighted average of hydrogen and diesel fuel lower heating values, weighted with the corresponding fuel consumption (kg/h).

The price of energy produced by the hydrogen P_{kWH2} and the diesel fuel P_{kWDF} can then be calculated using equations (7), (14), (15) and (16):

$$P_{kWH2} = P_{kW}\frac{N_{eH2}}{N_{e\Sigma}} = P_{kW}\frac{Q_{H2}}{Q_F} = P_{kW}\frac{H_{LH2}.B_{H2}}{H_{LF}.B_F}, \text{€/kWh}; \quad (17)$$

$$P_{kWDF} = P_{kW}\frac{N_{eDF}}{N_{e\Sigma}} = P_{kW}\frac{Q_{DF}}{Q_F} = P_{kW}\frac{H_{LDF}\,B_{DF}}{H_{LF}\,B_F}, \text{€/kWh}. \quad (18)$$

The issue of the carbon emissions (CO_2) is particularly relevant. Hence, an assessment of the monetary value of carbon emissions emitted by the engine is appropriate. For this purpose it is necessary to calculate the mass of carbon emissions in the engine exhaust gases (G_{CO2}, kg/h). If multiplied by the price of carbon emissions futures (€/ton, converted to €/kg) traded on the international stock exchanges, this gives us the expenses for the CO_2 emissions in €/h.

$$P_{CO2} = \frac{p_{CO2}\,G_{CO2}}{1000}, \text{€/h}, \quad (19)$$

where p_{CO2} is price of CO_2 emissions, €/ton; G_{CO2} – quantity of CO_2 emissions emitted by the engine, kg/h.

4. Experimental results

An advantage of the methodology shown above is that the data from the test bench experiment could be used to make an economic assessment of the engine operation worldwide, depending on the resource price in the corresponding region. The calculations in the current article are based on fuel prices in Germany. The main reason for this approach is that hydrogen is not a fuel of mass consumption. The traded volumes of this gas are small, supply and demand for this market are of much lower volumes than the ones for the diesel fuel market. Therefore, it is not likely to find reliable information about hydrogen prices [9]. Due to scarcity of public data, the hydrogen end-customer price data for Germany is retrieved from the German website of the major fuel producer and distributor Shell Oil (https://www.shell.de/) as well as the online service H2-Mobility (https://h2.live/), a commercial webservice by AIR LIQUIDE Deutschland GmbH, OMV Deutschland GmbH, Shell Deutschland Oil GmbH, TOTAL Deutschland GmbH, Linde AG, and Daimler AG. According to the data, the average price per kilogram of hydrogen for Germany is 9.5 €/kg. Hydrogen price in hydrogen stations in USA is similar to the one in Germany, fluctuating between 12 and 16 $/kg [10-11]. Data for the hydrogen price from documents for purchases of gas storage cylinders for the experiment is also at authors' disposal (Bulgaria). In this case, the price for a kilogram of hydrogen is 127.82 €/kg. The price is correct and reliable, does not fluctuate like prices of mass consumption fuels. Yet, the application of IC engines is mostly in road vehicles so it is appropriate to use the price of hydrogen at hydrogen stations for the calculations. Apart from that, the price for hydrogen in gas storage cylinders in Bulgaria is several times higher than the one at hydrogen stations in Germany and that would twist the results of the study.

In order to eliminate the impact of short-term fluctuations in the price of diesel, its average price for Germany for April 2019 is taken. The fuel end-customer price is retrieved from the Shell Oil website for Germany (https://www.shell.de/). According to the information acquired, the average price of diesel fuel in Germany for April 2019 is 1.2975 €/l. The measured fuel density during the experiments is 0.8233 kg/l. After dividing the price per litre of diesel fuel by the fuel density, the price per kilogram of diesel fuel is obtained: 1.576 €/kg. The fuel prices noted above are used in order to calculate the total fuel expenses (P_F), the diesel (P_{DF}) and hydrogen (P_{H2}) fuel expenses. Results for the fuel expenses are shown in figures 1, 5 and 6. The weighted average fuel price (P_{FA}) as a function of the K_{H2} coefficient is given in figure 4.

Historic data for the price of carbon emission futures is gathered from the major financial portal Investing.com (https://www.investing.com/). It is also taken as a weighted average for the month of April 2019, again to eliminate the impact of the financial instrument short term volatility. The cost is

24.26 €/ton (weighted average). The price is valid not only for Germany, as carbon emissions are traded on the international stock exchanges. The weighted average price of carbon emissions in August 2018 (the period when the experiment was carried out) was 19.62 €/ton. To illustrate the upward trend in carbon emissions price, calculations were made with both prices. Still, carbon prices are several times lower than fuel prices at the current time. Calculation results for the CO_2 emission expenses (P_{CO2}) are shown in figures 3, 9 and 10.

As stated above, the operating expenses will be calculated for two separate cases. First case – as a function of variable K_{H2} coefficient with engine running at constant speed and load (1500 min^{-1}, 1.5kW power output). Second case – as a function of variable engine load (power output from 1 to 4kW), with engine running at constant speed (1500 and 2000 min^{-1}) and constant K_{H2} coefficient (10%).

The price for the entire energy P_{kW} (total energy price) obtained at the engine crankshaft is compared to the price of electrical energy in Germany (P_E) for the month of April 2019 – 0.30 €/kWh. Results for the energy prices (P_{kW}, P_{kWH2}, P_{kWDF} and P_E) are shown in figures 2, 7 and 8.

4.1. Calculations at variable K_{H2} coefficient (0-50%), constant engine load (power output 1.5 kW) and speed (1500 min^{-1}).

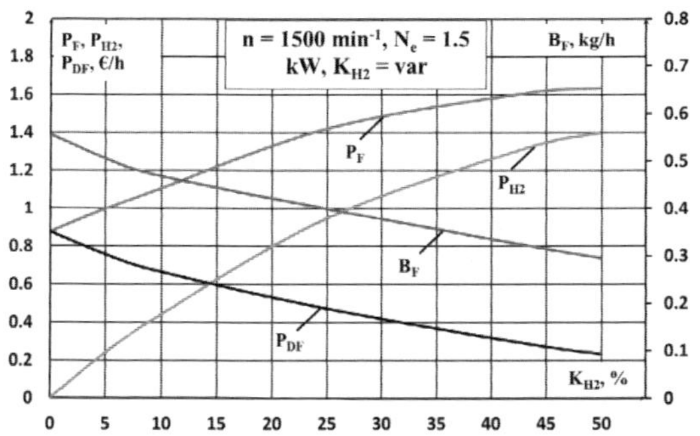

Figure 1. Variation of total fuel expenses, diesel fuel and hydrogen fuel expenses and total fuel consumption as a function of the K_{H2} coefficient.

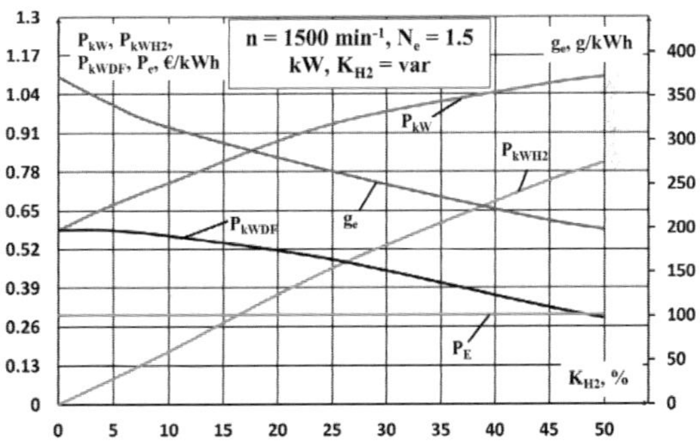

Figure 2. Variation of total energy price, diesel fuel energy price, hydrogen energy price and brake specific fuel consumption as a function of the K_{H2} coefficient; the price of electric energy in Germany for April 2019 is shown.

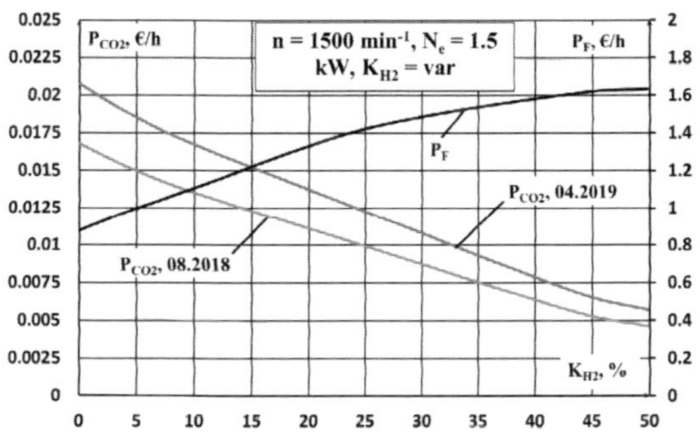

Figure 3. Variation of the expenses for CO_2 emissions at emission prices for August 2018 and April 2019 and the total fuel expenses as a function of the K_{H2} coefficient.

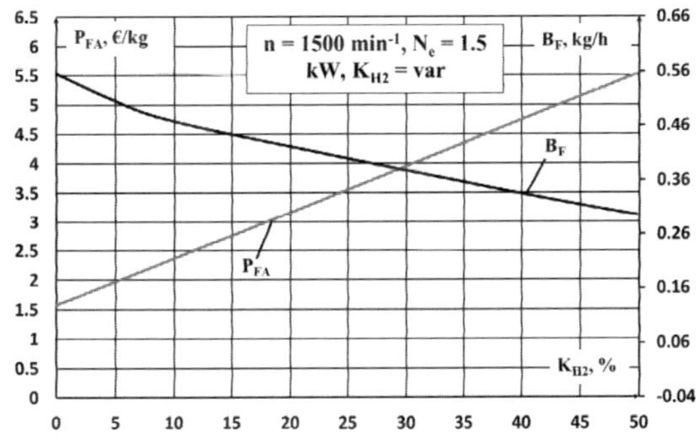

Figure 4. Variation of the weighted average fuel price and the total fuel consumption as a function of the K_{H2} coefficient.

IOP Conf. Series: Materials Science and Engineering **664** (2019) 012022 doi:10.1088/1757-899X/664/1/012022

4.2. Calculations at variable engine load (power output 1-4kW), constant engine speed (1500 min^{-1} and 2000 min^{-1}) and constant K_{H2} coefficient (10%).

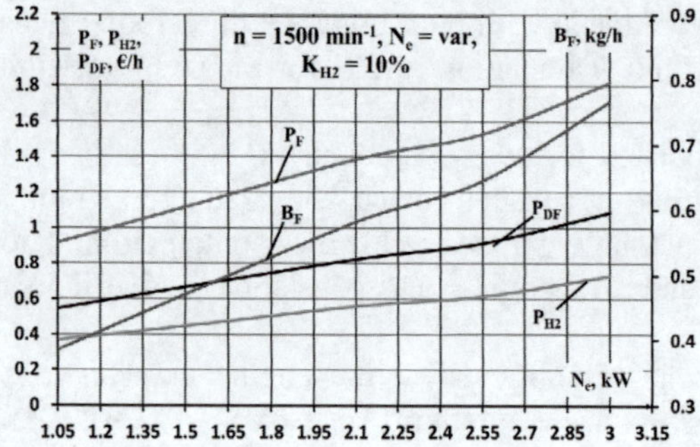

Figure 5. Variation of total fuel expenses, diesel fuel and hydrogen fuel expenses and total fuel consumption as a function of the engine load at engine speed of 1500 min^{-1}.

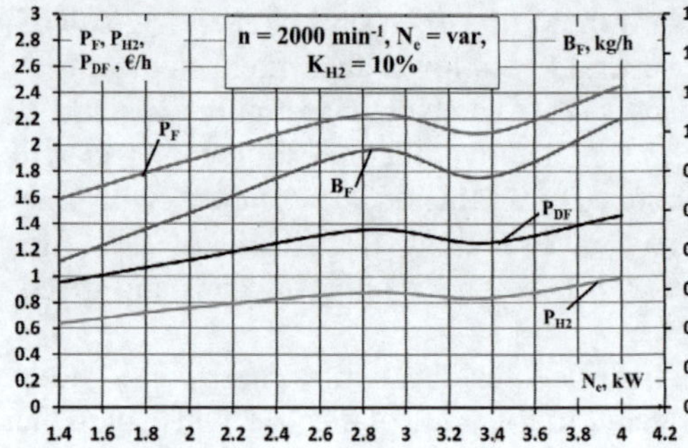

Figure 6. Variation of total fuel expenses, diesel fuel and hydrogen fuel expenses and total fuel consumption as a function of the engine load at engine speed of 2000 min^{-1}.

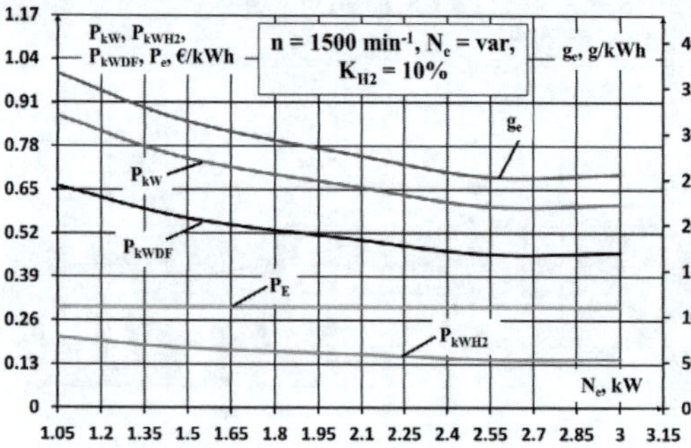

Figure 7. Variation of total energy price, diesel fuel energy price, hydrogen energy price and brake specific fuel consumption as a function of the engine load at speed of 1500 min^{-1}; the electric energy price in Germany for April 2019 is shown.

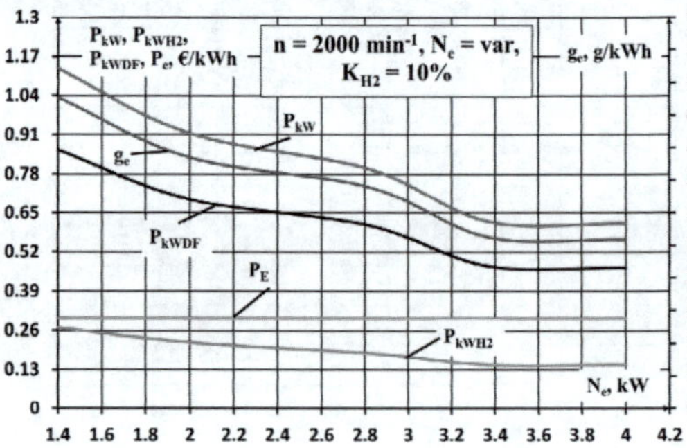

Figure 8. Variation of total energy price, diesel fuel energy price, hydrogen energy price and brake specific fuel consumption as a function of the engine load at speed of 2000 min^{-1}; the electric energy price in Germany for April 2019 is shown.

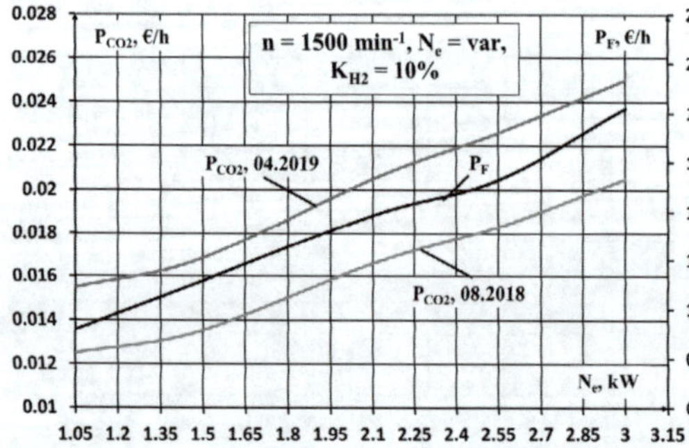

Figure 9. Variation of the expenses for CO_2 emissions at emission prices for August 2018 and April 2019 and the total fuel expenses as a function of the engine load at speed of 1500 min^{-1}.

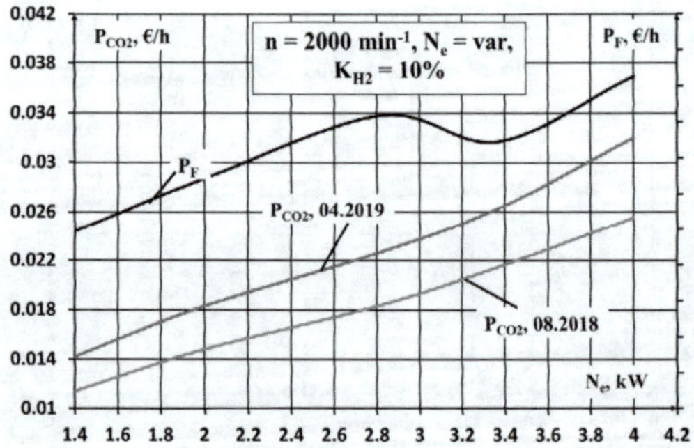

Figure 10. Variation of the expenses for CO_2 emissions at emission prices for August 2018 and April 2019 and the total fuel expenses as a function of the engine load at speed of 2000 min^{-1}.

5. Conclusions of the study

After performing an analysis of the gathered results, the following conclusions can be made:

- The total fuel consumption at a steady engine mode (constant speed and load) drops with the increase of the K_{H2} coefficient values due to the higher lower heating value of hydrogen. The curve of the total fuel expenses however rises due to the high price of hydrogen;
- The expenses for diesel fuel have an insignificant share in the total fuel expenses due to the much lower price of diesel fuel compared to the hydrogen price. The expenses for diesel fuel drop while the ones for hydrogen rise with the increase of the K_{H2} coefficient;
- The total energy price rises with the increase of the K_{H2} coefficient despite the fact that the brake specific fuel consumption drops. This is again due to the higher price of hydrogen. The diesel and hydrogen energy prices are equal for $K_{H2} = 26\%$;
- The expenses for CO_2 emissions drop almost 4 times with the increase of the K_{H2} coefficient from 0 to 50%. The rise of the price of CO_2 emission futures, traded at the international stock exchanges from August 2018 to April 2019, is significant and leads to the augmentation of the CO_2 emissions expenses;
- The weighted average fuel price, calculated in €/kg, understandably rises with the increase of the K_{H2} coefficient;
- Naturally, with constant values of the K_{H2} coefficient and constant engine speed, the fuel consumption rises with the increase of engine load. The curve of the total fuel expenses repeats the curve of the total fuel consumption;
- With the increase of engine load at constant values of the K_{H2} coefficient and at constant engine speed, the brake specific fuel consumption drops. As K_{H2} is *10%* at all engine loads, the weighted average fuel price and the weighted average lower heating value of fuel remain unchanged. Therefore, with the drop of the brake specific fuel consumption, the total energy price for the energy obtained from the engine drops;
- With the increase of engine load, the total fuel consumption and the amount of carbon emissions rise, hence the expenses for carbon emissions.

As a conclusion we can note that at this stage of hydrogen market development its use as a fuel for IC engines is not profitable. The expenses for carbon emissions drop several times, but not enough in order to compensate the drastic augmentation of the total energy price and the total fuel expenses with the increasing hydrogen share in the total fuel amount.

References

[1] Xu H, Luo Z, Wang N, Qu Z, Chen J and An L 2019 Experimental study of the selective catalytic reduction after-treatment for the exhaust emission of a diesel engine, *Applied Thermal Engineering*, **147** 198-204

[2] Serrano J, Jiménez-Espadafor F and López A 2019 Analysis of the effect of different hydrogen/diesel ratios on the performance and emissions of a modified compression ignition engine under dual-fuel mode with water injection. Hydrogen-diesel dual-fuel mode *Energy* **172** 702-11

[3] Chintala V, Benaerjee D, Ghodke P and Porpatham E 2019 Hydrogen rich exhaust gas recirculation (H2EGR) for performance improvement and emissions reduction of a compression ignition engine, *Int. J. Hydrogen Energy* (*Preprint* gr-qc/03603199)

[4] Dimitriou P, Tsujimura T and Suzuki, Y 2018 Hydrogen-diesel dual-fuel engine optimization for CHP systems *Energy* **160** 740–52

[5] Juknelevicius R, Rimkus A, Pukalskas S and Matijošius J 2018 Research of performance and emission indicators of the compression-ignition engine powered by hydrogen - Diesel mixtures *Int. J. Hydrogen Energy* **44(20)** 10129–38

[6] Dimitrov E, Gigov B, Pantchev S, Michaylov Ph and Peychev M 2018 A study of hydrogen fuel impact on compression ignition engine performance *MATEC Web of Conf.* **234** 03001

IOP Conf. Series: Materials Science and Engineering **664** (2019) 012022 doi:10.1088/1757-899X/664/1/012022

[7] Castro N, Toledo M, and Amador G 2019 An experimental investigation of the performance and emissions of a hydrogen-diesel dual fuel compression ignition internal combustion engine *Applied Thermal Engineering* **156** 660–7
[8] Dimitrov I P 1998 *Theory of Internal Combustion Engines* (Technical University – Sofia) p 138
[9] Fraile D, Lanoix J, Maio P, Rangel A and Torres A 2015 Overview of the market segmentation for hydrogen across potential customer groups, based on key application areas (CertifHy Grant agreement no 633107)
[10] Baronas J and Achtelik G 2018 Assessment of time and cost needed to attain 100 hydrogen refueling Stations in California (California Energy Commission CEC-600-2018-008)
[11] Alternative fuel price report April 2019 (US Department of energy, Energy Efficiency and Renewable energy)

IOP Conf. Series: Materials Science and Engineering **664** (2019) 012023 doi:10.1088/1757-899X/664/1/012023

Vibroacoustic analysis of marine engine supplied with mixture of fuels

Radosław Wróbel[1,3], Monika Andrych - Zalewska[1], Marcin Tkaczyk[1], Szymon Fryda[1] and Radostin Dimitrov[2]

[1] Wroclaw University of Science and Technology, 27 Wybrzeże Wyspiańskiego st. 50-370 Wrocław, Poland
[2] Technical University – Varna, 1 Studentska Str., 9010 Levski, Varna, Bułgaria

[3] E-mail: radoslaw.wrobel@pwr.edu.pl

Abstract. The main purpose of this paper is to obtain the differences in the vibrations of engine (Doosan MDT196TI) which was powered using two different fuel mixtures – liquefied natural gas with diesel pilot and diesel only. LNG and diesel oil differ in combustion temperature, density, and cetane number; the combustion process of various fuels proceeds differently, so it has an impact on the generated vibroacoustic vibrations (vibrations have a direct impact on the degree of machine degradation). In the research, the diagnostic system is based on piezoelectric accelerometers and Digital Signal Processing operations (windowing, Fast Fourier Transform). In the conclusions section the obtained results are analysed, indicating how the tested fuels have an effect on vibrations.

1. Introduction

From the very beginning of the use of internal combustion engines, the most optimal fuel combination has been searched. First the engines operated at excessive fuel to air ratio which generated great losses in the system of both fuel and power [1, 2].

At the same time with great automotive technological advancements (electronics control system for injectors) it was also crucial to design new types of fuel that would be in compliance in new standards of both emissions and power control. Engines needed high octane or cetane number fuels that provided enough heat to make them work.

Researchers at top crude oil companies are working on various eligible options like biodiesel, hydrogen, alcohols, compressed natural gas (CNG) and liquefied petroleum gas (LPG). However, with further development of oil chemistry new fuels have been introduced, which were both reliable in terms of preventing engine knocking (high octane) and ecological. The need for more engineered fuels instead of products of crude oil was also driven by a more green approach of society and politics. LNG gas is the last, widely tested and approved for popular use in heavy vehicles and ships [2-4]. The mixture of LNG and diesel (dual fuel) makes it possible to obtain more power (torque), which is used in engines that drive ships (table 1). We cannot overlook the ecological aspects of supplying with a mixture of these fuels.

2. Dual fuel

The values presented in table 1 are the reason why manufacturers are introducing dual fuel combustion into their engines. It is an indirect solution to the problems occurring when using diesel or LNG only.

Content from this work may be used under the terms of the Creative Commons Attribution 3.0 licence. Any further distribution of this work must maintain attribution to the author(s) and the title of the work, journal citation and DOI.
Published under licence by IOP Publishing Ltd

The tests that were conducted by leading massive engine manufacturers show that the use of methane in its liquidized form can be great addition to normal diesel combustion. It provides up-close combustion parameters that the diesel engine has to meet, but visibly limits fumes toxicity and flowing particles out of the exhaust system. Engines that were powered with LNG only lacked enough power and torque, but were much more eco-friendly then their diesel counterparts. By using double fuel systems operators can benefit from both solutions – diesel for power and torque, – LNG for price reduction and ecological aspects [1, 4]. With the reduction of costs and increased greenness of that solution more and more transport solutions providers will favour the combination. The adaptation for a dual fuel solution to a pre-era diesel engine is simple and many qualified workshops in Europe provide solutions for operators to install the system. As a diesel engine does not have spark plugs that could ignite the fuel mixture the Diesel working principle cannot be fully fulfilled. Thus the engine cannot run on LNG only and it needs to be replenished with gas oil.

Table 1. Power/Torque of different fuel composition in diesel engine [4].

	Maximum torque, Nm	Maximum power, HP
Diesel powered engine	1800-2300	400-550
LNG powered engine	1500	330-340
Dual powered engine	> 2000	> 500

On the other hand, there was an issue that can be observed when the engine runs on LNG only. Natural gas consists mostly of methane, which has a much bigger and stronger greenhouse effect than CO2. The problem that is widely addressed when powering the machine with LNG is called methane slip. It occurs when the unburned fuel leaves the engine through the exhaust. Methane poses a real threat to the environment when released as it has 86x more GWP (global warming potential). If the methane slip is not controlled properly the environmental benefits from using LNG powered engines are wasted and completely cancel the advantages. Another hazardous thing about using LNG on vessels or trucks is how the gas is stored as it needs to be cool enough to stay in its liquid state [4].

In spite of the many hazards that still need to be taken care of LNG powered engines lay a bright future of the ICE as the future means of powering transport. LNG implementation in future solutions depends on key factors:

- Gas availability;
- Demand for ships and heavy trucks;
- Emission limits (especially greenhouse gases);
- LNG tank installation and operation;
- Safety requirements.

To ensure that the climate will be taken into consideration future rules of emission compliance will be harsher and harder to fulfil. The use of LNG reduces sulphur oxides by nearly 100% and nitrogen oxide by about 90% [1, 4]. Such limitations should not pose a threat to manufacturers with great experience in dealing with much more complicated diesel engines and their equipment.

The mixture of fuels affects the engine power. It is associated with the combustion process, which has a direct impact on vibration (ergonomics). In many scientific works, vibrations have been shown as the factor of the degree of engine degradation. Therefore, it should be checked whether LNG, in addition to the above-mentioned beneficial economic and ecological features, does not affect the degradation of the machine by introducing redundant vibrations.

3. Research object

The object of the research was heavy duty marine diesel engine with 320 HP of power. Its characteristics are mentioned table 2 [5]. DOOSAN MD196TI is one of products of Korean based engine manufacturer which specializes in marine powering units and transmission. MD196TI is one of its key product with great capabilities and wide applicability – from small vessels to smaller cargo ships. Eleven litters of displacement in inline 6 type engine creates much power and torque which is further accompanied with turbocharger. Direct injection of fuel which is combined with precise computer management system allows for accurate control over fuel metering and exact point of injection in time. This type of fuel feed is great for optimal spray pattern which breaks the fuel droplets into atomized fog. With the use of direct injection the engine gets better power/emission ratio as more fuel is burned and the air/fuel mixture can stay at stoichiometric level for the whole combustion. What is more as the engine is a heavy duty type it was subjected to a test of fuel type change that consisted of direct feed of liquefied natural gas into the intake manifold. Further with the use of diesel pilot the combustion may take place and the engine was easily transformed into one with dual fuel se. This work focuses on a comparison of the engine characteristics after fuel change.

Table 2. Engine characteristics of DOOSAN MD196TI marine engine [5].

Engine type	4 cycle, in-line, direct injection, water cooled with turbo charger and inter-cooler
Rating output, kW/rpm	235 / 2000
Displacement, cc	11051
Cylinder number – bore (⌀) × stroke, mm	6- ⌀123×155
Low idling, rpm	725 ± 50
No load, max. Idling, rpm	2300
Compression ratio	16,5 : 1
Firing order	1 – 5 – 3 – 6 – 2 – 4
Govemor type of injection pump	Mechanical Variable Speed (R. S. V.)
Fuel consumption, l/h	58.61
Starting system	Electrical by motor

4. Research methodology

Measurements were taken using piezoelectric accelerometers and a National Instruments Data Acquisition – DAQ equipment. The accelerometer characteristic of gathered data is one of the most critical issues to be addressed when conducting vibration tests, because wrongly selected measurement device will cut out the most important data or the accelerometer can even be destroyed in the process.

For collecting the data from the accelerometer (PCB 353A – table 3) National instruments data acquisition card was used together with NI cDAQ-9215.The NI cDAQ-9215 is a 1-slot NI CompactDAQ USB chassis designed for small, portable sensor measurement systems.

Table 3. PCB353A accelerometer used in tests [6].

Sensitivity, mV	10.19 ± 10%
Measuring range, m/s^2	± 491
Frequency range, Hz	1-4000 ± 5%
Frequency range, Hz	0.7-6500 ± 10%
Resolution, g RMS	0.0005 (for 1-10 Hz)

The engine was placed on a measuring stand, which complies with the industry standards for testing equipment and mounting points. A stationary mounting stand prevents overemphasis of the engine movement and gives the accelerometers the opportunity to conduct the test correctly. The engine was supplied with every medium it needed to run through the whole operation.

As can be seen in the diagram (figure 1) the engine was supplied with all the media and accompaniments that made it easier for the data logging process to be correctly carried out. Engine load during operation was connected via clutch to electro wired brake steered directly by electronic controlled system. That allowed the operators to specify the load with great precision and get correct result, which

IOP Conf. Series: Materials Science and Engineering **664** (2019) 012023 doi:10.1088/1757-899X/664/1/012023

could later on be compared. Fumes from the exhaust system were discharged from the testing room using a pipe that was specially chosen for that kind of application.

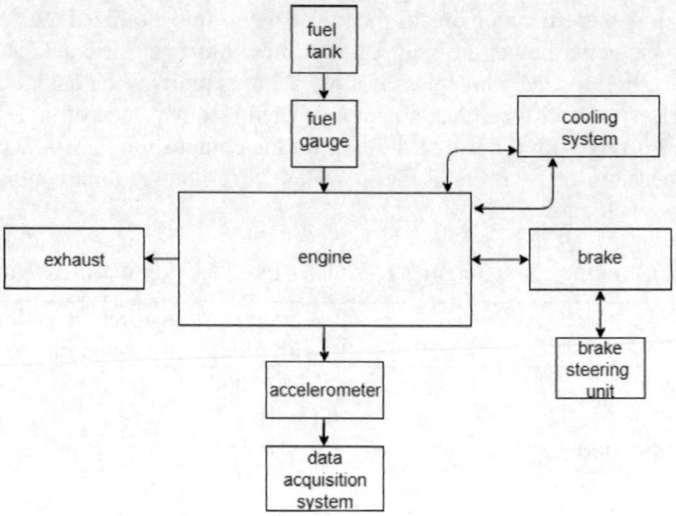

Figure 1. Diagram of the research stand.

The mounting point of the piezoelectric accelerometer has been chosen with great accuracy and several places have been considered, but as the engine characteristic shows the greatest amplitude of vibration should have come from the engine piston movement. Having that in mind and the purpose of this work, which is to determine the difference in the vibration characteristic with different fuel feed, it was determined that the best place was the engine head in one of the already threaded apertures. The sensor was placed on the engine head.

Four different states were taken from the original test results for comparison. The states differ from each other per rpm and load under which the engine was tested. States which were taken into consideration for vibration measurement were the standard states of the engine during operation taken from the characteristic given by the producer (1000 rpm / 500 Nm, 1400 rpm / 700 Nm, 1600 rpm / 1300 Nm, 2000 rpm / 1100 Nm [5, 7]).

Raw data from accelerators cannot be taken into consideration as valuable information data. Accelerometers data will be used to calculate the dominant harmonics of the engine with Fast Fourier Transform (FFT) [8]. In digital signal processing, filtration in the form of windowing (Hamming window) was used to extract the appropriate diagnostic information.

5. Results

The signal that was collected from the accelerometer had 8192 samples which lasted 2 seconds. That gives the 4 kHz as the sampling frequency of signal. From the Nyquist rule of frequency which states that for the maximum frequencies to be possible to be exact and correct they should not exceed half of the sampling frequency. Thus the frequency has been limited to 2 kHz. Most of the vibrations were already in that scale so it did not interfere with the final accuracy of the periodograms. What is more the ISO standard says that engine vibration analysis should cover the bandwidth of 0 Hz (DC component) to 1 kHz as those frequencies carry the most energy which could destroy the machine [9]. To obtain more accurate diagnostic information, a windowing operation was performed (figure 2).

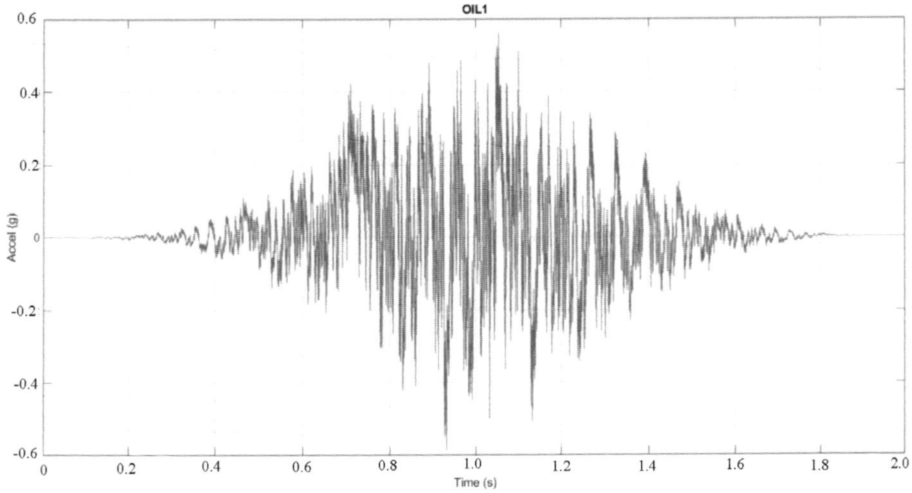

Figure 2. Example of measured (and conditioning via windowing) vibroacustic signal.

Figure 3 is an example of comparative periodogram of engine vibrations powered by various fuels.

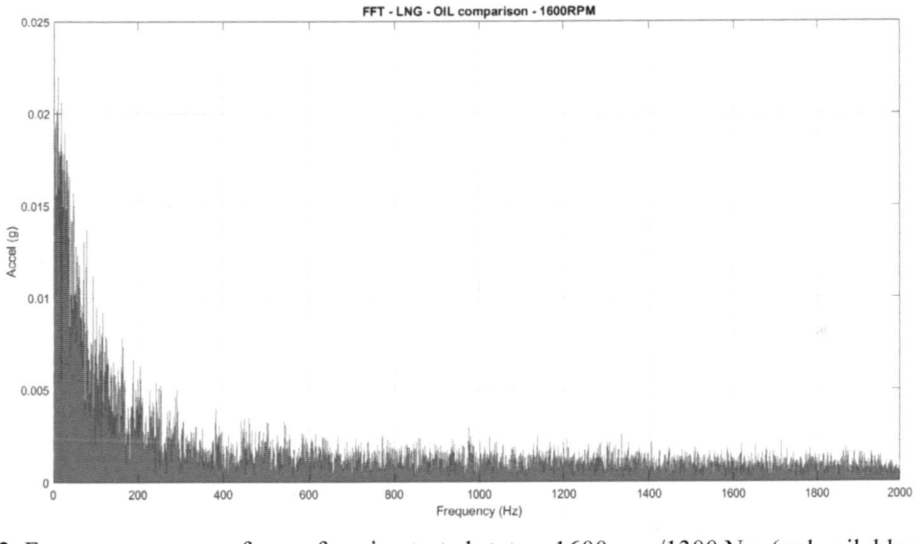

Figure 3. Frequency answer of one of engine tested state – 1600 rpm/1300 Nm (red –oil, blue- LNG).

The vibration amplitude value of the engine changed depending on the burning characteristics, engine load and rotations per minute. Each set of rpm and power has its own harmonic which is a result of the combustion in the cylinders [10, 11]. Dominant harmonics were calculated based on the principle of firing order that could be seen in the engine data characteristics. According to the literature inline 6 should have its harmonics that are connected to CF as 3rd order of the rotations per minute frequency. Table 4 presents the harmonics that were calculated directly from CF and rpm. As the engine in question was the inline 6 it has one of the best inherited balance. It is a result of the working principle and firing order of that kind of cylinder numbers. Each piston has its own pair and they work as a counter balance to each other – meaning that the vibration that comes from combustion force, the 2nd order vibration is well balanced in that engine. The periodograms corroborate with that statement. Further harmonics of the CF for each rpm and load could also be seen in the diagram [12].

Further harmonics that were not result of combustion force or engine order taken from rotation but further dominant frequencies of other oscillation are: 150, 200, 765 Hz and high frequent 1250 Hz.

Table 4. First engine harmonic depending on rpm and corresponding magnitude of vibration.

Engine speed, rpm	Fuel type	Dominant harmonic, Hz	Magnitude, abs
1000	LNG	50	0.01037
	Oil		0.01038
1400	LNG	70	0.00785
	Oil		0.00901
1600	LNG	80	0.01370
	Oil		0.00904
2000	LNG	100	0.00732
	Oil		0.00568

The values in table 5 show the magnitude of the engine harmonics at different loads and rpm. From a direct comparison of the rpm values it is clearly visible that the engine was at a higher level of vibrations that were in the frequencies from 1 Hz to 1 kHz and lower in the spectrum above those values. It means that lower frequencies that transfer greater energy are more present in the engine operation. The above statement was true for both LNG and diesel, but it is also viable to ascertain that LNG produced vibrations with greater frequency wave. It was observed that for all frequencies above 500 Hz the LNG produced more energetic oscillation, which may be a direct product of the fuel characteristics of combustion.

Table 5. Further engine harmonics depending on rpm and corresponding magnitude of vibration.

Engine speed, rpm	Fuel type	Harmonics, Hz			
		150	200	765	1250
1000	LNG	0.00494	0.00338	0.00077	0.00174
	Oil	0.00562	0.00433	0.00010	0.00032
1400	LNG	0.00324	0.00532	0.00103	0.00142
	Oil	0.00250	0.00312	0.00130	0.00071
1600	LNG	0.00254	0.00362	0.00099	0.00090
	Oil	0.00536	0.00563	0.00053	0.00118
2000	LNG	0.00546	0.00488	0.00286	0.00241
	Oil	0.00579	0.00649	0.00097	0.0083

As it was predicted in the introduction the engine had most vibrations in the low spectrum of frequencies, which is the outcome of the working principle of compression ignition engines. In comparison to the spark ignition engines they have much higher compression ratio, which is a main source for the great combustion force [11]. What is more, diesel engines differ from SI in that they add fuel to cylinders almost at the top dead centre of piston is achieved. Thus, fuel atomization and mixing with the compressed air is not complete. Such incompleteness generates huge force disturbances in the cylinder forces as the pockets of high concentrated fuel burn with greater energy in comparison to the lean mixture.

The root mean square values of the engine vibrations that were measured can be seen in figure 4. According to the ISO norm the RMS should be calculated at frequencies from 1Hz to 1 kHz to be viable for most destructive forces in the engine. Calculations were made in MATLAB software after limiting the frequency scale to the values mentioned above. The values of the obtained accelerations clearly determine the vibration profile of the engine. It is clearly seen that the LNG fuelled engine was at a higher level of vibration in 3 out of 4 measurements taken. Only for the operational speed at 2000 rpm the engine fuelled with liquidized natural gas was subjected to lower oscillation force in this frequency spectrum.

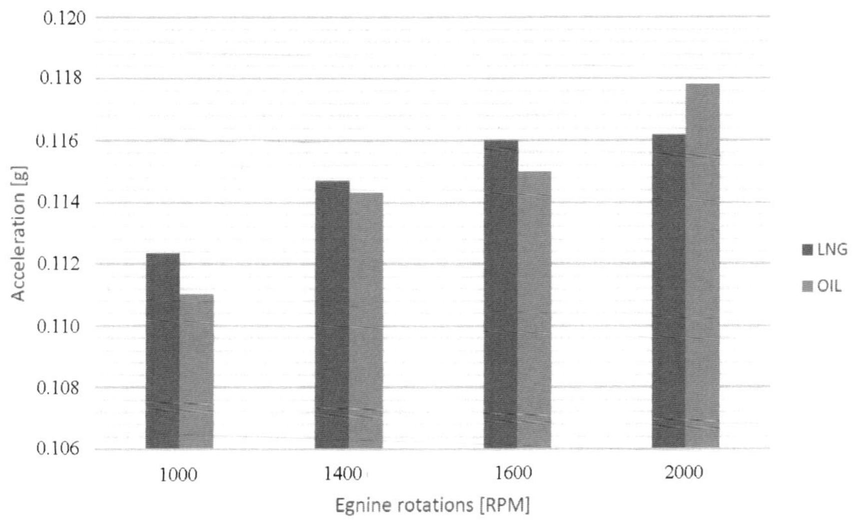

Figure 4. Root mean square values for different fuels and rotations per minute.

As this work focuses on the comparison of fuel blends it is viable to check which one of the fuels achieved lower magnitude in the harmonics that were created directly from the combustion force. The CF harmonics gives direct information about the combustion process and the forces that were induced from burning the fuel. As both diesel and LNG have different calorific values and different temperature of combustion the forces that were created in the combustion chamber were also much different. With each temperature point difference in the cylinder also the final pressure has changed which is directly connected to the working principle of the compression ignited engine. Table 6 shows that at 1000 rpm the engine achieved more or less the same vibration level that was created by the combustion forces. As the engine was designed and manufactured to operate on diesel fuel only it should not be a surprise if the engine would have a greater value of vibration when different fuel was supplied.

Table 6. Difference of vibration magnitude in the first harmonic of the obtained vibracustic signal.

Engine speed, rpm	Fuel type	Dominant harmonic, Hz	Lower Magnitude
1000	LNG; Oil	50	–
1400	LNG; Oil	70	LNG
1600	LNG; Oil	80	Oil
2000	LNG	100	Oil

Table 7 presents which of the fuel mixtures had lower magnitude of vibrations at different further (higher) harmonics.

Table 7. Difference of vibration magnitude in higher harmonics.

Engine speed, rpm	Fuel type	Harmonics, Hz			
		150	200	765	1250
1000	LNG; Oil	LNG	LNG	Oil	Oil
1400	LNG; Oil	Oil	Oil	LNG	Oil
1600	LNG; Oil	LNG	LNG	Oil	LNG
2000	LNG; Oil	LNG	LNG	Oil	Oil

During the research, it was examined how the vibration volume changes during load changes and rotation speed. It also studied how different fuels affect vibration. Most of the values were part of the G-force, which is inherent to the performance of the in-line engine 6, which is very balanced. It also shows that the engine has been correctly mounted because no resonance can be seen from the visible attachment points. Most vibrations were visible at 1000 rpm, which was the lowest rpm value recorded in the engine. For the operational rotational speed at which the motor normally operated, the amplitude of the vibrations at all four harmonics took the smallest values. The LNG driven engine had a higher level of vibration when it comes to high frequencies in all of the researches states.

6. Conclusions

Dual fuelled engines that use LNG as the fuel for the operational rotation speed are the future of the marine industry as their capability to generate the same power and lower the fuel consumption is great. The high frequency vibrations that were the result of combustion could easily be counteracted by specialized appliances and should not be problematic to solve. Further fuel consumption reduction and environmental causes should persuade the industry that even supplying existing diesel engine with LNG through the intake manifold, thus removing the need of specialized changes in the structure is a favourable thing. As emission controls and upcoming emission regulations both for marine and land vehicles will be more and more restrictive the engine manufacturers should take the dual fuel solution into consideration as it will greatly help to be in compliance with the standards. The present study clearly shows that there is great change of vibration level when it comes to lower rpm, which are below the operational level of the engine. Thus the engine should not be powered with just one type of fuel, but then after the engine is warm enough to provide better combustion parameters of the liquidized gas it would change. As it has been already mentioned the LNG is often the by-product of the transportation system and could easily be used in marine vessels powering units.

The only real issue that is still to be addressed is the not big enough access to the fuelling stations across the ports around the world. However, that should change when more and more engine manufacturers turn to dual fuel solutions.

Acknowledgments

The research has been financed by the Wroclaw University of Science and Technology statutory funds No. 0401/003/18.

References

[1] Ivanov Z, Mihaylov V and Chekelov G 2017 Exhaust gas recirculation parameters in a diesel engine *Proceedings of University of Ruse* **56(4)** 52–7
[2] Taylor C F 1968 *The Internal-Combustion Engine in Theory and Practice-Volume II: Combustion, Fuels, Materials, Design* (MIT Press)
[3] Hristov R, Bogdanov K and Dimitrov R 2018 Reseach the influence of spark plugs types on the performance of the engine operating on gaseous fuels *Mobility&Vehicle Machines* **44(1)** 45–52
[4] Chehroudi B 1993 Use of natural gas in internal combustion engines *International Non-renewable Energy Sources Congress* 1–5
[5] MD196TI MARINE ENGINE datasheet. https://doosan.nl/nl/download/file/fid/21235
[6] PCB353A accelerometer datasheet. http://www.pcb.com/Products/model/353A33
[7] *IMO Marine Engine Regulations* https://www.dieselnet.com/standards/inter/imo.php
[8] Van Drongelen W 2018 *Signal Processing for Neuroscientists* (Academic press)
[9] Grajales J A 2017 Engine diagnosis based on vibration analysis using different fuel blends *Diagnostyka* **18(4)** 27–36
[10] Junhong Z and Jun H 2006 CAE process to simulate and optimise engine noise and vibration *Mechanical Systems and Signal Processing* **20(6)** 1400–9
[11] Taghizadeh-Alisaraei A and Mahdavian A 2019 Fault detection of injectors in diesel engines using vibration time-frequency analysis *Applied Acoustics* **143** 48–58
[12] Ahirrao N S 2018 Dynamics and vibration measurements in engines *Procedia Manufacturing* **20** 434–9

Research of the impact of geometric unevenness of the railway track on the dynamic parameters of the railway rolling stock with two-stage spring suspension

A Kuzyshyn[1,4], S Kostritsa[1], L Ursulyak[1], A Batig[2], J Sobolevska[3] and O Voznyak[3]

[1] Dniprovsk National University of Railway Transport named after Academician V. Lazaryan, 49000 Dnieper, Ukraine
[2] Lviv Research Institute of Forensic Expertise, 79000 Lviv, Ukraine
[3] Lviv branch of Dniprovsk National University of Railway Transport named after Academician V. Lazaryan, 79052 Lviv, Ukraine

[4] E-mail: kuzyshyn1993@gmail.com

Abstract. The paper studies the possibility of using previously developed by the authors spatial mathematical model of the modern diesel train DPKr-2 for parameters determination, with the help of which the dynamic properties of the mechanical part are evaluated. The dependences of maximum accelerations of the body in the vertical and horizontal planes, the coefficients of vertical and horizontal dynamics both in the primary and in the secondary suspension on the speed of the car movement, the amplitude and the length of geometrical unevenness of the railway track were obtained. It is established at what speeds, depending on the parameters of geometrical unevenness of the railway track, the phenomenon of resonance and danger for the diesel train car movement appears. For a particular length of geometrical unevenness, the maximum permissible value of its amplitude over the entire range of speed movement, where the coefficients of vertical and horizontal dynamics in the primary suspension do not exceed permissible limits, has been studied. The adequacy of the mathematical model is checked by comparing the results obtained during calculations with the experimental ones.

1. Introduction

The study of the processes of interaction between the track and the railway rolling stock began with the birth of the railway engineering, since the results of this type of research were needed to create a new railway rolling stock and a track, setting standards for their maintenance and repair rules.

Practical needs demanded the immediate study and solution of interrelated tasks such as:

- studying the data and characteristics of the railway rolling stock and tracks during train movement;
- studying the magnitude and direction of the forces that arise between the rails and the wheels of the car, separate structural elements of the railway rolling stock;
- studying deformations, necessary structural dimensions of the elements of the track and the railway rolling stock, requirements for materials manufacture that provide sufficient strength, durability and reliability in operation.

The use of railway rolling stock is allowed only in case when the acceptable level of dynamic characteristics is ensured throughout its lifetime.

Content from this work may be used under the terms of the Creative Commons Attribution 3.0 licence. Any further distribution of this work must maintain attribution to the author(s) and the title of the work, journal citation and DOI.
Published under licence by IOP Publishing Ltd

On the domestic railways the following indices of dynamic qualities are adopted:

- indices that assess the dynamic properties of the mechanical part;
- traffic safety indices.

The main indices that assess the dynamic properties of the mechanical part of the railway rolling stock are:

- maximum acceleration of the body both in the vertical and in the horizontal plane;
- coefficients of vertical and horizontal dynamics of the primary and the secondary suspension.

At present, the modern diesel train DPKr-2, which is intended for provision of commuter passenger traffic, is gradually being put into operation on the territory of Ukraine.

The peculiarity of this diesel train is the presence of the two-stage spring suspension (figure 1):

- primary – with the use of cylindrical springs;
- secondary – with the use of pneumatic springs.

To ensure its safe operation, "Branch Research Laboratory of Railway rolling stock Dynamics and Strength" of Dniprovsky National University of Railway Transport named after Academician V. Lazaryan tested the train to determine the main indicators of the dynamic qualities of its mechanical part.

(a)

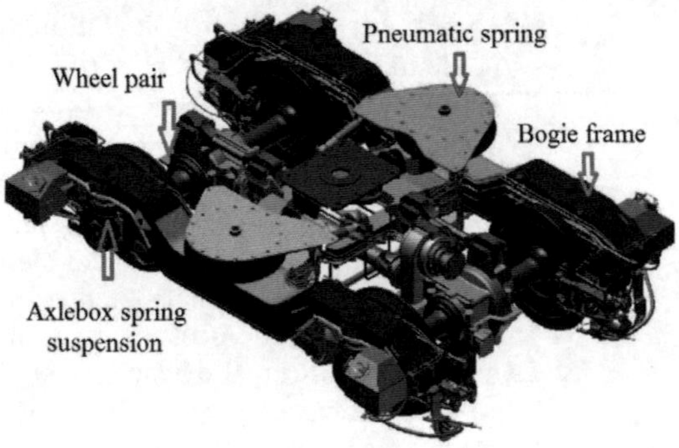

(b)

Figure 1. Diesel train DPKr-2: (a) general form of the train; (b) the main units of the bogie of the diesel train car DPKr-2.

It should be noted that the unevenness of railway lines is a source of forced fluctuations in the above spring structure of the freight car, which ultimately causes dynamic loads on the elements of the structure of the railway rolling stock and the railway track. This leads to an increased impact of the railway rolling stock on the track and, finally, to its derailment.

Studying the impact of a rail track geometric unevenness in the trail dynamics is not possible without the knowledge of the unevenness parameters, like its size, form, spectral properties, etc. Moreover, rail track geometry is subjected to deterioration during the track lifecycle. The most intensive deterioration usually takes place in high loaded parts of the track: railway turnouts and transition zones of engineering structures. The modelling of the track and turnout geometric unevenness during its lifecycle is presented in studies [1-2]. An intensive deterioration of track geometry after tamping works is studied in [3]. The impact interaction of a wheel and rails in the zone of common crossings during the lifecycle of turnouts is presented in experimental studies [4-6]. The studies demonstrate a significant variation in geometric unevenness and dynamic interaction depending on the track condition.

However, theoretical studies of diesel train DPKr-2 carriage for determining its basic parameters and assessing the dynamic properties of the mechanical part, namely, the maximum accelerations of the body, the coefficients of vertical and horizontal dynamics depending on the speed of motion and the parameters of geometric unevenness of the track have not been carried out. It should also be noted that the solution to this problem must be carried out with the help of computer simulation methods, which will allow for a more accurate assessment of railway transport safety [7-9].

The research of safety indicators of modern diesel-train DPKr-2 is a separate task that is considered and solved in works [10-11].

Therefore, to ensure traffic safety under different operating conditions, it is relevant to determine the indices that assess the dynamic properties of the railway rolling stock.

2. Purpose and objectives of the research

The purpose of the present study is to determine the indices that estimate the dynamic properties of the mechanical part of the modern diesel-train DPKr-2.

To achieve this goal, the following tasks must be solved:

- construct dependences of the maximum accelerations of the body in the vertical and horizontal plane, the coefficients of vertical and horizontal dynamics both in the primary and in the secondary suspension on the speed, amplitude and length of geometrical unevenness of the track;
- establish at which velocities of motion, depending on the parameters of geometric unevenness of the railway, the phenomenon of resonance and the danger for diesel train DPKr-2 carriage movement starts to occur;
- study the adequacy of the mathematical model by comparing results obtained through calculations with experimental ones.

3. Mathematical model for determining dynamic parameters

To determine the dynamic characteristics that are included in the indices, with help of which the dynamic properties of the mechanical part of the diesel train are estimated, analysis of the car design and its spatial mechanical model should be carried out [12].

The spatial mechanical model consists of 7 solids (body, two bogies, four wheel pairs), connected together by elastic-dissipative joints.

The spatial mathematical model, consisting of 30 differential equations of the second order, was obtained for describing diesel train car fluctuations [13].

The peculiarity of the mathematical model is the possibility of taking into account longitudinal and transverse forces of the creep that arise at the point of interaction between working surfaces of the wheel and the rail [14–16].

At each instant of time, by means of integrating the differential equations of motion, the values of all generalized coordinates were determined, which made it possible to find necessary values for calculating the parameters, such as forces in the primary and in the secondary suspension, frame forces, etc.

The maximum acceleration of the body both in the vertical and horizontal plane characterizes dynamic forces that act on the equipment of the car. To find them, the following equations were used:

$$\ddot{z}_k = \left(-4\beta_{21}\dot{z}_k - 4\beta_2\dot{z}_k - 4s_{21}z_k + 2s_{21}z_{b1} + 2s_{21}z_{b2} + 2\beta_{21}\dot{z}_{b1} + 2\beta_{21}\dot{z}_{b2} + 2\beta_2\dot{z}_{b1} + 2\beta_2\dot{z}_{b2}\right)/m_k \tag{1}$$

$$\ddot{y}_k = \frac{1}{m_k}\Big[-4s_{2\text{pt}}\left(y_k + a_5\theta_k\right) - 4\beta_{2\text{pt}}\left(\dot{y}_k + a_5\dot{\theta}_k\right) + 2s_{2\text{pt}}\left(y_{b1} - a_6\theta_{b1}\right) + $$
$$+ 2s_{2\text{pt}}\left(y_{b2} - a_6\theta_{b2}\right) + 2\beta_{2\text{pt}}\left(\dot{y}_{b1} - a_6\dot{\theta}_{b1}\right) + 2\beta_{2\text{pt}}\left(\dot{y}_{b2} - a_6\dot{\theta}_{b2}\right)\Big] \tag{2}$$

where s_{21} is equivalent stiffness of the pneumatic spring, kN/m; $s_{2\text{pt}}$ – horizontal stiffness of the pneumatic spring in the transverse direction, kN/m; β_{21} – equivalent coefficient of attenuation determined by the flow of air through the throttle, (kN·sec)/m; β_2 – damping coefficient of the vertical hydraulic quencher in the secondary suspension, (kN·sec)/m; $z_{b1,2}$, $y_{b1,2}$, $\theta_{b1,2}$, z_k, y_k – generalized coordinates of the diesel train movement obtained through calculations of the spatial mathematical model, m and radian; a_5, a_6 – linear parameters, m; m_k – body weight, tone.

The coefficients of vertical $K_{\text{d}}^{\text{ver}}$ and horizontal $K_{\text{d}}^{\text{hor}}$ dynamics can be determined on the basis of deflections Δ or forces F that arise in a separate set of spring hanging:

$$K_{\mathrm{d}}^{\mathrm{ver}} = \Delta_{\mathrm{d}}^{\mathrm{ver}} / \Delta_{\mathrm{st}} = F_{\mathrm{d}}^{\mathrm{ver}} / F_{\mathrm{st}} \qquad K_{\mathrm{d}}^{\mathrm{hor}} = \Delta_{\mathrm{d}}^{\mathrm{hor}} / \Delta_{\mathrm{st}} = F_{\mathrm{d}}^{\mathrm{hor}} / F_{\mathrm{st}} \tag{3}$$

The dynamic index corresponds to the lower index of "d", statics – "st". Vertical deflections are forces that have upper index "ver", horizontal – "hor".

The coefficients of vertical and horizontal dynamics in the primary suspension are determined:

$$K_{\mathrm{d\,c.b.}}^{\mathrm{ver}} = \left(z_{\mathrm{b}j} \pm a_1\varphi_{\mathrm{yb}j} \mp a_4\theta_{\mathrm{b}j}\right) - \left(z_{\mathrm{w.p.}i} \mp a_4\theta_{\mathrm{w.p.}i}\right) / \Delta_{\mathrm{st}} \tag{4}$$

$$K_{\mathrm{d\,c.b.}}^{\mathrm{hor}} = \left(y_{\mathrm{b}j} \pm a_1\psi_{\mathrm{b}j} - y_{\mathrm{w.p.}i}\right) / \Delta_{\mathrm{st}} \tag{5}$$

The coefficients of vertical dynamics in the secondary suspension are determined:

$$K_{\mathrm{d\,k.}}^{\mathrm{ver}} = \left(z_{\mathrm{k}} \pm a_2\varphi_{\mathrm{yk}} \mp a_3\theta_{\mathrm{k}}\right) - \left(z_{\mathrm{b}j} \mp a_3\theta_{\mathrm{b}j}\right) / \Delta_{\mathrm{st}} \tag{6}$$

In the paper [1-2] the quantitative distribution of various types of unevenness that occur on the railway track during railway rolling stock operation is presented, where the most common types are sinusoidal geometric unevenness. This type of unevenness makes up 60% of all other types of unevenness that are on the railroad tracks. Thus, in this work, sinusoidal geometric unevenness of the left and right rails was taken as perturbations both in the vertical and in the horizontal plane, which were determined by the formulas:

$$\eta_{\mathrm{ver}} = H_{\mathrm{ver}} \cdot \sin\left[2\pi vt / L_{\mathrm{ver}}\right] \qquad \eta_{\mathrm{hor}} = H_{\mathrm{hor}} \cdot \sin\left[2\pi vt / L_{\mathrm{hor}}\right] \tag{7}$$

where H_{ver}, H_{hor} are amplitude of vertical and horizontal unevenness, mm; L_{ver}, L_{hor} – length of vertical and horizontal unevenness, m; v – speed of movement, m/sec; t – time, sec.

It should be noted that according to regulatory norms used in Ukraine [17]:

- for the first degree of spring hanging:
  - the coefficient of vertical dynamics should not exceed 0.35;
  - the coefficient of horizontal dynamic – 0.3.
- for the second degree of spring hanging:
  - the coefficient of vertical dynamics should not exceed 0.25;
  - maximum acceleration in the vertical plane in fractions g should not be more than 0.25;
  - maximum accelerations in the horizontal plane in fractions g should not be more than 0.15.

According to European standards BS EN 14363: 2005 [18]:

- for the first degree of spring hanging:
  - the wheel unloading coefficient of the wheel coupling must not exceed 0.6;
- for the second degree of spring hanging:
  - maximum acceleration in the vertical plane should not be more than 3 m/s^2;
  - maximum acceleration in the horizontal plane should not be more than 3 m/s^2.

Comparing the level of admissible values that assess the dynamic properties of the mechanical part of the railway rolling stock, it can be concluded that the requirements of Ukrainian regulatory norms are more rigorous than the requirements according to the norms of European Union countries.

After calculating the mathematical model, we construct and analyse the dependences of the maximum accelerations of the body in the vertical and horizontal planes, the coefficients of vertical and horizontal dynamics both in the primary and in the secondary suspension on the speed of a car, the amplitude and the length of geometrical unevenness of the railway track (figures 2 – 5).

Analysing the results (figure 2) it can be stated that with the increase of the velocity of motion at any length of geometrical unevenness, the coefficients of vertical dynamics in the primary and in the secondary suspension increase at first and then fall for specific speed ranges.

The maximum value of the coefficient of vertical dynamics is observed when the frequencies of its own and forced oscillations coincide. For example, with the length of geometrical unevenness of the track for 20 m, the maximum value of the coefficient of vertical dynamics occurs at the speed of 35 m/s.

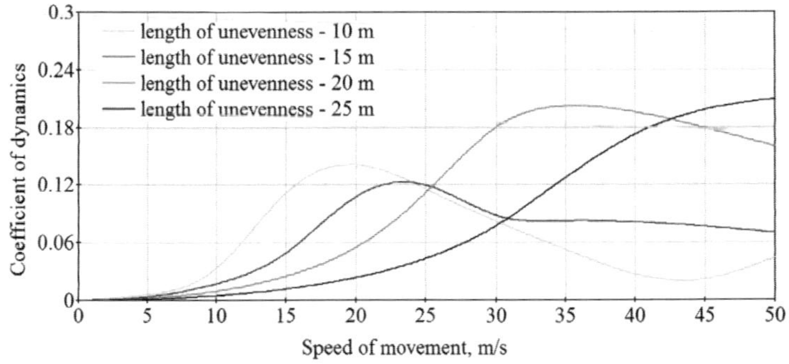

Figure 2. The dependences of the coefficients of vertical dynamics in the primary suspension on the speed of motion and the length of geometrical unevenness.

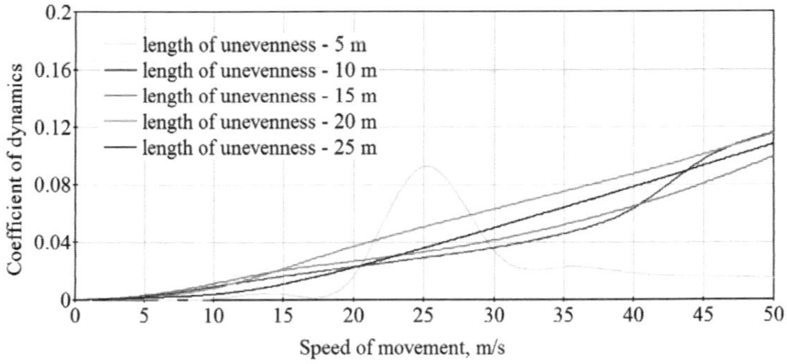

Figure 3. The dependences of the coefficients of horizontal dynamics in the primary suspension on the velocity of movement and the length of geometrical unevenness.

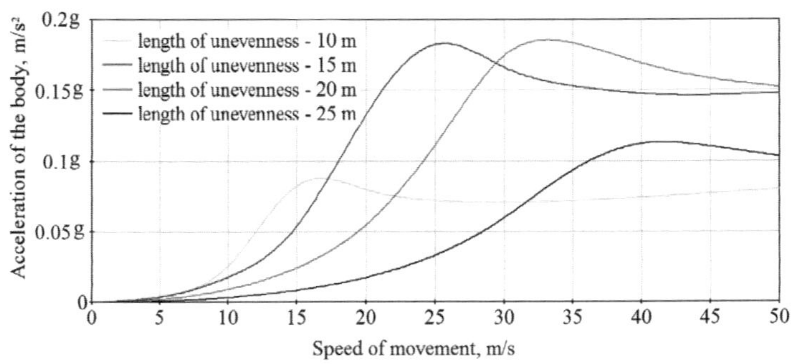

Figure 4. The dependences of maximum accelerations of the body in the vertical plane on the velocity and the length of geometrical unevenness.

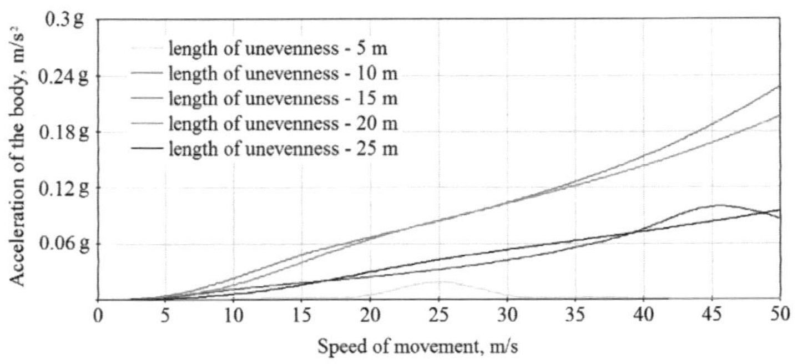

Figure 5. The dependences of maximum accelerations of the body in the horizontal plane on the velocity and the length of geometrical unevenness.

As can be seen from, with reduced geometrical unevenness the maximum value of the vertical dynamics coefficient will be shifted on figure 2-3 towards speeds decreasing. Since the frequency is directly proportional to the velocity of motion and is inversely proportional to the length of unevenness. Thus, for resonance appearance under condition of unevenness length decrease, a lower velocity of motion is required.

The analysis (figure 3) showed that the phenomena, which are characteristic for vertical dynamics, are observed for horizontal ones too. However, it should be noted that the stiffness of elements that connect solids of the spacious mechanical model in the horizontal plane is greater than the one in the vertical plane. Therefore, the resonance phenomenon will occur at higher frequencies, which make up about 5 Hz. For example, with geometrical unevenness of 5 m, the maximum value of the horizontal dynamics coefficient rises at the speed of 25 m/s. With the increase of the length of unevenness, the resonance zone will deviate towards the increase of speeds of motion and may not fall into the range of operating speeds.

The analysis of the results in figures 4 and 5 shows that the character of changes in the maximum accelerations of the body in both the vertical and horizontal planes is similar only in the other range of their changes.

The obtained dependencies of the coefficients of vertical and horizontal dynamics will further allow us to assess the safety of motion against wheel derailment [19–20].

Let us turn to the determination of the maximum allowable velocity of diesel train DPKr-2 carriage depending on the state of the railway track. According to paper [21], the state of the track is estimated by the degree of deviations from the standards of maintenance in terms of width, level and plan. There are five degrees for all deviations from the regulatory norms of track maintenance depending on their size and length.

On the basis of the analysis of the main dynamic factors of the diesel train carriage, obtained as a result of calculations, the maximum allowed speed of diesel train-DPKr-2 carriage was established, depending on the deviations on the level, including smoothness deviations, as well as distortions (deviations on a length of up to 20 m) (table 1).

These studies can be used in the analysis of traffic safety and in conducting forensic rail-transport expertise.

Table 1. Permissible speed of diesel-train DPKr-2 movement depending on the deviations on the level.

Degree of deviation	Smoothness deviations, mm	Allowable speed obtained as a result of calculations, km/h	Deviation on the length of up to 20 m, mm	Allowable speed obtained as a result of calculations, km/h
I	up to 6 including		up to 8 including	
II	over 6 up to 12 including	180	over 8 up to 12 including	180
III	over 12 up to 20 including		over 12 up to 16 including	
IV	over 20 up to 25 including	120	over 16 up to 20 including	150
	over 25 up to 30 including	110	over 20 up to 30 including	105
V	over 30 up to 35 including	75	over 30 up to 40 including	55
	over 35 up to 50 including	15	over 40 up to 50 including	movement stops
	over 50	movement stops	over 50	movement stops

4. Experimental determination of dynamic indices

To check the adequacy of the mathematical model of the diesel train car, we will compare the values of the coefficients of vertical and horizontal dynamics of the primary suspension, which are obtained by calculation, with the data from experimental studies [22].

Since during the test it was not known at what parameters of geometrical unevenness the diesel-train movement took place, the following comparison will be performed with the length of horizontal and vertical unevenness of 15 m and its amplitude of 0.006 m, which corresponds to the good state of the railway track.

As can be seen from figures 6 and 7, the experimental values of the coefficients of vertical and horizontal dynamics are in the range of values obtained by theoretical calculations according to formula (3).

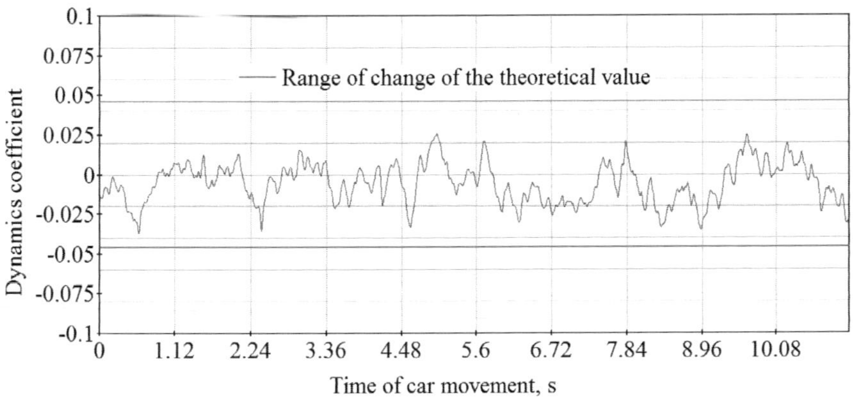

Figure 6. Comparison of experimental and theoretical values of the coefficient of vertical dynamics on the first wheel pair at speed of 35 m / s on the straight section of the track.

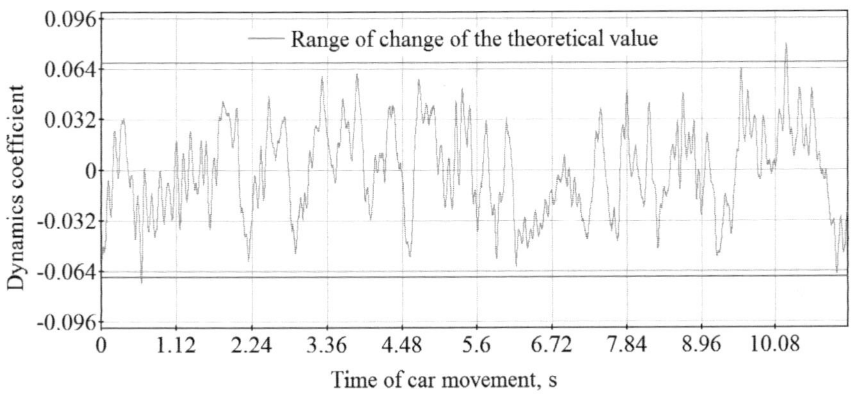

Figure 7. Comparison of experimental and theoretical values of the coefficient of horizontal dynamics on the first wheel pair at speed of 35 m / s on the straight section of the track.

The maximum experimental value of the coefficient of vertical and horizontal dynamics at a speed of diesel train carriage of 35 m / s is 0.038 and 0.08 respectively. The maximum theoretical value of these coefficients was 0.045 and 0.068. The difference in the results does not exceed 15%.

The comparison of the results of the theoretical research with the experimental ones shows that the developed spatial mathematical model provides sufficient accuracy of determination of the coefficients of vertical and horizontal dynamics and can be used in solving problems in the study of dynamic parameters of the mechanical part of the modern diesel train.

5. Conclusions

The paper shows the possibility of using the previously developed spatial mathematical model for solving problems related to research of the dynamic parameters of the mechanical part of the diesel train car. By means of the given mathematical model, forces were determined in the primary and in the secondary suspension, the frame forces that have major impact on the dynamic parameters.

It is established this is not the right verb to use, pls change that at any length of geometrical unevenness, the coefficients of vertical and horizontal dynamics increase at first, and then fall for specific speed ranges. This is due to the presence of resonance phenomena.

It is established that there is overlapping of the frequencies of its own and forced oscillations: in the vertical plane at the frequency of 2 Hz. and in horizontal plane at the frequency of 5 Hz. This allows us to establish that with the reduction of the length of geometrical unevenness, the maximum value of the coefficients of dynamics will be shifted towards the reduction of velocities.

While studying the dynamic properties of the mechanical part of the railway rolling stock, it was established that the requirements of the Ukrainian regulatory norms are more rigorous than the requirements according to the norms of European Union countries.

The maximum permissible speeds of diesel train DPKr-2 carriage, depending on the condition of the railway track, have been established, which allows us to determine safe conditions of its operation.

The comparison of the results of theoretical studies with experimental ones proves the adequacy of the developed model and the possibility of its use for solving problems in the study of dynamic diesel-train indices. The difference in the results does not exceed 15%.

References

[1] Nabochenko O, Sysyn M, Kovalchuk V, Kovalchuk Yu, Pentsak A and Braichenko S 2019 Studying the railroad track geometry deterioration as a result of an uneven subsidence of the ballast layer *Eastern-European Journal of Enterprise Technologies* **7(97)** 50–9

[2] Sysyn M, Gerber U, Gruen D, Nabochenko O and Kovalchuk V 2019 Modelling and vehicle based measurements of ballast settlements under the common crossing *European Transport / Transporti Europei - International Journal of Transport Economics, Engineering and Law* **71** 1–25

[3] Sysyn M, Gerber U, Kovalchuk V and Nabochenko O 2018 The complex phenomenological model for prediction of inhomogeneous deformations of railway ballast layer after tamping works *Archives of Transport* **46(3)** 91-107

[4] Sysyn M, Gerber U, Nabochenko O, Li Y and Kovalchuk V 2019 Indicators for common crossing structural health monitoring with track-side inertial measurements *Acta Polytechnica* **59(2)** 170–81

[5] Sysyn M, Kovalchuk V and Jiang D 2019 Performance study of the inertial monitoring method for railway turnouts *International Journal of Rail Transportation* **7(2)** 103–16

[6] Sysyn M, Gruen D, Gerber U, Nabochenko O and Kovalchuk V 2019 Turnout monitoring with vehicle based inertial measurements of operational trains: A machine learning approach *Communications - Scientific Letters of the University of Zilina*, **21(1)** 42–8

[7] Domin R, Domin IU and Cherniak G 2016 Investigation of some problems of running safety of railway rolling stock on the Ukrainian railways *Archives of Transport* **40(4)** 15–27

[8] Klimenko I, Černiauskaite L, Neduzha L and Ochkasov O 2018 Mathematical simulation of spatial oscillations of the "Underframe-Track" system interaction *12th Int. Conf. Intelligent Technologies in Logistics and Mechatronics Systems – ITELMS'2018 (Kaunas)* 105–14

[9] Kapitsa M, Kalivoda J, Neduzha L, Ochkasov O and Chernyaєv D 2018 Kompyuterne modelyuvannya zaliznichnikhtransportnikh zasobiv: metod. Vkazivki do vikonannya praktichnikh robit, kursovogo ta diplomnogo proektuvannya *DNUZT* 59

[10] Kuzyshyn A, Batig A, Kostritsa S, Sobolevska J, V Kovalchuk, Dovhanyuk S and O. Voznyak 2018 Research of safety indicators of diesel train movement with two-stage spring suspension *MATEC Web of Conferences* **234**

[11] Kovalchuk V, Kuzyshyn A, Kostritsa S, Sobolevska Ju, Batig A and Dovhanyuk S 2018 Improving a methodology of theoretical determination of the frame and directing forces in modern diesel trains *Eastern-European Journal of Enterprise Technologies* **6/7(96)** 19–25

[12] Kuzyshyn A and Batig A 2017 Construction of mechanical model of the diesel-train DTKr-2 car and its features *Science and Transport Progress. Bulletin of Dnipropetrovsk National University of Railway Transport* **6(72)** 20–9

[13] Kostritsa S, Sobolevska Y, Kuzyshyn A and Batih A 2018 Mathematical model of DPKr-2 dyzel train car *Science and Transport Progress. Bulletin of Dnipropetrovsk National University of Railway Transport* **1(73)** 56–65
[14] Dukkipati R 2000 *Vehicle Dynamics* (Alpha Science International)
[15] Santamaria J, Vadillo E and Gomez J 2009 Influence of creep forces on the risk of derailment of railway vehicles *Veh. Syst. Dyn.* **47** 721–52
[16] Shabana A, Zaazaa K and Sugiyama H 2007 *Railroad Vehicle Dynamics* (New York) pp 6-20
[17] Norms for calculating and evaluating the strength of load-bearing elements and the dynamic qualities of a motorized railway rolling stock etc 1997 (*VNIIZhT*) p 147
[18] BS EN 14363:2005 Railway applications. Testing for the acceptance of running characteristics of railway vehicles. Testing of running behavior and stationary tests.
[19] Zeng J and Guan Q 2008 Study on flange climb derailment criteria of a railway wheelset *Veh. Syst. Dyn.* **46(3)** 239–51
[20] Zeng J and Wu P 2008 Study on the wheel/rail interaction and derailment safety *Wear* **265(9–10)** 1452–9
[21] Rybkin V and Patlasov V 2012 Technical guidelines on the assessment of the condition of a rail track based on the readings of track measuring cars and ensuring the safety of train traffic during deviations from the norms of rail track content CP-0267 (Kiev) p 46
[22] Blokhin E, Granovsky R, Dzichkovsky E, Krivchikov A, Granovskaya N and Ciyupa A 2013 Results of running dynamic tests of a diesel train of type 630M produced by ao PESA *Bulletin of the DNUZT (Poland)* **4(46)** 18–29

IOP Conf. Series: Materials Science and Engineering **664** (2019) 012025 doi:10.1088/1757-899X/664/1/012025

Study of the prospect of k-Li lubricant for axle boxes of the railway rolling stock

Andrii Kravets[1,4], Andrii Yevtushenko[1], Leonid Kozar[1], Anna Kravets[2] and Maksym Kovalov[3]

[1] Construction, track and handling machines Department, Ukrainian State University of Railway Transport, Feuerbach sq. 7, 61050 Kharkiv, Ukraine
[2] Cargo and commercial work management Department, Ukrainian State University of Railway Transport, Feuerbach sq. 7, 61050 Kharkiv, Ukraine
[3] Structural Mechanics and Hydraulics Department, Ukrainian State University of Railway Transport, Feuerbach sq. 7, 61050 Kharkiv, Ukraine

[4] E-mail: kravets_am@ukr.net

Abstract. The use of various types of plastic lubricants (such as LZ-TsNII, ZhRO, etc.) in the axle boxes of locomotives, cars and track machines in the Ukrainian railway network causes technological inconveniences at the stage of procurement, supply and regulation of this process, moreover, their operational capabilities do not quite satisfy modern operating conditions of axle boxes of the railway rolling stock. The paper proposes to solve this problem by using a single brand of modern multi-purpose lubricant on lithium complex basis (k-Li) – MZT – in axle boxes of all types of railway rolling stock. The presented results of the comparative analysis of the physical and chemical properties of the mentioned lubricants demonstrate that the MZT lubricant completely meets the operating conditions of the axle bearings and can provide their high-quality lubrication under any conditions of operation. The tribological studies of lubricants on frictional testing machines, with simulation of their operation with rolling friction and sliding friction, demonstrated that anti-wear, antifriction and anti-cuffing properties of the MZT lubricant are much better than those of LZ-TsNII and ZhRO lubricants. It has been found out that the properties of the MZT lubricant are most effectively used under high load in friction pairs. The paper presents the results of testing the MZT lubricant in the actual conditions of operation of locomotives, railway cars and multiple unit rolling stock, which demonstrated its high efficiency and operation capacity.

1. Introduction

The reliable operation of the axle boxes of the railway rolling stock has a significant impact on its overall performance [1]. Trouble-free operation of the axle boxes can be achieved by using lubricants to grease the elements of the box to prevent their wear, significantly reduce the resistance during the operation of axle boxes and protect the surface of the friction pairs from corrosion [2].

Various types of rolling stock are operated in the Ukrainian railway, such as cars, locomotives, track vehicles of different manufacturers and years of production. Historically, different types of lubricants have been used to lubricate axle boxes of different types of rolling stock – predominantly, a Na-Ca based lubricant LZ-TsNII [3] and a Li-based lubricant ZhRO [4], and a few other brands of lubricant of limited use. This situation creates certain technological and economic problems in the operation of the railway rolling stock. Also, since these lubricants were developed 30-40 years ago, their functional properties

Content from this work may be used under the terms of the Creative Commons Attribution 3.0 licence. Any further distribution of this work must maintain attribution to the author(s) and the title of the work, journal citation and DOI.
Published under licence by IOP Publishing Ltd

do not always meet the modern requirements to box lubricants predetermined by the modern conditions of rolling stock operation with increasing speeds and train weight [5].

The above inconvenience can be solved by using of a single brand of modern multi-purpose lubricant on the lithium complex basis (k-Li) – MZT – in axle boxes of all types of railway rolling stock. As significant research shows, lubricants on k-Li basis have certain advantages over other lubricants, including those on an ordinary Li base, and even more so before Na-Ca lubrication [6, 7 and others].

2. Purpose and tasks of the paper

In order to substantiate the expediency of using a single brand of box lubricant in the rolling stock, a wide range of versatile studies was conducted with the task of comparing the properties and possibilities of common use of LZ-TsNII and ZhRO lubricants vs. the new lubricant MZT in different conditions.

For this purpose, comparative studies of the physicochemical and tribological properties of the named lubricants were conducted in the laboratory. Also, to check the performance of the studied lubricants in real operating conditions of the railway rolling stock, they were tested in the axle boxes of cars, locomotives and multiple unit rolling stock in the regular mode of operation of the latter.

3. Results of laboratory research

3.1. Analysis of physicochemical properties

The properties of MZT lubricant were studied using the standard methodologies (see table 1). The obtained results were compared with the requirements for the currently used LZ-TsNII and ZhRO lubricants, according to the normative documents and technical specification for their manufacture.

Table 1. Results of physicochemical study.

Quality indicator	LZ-TsNII [a]	ZhRO [a]	MZT [b]	Study method
Thickener	Ca-Na	Li	к-Li [a]	-
Temperature limits of application, °C	–60…+100	–50…+120 (for a short term +130)	–30…+130 (for a short term +150)[a]	-
Viscosity at an average shear rate gradient 10 s^{-1}, Pa·s	not more than 420 (at 0°C)	not more than 2000 (at –30°C)	1874 (at –30°C)	GOST 7163 [9]
Shear strength, Pa	not less than 220 (at 50°C)	not less than 350 (at 50°C)	600 (at 20°C) 410 (at 50°C) 180 (at 80°C)	GOST 7143 [10]
Melting point, °C	not lower than 135	not lower than 175	248	ISO 2176 [11]
Colloidal stability, % extracted grease	not more than 22	not more than 15	8.76	GOST 7142 [12]
Mass fraction of free alkali expressed as NaOH, %	not more than 0.2	not more than 0.25	0.16	DIN 51809-1 [13]
Evaporation at 100°C for 1 hour, %	not more than 7.0	not more than 2.5	0.53	ASTM D 2595 [14]
Penetration at 25°C, mm×10^{-1}	200 to 260	190 to 250	260	ISO 2137 [15]
Mass fraction of water, %	not more than 0.4	traces	traces	ASTM D95-13e1 [16]
Corrosive influence on metals	-	withstands	withstands (steel and bronze)	ISO 11007 [17]

[a] Information according to the normative documents for the material.
[b] The values actually determined during the study.

The analysis of the results of physicochemical study demonstrated that all the properties of lubricant MZT meet the requirements for currently used lubricants, and are better by most indicators.

The use of complex lithium soaps on a mixture of fatty acids as a thickener in MZT ensured its high flooding resistance, i.e., formation of stable emulsions with water. In case of heating to rather high temperatures, which occurs during the operation of bearing boxes of the rolling stock, flooding resistance, together with high adhesion properties and thermal stability, ensures the protection of the friction surfaces from corrosion wear.

The low temperature viscosity of MZT has a value that ensures the least possible resistance to the rotation of the bearing in the initial period of motion after a prolonged standstill of the rolling stock, especially in the cold season, when the temperature of the lubricant becomes low enough for transition to the softened state. This value of viscosity will also ensure that the lubricant reaches all pairs of friction of the axle box. This indicator is in some way related to the energy consumption of traction rolling stock for the train movement. The data obtained by studying the low-temperature viscosity are completely correlated with the results of other studies of lithium oils [8 and others]

The strength of the lubricant at the operating temperature of the box must not exceed (300 to 500) Pa, and its minimum value at the highest temperature in the working area should not be lower than (100 to 200) Pa. Since the axle box bearings operate in sealed conditions, and the box is filled with lubricant in sufficient quantity, the possibility of removing the lubricant from moving surfaces is out of the question, and insufficient lubrication of the surfaces is only possible with too high a limit of strength when the lubricant simply does not come to surfaces. The properties of the MZT fully meet these conditions, and it always reaches the lubricant of the friction areas of the axle bearings.

Due to the melting point, MZT has a sufficient margin of properties, which ensures its efficiency with substantial heating, which often happens in the axle boxes, especially during the initial period of operation of a new bearing.

Also, the data in table 1 suggest that MZT in comparison with other lubricants has:

- excellent resistance to separation of the lubricant base (breaking) during the storage and use;
- good anti-corrosion properties for non-ferrous and ferrous metals;
- low inclination to liquid fraction evaporation, i.e., not inclined to increase the concentration of thickener, dehomogenization and a decrease in plasticity;
- good ability to withstand the load and resist pulling out of the bearing.

Rather high physicochemical properties of MZT were the reason for further studies.

3.2. Results of tribological tests

Since the tribological characteristics for LZ-ZNII and ZhRO are not regulated with normative technical documentation, all three lubricants were tested to evaluate the performance of MZT. Anti-wear, anti-friction and anti-cuffing properties were studied on well-known friction and wear machines [18].

One of the machines used in the study was a four-ball extreme pressure tester (FBEPT), used to conduct testing in accordance with ASTM D2596-15 [19] and ASTM D2266-01(2015) [20] standard test methods. The tests were conducted according to the scheme pattern in figure 1, which implements the point contact of the friction surfaces and simulates the work of the higher kinematic pairs. This type of contact is characteristic for roller bearings, which are used in the axle boxes of some track vehicles.

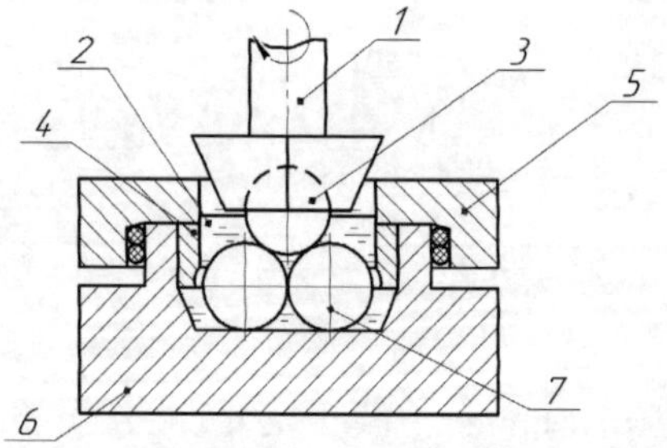

Figure 1. Testing pattern on FBEPT: 1 – collet clamp (cartridge); 2 – tested lubricant; 3 – the upper moving ball; 4 – clamping washer; 5 – nut; 6 – cup; 7 – the bottom fixed ball.

In FBEPT used for testing, standard Ø12.7 mm balls from ball bearings are used in the friction unit.

The results of the study on FBEPT (figure 2) demonstrated the following:

- by welding load (L_W), which determines the boundary capacity of the lubricant, MZT exceeds LZ-TsNII 1.4 times and exceeds ZhRO almost 1.5 times;
- by critical load (L_C), the characteristics of the lubricant are better than those of the other two lubricants by 12%;
- load wear index (I_W) that characterizes the properties of the lubricant at marginal friction, does not differ substantially (by 3.5%) in ZhRO and LZ-TsNII, but in MZT it is about ¼ higher than in other lubricants;
- The diameter of wear scar at loads of 196 N (D_{196}) and 392 N (D_{392}), which determines the anti-wear properties of lubricants, showed that in MZT it is by 27.3% and 30.5% better than in LZ-TsNII, depending on the load, and by 32.2% and by 33.3% better than in ZhRO, respectively.

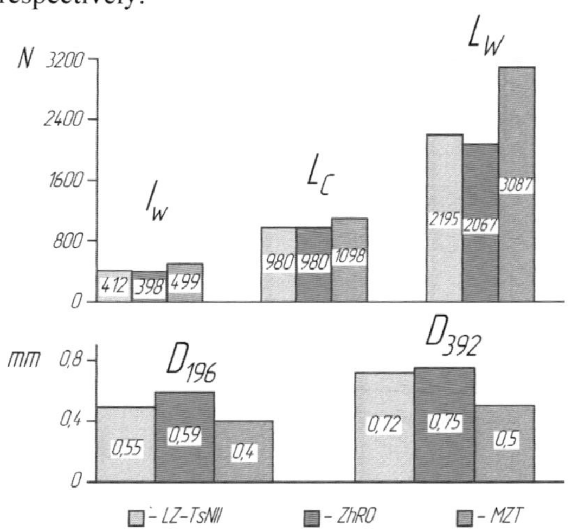

Figure 2. Results of tests on FBEPT.

The tribological properties of lubricants were also tested on a SMTs-2 friction and wear machine according to the generally accepted method [21] using two contact patterns of test specimens:

- the "roller-pad" pattern (figure 3 (a)), which implements the area contact of the tribocoupling and simulates the operation of the lower kinematic pairs. This pattern simulates the operation of tribocoupling with slipping friction. This type of contact is found in roller axle bearings between the end face of the rollers and the flange of the outer ring;
- the "roller-roller" pattern (figure 3 (b)), which implements the line contact of the tribocoupling and simulates the operation of higher kinematic pairs. This pattern simulates the operation of tribocoupling with rolling friction. This type of contact occurs between the working surfaces of rollers and bearing rings in roller axle bearings.

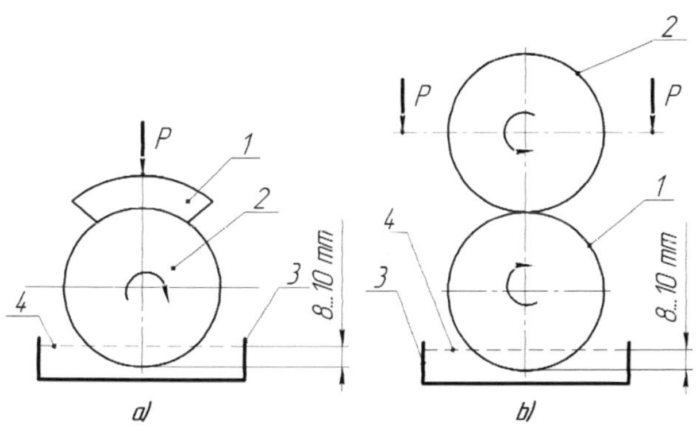

Figure 3. Test patterns on the friction and wear machine SMTs-2: (a) – "roller-pad"; (b) – "roller-roller"; 1 – pad; 2 – roller; 3 – lubricant tray; 4 – lubricant level; 5 – main roller; 6 – counterbody roller.

In these studies, the material of the pads and rollers was the same – steel with surface hardness of (55 to 58) HRC, which is close to the surfaces hardness of the parts of the axle bearings of the railway rolling stock. The maximum (working) load (*P*) in the friction pair of the SMTs-2 machine was selected based on the loads in the real friction units of the cars, locomotives, track vehicles, motor car rolling stock which were determined using the methods described in [22], taking into account the ratio of the dimension of the parts of the axle boxes of rolling stock and rollers and pads. All tests on SMTs-2 were repeated 4 times for each lubricant with each test pattern, and every time new parts of the friction pair and a fresh portion of lubricant were used, as recommended [21].

In the "roller-pad" tests, the loading mode in the friction pair shown in table 2 was selected. In this case, the rotation rate of the roller was set at 5 s^{-1} (300 min^{-1}). The result of the tests (the level of anti-wear properties of the lubricants) was determined by the weight loss of the friction pair parts during the test. In this case, the control of temperature of the roller surface was used to evaluate the antifriction properties of lubricants.

Table 2. "Roller-pad" test pattern.

Load, N	Time of testing, min.
0	10
50	20
100	30
500	30
800	240
	$\sum$= 360 (6 hours)

Tests on the SMTs-2 friction and wear machine demonstrated that the total wear (*W*) of both samples (figure 4 (a)) lubricated with MZT was by 30% lower than that of the ones lubricated with LZ-TsNII and by 47% lower than that of the ones lubricated with ZhRO. Similar results are also shown in temperature control (figure 4 (b)) – the antifriction properties of MZT are better than those of the other two lubricants. In samples that were lubricated with MZT, the average surface temperature was approximately by 12°C lower than that in the ones lubricated with LZ-TsNII, and by 17.6°C lower than that in the ones lubricated with ZhRO.

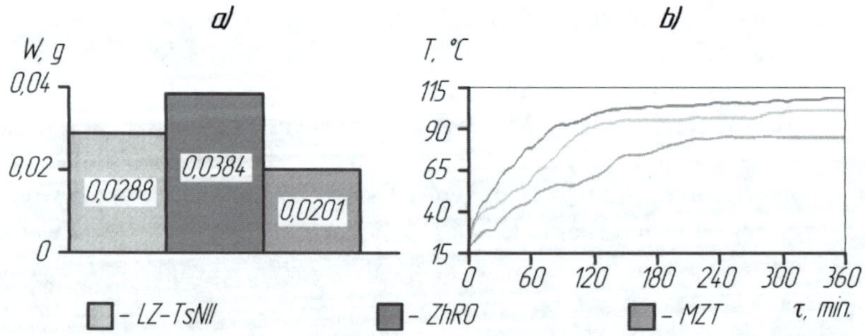

Figure 4. Results of tests on SMTs-2: (a) Total wear of pads and rollers; (b) Temperature of the roller surface during the tests.

The main type of wear in the second contact pattern ("roller-roller") is fatigue flaking (pitting), although there are also other types of wear. Since during rolling friction, the load in a friction pair is much lower, then taking into account the above principles of force calculation and simulation of operation of the friction pair, two test modes were selected on the basis of the "roller-roller" mode (table 3). Two modes were chosen because the estimated values of loads between the roller and the bearing ring in the real operating conditions of the axle boxes of cars, locomotives and motor car rolling stock differ quite significantly. The rest of the test conditions remained as described above.

Table 3. "Roller-roller" test pattern.

Load, N		Time of testing, min.
mode I	mode II	
0	0	10
100	100	10
300	500	20
500	800	20
1,000	1,400	30
1,500	1,800	30
2,000	2,500	240
		$\sum$= 360 (6 hours)

According to the results of the tests shown in figure 5, in the case of rolling friction, MZT provides better wear protection of the friction surfaces vs. LZ-TsNII by 26 to 36%, and vs. ZhRO by 34 to 41%. The same trend was observed in changing the temperature of the friction surfaces. Generally, the temperature of the friction surfaces during these tests was significantly lower than the one during the slipping friction, which is quite natural. The nature of the temperature change was similar to the one shown in the diagram (figure 4). During the tests, the rollers lubricated with MZT had a lower average temperature than those that were lubricated with LZ-TsNII by 9.3 to 11.9°C. In turn, the temperature of the samples lubricated with ZhRO was higher than that o f those lubricated with MZT by 16.9°C to 19.5°C.

The best tribological indicators of MZT lubricants can be explained not only by its k-Li base, but also by the content of certain active components such as sulfur [23].

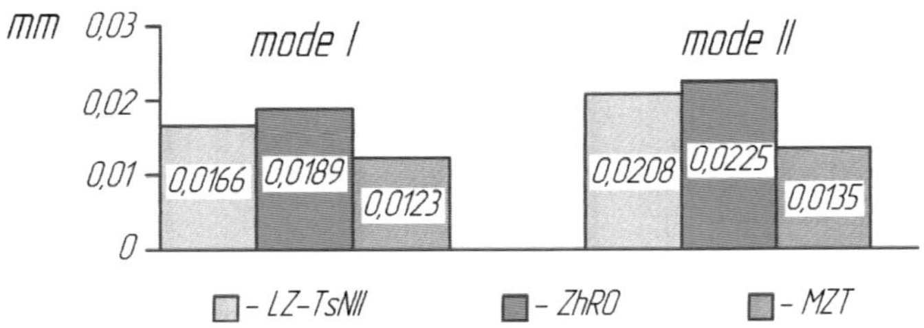

Figure 5. Total wear of rollers in testing on SMTs-2 in the "roller-roller" pattern.

4. Operational tests

Positive results of laboratory tests of MZT were followed by tests in the actual operational conditions of the axle boxes of railway rolling stock, namely, in freight cars, locomotives and electric trains.

Tests on a refrigerator car lasted until it had run 65706 km and were discontinued due to its major repair. During the periodic inspections of the axle boxes of the car, samples of the tested lubricants were taken and further analyzed in the laboratory using standard methods. The results of the study of the properties of all lubricant samples (table 4) showed that during the tests its quality slightly deteriorated, which is quite natural, but even at the end of the test, the values of the indicators are sufficient to ensure the normal operation of the axle box.

The following operational tests were carried out in the axle boxes of cars of the EP-9M No. 540 electric train until it had run 200000 km. During the tests, samples of the tested lubricant were taken from the axle boxes of the cars of the electric train, as in the previous case, to control their operational properties. The results of physicochemical studies (table 5) of these tests showed that MZT had stable properties throughout the test time. After the electric train ran for more than 200000 km, the chemotological parameters of the lubricant correspond to the operating conditions of the friction pairs of the axle boxes of the motor car rolling stock and have sufficient margin to achieve the boundary values.

Table 4. Quality indicators of MZT during the operational tests in the axle boxes of the freight car.

Indicator	Lubricant sample		
	after 20000 km run	after 40000 km run	final
Penetration at 25°C, m×10^{-4}	265	270	280
Melting point, °C	238	231	234
Colloidal stability, %	11.2	11.3	11.6
Mass fraction of free alkali expressed as NaOH, %	0.36	0.37	0.35
Fe content, %	0.018	0.028	0.04
S content, %	1.405	1.365	1.305
P content, %	0.245	0.235	0.230
Tribology on FBEPT at t= (20±5)°C:			
- load wear index	480	480	486
- wear scar at load 196 N, mm	0.4	0.37	0.38
- critical load, N	1.039	980	980

Table 5. Quality indicators of MZT in the operational tests in axle boxes of the electric train.

Parameter	Estimated run of the electric trains at sampling, km			
	more than 20000	more than 45000	about 100000	more than 200000
Colloidal stability, %	2.79	2.77	2.9	2.87
Melting point, °C	247	232	232	232
Penetration, mm^{-1}	266	264	267	271
Mass fraction of water, %	none	none	none	none
Mass fraction of mechanical impurities, %	none	none	none	none

Similar tests of MZT were carried out in the axle boxes of ten freight cars (gondola cars) that had been in operation for more than two years. At the end of the test period, the inspection of the axle boxes of the cars showed that the friction surfaces of the axle bearings were in a satisfactory condition, without apparent significant damage, and the lubricant in them appeared as homogeneous grease, was neither contaminated, nor contained any metal inclusions, traces of rust or flooding. Analysis of the physicochemical properties of lubricant samples, like in the previous case, showed that all the tested quality indicators were rather high, and the change in most of them in comparison to the initial values was negligible, which allows for prediction of the effective operation of lubricants and longer time of operation and runs.

Operational tests of MZT in the axle boxes of locomotives were carried out on locomotive 2TE116 and electric locomotive VL-80. The tests lasted until the locomotive had run 111428 km, and the electric locomotive – 100705 km. During the tests, the axle boxes of one section of the locomotive ran with MZT, and the second section – with the reference lubricant ZhRO (for comparison). During the testing period, there were no problems with the operation of the locomotive axle boxes. During the tests, samples of the tested and reference lubricant were taken to control their operational properties. The results of these tests showed that MZT had stable properties throughout the test time. After the locomotives had run for over 100000 km, the physicochemical and tribological indicators of the MZT corresponded to the operating conditions of the friction pairs of the axle box of locomotives, had sufficient margin to achieve the boundary values and wear at a much higher level than those of ZhRO.

5. Conclusion

In general, all conducted studies show rather high operation capacity of the MZT lubricant, which corresponds to the operation conditions of a pair of rolling bearings of axle boxes of cars, locomotives, motor cars and other railway rolling stock.

The results of separate studies (figures 2 and 5) show that the anti-wear properties of lubricant MZT manifest themselves more explicitly when the load in the friction pair increases. This data comply with the results obtained by other researchers [24], which noted the effect of friction conditions on tribological indicators of oils which have a complex lithium base.

High stability of MZT lubricant recorded during the operational tests provides basis for proposing to extend both operating periods of axle boxes between repairs and lubricant change intervals.

Testing of MZT in the laboratory for possible mixing with ZhRO gave positive results. The mixture of two lubricants has a homogeneous structure; its physicochemical properties are not much inferior to MZT and exceed those of the ZhRO. Partial mixing is allowed during the lubricant change in the friction unit. However, this cannot be the case for LZ-TsNII – it has a fundamentally different base and may not be mixed in operation with MZT, which should be taken into consideration.

For the Ukrainian railway it is especially favorable that MZT is manufactured by a domestic producer "Research and Production Enterprise Agrinol" Limited Liability Company (NVP Agrinol LLC). This significantly simplifies the procurement procedures and increases the stability of the supply of the lubricant.

The results presented in the paper allow for recommending the transfer of all rolling stock of the Ukrainian railways to the use of the single modern universal lubricant on the complex lithium base – Agrinol MZT – in the axle boxes.

References

[1] Yurov V H 2010 Bearing unit as an integral part of the railway rolling stock, ensuring traffic safety and increasing its standard service life *Vagonnyy Park* **9** 24–27

[2] Kleinlein E and Kroner H 2004 Lubrication of axle box roller bearings for different speeds: how to choose appropriate lubricants and prolong relubrication intervals *European Railway Review* **1** 59–65

[3] Instruction on the use and operation of lubricants in the axle boxes of freight car TsV-0071 2006 (Kyiv: Ukrainian Railways) p 35

[4] Instruction on the use of lubricants on the traction rolling stock of Ukrainian railways TsT-0060 2003 (Kyiv: Ukrainian Railways) p 54

[5] Bladon Keith, Rennison David, Izbinsky Grigory, Tracy Roger and Bladon Trevor 2004 Predictive Condition Monitoring of Railway Rolling Stock *Conf. Paper Research CORE 2004: New Horizons for Rail. Darwin, N.T. (Railway Technical Society of Australasia)*

[6] Xiaoqiang Fan, Wen Li, Hao Li, Minhao Zhu, Yanqiu Xia and Junjun Wang 2018 Probing the effect of thickener on tribological properties of lubricating greases *Tribology International* **118** 126–39

[7] De Laurentis N, Kadiric A, Lugt P and Cann P 2016 The influence of bearing grease composition on friction in rolling/sliding concentrated contacts *Tribology Int.* **94** 624–32

[8] Maciej Paszkowski 2012 Assessment of the effect of temperature, shear rate and thickener content on the thixotropy of lithium lubricating greases *Proc. Institution of Mechanical Engineers, Part J: Journal of Engineering Tribology* **227** 209–19

[9] GOST 7163-84 1984 Petroleum products. Method for mensurement of viscoslty by automatic cappillar viscosimeter (Moscow: Standards Publishing)

[10] GOST 7143-73 1973 Greases. Method for determination of ultimate strength and thermostrengthening (Moscow: Standards Publishing)

[11] ISO 2176:1995 Petroleum products. Lubricating grease. Determination of dropping point. Edition 2 (International Organization for Standardization)

[12] GOST 7142-74 1974 Lubricating greases. Methods for determination of oil separation (Moscow: Standards Publishing)

[13] DIN 51809-1:1978 Testing of lubricants. Determination of the free acidic or alkaline parts (neutralization number) in greases, colour indicator titration (German Institute for Standardisation)

[14] ASTM D2595-2017 Standard test method for evaporation loss of lubricating greases over wide-temperature range (ASTM International)
[15] ISO 2137:2007 Petroleum products and lubricants. Determination of cone penetration of lubricating greases and petrolatum. Edition 3 (International Organization for Standardization)
[16] ASTM D95-13e1 2013 Standard test method for water in petroleum products and bituminous materials by distillation (ASTM International)
[17] ISO 11007:1997 Petroleum products and lubricants. Determination of rust-prevention characteristics of lubricating greases. Edition 1 (International Organization for Standardization)
[18] Kombalov V S 2008 *Methods and Means of Testing for Friction and Wear of Structural and Lubricants: A Handbook* (ed Frolov K V and Marchenko E A, Moscow: Engineering) p 384
[19] ASTM D 2596-15 2015 Standard test method for measurement of extreme-pressure properties of lubricating grease (Four-Ball Method) (ASTM International)
[20] ASTM D 2266-01 2015 Standard test method for wear preventive characteristics of lubricating grease (Four-Ball Method) (ASTM International)
[21] Kravets A M, Kravets V H and Afanasov H M 2011 *Investigation of antiwear properties of lubricants using a friction machine SMTs-2: methodical instructions* (Kharkiv : UkrSURT) p 26
[22] Nevmerzhytskaia H V 2003 *Load Elements of Axle Components of Railway Rolling Stock and its Impact on the Reliability of the Axle Bearing* (Dissertation, Bryansk: BSTU) p 199
[23] Singh T and Bhan A 2001 Technical note: Extreme-pressure activity assessment of certain nitrogen and sulphur compounds in lithium-based grease *Lubrication Science* **13(3)** 291–7
[24] Sosulina L N and Volobuev N K 1989 Influence of friction conditions on tribochemical stability of lithium lubricants *Chemistry and Technology of Fuels and Oils* **6** 26–8

Comparative analysis of the results of theoretical and experimental studies of freight wagon Sdggmrss-twin

V Stoilov[1], G Simić[2], S Purgić[1,3], D Milković[2], S Slavchev[1], S Radulović[2] and V Maznichki[1]

[1] Department of Railway Engineering, Faculty of transport, Technical University – Sofia, 8 Kliment Ohridski Blvd., 1000 Sofia, Bulgaria
[2] Laboratory of rail vehicles, Faculty of Mechanical Engineering, University of Belgrade, 16 Kraljice Marije Str., 11120 Belgrade, Serbia

[3] E-mail: s_purgic@tu-sofia.bg

Abstract. This paper presents a comparative analysis based on the results from static strength calculation of wagon body, series Sdggmrss-twin, and on the results from the real wagon test. The verification of results from calculations and tests and their comparison was mandatory for client's commissioning of the wagon by notified body. Calculations based on the finite elements method were carried out in the Department of Railway Engineering at Technical University of Sofia. Experimental studies on real wagon construction were conducted at the facilities of Bulgarian National Transport Research Institute by testing team from Laboratory of rail vehicles at University of Belgrade. It was found that the obtained static stress results are similar, which proves that the proposed models are appropriate and they can help to solve a wide range of issues, for example those related to lightweight design of railway vehicles.

1. Introduction

The strength calculations and real tests of the freight wagons underframe and wagon body have to be carried out in accordance to procedures regulated in the relevant European standards, Technical Specifications for Interoperability – Freight Wagons (WAG TSI [1]) and/or UIC leaflets. Before the new freight wagon is allowed for operation, the strength calculations and real tests are mandatory. If new freight wagon structure passes both strength calculations and tests, notified body allows the commissioning of the wagon based on positive results. If there is a large discrepancy in results from both procedures, notified body can order either new calculations or tests or both. This causes loss of time and money, but also loss of confidence in calculation or testing teams by the wagon manufacturer. That's why the verification of results from calculations and tests and their comparison is important for the client and is mandatory for commissioning. On other hand, this is also very important for calculation teams, because they can prove precision and reliability of their models, as well as the appropriateness of calculation method and the level of their know-how. If the know-how of calculation team is not on enough high level, chances for future orders are small. This is valid also for the testing crew. Based on similar results from calculations and tests and in case of relatively low stresses, wagon structure could be built up from thinner profiles and beams. This helps to reduce the own weight of the wagon and to increase the maximum payload.

Chapter 2 of this paper describes the object of the study, load cases applied on the wagon structure and the properties of calculation model and methods. Testing and measuring equipment, testing and

Content from this work may be used under the terms of the Creative Commons Attribution 3.0 licence. Any further distribution of this work must maintain attribution to the author(s) and the title of the work, journal citation and DOI.
Published under licence by IOP Publishing Ltd

measuring methods and some specific features regarding tests are the subject of chapter 3. Chapter 4 presents the results from calculations and tests as well as their analysis. Paper ends with conclusions.

2. Object of the study and calculation model

The object of the study is a wagon series Sdggmrss-Twin, manufactured by Bulgarian wagon manufacturer KOLOWAG Plc. Sdggmrss-Twin is articulated wagon with two underframes that differ among themselves only in a small area of articulation. This wagon is designed for transportation of containers and semitrailers and is defined in category F-II (vehicles restricted in hump and loose shunting). The three-dimensional wagon model (section 2) is shown in figure 1.

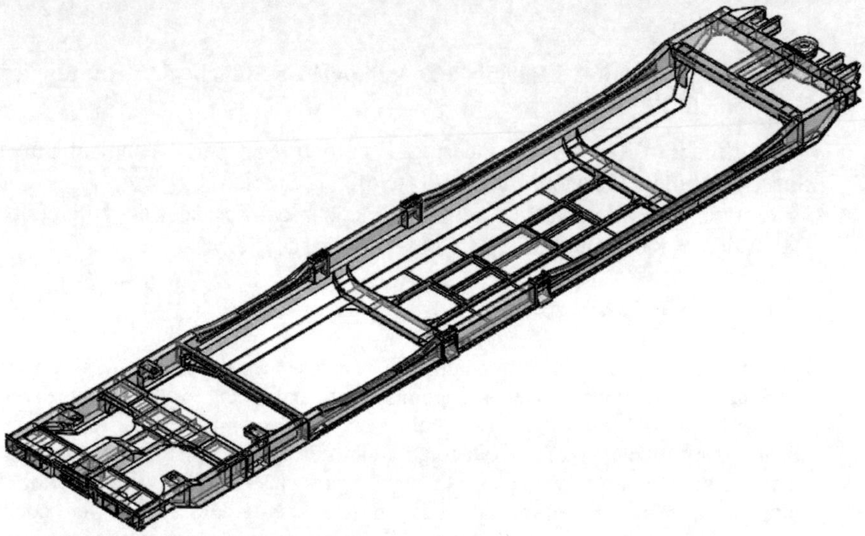

Figure 1. Three-dimensional model of wagon series Sdggmrss-Twin.

The main parameters of the wagon are: gauge G1, number of axles 6, axle load 22.5 t, payload capacity about 101 t, own weight 34 t, wagon length without buffers 33 m. The underframe of the wagon is made of steel S355J2 according to European standard EN 10025.

2.1. Calculation model

Strength calculations of the wagon are made by the finite element method. The software product SolidWorks is used for calculations. For the purposes of strength analysis, spatial computational models were developed by using 3D solid elements.

The models were optimized by analysing the approximation of the calculations. Optimized model is made up of 561722 notches and 289763 finite elements. The finite elements mesh is shown in figure 2. The maximum size of the finite elements is 80 mm and the minimum size (in the zones with expressed stress concentrators) is 10 mm.

2.2. Load cases

In accordance with international standards, each newly designed wagon is object of theoretical (by calculation) and experimental studies in order to determine stresses and deformations. Load cases and conditions are specified in European standard EN 12663-2:2010 [2], TSI subsystem "Rolling stock – Freight wagons" [1] and leaflet 577 [3] of the International Union of Railways (UIC). According to documents cited, loads are divided into the following groups: longitudinal static loads (HLC), vertical static loads (VLC), exceptional static loads (LLC) and superposition of static loads (CHLC). An overview of all load cases is shown in table 1. Methods for the application of loads from HLC, VLC, LLC and CHLC load cases groups are described in detail in [4-10] and for articulated wagons in [11].

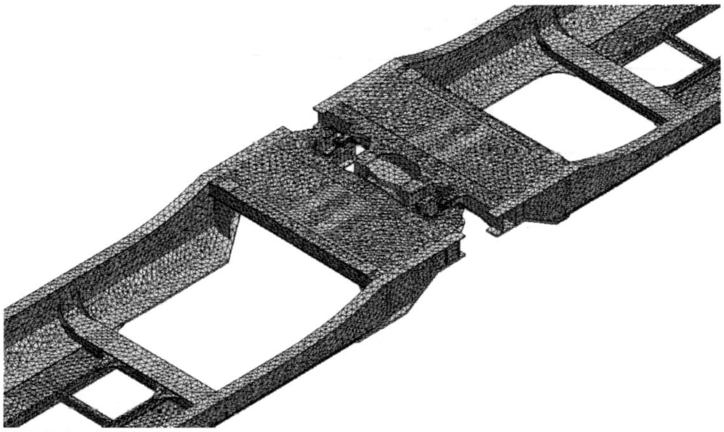

Figure 2. The finite elements mesh used in calculation model.

Table 1. Load cases applied to wagon construction.

Load case	Description
HLC 1	Compressive force at buffer axis height 1200 kN (table 2 of [2])
HLC 2	Compressive force 50 mm below buffer centre line 900 kN (table 3 of [2])
HLC 3	Compressive force applied diagonally at buffer level 400 kN (table 4 of [2])
HLC 4	Tensile force at coupler axis level 1500 kN (table 5 of [2])
VLC 1	Vertical load from weight of 2×20 ft. containers 1,3 g $(m_1 + m_3)$
VLC 2	Vertical load from weight of 40 ft. container 1,3 g $(m_1 + m_3)$
VLC 3	Vertical load from weight of 45 ft. container 1,3 g $(m_1 + m_3)$
VLC 4	Vertical load from weight of semitrailer 1,3 g $(m_1 + m_3)$
LLC 1	Lifting at one end of the vehicle (table 7 of [2])
LLC 2	Lifting the whole vehicle at 8 lifting positions (table 8 of [2])
LLC 3a	Lifting the whole vehicle at 8 lifting positions with one lifting point displaced 10 mm vertically (table 8 of [2]) – at wagon end
LLC 3b	Lifting the whole vehicle at 8 lifting positions with one lifting point displaced 10 mm vertically (table 8 of [2]) – in the middle of wagon
CHLC 1	Compressive force at buffer axis height 1200 kN combined with most unfavourable vertical load
CHLC 2	Compressive force 50 mm below buffer centre line 900 kN combined with most unfavourable vertical load
CHLC 3	Compressive force at buffer axis height 1200 kN combined with vertical load from own mass
CHLC 4	Tensile force at coupler axis level 1500 kN combined with most unfavourable vertical load
CHLC 5	Tensile force at coupler axis level 1500 kN combined with vertical load from own mass

For freight wagon's underframe and body, the methodologies and procedures for strength analysis and testing are defined in the European standard EN 12663-2: 2010 - Railway applications - Structural requirements of railway vehicle bodies - Part 2: Freight wagons [2]. In addition to the load cases it also defines the minimum requirements for wagon body strength, materials, permissible stresses and displacements and presents principles and methods that are used to validate the structure by strength analysis and testing.

The strength evaluation of the wagon underframe (acceptance criteria) is made using the safety factors S as given in equation (1):

$$S = \frac{\sigma_{\lim}}{\sigma_{Mises}} \geq 1.0, \quad (1)$$

where $\sigma_{\lim}$ is the permissible stress. For steel S355J2 the parameter $\sigma_{\lim} = R_e$ (yield limit of the material) = 355 MPa for parent material and $\sigma_{\lim} = R_e/1.1$ = 323 MPa for immediate vicinity of welds. The stress σ_{Mises} represents the maximum stress received under the corresponding load case.

IOP Conf. Series: Materials Science and Engineering **664** (2019) 012026 doi:10.1088/1757-899X/664/1/012026

3. Testing and measuring procedure and equipment

Current European regulations [1] and standards [2] require strength tests for every new rail vehicle structure. The first part of the test consists of a static strength test, which is dealt with in this document. The second part of the strength test includes buffing impact tests. Structural problems during operation may require additional strength testing under actual operating conditions [12].

The static strength test of the Sdggmrss wagon prototype was performed in the hall of the Bulgarian National institute for transport research (NIIT) in Sofia and in accordance with the "Program for static strength test of Sdggmrs-twin wagon carbody" which is previously approved by the manufacturer and Notified body.

As already mentioned above, the wagon has two carbodies that differ among themselves only in a small area of articulation. In order to include articulation influence, for the test was chosen one carbody and head part of the other carbody, resting on two bogies as shown in figure 3. In order to reduce lowering of the carbody during loading, the carbody was resting on two bogies with blocked suspension. The test was performed on the test rig (figure 4) with adjustable configuration. Different load cases according to the test program, i.e. application of the appropriate compression or tension forces as well as different combinations of vertical forces was provided by using hydraulic cylinders and auxiliary tools.

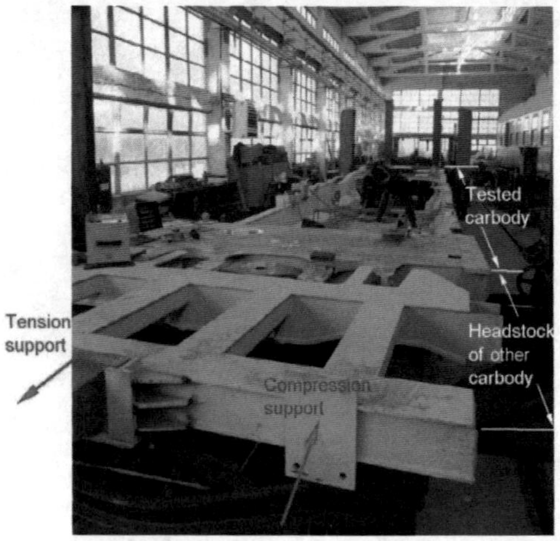

Figure 3. Arrangement of carbodies during the test.

Figure 4. Test rig overview.

Strain gauges were applied in the points according to test program. The positions of one part of those points were determined based on the locations of highest stresses obtained during the calculations. Other part of strain gauges is placed in some standard positions on wagon underframe, as well on the positions, for which the testing team proposed to be important based on their experience and know-how. Figure 5 shows example of applying the strain gauges on the testing object. Strain gauges were covered with glue for protection against mechanical damage as shown in figure 6.

The following quantities were measured:

- local strains ε_i, μm/m of the testing object and derived stresses σ_i, N/mm^2;
- forces for the vertical load application resulting from vertical cylinders pressures p_i;
- forces F_i, kN used for horizontal loads application;
- deflections f_i, mm of the superstructure under different load cases.

In order to measure all measuring quantities measuring chain presented schematically in figure 7 was used. Data acquisition system (DAQ), consisting of four Quantum X MX840A, one Quantum MX1615 and two SPIDER8 systems, was able simultaneously to record all signals of measuring quantities (strain gauges, forces and displacements). Readings of both DAQ systems were timely

IOP Conf. Series: Materials Science and Engineering **664** (2019) 012026 doi:10.1088/1757-899X/664/1/012026

synchronized, which was simply because of the static character of the test. Quantum X was connected via Ethernet connection to the PC. For the data acquisition and analysis the HBM CatmanEasy AP software was used. SPIDER8 was connected via USB-LPT adapter to the PC. For the data acquisition and analysis the HBM CatmanExpress software was used. The stress measurements were performed using 34 resistance strain gauges and 2 rosettes 0/45/90° with nominal resistance of 120 Ω and length of the measuring grid of 6 mm.

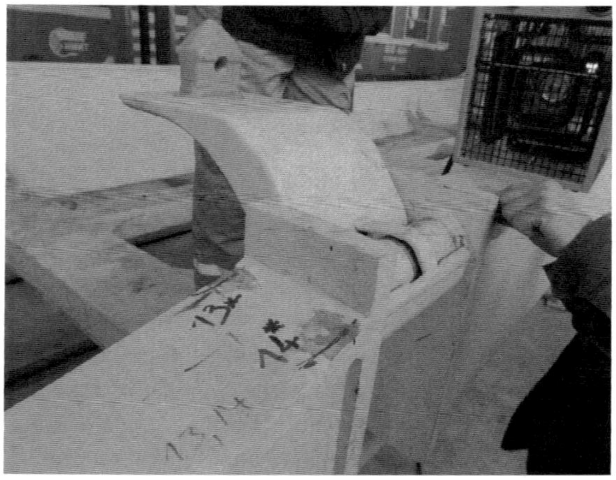

Figure 5. Applying the strain gauges on the testing object.

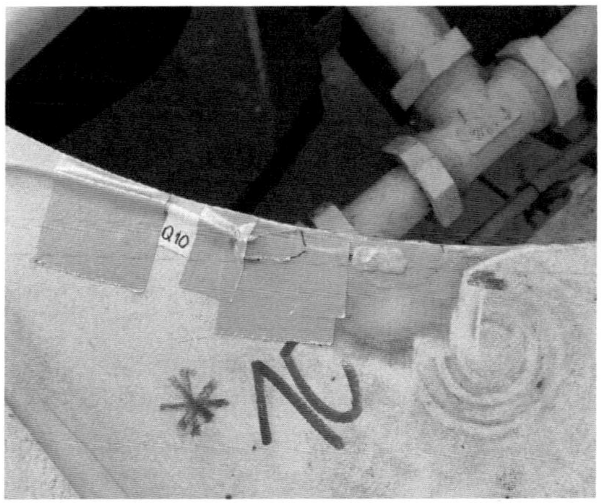

Figure 6. Protected strain gauge.

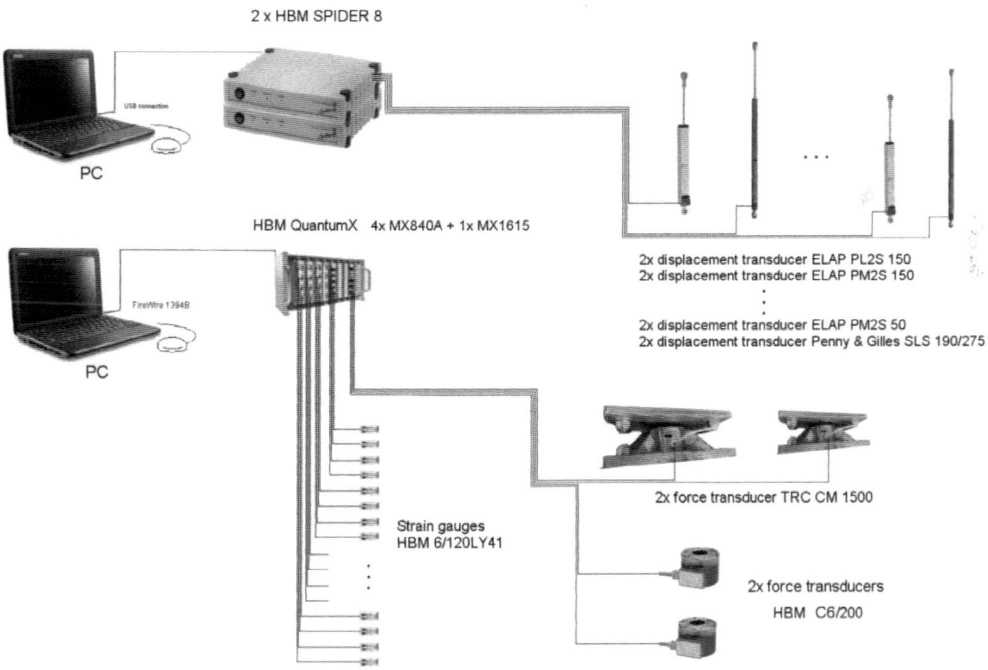

Figure 7. Measurement chain.

Longitudinal forces were measured using two force transducers CM1500. Vertical forces were measured at one hydraulic cylinder in each cylinder group connected in parallel using load cells HBM C6. Eight displacement transducers were used for measurement of carbody vertical deflections on each longitudinal beam (dz1 to dz4 on one side and dz5 to dz8 on the other side). Four displacement transducers were used for measurement of the longitudinal displacement of the both main longitudinal beams (dx1 to dx4). From eight displacement transducers used for vertical deflection measurements, two of them on each longitudinal wagon side are in the vertical plane of the supporting points on bogies. They are used to define referent straight line relative to which the deflections in the middle of the wagon and on the wagon end are determined. Horizontal displacement transducers were used for measurement of the main longitudinal beams change in length under compression and tension load.

4. Analysis of results from calculations and tests and their comparison

The discrepancies between the calculated and measured stress values arise for various reasons:

- geometrical idealization and simplifications of the calculation model;
- numerical calculation errors;
- mesh size, mesh adaptation to object geometry;
- characteristics of finite elements used;
- boundary conditions fidelity;
- tolerances in real material characteristics (R_{eH}, E, etc.);
- geometrical imperfections of the real structure. For example tolerances of profile and sheet thicknesses and shapes, not ideal symmetry of the real (especially welded) structure etc.;
- residual stresses from manufacturing (after welding, bending of sheets etc.);
- measurement uncertainty which includes among others strain gauge gluing imperfections, strain gauge tolerances, errors of positioning and aligning of strain gauges, measurement equipment accuracy etc.;
- averaging along measurement base of strain gauge;
- tolerances in the force introduction (intensities, symmetry, accuracy of the force application area etc.).

Relevant standards [2] prescribe that in order to stabilize the residual stresses due to manufacturing, before proceeding with the recording of the stresses, for all static tests preliminary loading shall be carried out. Preliminary loads are applied in stages, up to the stipulated maximum loads and after removal of the loads, the strains are considered to be zero. After applying the loads a second time up to the maximum value, the measurement should be considered as decisive.

The data used in the analysis is excerpted from reports of calculations and tests performed. It summarizes only the results registered in the strain gauges, which were recommended by calculation team to be used in the tests. It should be noted that the results obtained in calculation have been refined; taking into account the stresses at exactly the same points at which the strain gauges were positioned in the test. Only the so-called "clear" load cases have been taken into account, i.e. only the load cases for which the stress values were measured directly, without additional mathematical processing of the results, such as superposition, application of strength hypotheses for calculation of the equivalent stresses etc. Such load cases are: HLC2, HLC4, HLC6, HLC8, VLC2, VLC8 and LLC1 (table 1). The purpose of the above-mentioned features is to use the most accurate information possible to objectively assess the contrasting juxtaposition of results from calculations and tests. Namely for this reason, the comparative analysis was carried out according to criterion σ_1, which is actually measured by strain gauges in the wagon tests. Only the stress values for the positions where the highest stresses during the tests were registered are compared. The comparison of results obtained in calculation and tests for load cases mentioned above for some strain gauges is given in table 2. In table 2 abbreviation T stands for "Test" and C for "Calculation". All values are given in N/mm^2.

The analysis of the data from table 2 shows that there is a good match of the results for the stresses obtained in calculations and tests. Largest differences in measured and calculated stress values can be observed for resistant strain gauges Rosette 1 and Rosette 2 particularly for load case LLC 1. The discrepancies mentioned above can be explained visually in figures 8 and 9.

As it can be seen in the figure 8, maximal stress values are calculated in parent metal in the bending area of large metal sheet, divided in this way in vertical and horizontal sills with small bending radius between. It should be noted, that the calculated values of stress are valid for whole finite element independent on its size and present actually the average stress value. This means, that the average stress value is same e.g. for point at distance 1 mm from the edge and for point at 30 mm from the edge. In figure 9 the position of two corresponding measuring strain gauge rosettes is shown. Distance between them is 36 mm. This is because it is almost impossible to place the rosette so close to the edge because of restrictions for placing the strain gauges. This area between bended vertical side sill and horizontal plate also causes structural changes in the material because of bending during the manufacturing of the wagon and presents significant stress notch.

However, in this load case and for all tested load cases, the resulting stress values are lower than permissible.

Table 2. Comparison of results from calculations and tests.

Strain gauge No.		1	2	4	14	15	19	Rossette 1	Rossette 2
HLC 1	T	–124.2						66.2	67.0
	C	–117.8						48.3	58.0
HLC 2	T		–157.6						
	C		–145.3						
HLC 3	T		–50.7						
	C		–53.9						
HLC 4	T				–121.0				
	C				–122.8				
VLC 1	T			–223.3					
	C			–193.5					
VLC 4	T						–92.2		
	C						–105.8		
LLC 1	T					–120.1		190.9	197.6
	C					–95.4		259.7	276.9

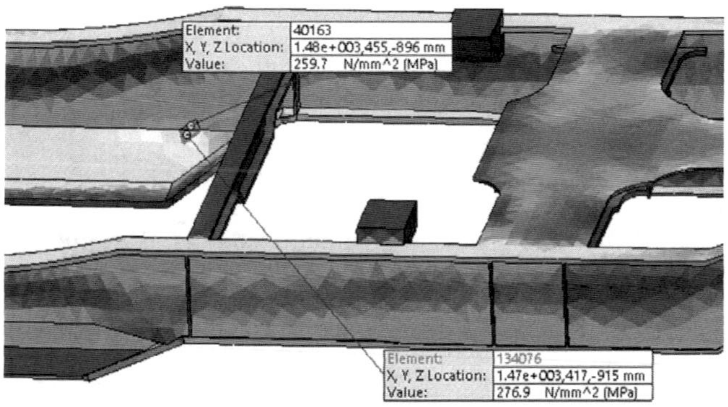

Figure 8. Calculated stress results for load case LLC 1.

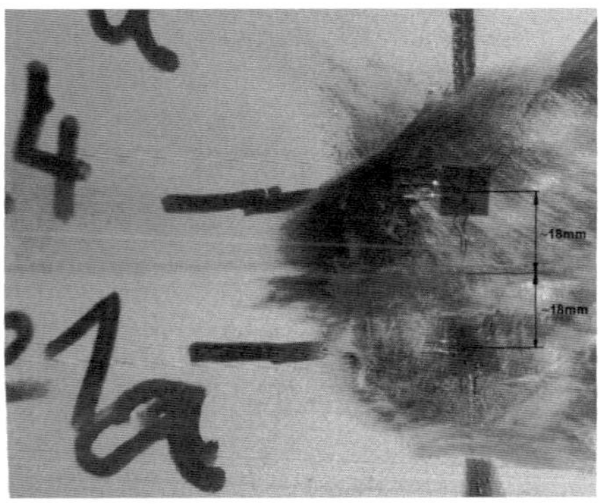

Figure 9. Position of strain gauge rosettes.

5. Conclusion

Summarizing the overall work on this study, it can be concluded that very good match of the results for the stresses obtained by calculation and those of the wagon test is present. This allows calculation models to be used for research and optimization of similar objects with similar construction. The performed tests show that the wagon withstands all prescribed loads without exceeding the permissible stresses and displacements. The developed models for static strength analysis of the articulated wagon series Sdggmrss-Twin structure can be used to optimize areas with considerably lower stresses. This will lead to a considerable reduction of the wagon's own mass and an increase in its load capacity.

Acknowledgements
The authors would like to thank to the Research and Development Sector at the Technical University of Sofia for the financial support. This work was also supported by the European Regional Development Fund within the Operational Programme "Science and Education for Smart Growth 2014 - 2020" under the Project CoE "National centre of mechatronics and clean technologies" BG05M2OP001-1.001-0008-C01.

References
[1] Commission Regulation (EU) No 321/2013 Technical specification for interoperability relating to the subsystem rolling stock - freight wagons (WAG TSI) of the rail system in the European Union and repealing Decision 2006/861/EC
[2] EN 12663-2:2010 *Railway applications - Structural requirements of railway vehicle bodies - Part 2: Freight wagons*
[3] UIC Leaflet 577, 5th edition, 2012 *Wagon stresses*
[4] Slavchev S, Georgieva K, Stoilov V and Purgic S 2015 Analysis of the results of theoretical and experimental studies of freight wagon Fals *University of Nis, Facta Universitatis, Series: Mechanical Engineering* **13(2)** 91
[5] Slavchev S, Stoilov V and Purgic S 2015 Static strength analysis of the body of a wagon series Zans *J. Balkan Tribol. Assoc.* **21(1)** 49
[6] Milovanovic V, Zivkovic M, Disic A, Rakic D and Zivkovic J 2014 Experimental and numerical strength analysis of wagon for transporting bulk material *J. IMK-14 – Research & Developement in Heavy Machinery* **20(2)** 61
[7] Baier A and Placzek M 2015 Computer-aided strength analysis of the modernized freight wagon *IOP Conf. Ser.: Mater. Sci. Eng.* **95** 012042
[8] Štastniak P, Moravčik M and Smetanka L 2019 Investigation of strength conditions of the new wagon prototype type Zans *MATEC Web of Conferences* **254** 02037
[9] Krason W and Niezgoda T 2014 FE numerical tests of railway wagon for intermodal transport according to PN-EU standards *Bulletin of the Polish Academy of Sciences Technical Sciences* **62(4)** 843–51
[10] Baykasoglu C, Sunbuloglu E, Bozdag S, Aruk F, Toprak T and Mugan A 2012 Numerical static and dynamic stress analysis on railway passenger and freight car models *Proc. of the Int. Iron and Steel Symp. (Karabuk University ISBN 6058708915*) 579–86
[11] Sunbuloglu E and Bozdag E 2018 A design strategy on the talbot type articulation of a freight wagon *Computers in Railways XVI: Railway Engineering Design and Operation (Ashurst Lodge: WIT Press ISBN 978-1-78466-285-1)* 145–56
[12] Lučanin V, Simić G, Milković D, Ćuprić N, Golubović S 2010 Calculated and experimental analysis of cause of the appearance of cracks in the running bogie frame of diesel multiple units of Serbian railways *Engineering Failure Analysis* **17(1)** 236–48

Analysis of some issues in the theoretical studies of unloading flaps strength of wagon series Falns

S Slavchev[1], V Stoilov[1], V Maznichki[1, 2] and S Purgic[1]

[1] Department of railway Engineering, Faculty of transport, Technical University – Sofia, 8 Kliment Ohridski Blvd., 1000 Sofia, Bulgaria

[2] E-mail: v.maznichki@tu-sofia.bg

Abstract. The self-unloading wagons with saddle-shaped floor, known under the serial designation Falns, are equipped with unloading flaps contributing to the rapid mechanized unloading of the wagon. In general, the theoretical and experimental studies of such wagons are standardized in EN 12663-2: 2010. Unlike many other nodes, no specific conditions (load forces, application locations, supports, etc.) have been described in the standard in order to properly investigate the strength of unloading flaps. This leads to the use of various methods for determination of their strength, which on other hand can lead to their improper design. The report analyzes the current used methods for the investigation of the unloading flaps strength, as well as their advantages and disadvantages. It is recommended to develop a specific methodology, reflecting the specifics of the construction and its way of operation. In this way, unification in the investigation of the unloading flaps strength will be achieved, as well as avoiding of the oversizing or undersizing of these responsible nodes in the Falns series wagons.

1. Introduction

Wagons of the Falns series (figure 1) are specialized self-unloading wagons designed for the transport of bulk goods. Numerous varieties are exploited in the world practice, differing in their basic parameters: axle load, pivot distance, width, length, console length, useful volume, number of unloading flaps, etc.

The common ground between them is that the floor is not horizontal but is inclined at a certain angle to the horizontal plane, which contributes to the rapid unloading of the vehicle. Additionally, the process is favored by the presence of unloading side flaps, a mechanism for opening and closing the flaps and a mechanism for locking the flaps during loading and transport. Those mechanisms are operated with pneumatic cylinders and suitably designed lever systems.

The theoretical study of the strength of wagons of this type is carried out in accordance with EN 12663-2:2010 Railway applications. Structural requirements of railway vehicle bodies. Freight Wagons [1]. Figure 2 shows the calculation model for testing the strength of the wagon body. The full list of load cases is given in [2-5]. In the stress-strain analysis, the unloading flaps are not presented in the wagon calculation model.

The reason for this is that they are an auxiliary bearing element and are hinged to the wagon body. Here the hypothesis is that if the structure stands without any problems the regulated loads in the absence of the unloading flaps, a more favorable distribution of stresses, deformations and displacements will be found upon their presence. In practice, complying with the requirements of [1, 6-8], the unloading flap and the unloading and locking mechanisms may remain unanalysed.

This is the reason why the manufacturers (or initiators of the theoretical analysis) of such wagons explicitly require the calculation teams to perform a stress-strain analysis of the mentioned parts.

Content from this work may be used under the terms of the Creative Commons Attribution 3.0 licence. Any further distribution of this work must maintain attribution to the author(s) and the title of the work, journal citation and DOI.
Published under licence by IOP Publishing Ltd

Figure 1. CAD model of wagon series Falns [5].

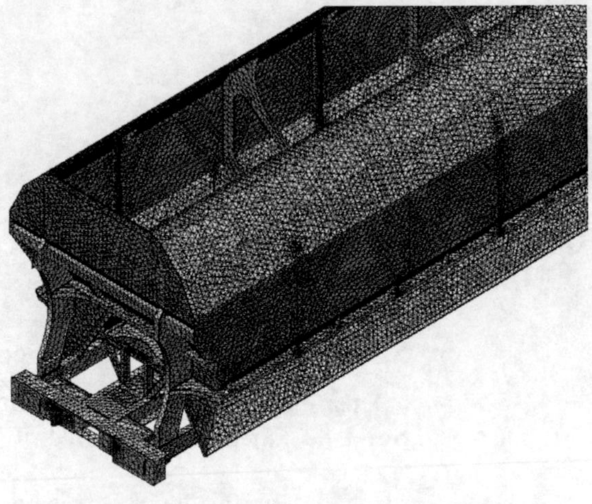

Figure 2. Finite elements mesh of calculation model [5].

2. Methods for determination of the loads acting on flaps

Extensive analysis of the normative documents shows that there are no regulated methodologies for determination of the values of the loads, the positions of their application, the methods of evaluation, etc. needed for the analysis of the flaps strength. The issue is dealt with in a number of publications [9-16], most fully analyzed in [9]. The methods of Jansen HA, Kleyn, GK, Mazarskiy SM, Putyato AV, Rusinek R., Zenkov RL, as well as the Standards of Eurasian Economic Union 1996 [17], GOST 33211. 2014 [18] and the German engineer's manual [19] were analysed. These publications offer different formulas for determining the pressure acting on the walls of the wagon (in this case, the wagon body). It has been found that the proposed functional dependencies are derived from the classic hydrostatic pressure formula. They mostly take into account the parameters of the repose angle, the angle of internal friction of the bulk load, the coefficient of friction between the bulk material and the wall studied, the angle of wall inclination, etc. and they allow simplifications (to alleviate the calculation process and avoid the demand for data specific to the material), which in practice lead to the use of the classic hydrostatic pressure formula. The proposed functional dependencies resulting in increased pressure values on the walls.

This is the reason in [19] to recommend following dependencies (equations (1) and (2)) to be used when analysing the strength of bulk storage vessels (including the Falns series flaps):

$$F = \frac{\rho\, g\, ({h_1}^2 - {h_2}^2)}{2 \sin \alpha}\, l; \tag{1}$$

$$e = \frac{(h_1 - h_2)\,(h_1 + 2h_2)}{3 \sin \alpha\, (h_1 + h_2)}. \tag{2}$$

The variables used in both equations are (as illustrated in figure 3): F – the force acting on the wall; ρ – density of bulk good; g – gravity acceleration; h_1 – height of the vessel; h_2 – the height of the unstudied part of the vessel; α – angle of inclination of the wall studied; l – length of the studied wall and e – point where the force F is applied.

3. Analysis of wagon's Falns unloading flap

The theoretical strength analysis of similar wagons was done in some previous works of calculation team [2-5]. Strength analysis of flap is done using the Finite Element Method [20, 23] in SolidWorks Simulation [21, 22]. For this purpose, the calculation model shown in figure 4 was used. The object geometry is modelled with 3D finite elements.

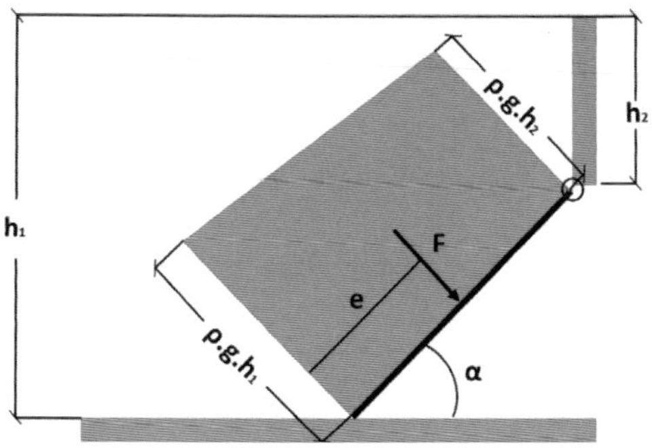

Figure 3. Onesided force acting on inclined wall [10].

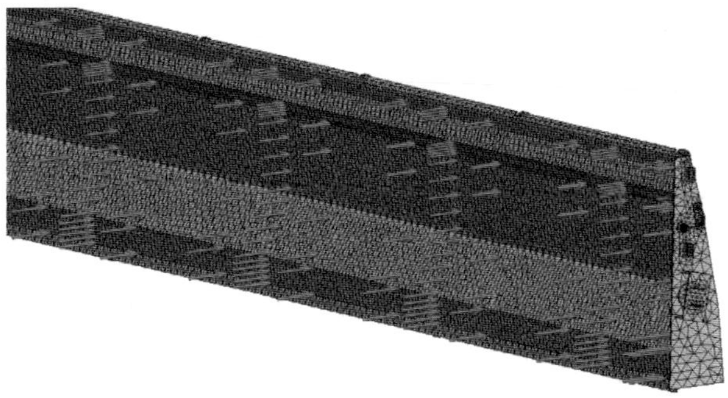

Figure 4. Finite elements mesh of calculation model [5].

The load is distributed unevenly over the entire surface of the flap. For this purpose, equation (3) is used at variable values of the distance h_{2z} from h_0 – the height of the fixed part of the wagon wall to h_1 – the height of the movable (flap) and the fixed part of the wall, i.e.:

$$h_{2z} = h_0 \div h_1. \quad (3)$$

The displacements are limited to the zones of the six hinges around which the valve rotates, two lateral hinges for the levers to open and close the flap and in the three zones of thumbs that lock the flap, as shown in figure 5.

The results of strength analysis are shown in figures 6 and 7.

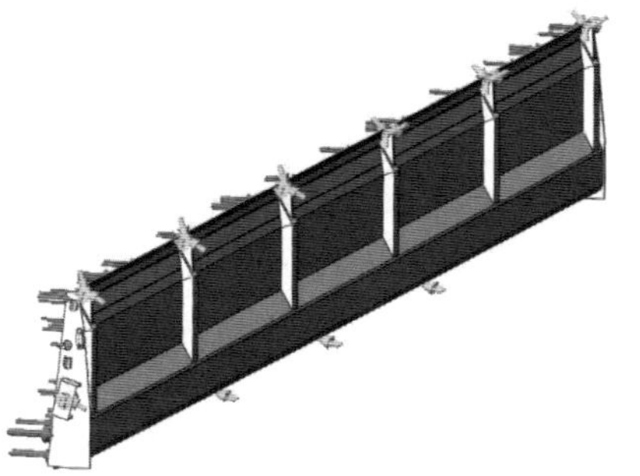

Figure 5. Limitation of displacements.

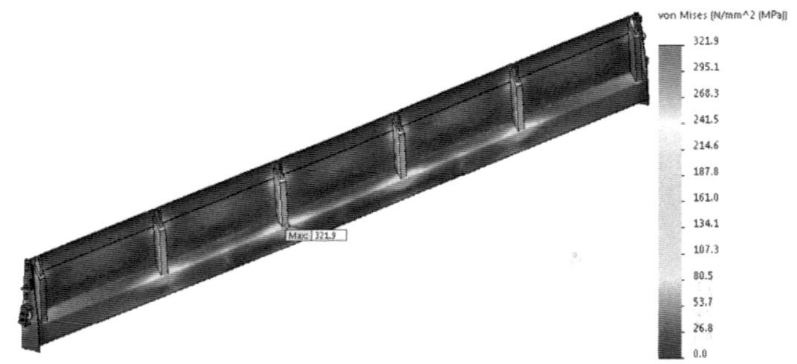

Figure 6. Distribution of stresses in flap.

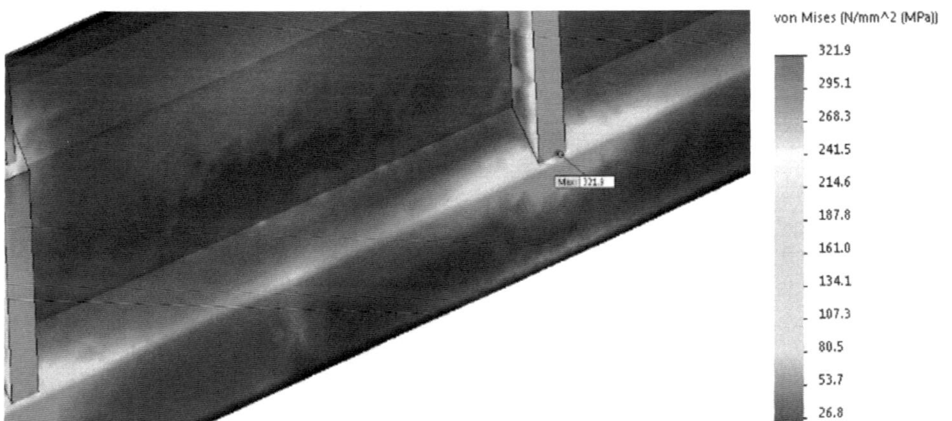

Figure 7. Maximal srtess values in flap.

The strength assesment is performed using permissible stresses in accordance to equation (4) [1, 7, 8, 24]:

$$\sigma_c \leq \sigma_{lim}, \tag{4}$$

where σ_c is calculated stress and σ_{lim} is permissible stress. The permissible stresses σ_{lim} depend on material used and analysed area. For steel S355J2 permissible stresses are 355 MPa for parent material and 323 MPa for parent metal in immidate vicinity to weld. The maximal stress value is 321.9 MPa and is obtained in the weld area. This value is very close to the permissible value and is characterised with safety factor as given in equation (5):

$$S = \sigma_{lim} / \sigma_c = 1.0034 > 1.0. \tag{5}$$

According to [1, 7] equation (5) allows the flap to be commissioned.

4. Additional calculations and test analysis of wagon's Falns flap

The minimal fullfillment of a condition in equation (5) has provoked the authors to continue the flap research in order to improve the calculation model by reflecting the design features, the modes of operation of the object, its behavior during operation. For this purpose, real observations were made of 80 Falns series wagons operating across Europe for more than 3 months. They serve for transporting powdered material, calcium carbonate $CaCO_3$, with a grain diameter of 0.1 to 5 mm. The following more important conclusions were made:

- The wagon as a whole (calculations done by the same team of authors) does not show any indications of insufficient strength. It is convenient for service. The mechanisms for opening and closing the flaps and for locking them operate flawlessly.
- Flaps are designed and manufactured to a high level. There is no spilling of the material transported.
- When the flaps are unlocked (without using the opening mechanism) there is also no spilling of the loaded material.
- No plastic deformations have been observed on the flap bottom sealing profile (position 2 in figure 8).
- The middle locking pin of the flap (figure 5, middle lower profile support) does not contact the lower profile both during the movement and when unloading. This is confirmed by the undamaged layer of paint on both elements.

The last four conclusions are extremely important as they show that, in the absence of locking, the flap seals seamlessly the wagon's cargo space and thus prevents leakage of the load.

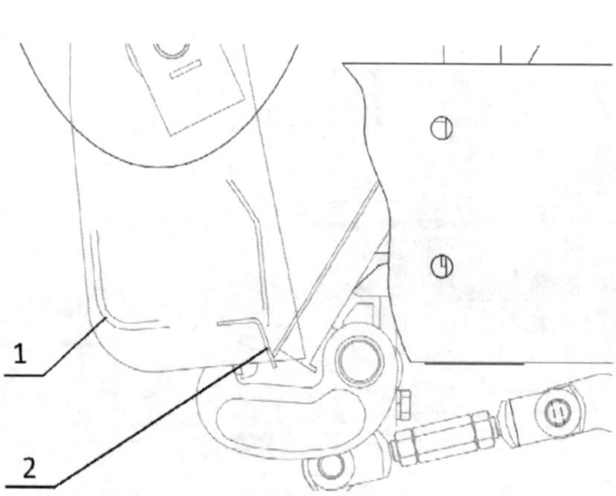

Figure 8. Flap bottom sealing profile.

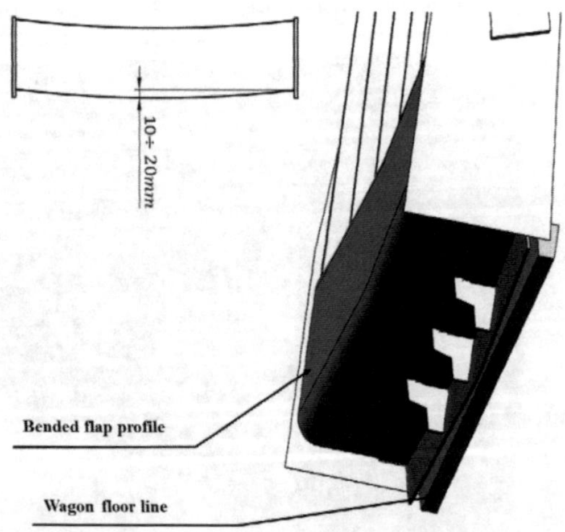

Figure 9. Schematic diagram of the contact between the wagon floor and the bottom flap profile.

This is assured as follows: Constructively, the bottom profile (position 1 of figure 8) is designed with a certain curvature. In this case, the bending arrow is 15 mm (figure 9). When the flap locking mechanism is actuated, the bottom profile travels the free distance to the floor, touches its convex part to the floor and, under the action of the applied forces, elastically deforms until contact is made across the entire floor line. Elastic deformations occur in both the sheet metal and the sealing profile 2 in figure 8. Deformations are elastic, which is confirmed by the observations made.

The described principle of flap closing operation required correction of the test method to maximize the simulation to the actual flap operation. For this purpose, the stress-strain analysis is done through the "Non-linear task" module, with contact between the floor and the valve. The results of the calculations are given in figure 10. In this case, the calculationl model and boundary conditions are preserved the same as in the initial study.

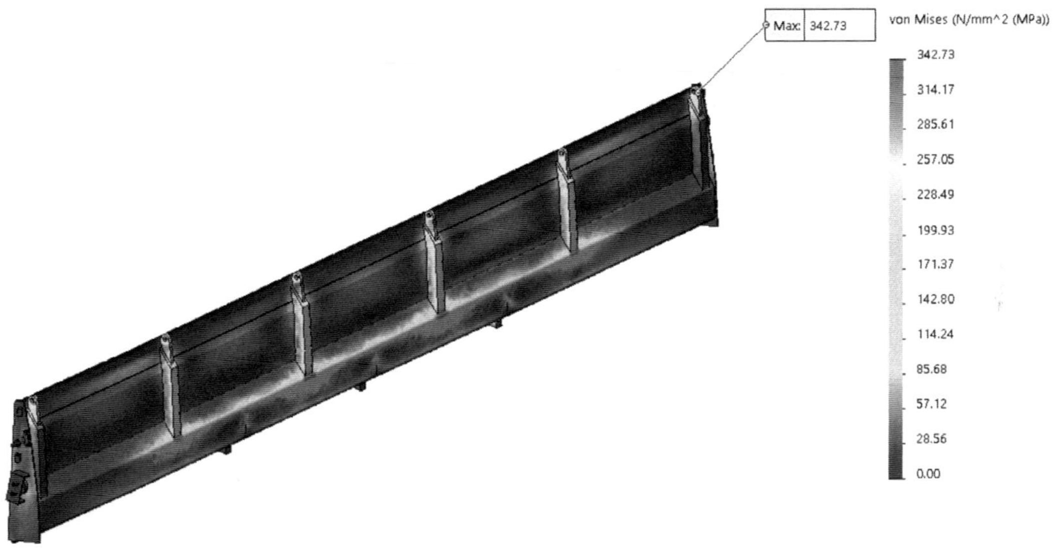

Figure 10. Distribution of stresses in the flap when calculated with "Non-linear task".

The analysis of the data from figures 6 and 10 allows the following conclusions to be made:

1. The stress distribution in the flap for the two calculation methods is similar.
2. When using the "Non-Linear Task" Module with Floor/Flap contact the maximum stresses are 342.73 MPa recorded in the upper end hinges of the flap.The stresses in the bottom area of the flap increase by an average of 21%. The stresses in the connections between the bottom profile and the vertical columns of the flap increase by 16.2% in the middle and by 43% in the intermediate columns adjacent to them. This is particularly worrying as these are welding zones and the allowable stresses are 323 MPa.

These data make it possible to conclude that the stress-strain analysis should be done through the "Non-linear task" module with floor-to-flap contact.

5. Analysis of the need for correction of methods for theoretical determination of the pressure on Falns series wagons

The conclusions from the Falns series wagon observations unambiguously showed that the flaps reliably seal the interior space, no gaps are detected when the locking thumbs are released, the middle locking pin does not contact the bottom profile of the flap during movement and unloading. These facts make it possible to assess the pressure-determining method used on the wagon flaps. For this purpose, the following hypothesis is formulated: The bulk load does not overcome the forces of elastic deformation of the bottom profile of the flap, the sealing profile 2 of figure 8 and the floor sheet. Consequently, the assessment of the working pressure can be carried out according to the following procedure:

- In the model shown in figure 5 the three displacement constraints of the bottom profile should be removed. This corresponds to the state "released locking thumbs".
- The flap shall be loaded in the manner described in chapter 2.
- The displacements are determined theoretically for the flap in the state "released locking thumbs".

If the flap displacements obtained are smaller than the bending arrow of the flap bottom profile (figure 9), the pressure is determined fairly accurately. Otherwise, it is necessary to adjust the pressure calculation methodology. The suggested hypothesis was tested in accordance with the methodology described. The results of the study are shown in figure 11.

Figure 11. Distribution of displacements in the flap for state "released locking thumbs".

Their analysis showed that the maximum displacement in the center of the flap was 54.61 mm. This indicates that when the locking thumbs are released, the bulk material should begin to flow through the gap between the flap and the floor. In experimental observations this was not found. Therefore, the displacements obtained under the theoretically imposed load do not correspond to real displacements during operation. The conclusion of the hypothesis is that there is significantly less pressure on the flaps and walls of the wagon. Therefore, the determination of the load with equations (1) and (2) results in substantially higher values. Similar findings have been made in [9], where it has been found that the forces applied in analysis of the walls of different series of wagons give rise to stresses which are about 2.5 times greater than the permissible.

As a sign of correctness, it should be noted that the proposed hypothesis is based on the complex application of theoretical and experimental studies. It can not be used alone at the design stage. It is possible to apply at the earliest after wagon prototype has been made and an test is conducted. This allows to see to what extent the theoretical values for Falns wagon flaps loading differ from actual.

6. Conclusions and recommendations

Summarizing the overall work on this study, the following main conclusions can be drawn:

- A comprehensive theoretical and experimental study of the flaps of a special wagon Falns was carried out.
- It is proven that if there is a pre-curvature in the bottom profile of the flap, the theoretical strength tests should be done through the "Nonlinear Task" module with floor-to-flap contact. This leads to a more accurate determination of the stresses in the subject and can prevent the occurrence of strength problems during operation.
- A hypothesis and a methodology for assessing the coincidence of flap loading forces obtained theoretically with those occurring during operation are proposed.

- After applying the hypothesis and the accompanying methodology, it has been found that for the particular object under consideration the theoretically determined flap loading forces are significantly higher than those acting on it during operation.

The conclusions made unequivocally suggest that it is expedient to develop a new methodology for determining Falns wagon flap loading forces to be used correctly in the stress-strain analysis of these objects.

Acknowledgements
The authors would like to thank to the Research and Development Sector at the Technical University of Sofia for the financial support. This work was also supported by the European Regional Development Fund within the Operational Programme "Science and Education for Smart Growth 2014 - 2020" under the Project CoE "National centre of mechatronics and clean technologies" BG05M2OP001-1.001-0008-C01.

References

[1] EN 12663-2:2010 Railway applications - Structural requirements of railway vehicle bodies - Part 2: Freight wagons
[2] Slavchev S, Georgieva K, Stoilov V and Purgic S 2015 Analysis of the results of theoretical and experimental studies of freight wagon *Fals Facta Universitatis, Series: Mechanical Engineering* **13(2)** 91
[3] Slavchev S, Stoilov V, Purgic S and Damyanov B 2015 Static strength analysis of the body of a wagon series Falns 176, *Proc. Int. Sci. Conf. BulTrans-2015 (Sozopol, Bulgaria)* **1** 163–6
[4] Slavchev S, Purgic S and Stoilov V 2015 Strength analysis of a wagon series Falns 274 with purpose to commissioning without testing, Proc. *Int. Sci. Conf. BulTrans-2015 (Sozopol, Bulgaria)* **1** 167–70
[5] Slavchev S, Maznichki V, Stoilov V and Purgic S 2018 Strength analysis of hopper freight wagon with regard to lightweight design. *Proc. Int. Sci. Conf. BulTrans-2018 (Sozopol, Bulgaria)* **1** 106–9
[6] UIC Leaflet 577, 5th edition, 2012 Wagon stresses
[7] Commission Regulation (EU) No 321/2013 of 13 March 2013 concerning the technical specification for interoperability relating to the subsystem rolling stock - freight wagons (WAG TSI) of the rail system in the European Union and repealing Decision 2006/861/EC, including amendments M1:2013 and M2:2015
[8] Report ERRI B 12/RP 17, 8th Edition, 1997 Programme of tests to be carried out on wagons with steel underframe and body structure (suitable for being fitted with the automatic buffing and draw coupler) and on their cast steel frame bogie
[9] Afanaskou P 2018 Loading the wagon body with bulk material in the calculation of strength (in Russian) *Proc. 21th Conf. for Junior Researchers "Science – Future of Lithuania"*, (*Transport Engineering and Management 4-5 May 2018 Vilnius Lithuania)*
[10] Jansen H 1995 Experiments on cereal pressure in silo cells. *Proc. 3rd European symp. storage and flow of particulate solids (Nuremberg)* **1** 1–6
[11] Kleyn G 1977 *Structural Mechanics of Bulk Goods 2nd Edition (in Russian)* (Moscow: Stroyizdat)
[12] Mazarskiy M 1971 *Silo Storage for Wood Chips (in Russian)* (Moscow: Sowiet Union Scientific Research and Design Institute of Economics)
[13] Putyato A 2010 On the issue of the regulatory loading of the wagon body with bulk cargo in the calculation of strength mechanics *Research and Innovations* **4(1)** 129–35 (in Russian)
[14] Rusinek R and Lukaszuk J 2004 Influence of moisture content on pressure ratio of rape seeds *Research Agricultural Engineering* **50(1)** 11–4
[15] Zenkov L 1964 *Mechanich of bulk goods (in Russian)* (Moscow: Mashinostroene)

[16] Russian Norms 1996 *Norms of calculation and designing of cars of railways mps goe 1520 mm (non-self-powered)* (in Russian)
[17] EN 1991-4. 2006. Eurocode 1: Actions on structures - Part 4: Silos and tanks [Authority: The European Union Per Regulation 305/2011, Directive 98/34/EC, Directive 2004/18/EC]
[18] Russian Standard GOST 33211 2014. Freight wagons. Requirements for strength and dynamic qualities (in Russian).
[19] Goris A 2012 Bautabellen für Ingenieure - mit Berechnungshinweisen und Beispielen, 20th Edition (Cologne: Werner Schneider)
[20] Kurowski P 2013 *Engineering Analysis with SolidWorks Simulation* (Mission: SDC Publications)
[21] Tenchev R 2010 *User guide for COSMOS/M* (Technical University – Sofia)
[22] Zenkiewicz O 2002 *The Finite Element Method in Engeneering Science* (London: McGraw-Hill)
[23] Segerlind L 1984 *Applied Finite Element Analysis 2nd edition* (New York: Wiley–Interscience)
[24] Slavchev S, Stoilov V and Valcheva T 2011 Application of a methodology for the assessment of stresses using numerical methods *Proc. Int. Sci. Conf. BulTrans-2011 (Sozopol, Bulgaria)* **1** 271–4

IOP Conf. Series: Materials Science and Engineering **664** (2019) 012028 doi:10.1088/1757-899X/664/1/012028

Reliability and availability assessment of a transport system using Dynamic Fault Tree and Monte Carlo simulation

Maciej Szkoda[1,2], Grzegorz Kaczor[1] and Magdalena Satora[1]

[1] Institute of Rail Vehicles, Faculty of Mechanical Engineering, Cracow University of Technology, Jana Pawla II 37 Ave., 31-864 Cracow, Poland

[2] E-mail: maciej.szkoda@mech.pk.edu.pl

Abstract. Ensuring a high level of reliability of the transport system is one of the most important issues related to the proper functioning of this system. It is concerned with the delivery of the transported goods on time and in accordance with recipient's order specification. The means of transport used in the delivery process are exposed to many external factors that may affect the reliability and safety of supply. Therefore, modelling and assessing the reliability of these systems requires the use of appropriate calculation methods that take into account complex scenarios reflecting the conditions of actual operation and maintenance. The presented paper concerns the application of Dynamic Fault Tree (DFT) together with Monte Carlo simulation to assess the selected reliability and availability indicators for the given transport system. The assessment of reliability was done based on the typical probabilistic models for analysis of the time to failure and the time to repair assigned to the selected components. The proposed approach allows detecting the weakest components of the transport system and is the basis for the development of preventive maintenance tasks. The presented model of the transport system can be further developed and adapted to needs of the specific system.

1. Introduction

One of the most basic aims of the transport system's proper functioning is providing the highest level of reliability in each phase of a transport process. Conclusion from current scientific research show the mentioned regularity regardless of the profile of an enterprise (figure 1). Reliability is the factor that improves the transport system's competitiveness. It comprises such delivery parameters as accuracy, completeness, and punctuality [1].

The reliability of a transport system can be evaluated by chosen measurements (indexes). They present the certainty level at which the given service is properly performed. A reliable realization of transport processes depends primarily on the strategy of management and organization of all elements of the transport system [2].

Due to a high level of complexity of real transport systems dynamic changes of technical states between their elements take place. The changes are connected with time relations, such as sequence dependent events, waiting for operation or repairing, and also with the lack of the possibility of taking spare parts and their priorities into account. Considering both the failure and repair characteristics in modelling of logic systems requires choosing the proper method of analysis. Classical methods of the reliability representation of the transport processes include Reliability Block Diagram (RBD) and Fault Tree Analysis (FTA). These techniques are successfully applied in industry. They are widely recognized as the best ones for the evaluation of reliability of technical systems, including transport systems [3, 4].

Content from this work may be used under the terms of the Creative Commons Attribution 3.0 licence. Any further distribution of this work must maintain attribution to the author(s) and the title of the work, journal citation and DOI.
Published under licence by IOP Publishing Ltd

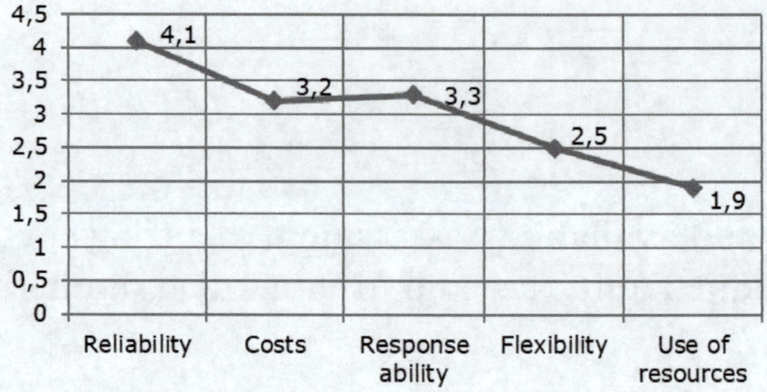

Figure 1. The main objectives of logistics – trade [5].

FTA is a technique of an analytical evaluation of reliability. It presents a set of independent events or processes in a graphical way. A certain combination of those events or processes leads to undesirable events. Based on the Boolean's gates (AND, OR, VOTING), FTA determines the so-called probability of the top event. Since FTA is a tool for qualitative and quantitative evaluation of reliability it is possible to identify critical events together with the probability of their appearance. Like other analytical tools, FTA has one major fault. It is the application of this method in systems with dynamic redundancy (e.g. load sharing redundancy or standby redundancy). Lack of the possibility for modelling the events whose only a determined sequence of occurrence can lead to the system's failure is an additional difficulty. A dynamic fault tree analysis (DFTA) method has been created to get rid of these limitations. It is complementary to a classical fault tree (FT) due to a few additional logic gates [6-9]. Several methods have been proposed to solve DFT; two most frequently are Markov models and the Monte Carlo simulation method. The first one can be applied only when the elements of the technical state have the exponential distributions of time through failure and repair time. Moreover, in the case of a complex system with a great number of elements the state space complicates the calculation procedure in a Markov process. Due to these drawbacks in the application of the Markov process, in the following work the Monte Carlo simulation method was used, which seems to meet these constraints. This method is based on a discrete simulation that allows to estimate the values of reliability indicators in the specific time [8, 10-12].

A modelling issue and a transport system reliability evaluation are discussed in numerous academic papers since there is a need of both limiting the occurrence of undesirable events and increasing the efficiency of functioning of these systems. Transport systems are complex systems. Their reliability functional indicators change at each stage of the process of load moving in a dynamic way [5, 11, 13-15].

The typical approach to fault tree analysis involves the use of Boolean algebra to present the relationship between undesirable events and their impact on system operation. Fault tree method is still widely used despite many limitations resulting from its idea. The most important limitation is the inability to take into account dynamic and complex time dependencies between the basic events. An evaluation of reliability of the power system using quantitative and qualitative approach underlines its valuable and application [16]. In the paper [10] the authors conducted a study related to the reactor regulation system (RRS) for the selected power plant. Solving a dynamic fault tree was accomplished by a numerical simulation.

In the selected papers, the application of FTA method was addressed to assess the risk of a hazard occurrence on the proper operation of a transport system. The possibility of using the Monte Carlo simulation to perform the necessary calculations in the air transport system was also considered. By combining both the methods, it was possible to generate alerts for air traffic controllers about the possibility of collisions in the air traffic [17]. This approach allowed to extend the use of the FTA method in transport-related issues.

New techniques which extend traditional fault trees are developed in order to eliminate the FTA limitations. One of them, the so-called timed fault trees (TFTs), allows an identification of faults that should be immediately eliminated. This technique also allows the determination of time needed for maintenance activities. One of the possibilities of using TFTs was the application referring to the selected rail transport system. [18].

To build transport systems also the Petri nets can be used. One of the examples of this method's application is an analysis of the reliability and efficiency of a real tram system. The performed research shows that models based on the Petri nets can also include time dependencies [19].

2. Description of the analysed system

The transport system analysed in this paper is presented in figure 2. It is an inter-modal transport system that consists of a few subsystems in which various means of transport (materials handling, road transport and rail transport) perform other transport tasks. The required reliability level of the transport system is achieved by assuming the number of backup elements. It has been assumed that each subsystem consists of two identical means of transport arranged in a cold standby redundant configuration. They are repairable objects which undergo strictly determined maintenance activities.

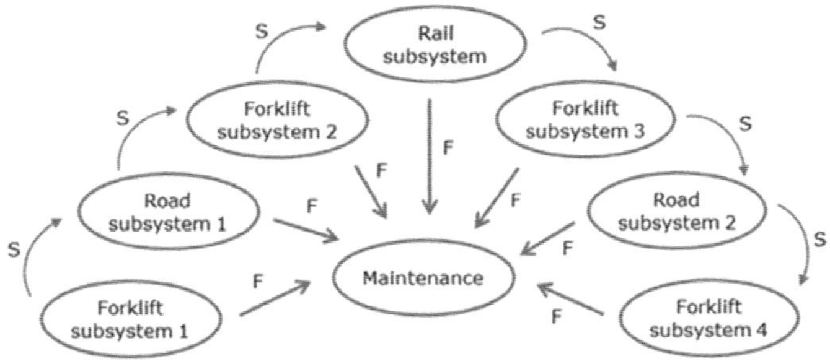

Figure 2. Scheme of transport system functioning: S – suspended, F – failed.

An analysis of transport system reliability should include not only the time of damaged elements repair but also a delay time that is connected with waiting for a standby object. A successful operation of the system requires all of its individual partial processes to be performed, i.e.:

a) Loading: formation of a unit load and preparation for road transport;
b) Road transport 1: moving the cargo to the desired railway terminal transport of a unit load to the rail terminal;
c) Reloading: changing the road transport to a railway transport;
d) Rail transport: moving the cargo to the nearest road terminal;
e) Reloading: changing the railway transport to a rail transport;
f) Road transport 2: moving the cargo to the final destination place;
g) Unloading: discharging the cargo and completing whole to process.

Inability to perform any of the elementary tasks is considered as incorrect operation of the system.

In order to make a model of the presented transport system by using a fault tree and to take into consideration the described assumptions one has to use gates with a dynamic dependency. They are, e.g. sequence enforcing gate (SEQ), spare gate (SPARE), priority AND (PAND) and functional dependency (FDEP), shown in figure 3. The rules for the gates are as follows [9]:

a) SEQ gate: it goes into a failure state only when all the assigned components occur in a given sequence. Every other combination of the input events cannot take place;
b) SPARE gate: it is associated with both active and spare components. The correct operation of the system described by this gate occurs if the required number of components is working correctly.
c) FDEP gate: it is used to show that all the components are functionally dependent on an additional factor, called the trigger;

d) PAND gate: the gate indicates the failure of a system when all the input components will fail in a pre-assigned order.

Unlike the SEQ gate, the PAND gate allows for other than pre-assigned sequence of events.

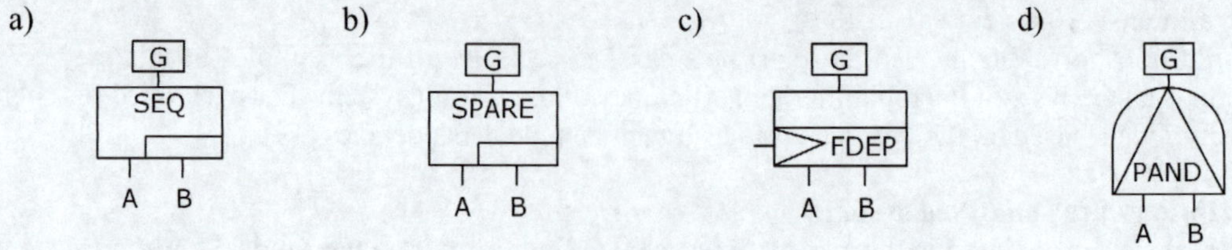

Figure 3. Dynamic gates: a) SEQ, b) SPARE, c) FDEP, d) PAND.

Fault tree model of the transport system taking into account in the presented study is presented in figure 4. The OR gate represents a top event. This event occurs if any of the partial process of transport is failed. Each process uses certain transport means. The case in which the primary vehicle is failed will result in a stop the operation and the need to take a corrective task. The damaged vehicle is under corrective maintenance at that time. After the corrective maintenance, the vehicle goes into a standby stage. If the standby vehicle gest damaged during repair activities of the primary vehicle, the whole partial process is regarded undone. Corrective maintenance is accomplished when the vehicle is back again in the as good as a new condition.

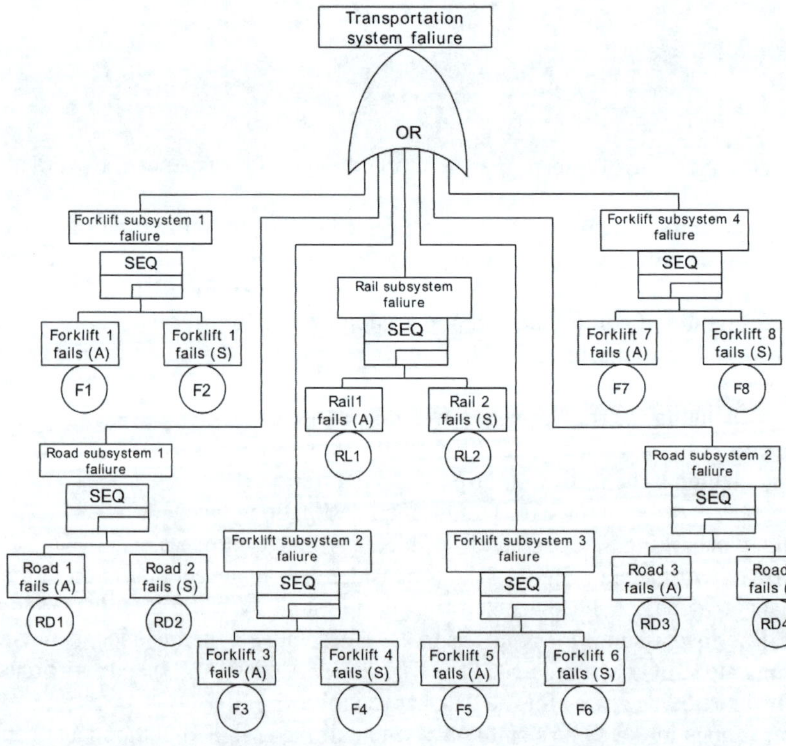

Figure 4. The fault tree model of the analysed system.

3. Reliability and availability analysis of the transport system

In the analysed transport system, the system elements have a normal distribution of time over failure (TTF) and a lognormal distribution of the repair time (TTR). The mean delay time (MDT), which is related to the substitution of spare components, is also taken into account. The detailed data are in table 1.

Table 1. Failure and repair data used for the analysis.

Component	Failure			Repair	Delay
	Distribution	Parameters		Parameter	Parameter
		Mean	Std, days	MTTR, hours	MDT, hours
F1, F2	NORMAL	60	5	2	2
RD1, RD2	NORMAL	95	5	4	4
F3, F4	NORMAL	70	6	2	2
RL1, RL2	NORMAL	187	11	6	6
F5, F6	NORMAL	65	5	2	2
RD3, RD4	NORMAL	100	10	4	4
F7, F8	NORMAL	55	7	2	2

The ReliaSoft's software, which allows to simulate of a discrete event, has been used to perform the Monte Carlo simulation. This specialized tool is widely applied in the different application of the industrial areas. The simulation performed in the following study is based on the Random Number Generator with Bays-Durham shuffle algorithm. The simulation requires an introduction of some input parameters, such as [20]:

a) Simulation End Time (SET);
b) Point Results Every (PRE);
c) Number of Simulations (NoF).

4. Results

The results were obtained using one hundred thousand simulations in the specified period: 0 to 1825 days. The PRE was assumed as 100000. When the simulation was ended the results were gathered in the System Overview table (table 2). The information in the table relies on the mean values.

Table 2. The simulated results for the 1825 days of operation.

System Overview, General	
Mean Availability (All Events):	0.9996
Std Deviation (Mean Availability):	0.000147
Mean Availability (w/o PM, OC & Inspection):	0.9996
Point Availability (All Events) at 1825:	0.9964
Expected Number of Failures:	2.3691
Std Deviation (Number of Failures):	0.5940
MTTFF, Day:	1148.7
MTBF, Total Time, Day:	770.3
MTBE, Total Time, Day:	543.3
System Uptime/Downtime	
Uptime, Day:	1824.4
CM Downtime, Day:	0.5992
MTTFF, Day:	1148.7
MTBF, Total Time, Day:	770.3
MTBE, Total Time, Day:	543.3

Mean availability A, is defined as a mean contribution of time in which the investigated vehicle remains in the state of serviceability. For an individual object the availability index is defined as [2]:

$$A = \frac{\sum_{i=1}^{N} TZ_i}{\sum_{i=1}^{N} TZ_i + \sum_{i=1}^{N} TUB_i + \sum_{i=1}^{N} TUP_i}, \tag{1}$$

where TZ_i – time of vehicle i in serviceability state; TUB_i – time of vehicle i in unavailability state due to corrective repairs; TUP_i – time of vehicle *i* in unavailability state due to preventive repairs; N – sample size of vehicles taken for tests.

It can be stated on the basis of the obtained results, that for a considered transport system a significant decrease in the system's availability occurs at certain intervals.

By analysing the participation of downing events and the number of failures of the system's elements the weakest component of the system for the simulated end time can be determined. The ReliaSoft Downing Event Criticality Index (RS DECI) and ReliaSoft Failure Criticality Index (RS FCI) are used to do that [20].

RS DECI is a relative index. It shows the amount of time of which the component contributed to the system downing event. (i.e. the number of the system downing events caused by the block divided by the total amount of the system downing events).

$$RS\ DECI = \frac{C_{NSDE}}{N_{ALLdown}}, \tag{2}$$

where C_{NSDE} is the Number of System Downing Events, this is the number of time when this component's downing causes the system to be down; $N_{ALLdown}$ – the number of downing events.

In the simulation results (figure 5), it can be observed that for the F1element the RS DECI = 59%. This implies that 59% of the time when the system was down were due to the component F1 being down. Besides, the conducted analysis allows for noticing that the greatest influence on the system's downtime have two elements: F1 and F2 (forklift 1 and forklift 2). It is connected with an assumed reliability structure together with an unloaded standby. If the primary element gets damaged then its functions are taken by the standby element.

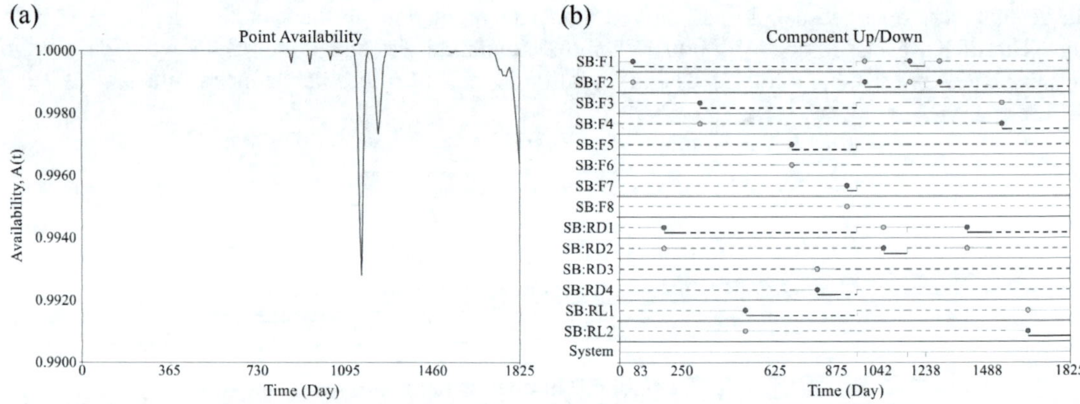

Figure 5. Simulation Results: (a) Point Availability chart; (b) Component Up/Down chart.

RS FCI is a relative index that shows the percentage of time when a failure of this component caused the system's failure. This is obtained from:

$$RS\ FCI = \frac{C_{NSDF}+F_{ZD}}{N_F}, \tag{3}$$

where F_{ZD} – a special counter of the system's failures not included in C_{NSDF}. This counter is not directly included in the calculations but it is used by the software. The reason for the presence of this counter is the that failures with zero duration time are not contained in C_{NSDF} since they really do not down the system. However, these failures are required to be taken into account when we are calculating the RS FCI; N_F – the number of system's failures.

An RS FCI chart for the selected components is shown in figure 6. For the RD1component, RS FCI = 43.5%. This implies that the component RD1 failure was responsible for 43.5% of the time when the system failed. It should be noted that the combined RS FCI of RD1 and F2 is almost 80%. In other words, RD1 and F2 contributed to about 80% of the system's total downing failures.

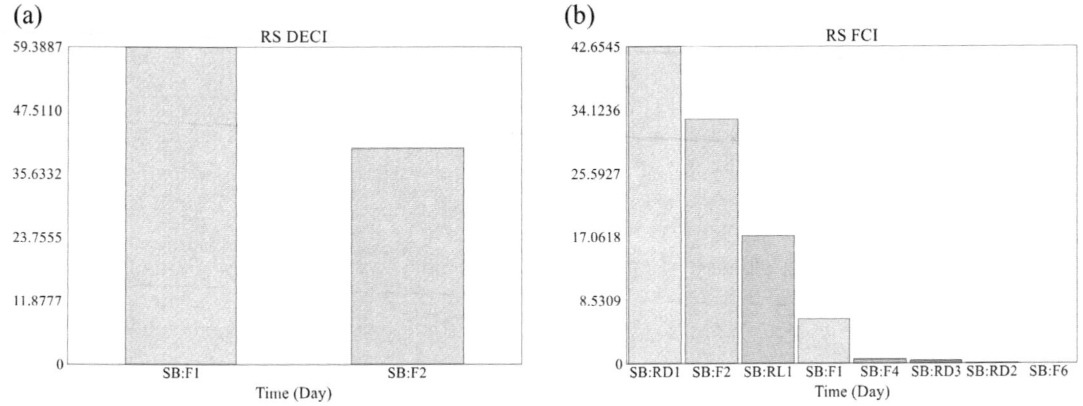

Figure 6. Simulation Results for the selected components: (a) RS DECI chart; (b) RS FCI chart.

It is worth to mention that for both RS DECI and RS FCI, with the overlapping events present, the component that contributed to the system downing event gets credited with this system event. Subsequent component that do not cause the system to go down (because the system might be already down) is not included in this metric [20].

5. Conclusion

As a result of the conducted analysis of the transport system's reliability with the use of a fault tree and the Monte Carlo simulation, the authors obtained the values of the selected indexes that may be used for the determination of the probability of this system's elements failures during operation. This approach allows for a qualitative and quantitative evaluation of reliability and an identification of weak components of the system. It can also constitute the basis for a preventive maintenance strategy. Such created model of the transport system can be further developed until it achieves a required level of detail. The applied ReliaSoft software package undoubtedly makes the usage of simulation techniques easier, especially in systems with a high level of complexity.

References

[1] Gajewska T and Kaczor G 2014 Analiza niezawodności dostaw w transporcie chłodniczym *Logistyka* **5** 453–62
[2] Szkoda M 2014 Assessment of reliability, availability and maintainability of rail gauge change systems *Eksploatacja i Niezawodność – Maintenance and Reliability* **16** 422–32
[3] Jacyna M and Merkisz J 2014 Proecological approach to modelling traffic organization in national transport system *Archives of Transport* **30** 31–41
[4] Nowakowski T 2004 Reliability model of combined transport system. Probabilistic safety assessment and management *Proc. Europ. Safety and Reliability Conference PSAM7-ESREL 2004 (London: Springer)*
[5] Nowakowski T 2011 *Niezawodność Systemow Logistycznych* (Wroclaw: Oficyna Wydawnicza Politechniki Wroclawskiej)
[6] Amari S, Dill G and Howald E 2003 A new approach to solve dynamic fault trees *Annual IEEE reliability and maintainability symposium* 374–89
[7] Cepin M and Mavko B 2002 A dynamic fault tree *Reliab. Eng. Syst. Safe.* **75** 83–91
[8] Manno G, Chiacchio F, Compagno L, D'Urso D and Trapani N 2012 An integrated FT and Monte Carlo Simulink tool for the reliability assessment of dynamic fault tree *Expert. Syst. Appl.* **39** 10334–42
[9] Faulin J, Juan A A, Matrorell S and Remirez-Marquez J E 2010 *Simulation Methods for Reliability and Availability of Complex Systems* (London: Springer-Verlag)

[10] Durga Rao K, Gopika V, Sanyasi Rao V V S, Kushwaha H S, Verma A K and Srividya A 2009 Dynamic fault tree analysis using Monte Carlo simulation in probabilistic safety assessment *Reliab. Eng. Syst. Safe.* **94** 872–83

[11] Marseguerra M and Zio E 2004 Monte Carlo estimation of differential importance measure: application to the protection system of a nuclear reactor *Reliab. Eng. Syst. Safe.* **86** 11–24

[12] Szkoda M and Kaczor G 2016 Reliability and availability assessment of diesel locomotive using Fault Tree Analysis *Archives of Transport* **40** 65–75

[13] Jacyna-Gołda I, Żak J and Gołębiowski P 2014 Models od traffic flow distribution for various scenarios of the development of proecological transport system *Archives of Transport* **32** 17–28

[14] Kaczor G 2015 Modelowanie i ocena niezawodnosci systemu transportu intermodalnego *Logistyka* **3** 2047–54

[15] Nowakowski T and Zając M 2005 Analysis of reliability model of combined transport system *Advances in Safety and Reliability – Proc. Europ. Safety and Reliability Conference (ESREL 2005)* 147–51

[16] Volkanowski A, Cepin M and Mavko B 2009 Application of the fault tree analysis for assessment of power system reliability *Reliab. Eng. Syst. Safe.* **94** 1116–27

[17] Stroeve S H, Blom H A P and Bakker G J 2009 Systematic accident risk assessment in air traffic by Monte Carlo simulation *Safety Science* **47** 238–49

[18] Peng A, Lu Y Miller A, Johnson C and Zhao T 2014 Risk assessment of railway transportation system using timed fault tree *Qual. Reliab. Eng. Int.* **32** 181–94

[19] Kowalski M, Magott J, Nowakowski T and Werbińska-Wojciechowska S 2011 Analysis of transport system with the use of Petri nets *Eksploatacja i Niezawodność – Maintenance and Reliability* **13** 48–62

[20] Reliasoft Corp. 2010 *System Analysis Reference* (Tucson, USA: ReliaSoft Publishing)

Development of a repair technology for locomotive units on the basis of the theory of decision

V Puzyr, Y Datsun[1], O Obozny and V Pyvo

Ukrainian State University of Railway Transport, Feuerbach Sq. 7, Kharkov, Ukraine

[1] E-mail: remlocomot@gmail.com

Abstract. The operative condition of locomotives is maintained by the existing repair system; it consists of the following elements: strategy, organization and technology. This repair technology for units and aggregates is based on disassembling and elimination of defects, and it can be inappropriate. The actual technical condition of the nodes requires an individual approach in determining the volume and content of technological impact during the repair process. Therefore, the purpose of this work is to research the task of developing a repair technology for locomotive units, depending on their actual technical condition. As far as probability distribution functions of the operating condition of units are known, a repair technology is chosen under risky conditions. Therefore, the decision-finding problem is framed in the theory of decision terms. The analysis of the decision tree of the choice of repair technology for fuel injectors of a diesel locomotive demonstrated that at least four variants of technologies (strategies) were possible. Operational costs were taken as the evaluation criterion for the technological process. The pay-off matrix obtained does not have a saddle point; therefore, the problem of choice of the technology cannot be solved only strategically. It substantiates a need to use flexible, adaptive repair technologies which consider the actual technical state of each unit. For a known distribution of states for fuel injectors, the greatest gain with minimum risks is the repair technology of dismantling and reseating the worn-out parts. Application of the described approach at locomotive repair enterprises will allow for improvement of the quality of repair and reduction of costs.

1. Introduction

The locomotive facility is one of the leading rail transport industries the key task of which is provision of the rail industry with sufficient amount of serviceable equipment and qualified personnel. The proper function of locomotives is maintained by the existing repair system; it consists of the following elements: strategy, organization and technology. Repair of machines and aggregates includes detection and elimination of trouble; it states the principal difference between the repair and machinery production facilities. The existing repair technology for units and aggregates of the rolling stock was formed before the active development and implementation of methods and tools to assess the technical condition of objects. It explains certain special features of the technological repair processes which require the disassembling of most units and aggregates to detect faults. Such a technology is unprofitable and sometimes unpractical.

The repair of units and aggregates of the rolling stock is characterized by a certain order of the key technological operations of cleaning, state controlling, disassembling, renewing, assembling and testing. Generally, these operations are carried out in a sequence [1-3]. Nowadays, the general technical deterioration of the rolling stock results in an increased number of panic repairs and

Content from this work may be used under the terms of the Creative Commons Attribution 3.0 licence. Any further distribution of this work must maintain attribution to the author(s) and the title of the work, journal citation and DOI.
Published under licence by IOP Publishing Ltd

replacement of units and aggregates [4]. Consequently, even a scheduled repair of single-type units can considerably differ due to their technical conditions. Thus, the established repair technology for units of different technical conditions can be an unreasonably time- and resource-consuming procedure. Besides, it causes a considerable deterioration of some details due to consequent running-in wears after the assembly-disassembly procedure [5-7].

Thus, the situation necessitates the development and introduction of measures to improve the quality and decrease the cost of the locomotive unit repair. And one of the resource-saving approaches is the development of individual technological repair procedures for units in different technical conditions. It implies that out of all possible types and kinds of technological operations, there is one which stipulates such a content and consequence of operations that ensures the required quality at minimal costs. In production, such a task is successfully solved by the network modelling methods [8, 9]. However, for the repair facility such an approach does not yield favourable results, as the technical conditions of objects can considerably differ and thus demand various technological approaches. Therefore, the purpose of this work is to research the task of forming a repair technology for locomotive units, depending on their actual technical condition.

2. Methodology

As far as the probability distribution functions of the operating condition of units can be known, a repair technology is chosen under risky conditions. Therefore, the decision-finding problem is commonly framed in the theory of decision terms [10, 11].

As far as the repair technologies for various units differ, there is a need to use the actual data for modelling. The study used the information on the repair of fuel injectors for diesel locomotive engines as the initial data. The established repair technology included the following operations:

- cleaning and checking of the injectors on a test bench;
- disassembling;
- washing, blowing and examining;
- cleaning and replacing nozzles;
- monitoring of the spring condition;
- replacing faulty parts;
- reseating the surfaces of bodies of injector, sprayer and nozzle;
- reseating the needle to the sprayer body;
- checking the needle lift;
- washing and blowing components;
- assembling;
- checking and adjusting on a test bench.

Such a technology is reasonable for cases of limiting wear or damages in injector basic elements. The application of a complete repair technology for injectors with partial wear can lead to unreasonable costs and reduce the operational life of the elements.

During the repair process, it should be possible to take successive solutions regarding the technological impact on the object in repair. Thus, it is reasonable to use a graphic method, the tree of decision, which allows for connection of the decision points, feasible strategies Ai, their consequences with possible factors and environmental conditions. The decision tree begins with an earlier decision, moves to possible events with their consequences, transfers to another decision (choice of the operation line), etc. until all logical consequences of the results are used up.

The required volume of technological impacts for injectors in different technical conditions was determined in building the decision tree (figure 1).

The analysis of the decision tree demonstrated that repair of fuel injectors might require at least four technological variants (strategies) according to the condition of the injectors: A_1 – disassemble, repair with replacement parts; A_2 – disassemble, repair with machined parts; A_3 – do not disassemble, adjustment; A_4 – do not disassemble, cleaning.

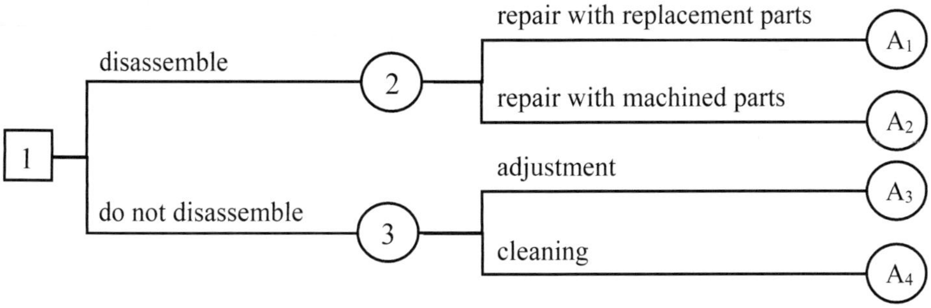

Figure 1. Decision tree of choice of a repair technology for the fuel injector of a diesel locomotive engine.

Here, the determining criteria are the general operative state and misalignment of the injector, and faults in certain parts. Probabilities of these states were determined from the statistical data on failures in the injectors collected by the repair facility of a locomotive depot (Table 1).

Table 1. Distribution of failures in fuel injectors of diesel locomotives under repair.

Type of failure in an injector	Number of injectors with faults	State probability
Damages and failures of certain components	654	0.26
Loss of tightness	830	0.33
Failures of adjustment	578	0.23
Dirty vents in a sprayer	453	0.18
Total	2515	1.0

As far as one of the assessment criteria for technological processes is the operational cost, it is crucially important to determine faults for the locomotive industry. The basic expenditure for a technological operation in the repair process of a locomotive unit can be determined according to:

$$C^{TO} = W + E_C + T_C + P_C \tag{1}$$

where: W is wages and salaries; E_C – operation and depreciation costs; T_C – costs of tool and equipment usage; P_C – costs of spare parts and materials.

If a technological repair process consists of k operations, the cost model of the process with consideration of (1) looks as:

$$C^{TP} = \sum_{k=1}^{K} C_k^{TO} \tag{2}$$

Calculations according to (2) make it possible to build an appropriate pay-off matrix in relative units. The rows of the matrix are the possible variants of the repair technology A_i (figure 1), and the columns are the possible conditions of the injectors (Table 1):

$$A = \begin{pmatrix} 122 & -7 & -953 & -1025 \\ -250 & 1072 & 1037 & 1055 \\ -156 & -1200 & 1166 & 1149 \\ -138 & -1088 & -1182 & 1184 \end{pmatrix} \tag{3}$$

On determining the strategies and building the matrix of gains and risks, the formalization of the problem is considered to be completed.

3. Results and discussion

The analysis of the pay-off matrix made it possible to determine a guaranteed gain, with the lower value of game $a = \max(a_i) = -250$ and the upper value of game $b = \min(b_j) = 122$. As $a \neq b$, the saddle point was absent, and the game value was within a range of $-250 \leq y \leq 122$. Therefore, the problem set did not have decisions in pure strategies. It was also determined, that no pure strategy was either dominating, or dominated. The technology currently used (strategy A_1) is characterized by the higher cost and is only appropriate for one condition of the injectors - with damage and parts failure. Considering that injectors with different technical conditions (Table 1) can come in for repair, it is rational to use flexible, adaptive repair technologies.

The pay-off matrix did not always reflect all existing conditions for taking decisions. Therefore, the model analysis should certainly include indications of "good" and "failed" choices of the strategy for each state of nature. The ease of state of nature Q_i is determined as the biggest gain at a given state, i.e. the largest element of the j-th column:

$$\beta_j = \max_i a_{ij}, \quad j = \overline{1,n} \tag{4}$$

The utility value of nature is used for determining risk r_{ij}, the characteristics of the utility for a player to use the strategy A_i for the state of nature Q_i:

$$r_{ij} = \beta_j - a_{ij} \tag{5}$$

According to (5) and the data on pay-off matrix (3) the risk matrix was obtained:

$$A = \begin{pmatrix} 0 & 1078 & 2118 & 2208 \\ 372 & 0 & 128 & 128 \\ 278 & 2272 & 0 & 34 \\ 260 & 2160 & 2348 & 0 \end{pmatrix} \tag{6}$$

Analysis of the risk matrix made it possible to reveal certain features of the model obtained. Thus, elements a_{22}, a_{23}, a_{24} were characterized by virtually equal gains in the pay-off matrix for strategy A_2. As for the risks, these elements were far from equal. For states Q_3 and Q_4 the risk was 128, and for the state Q_2 it was 0.

The decision-making techniques under risks use the choice theory called the utility theory. According to this theory, strategy A_i is chosen from the combination A_i $(i = 1 \ldots n)$ if it maximizes the expected cost of its utility function [11, 12]. The expected value for each alternative is determined as the average weighted gain $\overline{a_C}$ with consideration of probabilities w_i of all possible states of nature Q_j:

$$\overline{a_C} = w_1 a_{i,1} + w_2 a_{i,2} + \ldots + w_S a_{i,S} = \sum_{j=1}^{S} w_j a_{i,j} \tag{7}$$

The strategy with the maximum average weighted gain is optimal.

The average weighted risk was used for assessment of risk strategies.

$$\overline{r_C} = w_1 r_{i,1} + w_2 r_{i,2} + \ldots + w_S r_{i,S} = \sum_{j=1}^{S} w_j r_{i,j} \tag{8}$$

By comparing criteria $\overline{a_C}$ and $\overline{r_C}$ for strategies $A_{1\text{-}4}$ it was possible to determine a practical value in using these strategies for researched state probabilities (figure 2 (a)).

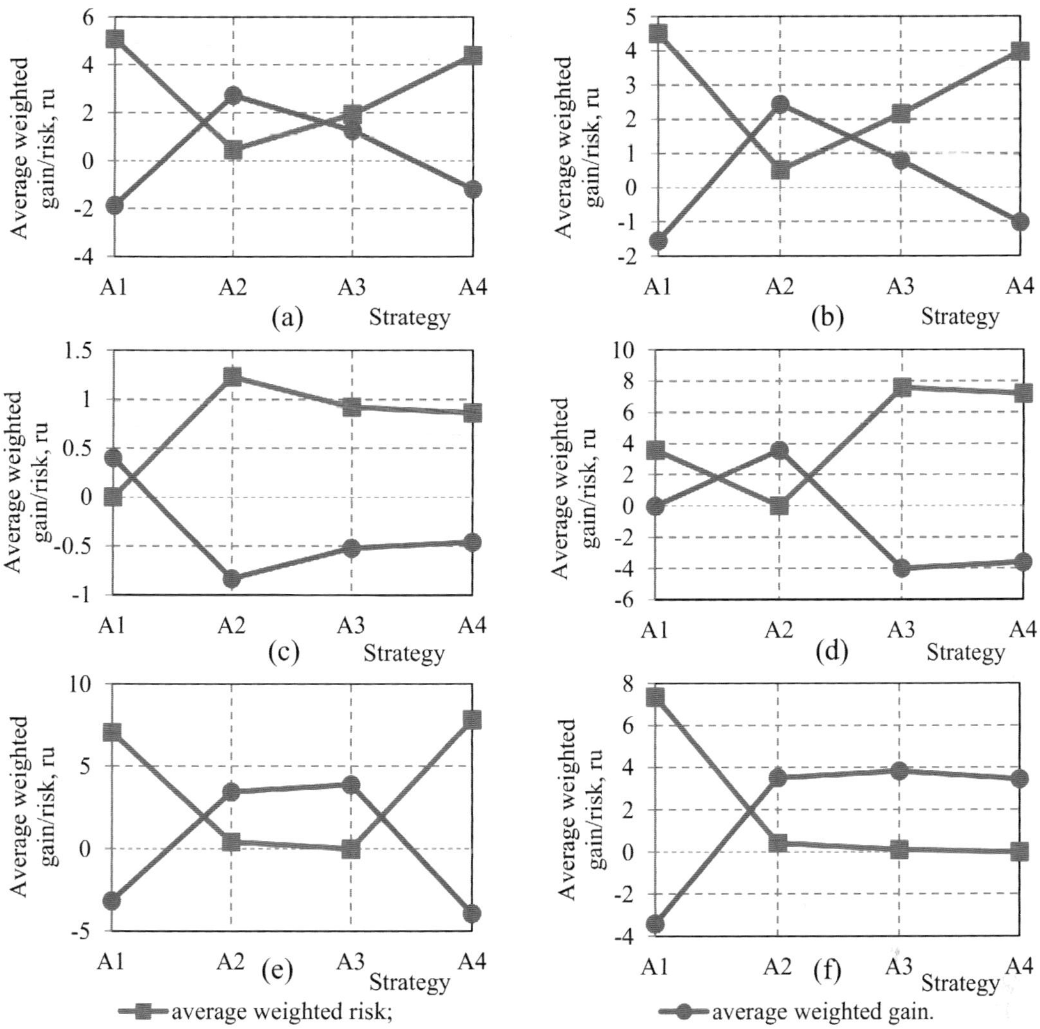

Figure 2. Comparison of the criteria $\overline{a_C}$ and $\overline{r_C}$ for the strategies $A_{1\text{-}4}$. (a) according to statistical data; (b) injector conditions probabilities are equal; (c) dominance of condition "Damages and failures of certain components"; (d) dominance of condition "Loss of tightness"; (e) dominance of condition "Failures of adjustment"; (f) dominance of condition "Dirty vents in a sprayer".

Strategy A_2 was characterized by the maximum average weighted gain (2.72) and, accordingly, by the minimal average weighted risk (0.47). By using the result to the problem, it should be mentioned that for a given distribution of state probabilities for fuel injectors, the biggest gain with minimal risks is a repair technology of disassembling and reseating worn-out parts. A simulation of the situation with equal probabilities of injector conditions showed a similar result (figure 2 (b)). Analysis of the simulation results for the cases of dominance of different injector conditions shows the maximum gain and the minimum risk of the most appropriate repair technology (figure 2 (c, d, e, f)). It confirms the need for an individual approach in determining the repair technology of injectors.

4. Conclusions

The authors proposed a model for development of a repair technology for locomotive units. The examples of fuel injectors made it possible to determine a guaranteed gain, with the lower value of game $a = \max(a_i) = -250$ and the upper value of game $b = \min(b_j) = 122$. Since $a \neq b$, the saddle point is absent, and the game value is within a range of $-250 \leq y \leq 122$. Therefore, the set problem does not have decisions in pure strategies.

The technology currently used (strategy A_1) is characterized by higher cost and is only appropriate for one condition of the injectors - damage and parts failure. Considering that injectors with different technical conditions can come in for repair, it is rational to use flexible, adaptive repair technologies.

For the considered probability distribution of the condition of the fuel injectors, the implementation of the repair technology with disassembly and grinding of worn parts will bring the greatest gain 2.72, with a minimum risk of 0.47.

Simulation of situations of dominance of different conditions of the injectors shows the maximum gain and the minimum risk of the most appropriate repair technology. It confirms the need for an individual approach in determining the repair technology of injectors.

References

[1] Kudryashov E and Stetsurin A 2008 More efficient repair of machine parts by a group method *Russ. Engin. Res.* **28(9)** 924–5

[2] Kheder M, Trigui M and Aifaoui N 2017 Optimization of disassembly sequence planning for preventive maintenance *Int. J. Adv. Manuf. Technol.* **90(5-8)** 1337–49

[3] Datsun Y 2016 Determination of the degree of influence of repair processes on serviceability of units of the traction rolling stock *East. Europ. J. Enterprise Techn.* **1(7(79))** 56–61

[4] Tartakovskyi E, Ustenko O, Puzyr V and Datsun Y 2017 Systems approach to the organization of locomotive maintenance on Ukraine railways *Rail Transport – Systems Approach* (Cham: Springer International Publishing) 217–39

[5] Rosmaini A and Shahrul K 2012 An overview of time-based and condition-based maintenance in industrial application *Computers & Industrial Engineering* **63(1)** 135–49

[6] Liu B, Wu S, Xie M and Kuo W 2017 A condition-based maintenance policy for degrading systems with age- and state-dependent operating cost *Europ. J. Operational Research* **263(3)** 879–87

[7] Heng A, Zhang S, Tan A and Mathew J 2009 Rotating machinery prognostics: State of the art, challenges and opportunities *Mechanical Systems and Signal Processing* **23(3)** 724–39

[8] Rao R V 2007 *Decision Making in the Manufacturing Environment: Using Graph Theory and Fuzzy Multiple Attribute Decision Making Methods* (London: Springer-Verlag) p 374

[9] Love J 2007 *Process Automation Handbook: A Guide to Theory and Practice* (London: Springer-Verlag) p 1093

[10] Jones A J 2000 *Game Theory* (Cambridge: Woodhead Publishing) p 300

[11] Ventsel E 1980 *Operation research* (Moskow: Nauka) p 206

[12] Goodwin P and Wright G 2014 *Decision Analysis for Management Judgment, 5th Edition* (New York: John Wiley and Sons) p 496

Determining traffic potential as an important part of sustainable railway passenger transport

Milan Dedík[1], Martin Kendra, Tomáš Čechovič and Martin Vojtek

Department of Railway Transport, University of Žilina, Univerzitná 8215/1, 010 26 Žilina, Slovakia

[1] E-mail: milan.dedik@fpedas.uniza.sk

Abstract. The main goal of EU transport policy in passenger transportation is to support public passenger transport and railway transport as a key transport mode, to know actual transport needs and to determine traffic potential properly. The paper considers particular current problems of public passenger transport, especially railway passenger transport. It proposes the basics of a methodical process that should increase its efficiency, attractiveness, a better image for the traveling public, but especially the increase of transport performance. A very important part of the methodical process is determining traffic potential. Particular methods of determining it are described and a proposal of a new methodology is included. The main aim of the methodology is to determine the value of the total traffic potential coefficient Kp. This value can be calculated using the proposed formula. It is possible to fit the values of the factors affecting traffic potential into the formula. As a result, there is practical application of the proposed methodology on real railway lines in Slovakia.

1. Introduction

At present, it is very important to ensure the optimal mobility of goods and people and the quality of the transport process. Freight and passenger transport is a significant sector of the national economy. Only a good transport system and transport infrastructure will contribute to the development of industry, agriculture and other manufacturing and non-productive sectors of the national economy. The speed and efficiency of transport, its impact on the environment and energy consumption significantly affect the sustainable development of the area and the development of society.

In recent years, however, public passenger transport performance has been reduced at the expense of individual motoring, which is much more popular among national citizens than mass passenger transport. However, the enormous increase in passenger cars leads to a number of negative facts, in particular congestion and adverse environmental impacts. Therefore, it is now necessary to adopt a number of effective measures to revitalize public passenger transport, which could motivate passengers to use public transport at the expense of individual transport. The European Commission has issued a number of strategy papers to help meet the objectives set.

The paper briefly analyzes and evaluates the current state of public passenger transport with emphasis on rail passenger transport. Furthermore, it proposes a methodological process for a sustainable status of traffic service provided by rail passenger transport linked to the extent of railway infrastructure. It includes sectional important steps which must be observed in the correct order.

Next, traffic potential as a very important part of the process is explained. Particular methods of determining it, including a proposal of a new method and its practical application, are described.

Content from this work may be used under the terms of the Creative Commons Attribution 3.0 licence. Any further distribution of this work must maintain attribution to the author(s) and the title of the work, journal citation and DOI.
Published under licence by IOP Publishing Ltd

2. Literature review and current status of solving the problems

Transport problems are also addressed by a large number of transport experts and scientists. For example, the book "Transport Modeling" [1] explains the basic principles of traffic planning and modelling. Another publication, "Railway Timetabling and Operations" [2] makes an analysis, modelling, simulation and performance evaluation. There is a lot of useful information about the rationalization and optimization of railway transport operation and railway transport management.

The scientific paper "Improving passenger transportation service as a part of tourism services complex in Russia" [3] explains the theoretical background of travelling motives, transport needs and structure of consumers by type of activity. It is an important part of the proposed methodical process. Another scientific contribution "Relationships between rail passengers′ satisfaction and service quality: a framework for identifying key service factors" [4] focuses on service quality and railway service and proposes a methodology aimed to establishing the importance of the various characteristics of a transit service for the overall service quality, in order to identify the key factors for the users of transit services.

Scientific contributions "Passenger- and operator-oriented scheduling of large railway projects" [5] and "Optimal train routing and scheduling for managing traffic perturbations in complex junctions" [6] include real-time railway traffic management problems, routing, scheduling, passenger perspective and railway infrastructure problems. Publication "The sustainable mobility paradigm" [7] contains measures for improving transport mobility and key elements of sustainable mobility. The diploma thesis "The proposal of the integrated cycle timetable in Slovakia in 2012" [8] proposes an integrated cycle timetable in Slovakia, but there are no general standards of rationalization which can be used all over Europe or the world.

However there are a lot of unexplored theoretical and practical principles in the field of planning, organizing and timetabling a railway passenger transport system and management. Therefore the goal of the study is methodological process proposal of sustainable railway transport and infrastructure and subsequently to define and determine the traffic potential, as an important part of the sustainable railway passenger transport.

2.1. Current issues in public passenger and railway passenger transport

At present, public passenger and non-motorized transport in some European countries is in a difficult situation. There was a sharp decrease in public passenger transport performance in the Slovak Republic from 1995 to 2010. While in 1995 the ratio of public passenger transport to individual car traffic was almost the same, this ratio is currently unfavorable for public passenger transport, because public passenger transport accounts for only 25% of total transport performance. For example, in the Slovak Republic, in 1995 there was 1 motor vehicle for more than 5 inhabitants of Slovakia, while in 2014 it was already 2.8 inhabitants.

In addition to the decrease of passengers, public passenger transport is currently struggling with various problems of different nature. These are financial problems, organizational problems, problems with transport infrastructure, rolling stock, maintenance, underestimated operating frequency, low vehicle capacity, as well as a negative public image of public transport. Also, the demands of the public traveler are constantly growing, because passengers no longer need to be transported between two selected locations but they are interested in travelling in the required quality. As public passenger transport is not capable of meeting these requirements, it is easier and more efficient for them to address their transport needs through individual transport. However, while the current trend of increasing the number of passenger cars continues, individual passenger motoring will gradually cease to be attractive because of congestion and few parking places. At the same time, there is a further negative effect, namely, the heavy burden on road infrastructure and in particular the negative impact on the environment.

In a number of post-socialist countries, especially in eastern, south-eastern and central Europe, public passenger transport and railway passenger problems appear to be much more intense than in more developed Western countries. The current negative situation of public passenger transport is particularly obvious in the area of rail passenger transport. The main advantages of this transport mode are greatly underestimated by the advantages of individual motoring. In many of these countries, after 1989 individual motoring has become highly preferred as modern democratic right for free transport mode Rail transport was ignored as a remnant of the old regime and gradually created various problems related to the organization, operation, problems with rail infrastructure and rolling stock.

The most significant deficiencies in the organization of rail passenger transport are related in particular to the construction of railway time table. Mostly it is constructed with a priority for long-distance transport. This results in insufficient satisfaction of regional train needs. Self-governing regions order bus connections mostly at the same time and almost on the same routes as operated regional trains. It creates the parallel connections of trains and buses.

There are also significant deficiencies in the area of transport planning. These often do not take into account the results of various transport analyses, transport research or ticket sales. Insufficient train connection frequency, weaker time and space availability, i.e. long distance of railway stations (stops) from the centers of villages (towns) have contributed to the weaker attractiveness of rail transport. Small accessibility of long-distance rail transport as well insufficient intervals in regional rail transport (2-hour interval) do not correspond to the basic tasks of rail transport system.

Other problems with rail passenger transport are caused by infrastructure quality. In many EU countries the infrastructure is in adequate with low line speeds and old safety devices. In some countries railway passenger transport rolling stock is often not well maintained and not renewed.

3. Methodology of research

The complex proposal of a methodological process for managing sustainable railway passenger transport consists of several sectional proposals that are closely interrelated. The purpose of the proposal will be a new characterization and classification of transport services provided by rail transport according to various criteria, as well as a definition of the factors affecting traffic service. However, it is essential to take into account current transport needs and current travel motives.

Subsequently, a relationship will be proposed, within the individual aspects of the traffic service, to determine the traffic potential that can be calculated on any selected transport route. It will also be necessary to define the railway infrastructure and its optimum range as an indispensable element needed to carry out the required traffic service. Consequently, it will be important to find the optimal connection between the required range of traffic services and the range of railway infrastructure. If this step is completed, it will be possible to proceed with the proposal of standards within the particular transport systems as well as the standards within the scope of the transport infrastructure range. These standards will then be the basis for the final construction of a diagram of train traffic.

This methodological process will be applicable to any selected railway transport route in particular countries. This process is graphically presented in figure 1as a flowchart.

The benefits of this methodological process should be significant in theoretical and practical terms, as the proposed steps and measures should offer a conceptual solution to the planning and organization of rail passenger transport, which should lead to its increased attractiveness and, in particular, to increased traffic flows. Every step of the process is important and irreplaceable. Achieving of the detailed individual results of particular steps requires detailed analysis. As an example, the step "determining traffic potential" is described in greater detail in the next chapter.

3.1. Determining traffic potential as an important part of the proposed methodical process

Determining traffic potential is a necessary part of the methodical process. However it is very complicated to establish it exactly. For example, empirical models can be used for its optimal determination. The most appropriate one is Lill's gravity model. It is described below and subsequently there are opportunities for extension of Lill's gravity model and also foe developing a new method for determining traffic potential.

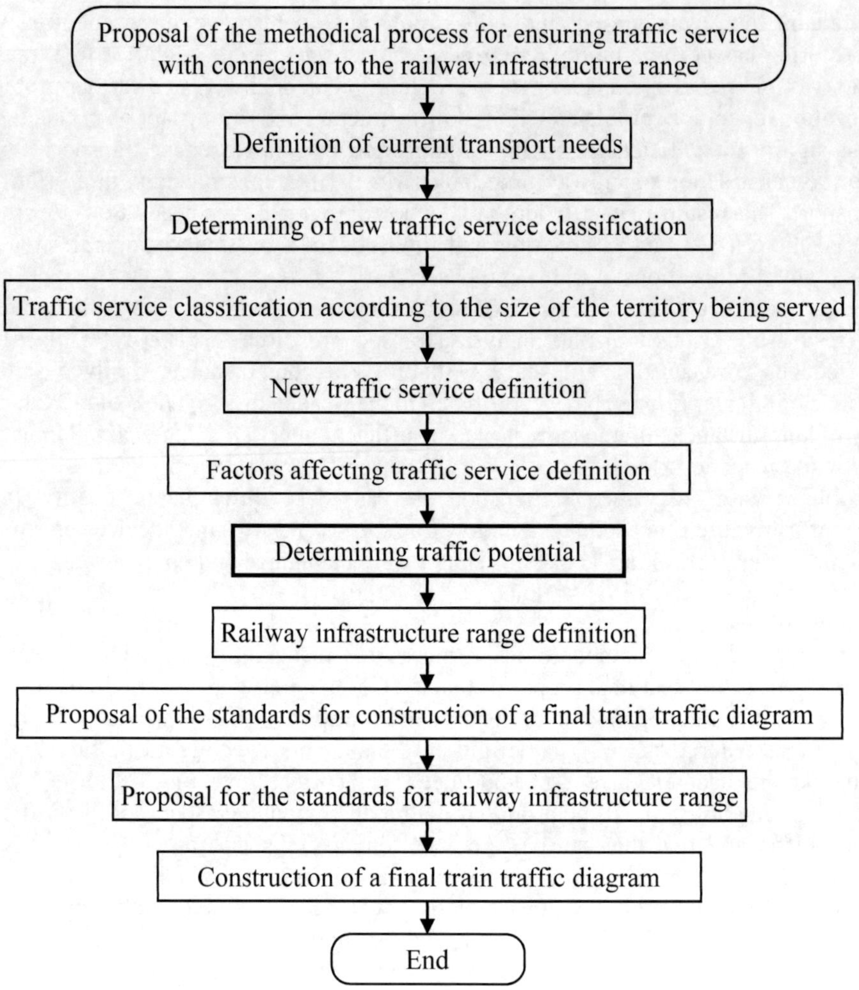

Figure 1. Methodological process for ensuring traffic service linked to railway infrastructure range.

3.1.1. Lill's gravity model. Lill's gravity model serves to determine the optimum number of return journeys of all types of public passenger transport between two selected traffic points. The model considers the number of inhabitants of these transport points and the distance between them. Calculated optimal number is directly proportional to the number of inhabitants of both transport points, the K coefficient and the inversely proportional distance of these traffic points. The model has the following formula, the result is always rounded up:

$$j_{1,2} = \frac{A_1 A_2}{d^2} K, \tag{1}$$

where $j_{1,2}$ is the optimal number of journeys (connections) between selected traffic points over a certain period of time; $A_{1,2}$ – population number (current, in thousands) of selected toll stations representing traffic points; d – distance between transport points, km; K – coefficient (depends on the character and boundaries of selected areas).

3.1.2. Another method for determining traffic potential. To ensure an attractive railway passenger transport system, it is necessary to analyze, among other things, the traffic potential of particular railway lines or transport routes. On the basis of the results of these analyses, it will then be easier to find bottlenecks on the railway lines or routes and then set the optimum service concept for them.

To determine traffic potential, a methodology is proposed that takes into account all relevant factors affecting this potential. This method is not an empirical model. Its basis is the formula, where is possible to fit the values of the factors affecting the traffic potential. The main objective is to determine the value of the total potential coefficient K_p. The value of the coefficient is a non-dimensional number that is directly proportional to transport potential, which means the higher the value, the higher the potential of the chosen railway transport route and the need to set up a service concept with maximum use of rail transport. The directly proportional factors of the resulting value are placed in the numerator of the formula, inversely proportional in the denominator. The overall formula is as follows:

$$K_p = \frac{A\,C_A}{L\,D\,C_0}, \qquad (2)$$

where K_p is total transport potential coefficient – the final value is directly proportional to traffic potential; A – average value of population of the each railway station or stop; current, in thousands; L – average value of transport distance between particular railway stations and stops in a chosen area, km; D – railway station or stop availability – average distance of all stations or stops from the center of adjacent towns or villages, km; C_O – overlapping of the railway line with road coefficient – comparison the total journey time by road and by rail on the chosen transport route, ratio of the running time by rail and by road; the final value is also average value of particular values.C_A is the average value of attractiveness coefficient of a chosen transport route. It expresses the extent to which the railway is attractive for tourism, employment and other. Coefficient C_A consists of three proposed sub-factors, which are expressed by additional coefficients, because the attractiveness of the monitored area is influenced mainly by the number of jobs, the number of different institutions for meeting the basic needs of citizens and the number of tourist centers. For the exact expression of these subfactors are proposed particular subcoefficients – subcoefficient of city size C_E, tourism subcoefficient C_{CR} and subcoefficient of unemployment rate C_N. The final calculation of C_A then will be:

$$C_A = \frac{C_E\,C_{CR}}{C_N}. \qquad (3)$$

The values of the particular subcoefficients were determined using the method of creative thinking and based on consultation with transport experts, it takes values from 1 to 5 where the higher the final value, the more attractive the selected territory and the higher the transport potential.

4. Results and discussion

Subsequently, the proposed methodology is tested on 2realrailway lines in Slovakia. The first example is Žilina-Vrútky railway line (shown in Picture 3 in red) and Hronská Dúbrava - Banská Štiavnica railway line (shown Picture 3 in blue).

Subsequently, the final values are inserted into the formula. Particular values including the final K_p value are shown in table 1.

In the second case, the particular values were changed except for transport distance. The changes were done intentionally, because it is important to prove that increasing the directly proportional values in the numerator of the formula causes an increase in the final K_p value and on the contrary indirectly proportional values in the denominator of the formula causes a decrease of the final K_p value. It follows that a decrease of the factor values that favourably influence the increase of K_p causes the decrease of the final K_p value and vice versa. An example is given in table 2.

The final K_p values are not limited. Based on the values fit into the formula, the final value may be within the first extreme, for example 0.10, or within the second extreme, for example 50 and more. It expresses the traffic potential on the chosen transport route and its comparison with other transport routes, for example how many times it is higher or lower than the other line and so on. Consequently, it will then be possible to calculate optimal number of trains that should by operated in the railway line. The exact procedure of this calculation will be the subject of further scientific and research work of the authors.

Table 1. Example on Žilina-Vrútky railway line.

	Žilina	Varín	Nezb. Lúčka - Strečno	Vrútky	Final average values
A	82.92	3.82	2.56	7.76	24.27
C_A	8	2	2	6	4.50
C_E	4	1	1	2	2
C_{CR}	4	4	4	3	3.75
C_N	2	2	2	1	1.75
L	8	8	4	9	7.00
D	0.5	1.8	1.3	0.5	1.03
t railway, min	7	7	3.5	7.5	6.00
t road, min	13	13	6.5	15	11.50
C_O	0.54	0.54	0.54	0.50	0.53
K_p				28.95	

Table 2. Example on Hronská Dúbrava - Banská Štiavnica railway line.

	Hronská Dúbrava	Kozelník	Banská Belá	Banský Studenec	Banská Štiavnica	Final average values
A	0.42	0.16	1.17	0.46	10.03	2.45
C_A	1	1	1	1	3	1.40
C_E	1	1	1	1	3	1.4
C_{CR}	3	3	3	3	3	3
C_N	3	3	3	3	3	3
L	8	8	7	3	2	5.00
D	0.3	0.7	1.3	5	1.1	1.68
t railway [min]	11.5	11.5	10	4	4	7.38
t road [min]	9	9	8.5	13	9	9.88
C_O	1.28	1.28	1.18	0.31	0.44	0.80
K_p					0.51	

In the case of the two monitored railway lines it means that on Žilina-Vrútky railway line traffic potential will be around 56 times higher than on Hronská Dúbrava - Banská Štiavnica railway line. It means that on Žilina-Vrútky railway line there should be 56 times more seats available on a train than on Hronská Dúbrava - Banská Štiavnica railway line. It could correspond to reality, because the Žilina-Vrútky railway line is part of the main corridor railway line Bratislava - Košice and Hronská Dúbrava - Banská Štiavnica is a short regional railway line in middle Slovakia. Nowadays there are approximately 10 regional trains operated daily in both directions on Hronská Dúbrava - Banská Štiavnica railway line with approximately 600 seats on the trains. There are operated approximately 90-100 trains (regional and long distance) in both directions on railway line Žilina-Vrútky approximately with 34000-40000 places to sit on the trains. It follows that the capacity of seats on the trains is about 60 times higher on Žilina-Vrútky railway line than in the case of the Hronská Dúbrava - Banská Štiavnica railway line. This value can approximately correspond to calculated traffic potential.

4.1. Variant calculations of chosen railway lines

However, if particular parameters are changed in the formula, the final K_p value will be changed too. Several variations of chosen railway lines are shown in tables 3 and 4.

The parameters in brown were changed in the tables. It can be seen how the final K_p value changed. In the case of Žilina-Vrútky railway line the marked in brown parameters were changed for the worse, but they were change for the better in the case of Hronská Dúbrava - Banská Štiavnica railway line. The final K_p value will then be only approximately 2.64 times higher in the first case than in the second case after these changes.

Table 3. Example on Žilina-Vrútky railway line after some parameter changes.

	Žilina	Varín	Nezb. Lúčka - Strečno	Vrútky	Final average values
A	62.92	3.82	2,56	7.76	19.27
C_A	4	1	1	1,33	1.83
C_E	4	1	1	2	2
C_{CR}	3	3	3	2	2,75
C_N	3	3	3	3	3
L	8	8	4	9	7.00
D	0.5	1.8	1.3	0.5	1.03
t railway, min	7	7	3.5	7.5	6.00
t road, min	13	13	6.5	15	11.50
C_O	0.54	0.54	0.54	0.50	0.53
K_p				9.36	

Table 4. Example on Hronská Dúbrava - Banská Štiavnica railway line after some parameter changes.

	Hronská Dúbrava	Kozelník	Banská Belá	Banský Studenec	Banská Štiavnica	Final average values
A	0.42	0.16	1.17	0.46	10.03	2.45
C_A	4	2	2	2	6	3.20
C_E	2	1	1	1	3	1.6
C_{CR}	4	4	4	4	4	4
C_N	2	2	2	2	2	2
L	8	8	7	3	2	5.00
D	0.3	0.2	0.4	2.5	0.5	0.78
t railway, min	8	8	7	3	3	5.25
t road, min	9	9	8.5	13	9	9.88
C_O	0.89	0.89	0.82	0.23	0.33	0.57
K_p					3.54	

4.2. Comparison of the K_p calculation with Lill's formula

For comparison it would be useful to calculate the optimal number of connections on the chosen railway lines using Lill's gravity model too. Only the population of the starting and destination station and distance between them are fit into the formula. However, it is very difficult to determine the K coefficient. Its value for Slovakia should be 100.

In the case of Žilina-Vrútky railway line the final value is as follows:

$$j_{1,2} = \frac{82.92 \cdot 7.76}{21^2} 100 = 145.91.$$

In the case of Hronská Dúbrava - Banská Štiavnica railway line the final value is as follows:

$$j_{1,2} = \frac{0.42 \cdot 10.03}{20^2} 100 = 1.05.$$

It follows from the calculated values that the final value for the first monitored railway line is more than 138 times higher than the final value of the second monitored railway line. It is too high and inaccurate. The calculation of transport potential using the proposed formula for K_p calculation is more accurate and objective. Therefore, it is more suitable to use the proposed methodology for other railway lines too.

5. Conclusion

This contribution addresses the current state of public passenger transport, highlighting its current problems and weaknesses, especially in the field of its operation and organization. The main proposal consists of the methodical process, which will be the basis for ensuring optimal transport service by rail passenger transport in the monitored area as well as the optimal range of railway infrastructure. The proposed methodological process contains a number of sub-steps and objectives, which when fulfilled will help to achieve the main objective. The most important part of the process is determining traffic potential. The paper explained current methods and also the new proposed methodology. The main aim of the proposed methodology is to determine the value of the total traffic potential coefficient K_p. This value can be calculated by using the proposed formula. Finally, there was practical application of the proposed methodology on real railway lines in Slovakia.

Generally, the calculated values of the traffic potential do not have to be exact values, it is only recommended mathematical value which should be taken into account when constructing train traffic diagram. Other detailed solutions of the proposed method and the specific solutions of other mentioned partial steps as well as the overall methodology proposal will be the subject of further scientific and research work of the authors.

Acknowledgments
This work was supported by the grant system of the University of Žilina.

References

[1] De Dios O J and Willumsen L G 2011 *Modelling Transport* (John Wiley & Sons.)
[2] Hansen I A and Pachl J 2014 *Railway Timetabling &Operations* (Eurailpress, Hamburg)
[3] Sokolov Y, Ivanova E, Anikeeva-Naumenko L and Lavrov I. 2019 Improving passenger transportation service as a part of tourism services complex in Russia *Management Science Letters* **9(1)** 53–60
[4] Eboli L and Mazzulla G 2015 Relationships between rail passengers' satisfaction and service quality: a framework for identifying key service factors *Public Transport* **7(2)** 185–201
[5] Kidd M P, Lusby R M and Larsen J 2019 Passenger-and operator-oriented scheduling of large railway projects *Transp. Res. Part C: Emerging Technologies* **102** 136–52
[6] Pellegrini P, Marlière G and Rodriguez J 2014 Optimal train routing and scheduling for managing traffic perturbations in complex junctions *Transp. Res. Part B: Methodological* **59** 58–80
[7] Banister D 2008 The sustainable mobility paradigm *Transport policy* **15(2)** 73–80
[8] Kuník P 2012 *The Proposal of the Integrated Cycle Time Table in Slovakia in 2012* (Diploma thesis, University of Pardubice, The Czech Republic) p 91

A multi-criteria selection of the transport plan of intercity passenger trains

Svetla D Stoilova

Department of Railway Engineering, Technical University – Sofia, Bulgaria

E-mail: stoilova@tu-sofia.bg

Abstract. The study proposes a methodology based on Sequential Interactive Modelling for Urban Systems (SIMUS) method for selection the transport plan of intercity passenger trains. The methodology uses linear programming to assess different transport plans. The research includes five steps. First, the alternatives transport plans of movement of passenger trains are determined. Second, the criteria for multi-criteria optimization of transport plan are defined. The SIMUS method is applied in the third step for ranking the alternatives. New approaches of integrated SIMUS-VIKOR and SIMUS-TOPSIS procedures have been proposed and compared. The verification of the results has been conducted in the fourth step by applying the Multi-criteria Optimization and Compromise Solution method (VIKOR) and Technique for Order of Preference by Similarity to Ideal Solution method (TOPSIS). The fifth step includes the determination of the suitable transport plan according the changes in passenger flows. This step takes into account of uncertainty of process. The methodology has been applied for study the transport plan of passenger trains in Bulgaria's railway network. The suitable organization of intercity passenger trains has been proposed. The elaborated new approach could be applied to make decision on a separate railway line or throughout the railway network, as well as for international passenger transport.

1. Introduction

The passenger trains planning are an important element in the railway operation process, which influences to determination of the efficiency of the railway system. It is a basis for draw up the train timetable, crew scheduling, rolling stock planning, and is related to the operating costs and economic benefits. The determination of the transport plan is an optimization problem related to obtain the trains frequencies, categories of trains, stop-schedules and others operating parameters.

The optimization and multi-criteria methods have been applied for studding the choosing of the transport plan of trains. The mixed-integer linear programming models of traffic management in railway have been proposed in [1, 2]. In [1] the authors elaborated model to optimize the train routes and train scheduling using mixed-integer linear programming. The research was conducted to the Chinese railway network. A two-layer optimization model for high-speed railway is elaborated in [3]. The first model determines the service frequencies and achieves an optimal stop-schedule. The nonlinear programming and genetic algorithm are applied to minimize the total operating cost and unserved passenger volume. The second model is based on mixed integer programming to maximize the passenger volume and minimize the total travel time for all passengers. The model has been experimented on Taiwan high-speed railway and also in Beijing-Shanghai high-speed railway line in China. In [4] are proposed an integrated line planning and timetabling model. The objective function presents a minimization of the total time spend by passengers waiting at origin and transfer stations and the operational costs. The

Content from this work may be used under the terms of the Creative Commons Attribution 3.0 licence. Any further distribution of this work must maintain attribution to the author(s) and the title of the work, journal citation and DOI.
Published under licence by IOP Publishing Ltd

model is used of Israel Railways. The binary integer programming models to optimize the stop planning according the passenger demand is elaborated in [5]. In [6] is proposed a multiobjective model for optimization the transport plan of passenger trains. The objective function consists minimizing the total passenger train stop times and minimization of the transportation costs. The mathematical model that takes into account of the operating costs and the costs of passengers are presented in [7]. The costs of passengers include the transport costs and the costs for waiting the train, train changing and others. The revenue-maximization model for transport planning is elaborated in [8]. The model considers the operating costs and costs of travel time. It accounts of the train scheduling and the trains stop, and also the passenger demand to ticket pricing.

The multi-criteria methods have been applied in [9-12].The multi-criteria analysis have been applied in [9] to choice and planning the suitable urban transportation. The authors used main criteria and sub-criteria to assess different alternatives. The main criteria that were applied are: economic, social impact, engineering, environmental impact. The model was experimented for monorail planning and route selection for Ankara. In [10] is elaborated a model for railway route planning based on multi-criteria methods. The criteria for assessment the different transport plans that were studied are the investment needed for railway construction, railway operating and maintenance costs, number of train and others. The methodology is applied for four railway routes on the Corridor X. In [11] the efficiency of the highway passenger transport enterprises is analysed according the criteria as the time, comfort, safety and rapidity.

The fuzzy linear optimization model for determination the number of passengers trains in railway network have been elaborated in [12]. As the optimization criterion has been used the minimum of operating costs. The model is integrated with multi-criteria analysis to choose the suitable transport plan of passenger railway transportation.

It can be concluded that the operating costs and the time travel are the main factors that influenced on the transport planning. Different techniques are applied to select the best transport plan as optimization models, multi-criteria analysis. The application of the multi-criteria analysis requires the use of experts to evaluate the criteria connected to transport planning. A major disadvantage of these approaches is subjectivism in decision-making. The multi-criteria method that does not use the expert's assessment is the SIMUS (Sequential Interactive Modelling for Urban Systems) procedure. It is hybrid approach that uses Liner programming, Weighed Sum Method (WSM) and Outranking, [13-15]. The SIMUS is use in different areas. It is applied in [16] to make decision about the carriage of containers in intermodal transport.

The aim of this study is to elaborate a methodology based on SIMUS application for assessment the alternative transport plans for passenger trains taking into account of different criteria that influence of transportation process.

2. Methodology

The methodology of research includes the following steps, figure 1:

Step 1: Determination the alternatives for transport plan of intercity trains. Determination the number of trains for each transport plans by using linear optimization according the criterion minimum direct operating costs.

Step 2: Defining the criteria for assessment of the alternatives.

Step 3: Application the SIMUS Method for ranking the alternatives. Experimentation a new approaches of integration of SIMUS and VIKOR methods, and also of SIMUS and TOPSIS methods for ranking the alternatives.

Step 4: Verification of results. This step includes a comparison of results between SIMUS, VIKOR and TOPSIS methods.

Step 5: Determination of the suitable alternative according the changes in passenger flows. This step takes into account of uncertainty of process.

2.1. Determination the alternatives of the transport plan in railway network
The transport plan of passenger trains includes the number of categories trains, number of trains in each category, number of itineraries of trains by categories, number of wagons in train composition. The determination of the parameters of transport plan contains the following steps: first, the itineraries and the categories of intercity trains in each transport plan have to be set; second, the linear optimization model to determine the number of trains of each category is applied for each transport plan. The study applies the fuzzy linear optimization model given in [12]. The number of transport plans defines the number of alternatives, and depends on values of passenger flows, possibilities of the railway infrastructure and the railway operator. The decision maker determines the number of alternatives.

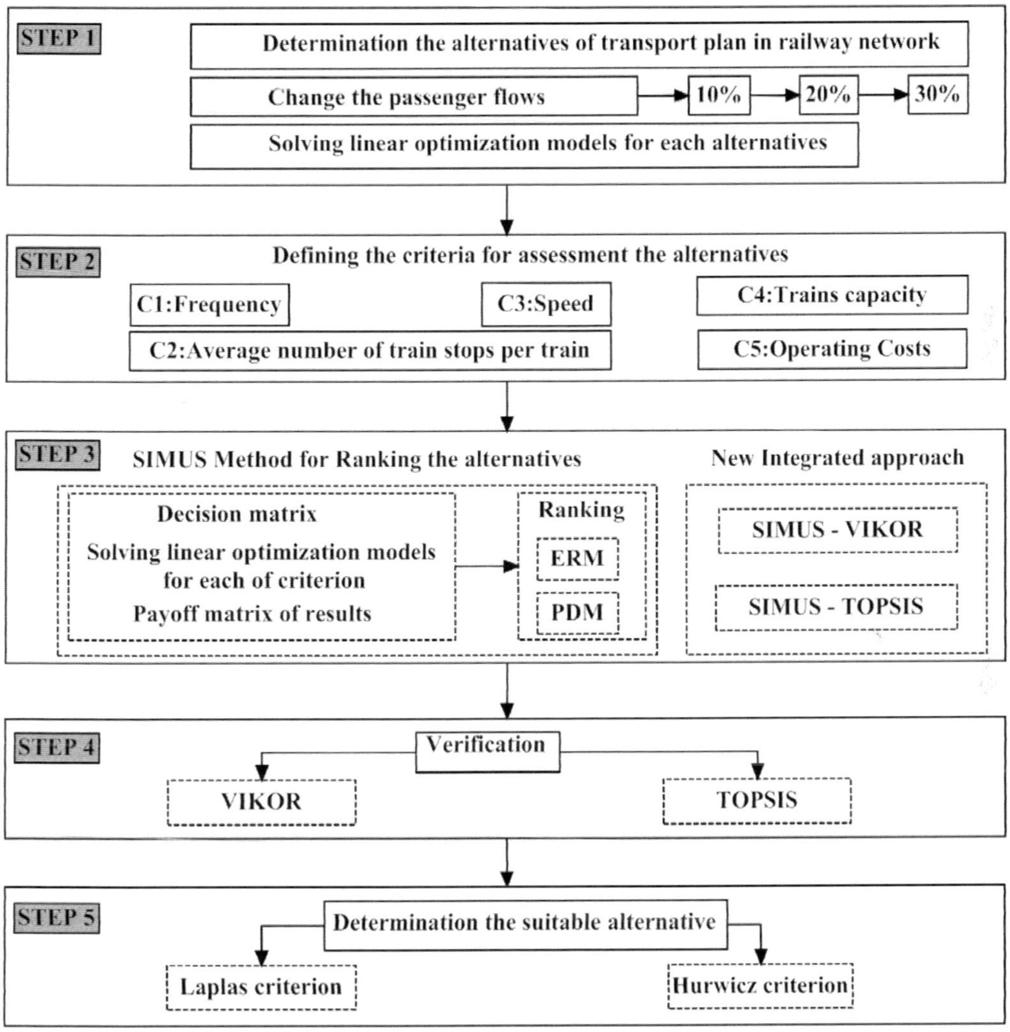

Figure1. Scheme of methodology.

2.2. Defining the criteria for assessment of the alternatives.
The second step of the methodology determines the criteria to assess the alternatives of transportation of intercity trains. The following criteria are proposed in the study:

- C_1 – Frequency, trains/day. This criterion shows the number of the trains in transport plan.
- C_2 – Average number of train stops per train. The organization of the category of intercity trains with reduced number of stops allows decreasing the travel time.

- C_3 – Average operating speed, km/h. This criterion presents the average speed of transport services of all categories intercity trains in the transport plan.
- C_4 – Trains capacity, seats/day. This criterion indicates the number of seats offered by the transport plan per day. The number of seats depends on the number of trains and the number of wagons in the train composition.
- C_5 – Direct operating costs, BGN/day. This criterion is important for railway operators, the purpose of which is to make transport with minimal operating costs.

2.3. SIMUS method for ranking the alternatives

The SIMUS procedure consist the following steps, [14]:

First, the decision matrix of criteria and alternatives is made. The columns present the alternatives, and the rows show the criteria. The type of optimization for each criterion is defined.

Second, the normalized matrix is calculated. It could be used different normalization methods, as total sum in row, maximum value in row, Euclidean formula and others. The choice of the method depends of decision maker. The results of the normalization have not influenced of the ranking. The limit to each criterion is determined. The limits can be set by the decision maker or determined by values given in normalized matrix. In the second case when the objective function is of maximum, the value of limit is set as equal of maximum normalized value of the row. In the case of minimum the value of limit is equal of minimum normalized value of the row.

Third, the procedure of SIMUS method is formed based on linear programming. The normalized matrix, the limits, and the type of optimization are used to define the optimization models for each of criterion. The first optimization model is made by using the first criterion as the objective function which. This criterion is removed from decision matrix. The others row of the matrix present the restrictive conditions of the linear model. The procedure is repeated with the other criteria. The number of linear optimization models is equal to the number of criteria. The results of linear models are recorded in Efficient Results Matrix (ERM). The values show the score of each alternative for each linear model.

Fourth, the ranking of alternatives is made. The SIMUS method is based on two approaches to make decision – Weighed Sum Method (WSM) and Outranking. The ranking are named ERM Ranking and PDM Ranking.

The procedure of ERM ranking consist the following steps: normalization of the ERM; summing the elements in each column (SC); determination the number of participation of each alternative in the each column of normalized ERM (PF); normalization the given results by dividing to the number of criteria (NPF); determination the score by multiplication the NPF and SC; ranking according the maximal value of the score.

The procedure of PDM ranking consist the following steps: starting from the highest value of the first row is determined the differences between values in the same row of normalized ERM; summing the elements of each row; summing the elements of the each column; determination the score of each alternative as the difference between row and column values; ranking according the maximal value of the scores.

2.4. New integrated approach

The study proposes new integrated approach for ranking the alternatives by using a combination of both SIMUS and VIKOR methods, and SIMUS and TOPSIS methods.

The VIKOR and TOPSIS methods are multi-criteria decision making approaches. These methods are distance-based. The both methods used the matrix of results of linear optimization models for each of criteria (ERM matrix) as a decision matrix. The calculations are applied to ERM matrix. The new approach could be applied also to verify the results. VIKOR and TOPSIS can be applied also to verify the results by SIMUS procedure.

2.4.1. VIKOR method for ranking the alternatives. The VIKOR method is based on a measure to the ideal solution, [17]. The main steps can be summarized as follows:

- Step1: Determination the decision matrix $(x_{ij})_{mxn}$ consisting of n alternatives and m criteria, $i = 1, \dots, n; j = 1, \dots, m$. Calculation the best and the worst values of all criterion functions.

The best values are calculated when j-th criterion represents a benefit; the worst values are calculated for non-beneficial criteria.

$$f_j^* = \max_i f_{ij} \; ; f_j^- = \min_i f_{ij}, \tag{1}$$

where f_j^* is the best value; f_j^- – the worst values of all criterion functions.

- Step 2: Calculation of the distance from each alternative to the positive ideal solution.

$$S_i = \sum_{j=1}^{m} \frac{w_j(f_j^* - f_{ij})}{f_j^* - f_j^-}; \quad R_i = \max_i \frac{w_j(f_j^* - f_{ij})}{f_j^* - f_j^-}, \tag{2}$$

where S_i is the distance of i-th alternative to the positive ideal solution; R_i – the maximal regret of each alternative; w_j – the weights of the criteria.

- Step 3: Determination of the index value Q_i

$$Q_i = \frac{v(S_i - S^*)}{S^- - S^*} + \frac{(1-v)(R_i - R^*)}{R^- - R^*}; \tag{3}$$

$$S^* = \min_i S_i; S^- = \max_i S_i; R^* = \min_i R_i; R^- = \max_i R_i, \tag{4}$$

where v is the coefficient of the decision making strategy, that shows the optimism level of the decision maker; $0 \le v \le 1$.

The value $v = 1$, and $v = 0$ indicate the choice of a strategy with maximum or respectively minimum group utility. The value shows a strategy of minimum individual regret. Usually, the value $v = 0.5$ is chosen to make research.

The minimum value of index Q_i shows the best alternative.

- Step 4. Determination the stability of the results

The best alternative has to satisfy two conditions. The first condition is:

$$Q((A^{(2)}) - Q((A^{(1)}) \ge \frac{1}{n-1}, \tag{5}$$

where $A^{(1)}$ is the best ranked alternative by the minimum of index Q_j, $A^{(2)}$ – the alternative with second position in the ranking.

The second check is about the ranking for the alternative $A^{(1)}$ in respect to parameters S or/and R for all values of v.

If one of both conditions is not satisfied, the following set of compromise solutions is proposed. $A^{(1)}A^{(2)}$,..., $A^{(N)}$. $A^{(N)}$ is determined according the satisfaction of the following condition:

$$Q((A^{(N)}) - Q((A^{(1)}) < \frac{1}{n-1}. \tag{6}$$

If only the condition about the ranking according the parameters S or/and R is not met the alternatives $A^{(1)}$ and $A^{(2)}$ are the best.

2.4.2. TOPSIS method for ranking the alternatives. The Technique for Order of Preference by Similarity to Ideal Solution (TOPSIS) is based on the principle of the shortest distance from the ideal solution and farthest distance from the negative ideal solution. The main steps are as follow, [18]:

- Step 1: Determination the decision matrix $(x_{ij})_{mxn}$ consisting of n alternatives and m criteria. Calculation of the normalization matrix $(r_{ij})_{mxn}$.

$$r_{ij} = \frac{x_{ij}}{\sqrt{\sum_{i=1}^{n} x_{ij}^2}}, \; i = 1, \dots, n; j = 1, \dots, m, \tag{7}$$

where $i = 1, \dots, n$ is the number of alternatives; $j = 1, \dots, m$ – the number of criteria.

- Step 2: Determination the weighted normalized matrix $(v_{ij})_{mxn}$.

$$v_{ij} = r_{ij}\, w_j; \;\; \sum_{j=1}^{n} w_j = 1 \tag{8}$$

where: w_j is the weight of criterion j.

- Step 3: Determination the ideal best v_j^+ and ideal worst v_j^- value for each criterion j.

$$v_j^+ = \min_i v_{ij} \text{ for no benefits criteria; } v_j^+ = \max_i v_{ij} \text{ for benefits criteria;} \tag{9}$$

$$v_j^- = \max_i v_{ij} \text{ for no benefits criteria; } v_j^- = \min_i v_{ij} \text{ for benefits criteria.} \tag{10}$$

- Step 4: Calculation the Euclidean distance from the ideal best D_i^+ solution and the Euclidean distance from the ideal worst D_i^- solution.

$$D_i^+ = \sqrt{\sum_{j=1}^{m}(v_{ij} - v_j^+)^2}\,; \;\; D_i^- = \sqrt{\sum_{j=1}^{m}(v_{ij} - v_j^-)^2}. \tag{11}$$

- Step 5: Determination the Performance score C_i . It presents the relative closeness of each alternative i with reference to negative ideal solution D_i^- .

$$C_i = \frac{D_i^-}{D_i^+ + D_i^-}\,; \;\; 0 \le C_i \le 1 \tag{12}$$

The maximum value of performance score C_i shows the best alternative.

2.5. Decision making in the state of variation of passenger flows.

The theory of Decision has been used to determine the best alternative taking into account of the variation of passenger flows. In the study we propose Laplace's criterion and Hurwicz's criterion to make decisions, [19].

- Laplace's criterion (Equal Likelihood). This is based on the principle that all events are equally. The criterion of choosing an optimal alternative depends of the type of elements of payoff matrix, i.e.:

$$L_i = \min_i \left\{\frac{1}{n}\sum_{i=1}^{n} a_{ij}\right\}, \text{ when } a_{ij} \text{ presents the costs;} \tag{13}$$

$$L_i = \max_i \left\{\frac{1}{n}\sum_{i=1}^{n} a_{ij}\right\}, \text{ when } a_{ij} \text{ presents the benefits.} \tag{14}$$

The main advantage of Laplace's criterion is its easy way of determining.

- The Hurwicz's criterion. This criterion uses an additional parameter – coefficient of optimism α that allows making decision according different decision approaches. The value of α can be set between 0 and 1. Generally, $\alpha = 0.5$. While α=1 represents an optimistic approach, $\alpha = 0$ represents a totally pessimistic approach.

The optimal alternative is determined as:

$$H_i = \min_i \left\{\alpha \min_j a_{ij} + (1 - \alpha) \max_j a_{ij}\right\}, \text{ when } a_{ij} \text{ presents the costs} \tag{15}$$

$$H_i = \max_i \left\{\alpha \min_j a_{ij} + (1 - \alpha) \max_j a_{ij}\right\}, \text{ when } a_{ij} \text{ presents the benefits} \tag{16}$$

The main advantage of Hurwicz's criterion is the application of a parameter that allows the variability of the results to be examined.

3. Results and discussion

3.1. Application of SIMUS method for ranking the alternatives

The proposed methodology is applied to make decision about transport plan of intercity trains in Bulgarian network. 17 routes between initial and finals points are studied. The categories of intercity trains have been set for these routes. In this study we investigated three categories of intercity trains – direct express trains (TC1) with reduced number of stops in comparison of the current train timetable (new category), accelerate fast trains (TC2) and faster trains (TC3) that stopped in the stations defined in the current train timetable. The direct intercity trains are studied for itineraries Sofia-Plovdiv, Sofia-Plovdiv-Burgas, and Sofia-G. Oryahovitsa-Varna taking into account of passenger flows, [12]. It was accepted that the all train compositions have 4 wagons. The direct trains Sofia-Plovdiv do not stop in other stations; the direct trains Sofia-Plovdiv-Burgas stop in stations Plovdiv and Stara Zagora; the direct trains Sofia-G. Oryahovitsa-Varna stops in stations Mezdra, Pleven, G. Oryahovitsa and Varna. The scheme of itineraries is presented in [12].

The alternatives have been formed according the number of train categories that are included in transport plan. Table 1 presents the main parameters of alternatives.

Table 1. Parameters of alternatives.

Alternative	Categories of train		Number of itineraries			
	Number	Type	TC1	TC2	TC3	Total
A1	3	TC1, TC2, TC3	3	7	17	27
A2	2	TC1, TC3	3		17	20
A3	2	TC2, TC3		7	17	24

The number of trains in each category for the alternatives and the values of criteria have been determined using fuzzy linear optimization model given in [12].

The influence of change the passenger on the ranking the alternatives flows has been investigated. The following variants of the increase of passenger flows have been studied: Variant 1 – current situation; Variant 2 – 10%; Variant 3 – 20%; Variant 4 – 30%. Tables 2-5 consist two parts. In the first part are presented the decision matrices that contain the values of criteria for each alternative. The normalized matrices, type of optimization of criteria, type of operator, and limits of criteria determined by SIMUS method are shown in the second part. The last columns of the tables present the limits that are set as equal of the maximum normalized value of the row when the criterion is of maximum and the minimum value of the row if the criterion is of minimum.

Table 2. Parameters of Variant 1.

Criterion	Decision matrix			Normalized matrix			Action	Operator	Limits
	A1	A2	A3	A1	A2	A3			
C1	38.00	35.00	37.00	0.60	0.55	0.58	max	≤	0.60
C2	15.45	16.17	16.19	0.56	0.59	0.59	min	≥	0.56
C3	63.00	63.00	63.00	0.58	0.58	0.58	max	≤	0.58
C4	10640.00	9800.00	10360.00	0.60	0.55	0.58	max	≤	0.60
C5	51583.00	48772.00	49716.00	0.60	0.56	0.57	min	≥	0.56

Table 3. Parameters of Variant 2.

Criterion	Decision matrix			Normalized matrix			Action	Operator	Limits
	A1	A2	A3	A1	A2	A3			
C1	40.00	36.00	38.00	0.61	0.55	0.58	max	≤	0.61
C2	14.85	16.36	16.47	0.54	0.59	0.60	min	≥	0.54
C3	64.00	63.00	63.00	0.58	0.57	0.57	max	≤	0.58
C4	11200.00	10080.00	10640.00	0.61	0.55	0.58	max	≤	0.61
C5	55336.00	50687.00	51969.00	0.61	0.56	0.57	min	≥	0.56

Table 4. Parameters of Variant 3.

Criterion	Decision matrix			Normalized matrix			Action	Operator	Limits
	A1	A2	A3	A1	A2	A3			
C1	44.00	42.00	42.00	0.60	0.57	0.57	max	≤	0.60
C2	15.11	15.55	15.38	0.57	0.58	0.58	min	≥	0.57
C3	64.00	64.00	63.00	0.58	0.58	0.57	max	≤	0.58
C4	12320.00	11760.00	11760.00	0.60	0.57	0.57	max	≤	0.60
C5	60696.00	59227.00	55197.00	0.61	0.56	0.57	min	≥	0.55

Table 5. Parameters of Variant 4.

Criterion	Decision matrix			Normalized matrix			Action	Operator	Limits
	A1	A2	A3	A1	A2	A3			
C1	48.00	44.00	45.00	0.61	0.56	0.57	max	≤	0.61
C2	15.00	15.36	15.58	0.57	0.58	0.59	min	≥	0.57
C3	64.00	64.00	63.00	0.58	0.58	0.57	max	≤	0.58
C4	13440.00	12320.00	12600.00	0.61	0.56	0.57	max	≤	0.61
C5	65518.00	61824.00	60134.00	0.60	0.57	0.56	min	≥	0.56

When the criterion is of maximum the type of operator is "≤", in the case of minimum the operator is "≥".

Table 6 presents the optimization models by SUMUS method for Variant 1. The scores of each alternative are labelled accordingly *x1*, *x2* and *x3*. The similar models have been formed also for the other variants.

Table 6. Optimization linear models by SIMUS method for Variant 1

Optimization model for Criterion C1	Optimization model for Criterion C2	Optimization model for Criterion C3
0.60$x1$+0.55$x2$+0,58$x3$→ max	0.56$x1$+0,59$x2$+0.59$x3$→ min	0.58$x1$+0.58$x2$+0.58$x3$→ max
0.56$x1$+0,59$x2$+0.59$x3$≥0.56	0.60$x1$+0.55$x2$+0.58$x3$≤0.60	0.60$x1$+0.55$x2$+0.58$x3$≤0.60
0.58$x1$+0.58$x2$+0.58$x3$≤0.58	0.58$x1$+0.58$x2$+0.58$x3$≤0.58	0.56$x1$+0.59$x2$+0.59$x3$≥0.56
0.60$x1$+0.55$x2$+0.58$x3$≤0.60	0.60$x1$+0.55$x2$+0.58$x3$≤0.60	0.60$x1$+0.55$x2$+0.58$x3$≤0.60
0.60$x1$+0.56$x2$+0.57$x3$≥0.56	0.60$x1$+0.56$x2$+0.57$x3$≥0.56	0.60$x1$+0.56$x2$+0.57$x3$≥0.56
0 ≤$x1$,$x2$,$x3$≤1	0 ≤$x1$,$x2$,$x3$≤1	0 ≤$x1$,$x2$,$x3$≤1
Optimization model for Criterion C4	**Optimization model for Criterion C5**	
0.60$x1$+0.55$x2$+0.58$x3$→ max	0.60$x1$+0.56$x2$+0.57$x3$→ min	
0.60$x1$+0.55$x2$+0.58$x3$≤0.60	0.60$x1$+0.55$x2$+0.58$x3$≤0.60	
0.56$x1$+0,59$x2$+0.59$x3$≥0.56	0.56$x1$+0.59$x2$+0.59$x3$≥0.56	
0.58$x1$+0.58$x2$+0.58$x3$≤0.58	0.58$x1$+0.58$x2$+0.58$x3$≤0.58	
0.60$x1$+0.56$x2$+0.57$x3$≥0.56	0.60$x1$+0.55$x2$+0.58$x3$≤0.60	
0 ≤$x1$,$x2$,$x3$≤1	0 ≤$x1$,$x2$,$x3$≤1	

Tables 7 and 8 show the results of SUMUS procedure for Variant 1. The ERM matrix in Table 7 presents the results of optimization. The results of both ranking are equally. It can be seen that the alternative A1 is the best. It is the best for three of criteria when they are given as objective function. This includes: maximum of frequency, minimum of average train stops for transport plan and maximum of train capacity. The scores of criteria for alterative A3 are equal to zero. Figure 2 shows the results of individual linear optimization by each criterion that represents the scores of criteria by alternatives. The summarized results of ranking for all studied variants are given in table 9. The alternative A1 is the best according the SIMUS procedure for all cases.

Table 7. ERM Ranking for Variant 1.

Criterion	ERM			Objective function	ERM Normalized		
	A1	A2	A3		A1	A2	A3
C1	1.00	0.00	0.00	0.60	1.00	0.00	0.00
C2	0.95	0.00	0.00	0.53	1.00	0.00	0.00
C3	0.00	1.09	0.00	0.63	0 .00	1.00	0.00
C4	1.00	0.00	0.00	0.60	1.00	0.00	0.00
C5	0.00	0.96	0.00	0.54	0.00	1.00	0.00
Sum of Column (SC)					3.00	2.00	0.00
Participation Factor (PF)					3	2	0
Norm. Participation Factor (NPF)					0.60	0.40	0.00
Final Result (SC x NPF)					1.80	0.80	0.00
ERM Ranking: A1 - A2 - A3							

Table 8. PDM Ranking for Variant 1.

Dominant	PDM Subordinated			Row sum	Net dominance
	A1	A2	A3		
A1	0.0	3.0	3.0	6.0	4.0
A 2	2.0	0.0	2.0	4.0	1.0
A 3	0.0	0.0	0.0	0.0	-5.0
Column Sum	2.0	3.0	5.0		
PDM Ranking: A1 - A2 - A3					

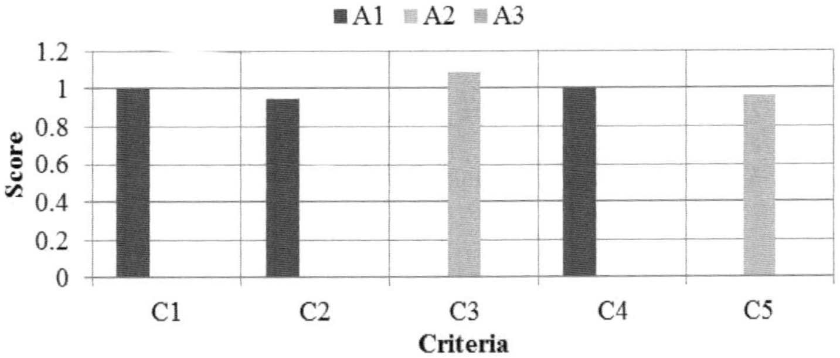

Figure 2. Score of criteria by alternatives.

Table 9. Ranking SIMUS (ERM, PDM).

Alternative	Variant 1			Variant 2			Variant 3			Variant 4		
	ERM	PDM	rank	ERM	PDM	rank	ERM	PDM	rank	ERM	PDM	rank
A1	1.8	4	1	1.8	4	1	1.8	4	1	1.8	4	1
A2	0.8	1	2	0.8	1	2	0.2	-2	2	0.2	-2	2
A3	0.0	-5	3	0.0	-5	3	0.2	-2	3	0.2	-2	3

3.2. New integrated approaches SIMUS-VIKOR, SIMUS-TOPSIS

Tables 10 and 11 present the results of new integrated approaches SIMUS-VIKOR and SIMUS-TOPSIS proposed in this study. In this case the decision matrix is the ERM matrix of results of SUMUS linear optimization models for all criteria. The calculations by VIKOR and TOPSIS methods have been conducted taking into account that the weights of the criteria are equal (in this case w_j =0.2). We made this acceptance because SIMUS does not use weightings on the criteria. The results by VIKOR method are for value of parameter $v = 1$ that indicates the choice of a strategy with maximum group utility. The other values of the parameter v are not applicable due to the specification of the data. The results show that according to the conditions of the stability of the results, the all three alternatives are suitable. The results of TOPSIS method present the same ranking as the ERM and PDM ranking by the SIMUS procedure.

The comparison of the results by new approaches and SIMUS is shown in table 12. The type of optimization is different. The VIKOR method use a minimization of the score as a criterion for ranking, while SIMUS (ERM, PDM) and SIMUS – TOPSIS applied the maximum of the score of the ranking's criterion. Figure 3 presents the ranking by the different approaches. It was found that the procedure SIMUS-TOPSIS is more appropriate than the procedure SUMUS-VIKOR due to the specification of the data in the decision matrix. Therefore, we recommend the SIMUS-TOPSIS integrated approach.

Table 10. Ranking SIMUS-VIKOR.

Alternative	Variant 1							Variant 2						
	Decision matrix					Q_i	rank	Decision matrix – 10%					Q_i	rank
	C1	C2	C3	C4	C5			C1	C2	C3	C4	C5		
A1	1.00	0.95	0.00	1.00	0.00	0.00	1	1.00	0.92	0.00	1.00	0.00	0.67	1
A2	0.00	0.00	1.09	0.00	0.96	0.33	2	0.00	0.00	1.11	0.00	0.00	1.00	2
A3	0.00	0.00	0.00	0.00	0.00	1.00	3	0.00	0.00	0.00	0.00	0.00	1.00	3
Alternative	**Variant 3**							**Variant 4**						
	Decision matrix – 20%					Q_i	rank	Decision matrix – 30%					Q_i	rank
	C1	C2	C3	C4	C5			C1	C2	C3	C4	C5		
A1	1.00	0.91	0.00	1.00	0.00	0.67	1	1.00	0.92	0.000	1.00	0.00	0.67	1
A2	0.00	0.00	1.05	0.00	0.00	1.00	2	0.00	0.00	1.09	0.00	0.00	1.00	2
A3	0.00	0.00	0.00	0.00	0.98	1.00	3	0.00	0.00	0.00	0.00	0.96	1.00	3

Table 11. Ranking SIMUS-TOPSIS – Ranking SIMUS (ERM).

Alternative	Variant 1				Variant 2				Variant 3				Variant 4			
	D_i^+	D_i^-	C_i	rank	D_i^+	D_i^-	C_i	rank	D_i^+	D_i^-	C_i	rank	D_i^+	D_i^-	C_i	rank
A1	1.41	1.73	0.55	1	1.00	1.73	0.63	1	1.41	1.73	0.55	1	1.41	1.73	0.55	1
A2	1.73	1.41	0.45	2	1.73	1.00	0.37	2	2.00	1.00	0.33	2	2.00	1.00	0.33	2
A3	2.24	0.00	0.00	3	2.00	0.00	0.00	3	2.00	1.00	0.33	3	2.00	1.00	0.33	3

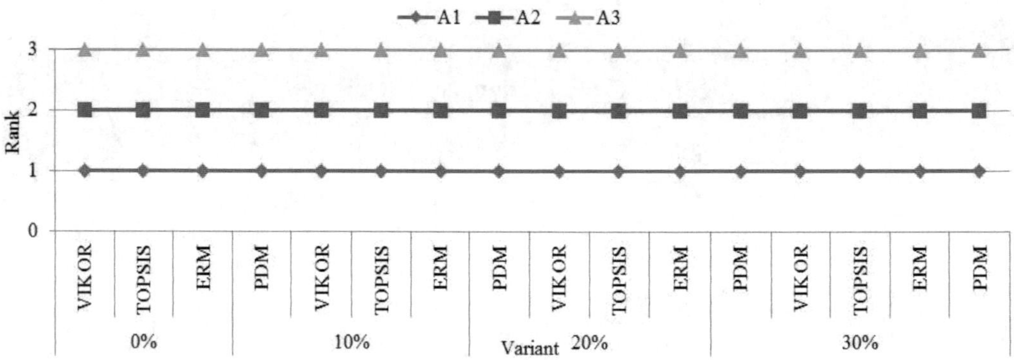

Figure 3. Comparison of the ranking by SIMUS-VIKOR, SIMUS-TOPSIS and SIMUS procedures.

Table 12. Comparison of ranking SIMUS-VIKOR, SIMUS-TOPSIS, SIMUS (ERM, PDM).

Alternative	Variant 1				Variant 2				Variant 3				Variant 4			
	SIMUS-VIKOR	SIMUS-TOPSIS	ERM	PDM	SIMUS-VIKOR	SIMUS-TOPSIS	ERM	PDM	SIMUS-VIKOR	SIMUS-TOPSIS	ERM	PDM	SIMUS-VIKOR	SIMUS-TOPSIS	ERM	PDM
A1	0.00	0.55	1.80	4	0.00	0.63	1.80	4	0.67	0.55	1.80	4	0.67	0.55	1.8	4
A2	0.33	0.45	0.80	1	0.67	0.37	0.80	1	1.00	0.33	0.20	-2	1.00	0.33	0.2	-2
A3	1.00	0.00	0.00	-5	1.00	0.00	0.00	-5	1.00	0.33	0.20	-2	1.00	0.33	0.2	-2

3.3. Verification the results

The verification of results given by SIMIS method has been conducted by using VIKOR and TOPSIS methods. The decision matrix in this case coincides with the initial decision matrix of values of criteria for each alternative. The results of all studied variants have been verified. Table 13 shows the results by VIKOR method for value of parameter $v = 1$ that is applicable. Table 14 shows the parameters of

TOPSIS method. Table 15 presents the comparison of the scores. The ERM Ranking is used for SIMUS procedure to compare the results. The ranking of the alternatives by both applied methods are the same as those given by SIMUS. Figure 4 shows the rank by the three methods. It can be seen that the results are equal.

Table 13. Ranking VIKOR.

Alternative	Variant 1							Variant 2						
	Decision matrix					Q_i	rank	Decision matrix – 10%					Q_i	rank
	C1	C2	C3	C4	C5			C1	C2	C3	C4	C5		
A1	38	15.45	64	10640	51583	0.00	1	40	14.85	64	11200	55336	0.00	1
A2	35	16.17	63	9800	48772	1.00	3	36	16.36	63	10080	50687	1.00	3
A3	37	16.19	63	10360	49716	0.44	2	38	16.47	63	10640	51969	0.56	2
Alternative	Variant 3							Variant 4						
	Decision matrix – 20%					Q_i	rank	Decision matrix – 30%					Q_i	rank
	C1	C2	C3	C4	C5			C1	C2	C3	C4	C5		
A1	44	15.11	64	12320	60696	0.00	1	48	15.00	64	13440	65518	0.00	1
A2	42	15.55	64	11760	59227	0.37	2	44	15.36	64	12320	61824	0.83	2
A3	42	15.38	63	11760	55197	1.00	3	45	15.58	63	12600	60134	1.00	3

Table 14. Ranking TOPSIS.

Alternative	Variant 1				Variant 2				Variant 3				Variant 4			
	D_i^+	D_i^-	C_i	rank	D_i^+	D_i^-	C_i	rank	D_i^+	D_i^-	C_i	rank	D_i^+	D_i^-	C_i	rank
A1	0.83	1.00	0.55	1	0.81	1.00	0.55	1	1.41	1.73	0.55	1	0.82	0.94	0.53	1
A2	0.97	0.83	0.46	2	0.93	0.81	0.47	2	2.00	1.00	0.33	2	0.64	0.67	0.51	2
A3	1.04	0.56	0.35	3	1.02	0.59	0.36	3	2.00	1.00	0.33	3	0.94	0.82	0.47	3

Table 15. Comparison of ranking by VIKOR, TOPSIS, SIMUS.

Alternative	Variant 1				Variant 2				Variant 3				Variant 4			
	VIKOR	TOPSIS	SIMUS (ERM)	SIMUS (PDM)	VIKOR	TOPSIS	SIMUS (ERM)	SIMUS (PDM)	VIKOR	TOPSIS	SIMUS (ERM)	SIMUS (PDM)	VIKOR	TOPSIS	SIMUS (ERM)	SIMUS (PDM)
A1	0.00	0.55	1.80	4	0.00	0.55	1.80	4	0.00	0.55	1.80	4	0.00	0.55	1.80	4
A2	1.00	0.46	0.80	1	1.00	0.47	0.80	1	0.37	0.33	0.20	-2	0.83	0.33	0.20	-2
A3	0.44	0.35	0.00	-5	0.56	0.36	0.00	-5	1.00	0.33	0.20	-2	1.00	0.33	0.20	-2

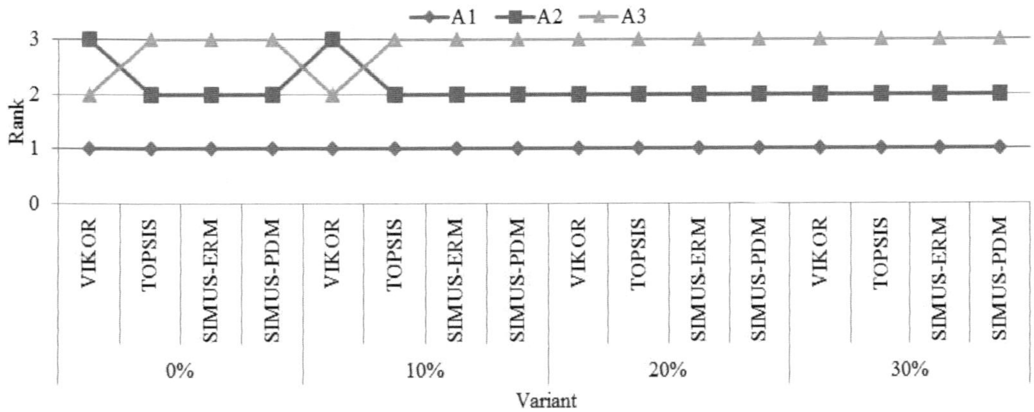

Figure 4. Comparison of the ranking by SIMUS, VIKOR and TOPSIS methods.

3.4. Decision making in variation of passenger flows

The Laplas's criterion and Hurwicz's criterion according formulas (13-16) have been applied to make decision. We used five values of the coefficient of optimism α, that are presented in figure 4. The results of both criteria are equal that verify the solutions. Figure 5 and table 16 present the ranking. The type of decision making by Laplas's criterion and Hurwicz's criterion for scores of TOPSIS and SIMUS (ERM ranking) is of minimum according formulas (13) and (15). When are used the scores of VIKOR to assess the changes in passenger flows we used formulas (14) and (16). In this case the maximum values of Laplas's criterion and Hurwicz's criterion determine the best solution. It can be seen in figure 5. Finally, it could be conclude that the best alternative of transport plan between studied ones in railway network is A1. It proposes three categories of intercity trains, as the direct trains suggested for itineraries Sofia – Plovdiv, Sofia – Plovdiv – Burgas, and Sofia – G. Oryahovitsa – Varna.

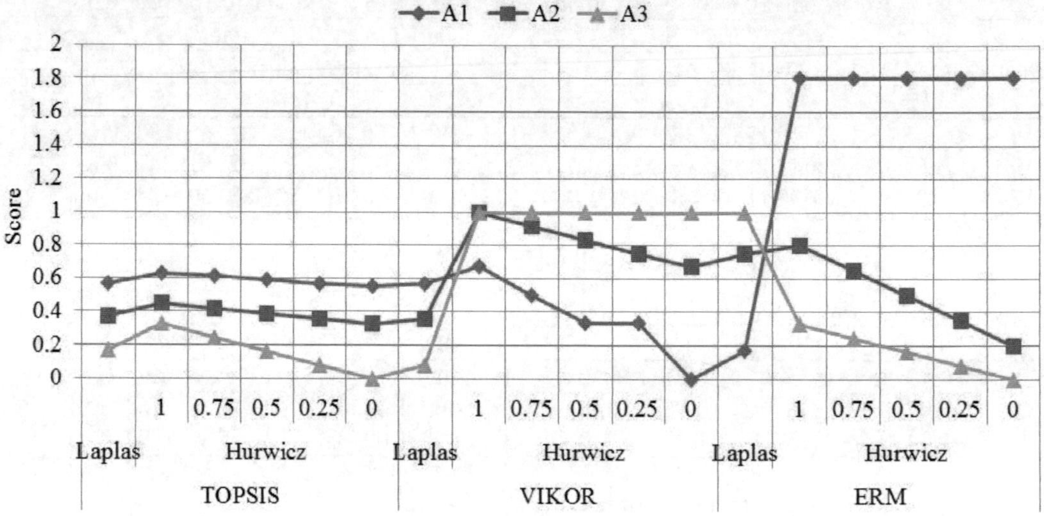

Figure 5. Comparison of SIMUS-VIKOR, SIMUS-TOPSIS and SIMUS (ERM).

Table 16. Comparison of the ranking by SIMUS-VIKOR, SIMUS-TOPSIS, SIMUS.

Alternative	TOPSIS						VIKOR						SIMUS (ERM Ranking)					
	Laplas		Hurwicz				Laplas		Hurwicz				Laplas		Hurwicz			
	L_i	rank	α 0.5	0.75	0.25	rank	L_i	rank	α 0.5	0.75	0.25	rank	L_i	rank	α 0.5	0.75	0.25	rank
A1	0.57	1	0.59	0.61	0.57	1	0.33	1	0.34	0.50	0.16	1	1.8	1	1.80	1.80	1.80	1
A2	0.37	2	0.39	0.42	0.36	2	0.75	2	0.84	0.92	0.75	2	0.5	2	0.50	0.65	0.35	2
A3	0.17	3	0.16	0.25	0.08	3	1.00	3	1.00	1.00	1.00	3	0.1	3	0.16	0.24	0.08	3

4. Conclusions

The methodology for selection the transport plan of intercity passenger trains that is based on integration of SIMUS and VIKOR, and SIMUS and TOPSIS methods has been elaborated. It was found that the procedure SIMUS-TOPSIS is more appropriate than the procedure SUMUS-VIKOR due to the specification of the data in the decision matrix. Therefore, we recommend the SIMUS-TOPSIS integrated approach as new procedure of decision making.

The indices for multi-criteria optimization of transport plan of passenger trains have been defined.

The methodology has been experimented for passengers transport in Bulgaria's railway network. Three transport plans of intercity trains have been proposed and studied.

Each criterion is applied as objective function of linear optimization according SIMUS procedure and the solution has been determined. The alternative A1 that includes accelerate fast trains, fast trains and additional service with direct trains with reduced number of stops are chosen as the suitable. It is the best alternative for three of criteria when they are given as objective function. This includes: maximum of frequency, minimum of average train stops for transport plan and maximum of train capacity.

The verification of results shows the applicability of the developed methodology. The new approach could be used also for verification the results.

An approach based on application of Laplas's criterion and Hurwicz's criterion to make decision in the state of variation of passenger flows is proposed. The results show that the proposed alternative retains its position as the best transport plan.

The proposed methodology and defined criteria could be used to investigate the different transport alternatives in railway network. It can be applied also to assess the organization of passenger trains of different categories and different number of wagons in train composition. The separated railway lines or the whole railway network can be explored to propose a suitable transport plan.

Acknowledgments
This research has been supported by contract №КП-06-Н27/12 of 11.12.2018 "Modelling and elaboration of complex system for selection of transport technology in transport network" funded by the National Science Fund of the Ministry of Education and Science of Bulgaria.

References

[1] Luan X, Miao J, Meng L, Corman F and Lodewijks G 2017 Integrated optimization on train scheduling and preventive maintenance time slots planning *Transportation Research Part C* **80** 329–59
[2] Pellegrini P, Marliere G and Rodriguez J 2014 Optimal train routing and scheduling for managing traffic perturbations in complex junctions *Transportation Research Part B* **59** 58–80
[3] Wang L, Jia L, Qin Y, Xu J and Mo W 2011 A two-layer optimization model for high-speed railway line planning *J. of Zhejiang University-SCIENCE A (Applied Physics & Engineering)* **12(12)** 902–12
[4] Kaspi M and Ravi T 2013 Service-oriented line planning and timetabling for passenger trains *Transportation Science* **47(3)** 295–311
[5] J Yung-Cheng Lai, Mei-Cheng Shih and Guei-Hao Chen 2016 Development of efficient stop planning optimization process for high-speed rail systems *J. Adv. Transp.* **50** 1802–19
[6] Liu P, Han B and Zhang Q 2014 Optimization model of passenger train service plan based on the degree of passenger satisfaction *Applied Mechanics and Materials* **505-506** 577–82
[7] Lingaitisa V and Sinkeviuius G 2014 Passenger transport by railway: evaluation of economic and social Phenomenon *Procedia - Social and Behavioral Sciences* **110** 549–59
[8] Zhang X, Li L, Le Vine S and Liu, X 2019 An integrated pricing/planning strategy to optimize passenger rail service with uncertain demand *J. of Intelligent & Fuzzy Systems*, **36(1)** 435–48
[9] Hamurcu M and Eren T 2019 An application of multicriteria decision-making for the evaluation of alternative monorail routes *Mathematics* **7** 16
[10] Kosijer, M, Ivic M, Markovic M and Belosevic I 2012 Multicriteria decision-making in railway route planning and design *Gradevinar* **64** 195–205
[11] Wen H and Lin S 2011 Performance evaluation of highway passenger transport enterprises' operation based on the model of AHP-DEA *Industrial Engineering and Engineering Management (IEEE 18Th International Conference Wuhan, China)* **2** 811–5
[12] Stoilova S 2018 An integrated approach for selection of intercity transport schemes on railway networks. *Promet-Traffic&Transportation* **30(4)** 367–77
[13] Munier N 2011 *A Strategy for Using Multicriteria Analysis in Decision-making: A Guide for Simple and Complex Environmental Projects* (Springer Science & Business Media) p 298
[14] Munier N. 2016 A new approach to the rank reversal phenomenon in MSDM with the SIMUS method *Multiple criteria decision making* **11** 137–52

[15] Munier N, Hontoria E and Jimenez-Saez F 2019 *Strategic Approach in Multi-Criteria Decision Making. A Practical Guide for Complex Scenarios* (Springer Nature Switzerland AG) p 273
[16] Stoilova S 2019 Application of SIMUS method for assessment alternative transport policies for container carriage *Proc. of 18th International Scientific Conference Engineering for Rural Development (Jelgava, Latvia)* 898–906
[17] Opricovic S and Tzeng GH 2007 Extended VIKOR method in comparison with outranking methods *European Journal of Operational Research* **178(2)** 514–29
[18] Srinivasa R K 2014 *Multicriterion Analysis in Engineering and Management* (PHI Lerning Private Limited, Delhi) p 266
[19] Hamdy A. Taha 2011 *Operations Research: An Introduction, 9th Edition* (Prentice Hall)

Choosing the container handling equipment in a rail-road intermodal terminal through multi-criteria methods

Svetla D Stoilova[1] and Svetoslav V Martinov

Department of Railway Engineering, Technical University – Sofia, Bulgaria

[1] E-mail: stoilova@tu-sofia.bg

Abstract. The study proposes a methodology based on the combination of multi-criteria methods for choosing the yard handling equipment in a rail-road container terminal. The methodology contains four steps. The alternative variants of the handling equipment have been determined in the first step. Three types of container handling equipment – electric Rail Mounted Gantry crane (RMG), diesel driven Rubber Tyred Gantry crane (RTG) and mobile Reach Stacker (RS), have been considered. The second step is based on a Technical, Economical, Technological and Ecological (TETE) analysis. The sub-criteria for each main TETE group to assess the alternatives have been defined. The alternatives have been assessed by applying of twenty sub-criteria. The weights of the main criteria and the sub-criteria have been determined in the third step by applying Analytic Hierarchy Process (AHP) method. The results of weights given in this step are used in the next fourth step to ranking the alternatives by using Preference Ranking Organization Method for Enrichment of Evaluations (PROMETHEE) method. The sub-criteria that have main impact of the choice of an alternative are procurement costs (11%), annual operation costs (9%), stacking capacity (9%), annual equipment maintenance costs (6%), flexibility (6%) and carbon dioxide emissions (6%). Three variants according the type of terminal as small, medium and large are investigated. The results have been verified by using Technique for Order of Preference by Similarity to Ideal Solution (TOPSIS) method. It was found that the RS is the suitable alternative for small rail-road terminals, and the RMG is the best one for the medium and large rail-road terminals. The proposed methodology could be applied also to assess other types of handling equipment.

1. Introduction

A part of the freight transport system is the intermodal transport system. It provides an opportunity to transportation of intermodal transport units in an intermodal supply chain in combining different modes of transport – railway, road and waterways. An important element of intermodal hinterland transport of containers to and from ports is the intermodal rail transport. It also is a major transport for continental transport of containers and the other continental load units – swap bodies and semitrailers. A standard intermodal supply chain usually consist a few stages of transportation by using different modes of transport. These stages are: pre-haulage transport to an intermodal terminal, handling of intermodal units in the intermodal terminal, transportation of intermodal transport units by rail and/or ship and end-haulage. In the intermodal supply chains the major stages are transportation by rail for a land-based intermodal supply chain and transportation by ship for a waterway intermodal supply chain.

The intermodal terminals are essential elements of the intermodal transport system. They are the connecting elements, a part of the intermodal system, where the different modes of transport interacts. There are two main groups intermodal terminals in depends on the modes of transport that the terminal

Content from this work may be used under the terms of the Creative Commons Attribution 3.0 licence. Any further distribution of this work must maintain attribution to the author(s) and the title of the work, journal citation and DOI.
Published under licence by IOP Publishing Ltd

serviced. The first group includes intermodal terminals that serve only railway and road transport. The second group intermodal terminals are the trimodal terminals that are situated in an inland waterway or in a maritime port. The three modes of transport – railway, road and ship are serviced in the trimodal intermodal terminals.

The various possibilities for handling of intermodal transport units are determined by a large number of technological and technical characteristics of the intermodal terminals [1, 2]. Various container handling technologies at the terminals are applied globally. The most widely used technologies in the rail-road intermodal terminals [3] are Reach Stackers (RS), Rail Mounted Gantry cranes (RMG), and Rubber Tyred Gantry cranes (RTG). The choice of a handling technology is related to multiple criteria and constraints that have to take into account. In many scientific studies, different criteria for selecting the appropriate intermodal handling technology are used by the authors. Cost analysis for choosing the handling equipment is used by some authors [4-6]. In [4, 7] the authors described the main characteristics of container handling systems in terminals and analysed the factors that affect the total cost of the handling technologies. The design and construction of intermodal terminals from the point of view of expenditures for handling equipment are described in [6].

Different operational attributes, economical cost attributes and management characteristics are used by authors as selection criteria in [8]. In [9] the container loading problem in rail-truck intermodal terminals is discussed by authors considering energy consumption.

Numerous parameters and features influence the choice of technology at intermodal terminals. These parameters could be classified into different groups – economic, technical, operational, etc. As a main characteristic of the intermodal terminals could be qualified the handled volume of containers. It could be determined by the quantity of containers or other intermodal transport units that are handled into the terminal over a certain period of time [1].

In the literature, the intermodal terminals are classified into several main groups according to the handled volume of containers [1, 2, 10, 11]. In [1, 2] the container terminals are classified by the authors in five main groups. The terminals up to 10000 TEU per year and 9000 m^2 terminal area are classified as smallest terminals. And opposite, the terminals over 500000 TEU per year and terminal area 400000 m^2 are classified as large terminals. The terminals are classified by authors [10] in three main groups – small, medium and large terminals with handled volume more than 350 intermodal transport units per day. The terminals are classified [11] in three groups – small, medium and big terminals that handled respectively up to 20000 TEU/year, 20000-100000 TEU/year and over 100000 TEU/year.

The main technical and technological characteristics of the intermodal terminals discussed in [4-6] are cycle time, productivity of handling equipment, fuel consumption of combustion engine handling equipment, energy consumption for electric handling equipment, etc.

The multiple of criteria and constraints in choosing of container handling equipment requires application of multi-criteria methods. Various models based on Multi-Criteria Decision Making (MCDM) methods [12] are used in the literature for selection of container handling equipment in intermodal terminals. In [8, 13] the authors apply AHP analysis for selecting yard cranes in marine container terminals.

The MCDM methods are an appropriate basis for decision-making when many criteria are applied for select an alternative. In general, there are weighting based, distance based and outranking multi-criteria methods to make decision.

The purpose of this study is to elaborate a methodology for choosing the container handling equipment in a rail-road intermodal terminal taking into account different criteria by its nature.

2. Methodology of research

This study proposes a new methodology based on the combination of TETE analysis, AHP [14] and PROMETHEE [15] methods for selecting the most appropriate handling equipment for a rail-road intermodal terminal.

The methodology includes the following steps:

Step 1. Determination the alternatives of handling equipment for a rail-road intermodal terminal.

Step 2. Definition of quantitative and qualitative criteria for the assessment of alternatives. This step of the model proposes a TETE analysis including Technical, Economics, Technological and Ecological criteria to assess the handling equipment. The sub-criteria for each main criterion of TETE group are defined for the evaluation of the alternatives. The sub-criteria are systematized in four main groups – Technical (sub-criteria C1, C2, C3 and C4), Economics (sub-criteria C5, C6, C7, C8, C9 and C10), Technological (sub-criteria C11, C12, C13, C14 and C15) and Ecological (sub-criteria C16, C17, C18, C19 and C20). In this research the following sub-criteria are proposed:

- C1 – Cycle time, min/move – it is a key characteristic of container handling equipment in the container terminals. It depends of different technical and technological factors. The value of the cycle time is equal of the average time for one move of a container in terminal;
- C2 – Productivity, moves/h – it is a parameter of handling equipment that express the number of movements that could be done per an hour;
- C3 – Energy consumption, kgoe/cont. – it expresses the amount of electricity or diesel fuel consumed to handling of a single container in container terminal. The energy that is consumed is measured by kilograms of oil equivalent per container;
- C4 – Equipment safety, rate – in this study the equipment safety is measured by a rate that indicates the presence of dangerous routes when handling equipment is moving into the container terminal. It is considered that the rail mounted gantry cranes have a higher rate of safety than mobile handling equipment as reach stackers;
- C5 – Annual operation costs, cost ratio – they are expressed by cost ratio between different types of container handling equipment;
- C6 – Annual equipment maintenance costs, % – they are expressed as percent of the initial procurement costs for handling equipment;
- C7 – Container yard development costs, EUR/m^2 – they include costs for engineering, designed and construction of container yard;
- C8 – Container yard maintenance costs, % – the annual maintenance costs of container yard are percent from container yard development costs;
- C9 – Procurement costs, EUR – total procurement costs for container handling equipment in the container terminal;
- C10 – Economic life, years – the expected economic life of the container handling equipment;
- C11 – Number of handlings of a container per number of rows high in the stacking yard, moves/row – relationship between average number of handlings per container at the yard and the average number of rows high of containers in the container yard;
- C12 – Flexibility, rate – it is measured by a rate that indicates the potential for replacement of the container handling equipment if it is necessary. The flexibility of the technology evaluates the ability to replace an existing type of handling equipment in the terminal with another type. In this study the flexibility assesses the existence or absence of a possibility of replacing the handling equipment with another similar or more advanced technology. Restrictions in this respect are imposed by the accompanying infrastructure for RMG cranes (rail-track and power supply). Such restrictions do not exist for the mobile reach stacker;
- C13 – Mobility, rate – it reflects the possibilities of the container handling equipment to handles containers in areas, different from the container stacking area;
- C14 – Stacking capacity, cont./ha – this indicator reflects the average number of containers that could be placed in the terminal stacking area at the same time;
- C15 – Average number of handlings at yard, moves/cont. – average number of handlings per container at the container yard excluding the operations of loading and unloading of the containers from the trucks and the container wagons;
- The ecological sub-criteria in the study are CO_2 emissions (C16), CO emissions (C17), NO_2 emissions (C18), SO_2 emissions (C19) and Particulate matter (PM) emissions (C20). All emissions are equal in kg per container.

Step 3. Determination the weights of criteria. In this step the AHP method is applied.
Step 4. Ranking the alternatives. The PROMETHEE multi-criteria method is proposed to make a decision.
Step 5. Verification of the results. To verify the results obtained by PROMETHEE method we propose other approach of multi-criteria analysis – TOPSIS method [16] that is distance-based.

2.1. Determination the values of the criteria

The study includes quantitative and qualitative criteria. The values of the qualitative criteria are set with values of 1, 2 or 3. The value of 3 shows the best score for the criterion and the value of 1 the lowest. The criteria equipment safety (C4), flexibility (C12) and mobility (C13) are qualitative ones.

A methodology to determine the values of parameters connected to the quantitative criteria has been developed.

The fuel consumption expresses energy consumption of different handling equipment in kilograms of oil equivalent per container (kgoe/cont.). The average energy consumption for engine driven RS and RTG cranes could be equal through:

$$EC = PH\ FC\ 0.83\ 1.01, kgoe/cont., \quad (1)$$

where EC is the energy consumption, kgoe/cont.; PH – the average period for handling of one container in the intermodal terminal, h/cont.; FC – average fuel consumption of one handling equipment per an hour, l/h; 0.83 – the weighs of one litre diesel fuel, kg/l; 1.01 is energy content of 1 kg diesel fuel, kgoe/kg [17].

The average number of container handling equipment in terminal could be equal through:

$$NHE = (PH\ N_{CONT})/WT, number, \quad (2)$$

where NHE is the average number of container handling equipment in the terminal; PH – the average period for handling of one container in the terminal, h/cont.; N_{CONT} – the average number of containers handling in the terminal per day, cont./day; WT – the average working time of the container terminal, h/day.

It is assumed that the all container handling equipment are identical and have the same productivity.

$$PH = CT\ MC, h/cont., \quad (3)$$

where CT is the average time cycle of container handling equipment in the terminal, h/move; MC is the average total number of handlings per container in the terminal yard, moves/cont.

$$MC = HCY + 2, moves/cont., \quad (4)$$

where HCY is the average number of handlings per container in the yard of the terminal, moves/cont.

The average number of handlings at yard is determined as follow:

$$HCY = (SH/MSH)\ NH, moves/cont., \quad (5)$$

where SH is the average height of stacking of the containers in container yard, rows; MSH is the maximum stacking height of container handling equipment, rows; NH is average number of handlings per container at yard in depends of container handling equipment [4], moves/cont.

The value of HCY does not include the moves of loading and unloading of a container from trucks and container wagons.

The required container capacity of the intermodal terminal is:

$$RCC = N_{CONT}\ DW, cont., \quad (6)$$

where RCC is required container capacity of the stacking area of the intermodal terminal, cont.; DW – the average dwell time of one container in the yard of the intermodal terminal, days.

The average stacking height in container yard could be determined by:

$$SH = RCC/N_{GS}, rows, \quad (7)$$

where N_{GS} is the number of ground slots in container yard, ground slots.

The land utility of the intermodal terminal could be determined through:

$$SC = RCC/SA\,, cont./ha, \tag{8}$$

where SC is stacking capacity of container yard, cont./ha; SA – stacking area of the container terminal, ha.

The ratio between the average number of handlings per a container and the average stacking height of containers in the stacking area of the terminal is determined through:

$$HHR = HCY/SH\,, moves/row, \tag{9}$$

where HHR is the ratio between average number of handlings at terminal yard per container and the stacking height of containers, moves/row.

The average amount of emissions by pollutants (CO_2, CO, SO_2, NO_2 and PM) released in the atmosphere per container that is handled by an electric container handling equipment is determined through:

$$EM_E = MP\ EL, kg/cont., \tag{10}$$

where EM_E are emissions by pollutants per container, kg/cont.; MP – the mass of the pollutants (CO_2, CO, SO_2, NO_2 and PM) released into the atmosphere for production of electricity, t/kWh; EL – the average value of the electricity needed to handling a container, kWh/cont.

The average amount of emissions by pollutants (CO_2, CO, SO_2, NO_2 and PM) released in the atmosphere from an internal combustion engine of the container handling equipment in the intermodal terminal is:

$$EM_C = PH\ EP\ EF\ LF\ 10^{-3}, kg/cont., \tag{11}$$

where EM_C are non-road engine exhaust emissions by pollutants, kg/cont.; EP – the engine power of non-road engine of the container handling equipment, kW; EF – the emission factor by pollutants, g/kWh; LF – load factor that is the ratio of average load used during normal operations as compared to full load at maximum rated horsepower [18, 19], coef.

To determine the emission CO_2, SO_2, CO, NO_2 and PM of the container handling equipment with diesel combustion engine it is necessary to determine emission factors of different pollutants.

Emission factor for CO_2 of a non-road container handling equipment with diesel combustion engine [20] could be determined through:

$$EF_{CO_2} = (BSFC - HC)\ 0.87\ (44/12), g/kWh, \tag{12}$$

where EF_{CO_2} is emission factor for CO_2, g/kWh; $BSFC$ – the in-use adjusted fuel consumption, g/kWh; HC – the in-use adjusted hydrocarbon emissions, g/kWh; 0.87 – the carbon mass fraction of diesel fuel (87% of carbon by mass), coef.; 44/12 – ratio of CO_2 mass to carbon mass (44 g CO_2 / 12 g C), coef.

Emission factor for SO_2 of a non-road container handling equipment with diesel combustion engine could be determined through [20]:

$$EF_{SO_2} = [BSFC\ (1 - FFS) - HC]\ 0.01\ WPS\ 2, g/kWh, \tag{13}$$

where FFS is the fraction of fuel sulfur converted to direct particulate matter, coef.; 0.01 – the conversion rate from weight percent to weight fraction, coef.; WPS – the weight percent of sulfur in diesel fuel, %; 2 – SO_2 (in grams) formed from a gram of sulfur, g.

Formulas (1-13) are used to determine the values of criteria for each of alternatives. The value of sub-criterion C3 (energy consumption) of technical mean group is determined according formula (1). The values of sub-criterion C9 (procurement costs) of economics main group and sub-criterion C15 of technological main group are determined using formulas (2-5). The values of sub-criteria C11 (number of handlings of a container per number of rows high in yard) and C14 (Stacking capacity) of technological main group are determined by formulas (6-9). The values of ecological criteria are determined by formulas (10-13).

2.2. Determination the weights of criteria by AHP Method

In the study we used the AHP method to determine the weights of criteria. This method is based on experts' assessments and pair-wise comparison of criteria by applying the fundamental Saaty's scale. Table 1 presents Saaty's scale for pair-wise comparison [14, 21].

The experts' assessments are verified by AHP consistency index CR. The value of consistency of the decision-maker has to satisfy the following condition: $CR \leq 0.1$.

Table 1. Saaty's scale for pair-wise comparison.

Explanation	Intensity of importance	Reciprocal values
Equal importance	1	1
Moderate importance	3	1/3
Strong importance	5	1/5
Very strong importance	7	1/7
Extreme importance	9	1/9
Average intermediate values between two close judgments	2; 4; 6; 8	1/2; 1/4; 1/6; 1/8

2.3. Ranking the alternatives of handling equipment by using PROMETHEE Method

The PROMETHEE multi-criteria analysis method is an outranking approach to prioritize the alternatives. In the study, the weights are determined in advance using the AHP method. For each criterion it is necessary to be set the type of optimization – maximum or minimum; the weight; the preference function which characterizes the difference for a criterion between the evaluations obtained by two possible decisions into a preference degree ranking from 0 to 1. The PROMETHEE method uses six basic preference functions: usual criterion; quasi criterion; criterion with linear preference; level criterion; criterion with linear preference and indifference area; Gaussian criterion.

The best alternative is determined according the maximum value of net outranking flows φ_i, which corresponds to the alternative with highest priority. The net outranking flows φ_i of alternative i $(i = 1, \dots, n)$ is determined as a difference between the positive outranking flow φ_i^+ and the negative outranking flow φ_i^-. The positive outranking flow shows how much an alternative – outranks the others. The negative outranking flow presents how much an alternative is outranked by the others.

$$\varphi_i = \varphi_i^+ - \varphi_i^-; \varphi_i \in [-1; 1]; \sum_{i=1}^{m} \varphi_i = 0 \quad (14)$$

The explanation and mathematical calculation steps of the PROMETHEE method are given in [15].

2.4. Verification the results by TOPSIS method for ranking the alternatives

To verify the results we used the distance-based multi-criteria decision analysis method. The Technique for Order of Preference by Similarity to Ideal Solution (TOPSIS) is based on the principle that best alternative should have the shortest distance from the ideal solution and farthest distance from the negative ideal solution [16]. TOPSIS consists the following steps:

- Step 1: Determination the decision matrix $(x_{ij})_{mxn}$ consisting of n alternatives and m criteria. Calculation of normalization matrix $(r_{ij})_{mxn}$. The values of normalization matrix are:

$$r_{ij} = \frac{x_{ij}}{\sqrt{\sum_{i=1}^{n} x_{ij}^2}}, i = 1, \dots, n; j = 1, \dots, m, \quad (15)$$

where $i = 1, \dots, n$ is the number of alternatives; $j = 1, \dots, m$ – the number of criteria.

- Step 2: Calculate weighted normalized matrix $(v_{ij})_{mxn}$. The elements of this matrix are:

$$v_{ij} = r_{ij}\, w_j; \sum_{j=1}^{n} w_j = 1, \quad (16)$$

where w_j is the weight of criterion j.

- Step 3: Calculate the ideal best v_j^+ and ideal worst v_j^- value for each criterion j.

$$v_j^+ = \min_i v_{ij} \text{ for no benefits criteria; } v_j^+ = \max_i v_{ij} \text{ for benefits criteria;} \quad (17)$$

$$v_j^- = \max_i v_{ij} \text{ for no benefits criteria; } v_j^- = \min_i v_{ij} \text{ for benefits criteria.} \quad (18)$$

- Step 4: Determination the Euclidean distance from the ideal best D_i^+ solution and the Euclidean distance from the ideal worst D_i^- solution.

$$D_i^+ = \sqrt{\sum_{j=1}^{m}(v_{ij} - v_j^+)^2}\ ;\ D_i^- = \sqrt{\sum_{j=1}^{m}(v_{ij} - v_j^-)^2} \quad (19)$$

- Step 5: Calculate Performance Score C_i that presents the relative closeness of each alternative i with reference to negative ideal measure D_i^- as follow:

$$C_i = \frac{D_i^-}{D_i^+ + D_i^-}\ ;\ 0 \leq C_i \leq 1. \quad (20)$$

The ranking of the alternatives is based on the C_i values. The higher is the C_i value, the better the alternative.

3. Results and discussion

The object of this study is a rail-road intermodal terminal. It is accepted that the handling of the containers in the terminal is done with identical container handling equipment and all cranes have the same productivity. Three variants for manipulating of containers in the terminal are discussed in the study – Reach Stacker (RS), Rubber Tyred Gantry crane (RTG) and Rail Mounted Gantry crane (RMG). The Reach Stacker and Rubber Tyred Gantry crane are diesel driven by a diesel combustion engine. The Rail Mounted Gantry crane is electricity driven and the crane moves on the fix rail-tracks.

Three types of container terminals have been discussed in the study according to volume – small, medium and large terminal. In the first variant up to 20000 TEU per year (V1) are handled in the terminal. In the second (V2) and third variant (V3) the containers handled in the terminal are between 20000 to 50000 TEU per year and between 50000 to 100000 TEU per year respectively.

The cycle time is a main characteristic of container handling equipment in container terminals. It could be expressed by the average time for one move of a container. The average cycle time for handling equipment is approximately 3-3.5 min for RS, 2.5-3 min for RTG and 2-2.5 min for RMG [4].

In the study we assume an average value of cycle time 3 min/move for RS, 2.5 min/move for RTG and 2 min/move for RMG crane [2, 4]. Productivity expresses the average number of the moves that the container handling equipment performs per an hour. Productivity of RS in container handling is approximately 25 moves/h for loading or unloading of railcars, 15 moves/h for stacking and 10 moves/h for loading or unloading of road trucks. Productivity of RMG cranes in container handling is approximately 30 moves/h for loading or unloading of railcars, 25 moves/h for stacking and 20 moves/h for loading or unloading of road trucks. Productivity of RTG cranes in container handling is approximately 30 moves/h for loading or unloading of railcars, 20 moves/h for stacking and 15 moves/h for loading or unloading of road trucks. The actual productivity of the intermodal terminal handling equipment depends on the work conditions and can be varied. In variant V3 for handled volumes over 55000 TEU/year, 65000 TEU/year and 80000 TEU/year two RS, RTG and RMG cranes are needed in the terminal respectively. The average number of handling equipment that is needed for handling containers at intermodal terminal is shown in table 2.

Table 2. Average number of the handling equipment.

Variant	Volume of terminal, TEU/year	RS	RTG	RMG
V1	up to 20000	1	1	1
V2	20000-50000	1	1	1
V3	50000-100000	2	2	1

The land utility of the stacking area in an intermodal terminal is assessed through storage capacity. The static storage yard capacity of a rail-road intermodal terminal could be determined by the density of the storage area (in cont./ha or TEU/ha) and the average stacking height (rows high).

The different handling systems [3, 7] can be achieved storage yard capacity up to 500 TEU/ha, based on 3-high stacking for RS systems, up to 1100 TEU/ha for RMG cranes and up to 1000 TEU/ha for RTG systems, based on 4-high stacking.

Information about land utility in the stacking area of the intermodal terminal, the number of ground slots and the average stacking height of containers in the terminal for different types of yard handling equipment is shown in table 3. The stacking yard of the terminal is 500 m length at a width of 22 m for RS and between 21 m and 72 m wide for the RTG crane and RMG crane. It is assumed that 20 foot and 40 foot containers are manipulated in the terminal and they are distributed in ratio 50% to 50%.

Table 3. Land utility in the stacking area for Variants 1, 2 and 3 (V1, V2 and V3).

Type of handling equipment	RS			RTG			RMG		
Variant	V1	V2	V3	V1	V2	V3	V1	V2	V3
Stacking height (up to rows)	3	3	3	4+1	4+1	4+1	4+1	4+1	4+1
Total number of ground slots	44-88	88-264	264-396	300	300	300	300	300	300
Number of ground slots per ha (ground slots/ha)	49-84	84-110	93-100	273	273	273	273	273	273
Average stacking height (rows)	1.8	2.3	2.5	0.4	1.3	2.8	0.4	1.3	2.8
Average storage capacity (cont./ha)	88-151	193-253	233-275	109	355	764	109	355	764

Gross electricity generation of Bulgaria was 45.28 TWh in 2016 [17]. In production was being emitted 25.42 million tons CO_2 [22]. The thermal power plant are major sources of pollutions. The main part of the electricity production in Bulgaria is from thermal power plants (over 45%) [23, 24]. On the basis of the produced electricity and the total emissions released by the thermal power plants in 2016 in Bulgaria are defined the values of pollutant emissions from the electricity: CO_2 – 0.56 t/MWh; SO_2 – 0.013 t/MWh; NO_2 – 0.0006 t/MWh; CO – 0.00004 t/MWh; PM – 0.00003 t/MWh.

The values of economics sub-criteria are defined according data given in [2, 4-6]. The values of qualitative criteria equipment safety (C4), flexibility (C12) and mobility (C13) are set with 1, 2 and 3, when 3 is the maximal score.

Studies related to energy consumption for container handling in container terminals indicate different average energy consumption. In [25] authors reported energy consumption of 7.25 kWh per move for an electric rail mounted stacking crane. The values of energy consumption reported in the other studies are 4.4 kWh per TEU [26, 27], 5-7.25 kWh per move [28] and 3.1-4.2 kWh per container [29]. In this study we assume average energy consumption 6 kWh per move for electric RMG crane.

The average fuel consumption of diesel driven container handling equipment depends from the work conditions, the load factor and the type of crane. The fuel consumption of diesel driven mobile cranes and Reach Stacker reported in different studies are 5 litres per km [25], average 3.34 litres per box [30] and 16 litres per an hour [28]. The consumption of diesel fuel for RTG is 24 litres per an hour or 2.4 litres per move at 10 moves per an hour [31]. We assumed 16 litres per hour average fuel consumption for RS and 22.2 litres per an hour for RTG crane. The values of emission factors of CO, NO_2 and PM pollutants have been defined by the limitations of Stage III A/B emission standards for non-road diesel engines category L in the power range between 130 and 560 kW [20, 32]: CO=3.5 g/kWh, NO_2=2.0 g/kWh and PM=0.025 g/kWh. To determine EF_{SO_2} we accept WPS=0.33 % and FFS=0.022.

The safety of the equipment is a parameter that expresses the absence of hostile routes when the handling equipment operating. The RMG crane has the maximum rate from the three alternatives due to the movement on a separate fix rail track. The values of sub-criteria for intermodal terminal with handled volume up to 20000 TEU/year, 20000-50000 TEU/year and 50000-100000 TEU/year are given respectively in table 4.

Table 4. Values of sub-criteria for Variants 1, 2 and 3 (V1, V2 and V3).

Sub-criteria	Dimension	RS			RTG			RMG			Type of optimization
		V1	V2	V3	V1	V2	V3	V1	V2	V3	
C1	min/move	3	3	3	2.5	2.5	2.5	2	2	2	min
C2	moves/h	20	20	20	25	25	25	30	30	30	max
C3	kgoe/cont.	2.21	2.48	2.55	1.78	2.33	3.18	1.19	1.55	2.12	min
C4	rate	1	1	1	2	2	2	3	3	3	max
C5	cost ratio	0.7	0.7	0.7	1	1	1	0.5	0.5	0.5	min
C6	%	5.5	5.5	5.5	6.5	6.5	6.5	4	4	4	min
C7	EUR/m^2	28	28	28	53	53	53	70	70	70	min
C8	%	0.6	0.6	0.6	0.75	0.75	0.75	0.5	0.5	0.5	min
C9	EUR	380000	380000	760000	520000	520000	1040000	700000	700000	700000	min
C10	years	10	10	10	15	15	15	20	20	20	max
C11	moves/row	0.73	0.74	0.72	0.75	0.77	0.75	0.75	0.77	0.75	min
C12	rate	3	3	3	2	2	2	1	1	1	max
C13	rate	3	3	3	2	2	2	1	1	1	max
C14	cont./ha	122	224	262	111	354	758	111	354	758	max
C15	moves/cont.	1.3	1.7	1.8	0.3	1.0	2.1	0.3	1.0	2.1	min
C16	kg/cont.	10.7702	12.0757	12.4020	11.1769	14.5786	19.9241	7.7280	10.0800	13.7760	min
C17	kg/cont.	0.0725	0.0812	0.0834	0.0660	0.0861	0.1177	0.0006	0.0007	0.0010	min
C18	kg/cont.	0.0414	0.0464	0.0477	0.0377	0.0492	0.0672	0.0083	0.0108	0.0148	min
C19	kg/cont.	0.0218	0.0244	0.0251	0.0226	0.0295	0.0403	0.1794	0.2340	0.3198	min
C20	kg/cont.	0.0005	0.0006	0.0006	0.0005	0.0006	0.0008	0.0004	0.0005	0.0007	min

3.1. Determination the weights of criteria

This research uses 7 experts with experience in intermodal transport (academics specialists in freight transportation and managers of transport companies). The experts were asked to perform pairwise comparisons of the main criteria and the sub-criteria. They are given an overall assessment using Saaty's scale. Table 5 shows the results of expert assessment for main TETE group. Tables 6 and 7 present expert assessments for sub-criteria for each main group, and the local weights (LW) and the global weights (GW) of sub-criteria. The local weights are the normalised values that show the weight of each sub-criterion in the group. The global weights show the priority of all sub-criteria taking into account of the weights of main TETE criteria.

Table 5. Pair-wise comparison of main TETE criteria.

Main Criteria	CR=0.05				
	Technical	Economics	Technological	Ecological	Weight
Technical	1	1/2	1/2	1/2	0.14
Economics	2	1	2	2	0.39
Technological	2	1/2	1	2	0.28
Ecological	2	1/2	1/2	1	0.19

Table 6. Pair-wise comparison of sub-criteria of Technical and Economics group criteria.

	Technical group; CR=0.05					
	C1	C2	C3	C4	LW	GW
C1	1	2	1/2	1/2	0.20	0.03
C2	1/2	1	1/2	1/2	0.14	0.02
C3	2	2	1	2	0.39	0.05
C4	2	2	1/2	1	0.27	0.04

	Economics group; CR=0.07							
	C5	C6	C7	C8	C9	C10	LW	GW
C5	1	1	4	2	1/2	3	0.23	0.09
C6	1	1	4	1	1/3	1	0.16	0.06
C7	1/4	1/4	1	1/3	1/2	1/2	0.06	0.03
C8	1/2	1	3	1	1/2	1/2	0.12	0.05
C9	2	3	2	2	1	2	0.29	0.11
C10	1/3	1	2	2	1/2	1	0.14	0.05

Table 7. Pair-wise comparison of sub-criteria of Technological and of Ecological group criteria.

	Technological group; CR=0.04								Ecological group; CR=0.05						
	C11	C12	C13	C14	C15	LW	GW		C16	C17	C18	C19	C20	LW	GW
C11	1	1/2	1	1/2	2	0.17	0.05	C16	1	3	2	2	1	0.31	0.06
C12	2	1	1	1/2	1	0.20	0.06	C17	1/3	1	3	2	1/2	0.19	0.04
C13	1	1	1	1/2	1	0.16	0.04	C18	1/2	1/3	1	1	1/2	0.11	0.02
C14	2	2	2	1	2	0.32	0.09	C19	1/2	1/2	1	1	1/2	0.12	0.02
C15	1/2	1	1	1/2	1	0.15	0.04	C20	1	2	2	2	1	0.27	0.05

The results show that the consistency index CR for all main criteria and sub-criteria satisfy the condition: $CR \leq 0.1$. It was found that the main importance of TETE group has the economics group criteria (39%). The sub-criteria that have main impact of the choice of an alternative are procurement costs C9 (11%), annual operation costs C5 (9%), stacking capacity C14 (9%), annual equipment maintenance costs C6 (6%), flexibility C12 (6%) and CO_2 emissions C16 (6%).

3.2. Ranking the alternatives by using PROMETHEE Method

After determining the weights of the criteria the PROMETHEE method is applied for ranking the handling equipment. The weights of the all sub-criteria determined by AHP method are used in the PROMETHEE method to estimate the alternatives. The study applied the Visual PROMETHEE software to make research [33]. The type of optimization of sub-criteria C2, C4, C10, C12, C13 and C14 are of maximum; for others sub-criteria the optimization is of minimum. A usual preference functions have been set for the sub-criteria C4, C12 and C13; for others sub-criteria have been set the linear preference.

Figures 1, 3 and 5 show the ranking of the handling equipment for studied variants. In the first part of the figures in the ordinate axis are presented the ranking according values of net outranking flows; the second part shows the global weights of the sub-criteria.

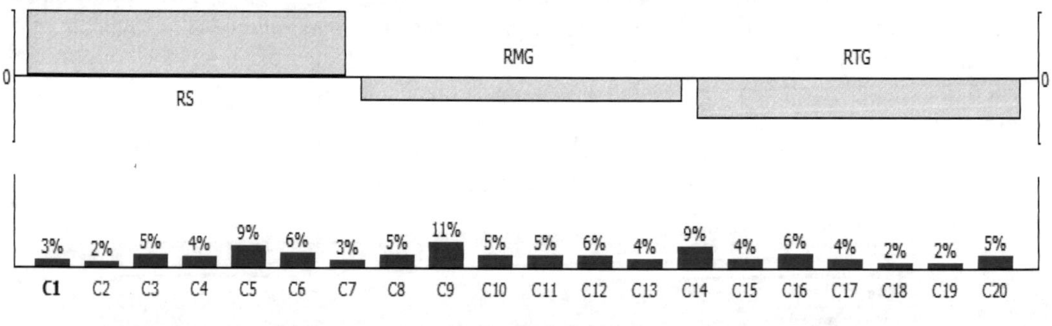

Figure 1. Ranking of alternatives for variant 1 in Visual PROMETHEE Software.

It can be concluded that the RS is the best alternative of handling equipment for variant 1 (up to 20000 TEU/year). The RMG is the best handling equipment for variants 2 and 3 (20000-50000 TEU/year and 50000-100000 TEU/year).

Figure 2 presents the net outranking flows for the criteria according the best alternative RS for variant 1. The main impact of choosing the best alternative has the criteria container yard development costs (C7), procurement costs (C9), flexibility (C12), mobility (C13) and stacking capacity (C14). Figure 4 presents the net outranking flows for the criteria according the best alternative RMG for variant 2. The main impact of choosing the best alternative has the criteria productivity (C2), equipment safety (C4), annual equipment maintenance costs (C6), economic life (C10), stacking capacity (C14) and CO_2 emissions (C16).

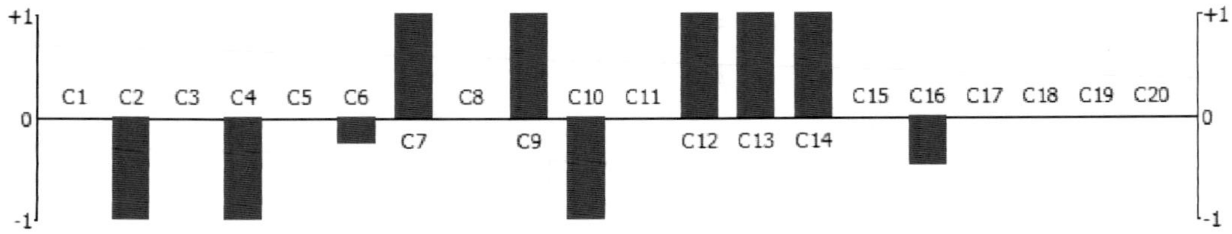

Figure 2. Net outranking flows for all criteria for alternative RS, variant 1.

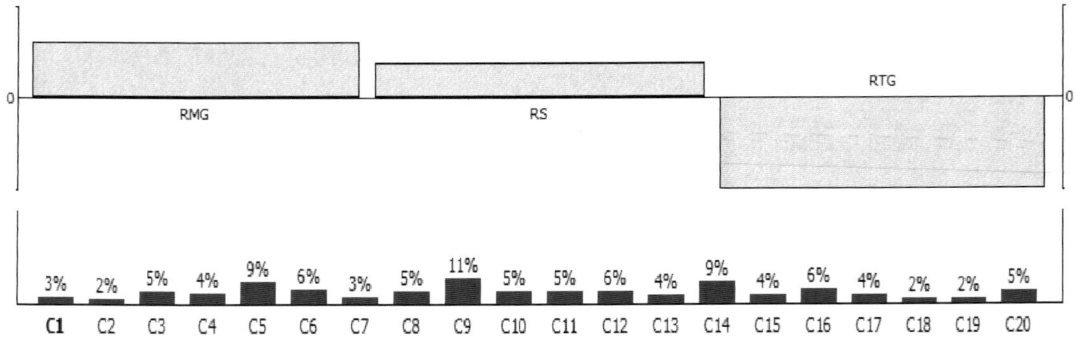

Figure 3. Ranking of alternatives for variant 2 in Visual PROMETHEE Software.

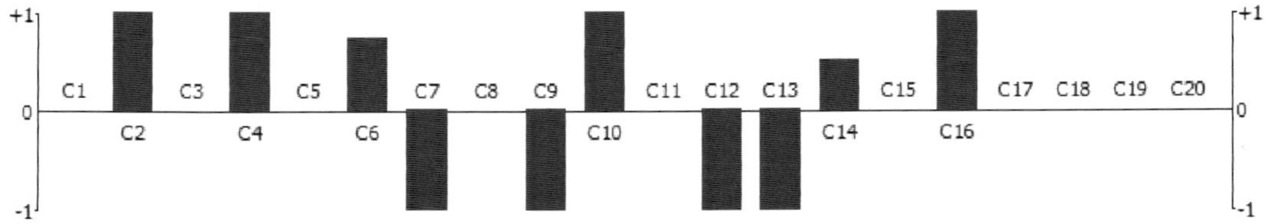

Figure 4. Net outranking flows for all criteria for alternative RMG, variant 2.

Figure 6 presents the net outranking flows for the criteria according the best alternative RMG for variant 3. The main impact of choosing the best alternative has the criteria productivity (C2), equipment safety (C4), annual equipment maintenance costs (C6), procurement costs (C9), economics life (C10), staking capacity (C14) and CO_2 emissions (C16).

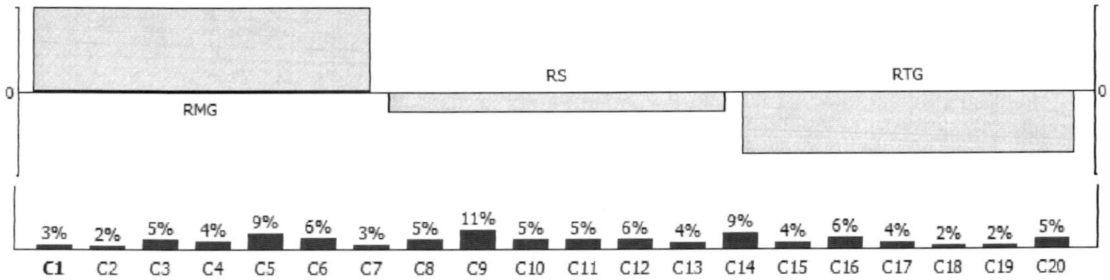

Figure 5. Ranking of alternatives for variant 3 in Visual PROMETHEE Software.

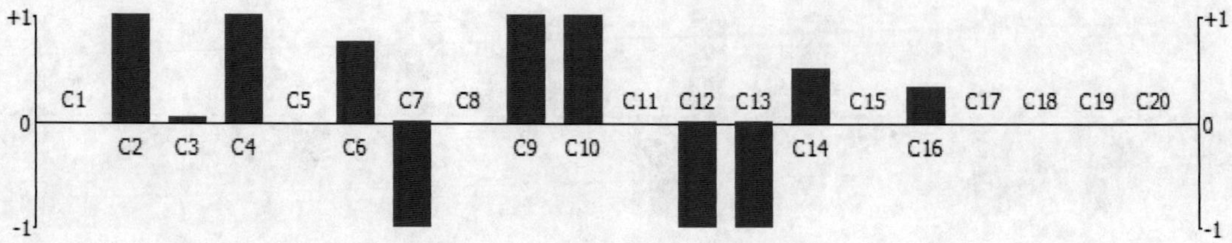

Figure 6. Net outranking flows for all criteria – alternative RMG, variant 3.

The sensitivity analysis was made for the studied variants. Table 8 presents the stability intervals of the weights by variants. The criteria with stability intervals [0-100%] are not given in the table.

Table 8. Stability intervals.

	Criteria	C2	C3	C4	C6	C10	C16				
Variant 1	From, %	0	0	0	0	0	0				
	To, %	11.8	92.2	13.6	23.2	14.5	18. 1				
	Criteria	C2	C4	C6	C7	C9	C10	C12	C13	C14	C16
Variant 2	From, %	1.49	3.34	5.01	0	0	4.50	0	0	8.36	5.0
	To, %	100	100	100	3.73	11.46	100	6.09	4.50	100.0	100.0
	Criteria	C7	C12	C13	C16						
Variant 3	From, %	0.0	0.0	0.0	0.0						
	To, %	14.4	16.2	15.3	45.5						

3.3. Verification the results by TOPSIS method

The results of ranking by PROMETHEE method have been verified by TOPSIS according the methodology. Table 9 presents the values of Euclidean distance from the ideal best D_i^+ solution, the Euclidean distance from the ideal worst D_i^- solution and the values of performance score C_i according formulas (19-20).

Table 9. Ranking by TOPSIS method.

Variant 1					Variant 2					Variant 3				
	D_i^+	D_i^-	C_i	rank		D_i^+	D_i^-	C_i	rank		D_i^+	D_i^-	C_i	rank
RS	0.058	0.073	0.555	1	RS	0.064	0.074	0.534	2	RS	0.079	0.075	0.488	2
RTG	0.080	0.048	0.376	3	RTG	0.080	0.056	0.412	3	RTG	0.079	0.072	0.479	3
RMG	0.061	0.072	0.539	2	RMG	0.062	0.071	0.535	1	RMG	0.051	0.090	0.639	1

The RS is the best alternative for variant 1; the RMG is the best handling equipment for variants 2 and 3. The results for ranking of RS and RMG for variant 2 are close.

Figure 7 presents the comparison of score for alternatives by the both methods for each variant. The scores for PROMETHEE method are the values of net outranking flows, the scores for TOPSIS method are the performance scores. Figure 8 shows the ranking by both methods. It can be concluded that the results are similar.

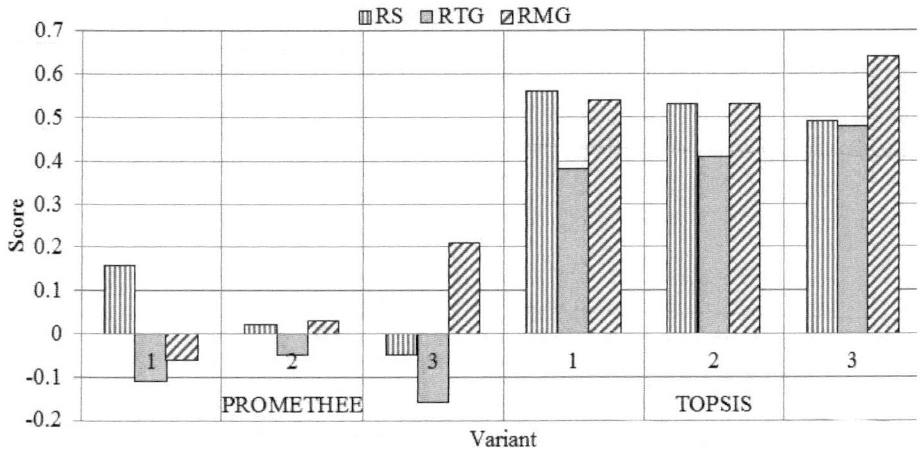

Figure 7. Comparison of results by PROMETHEE method and TOPSIS method.

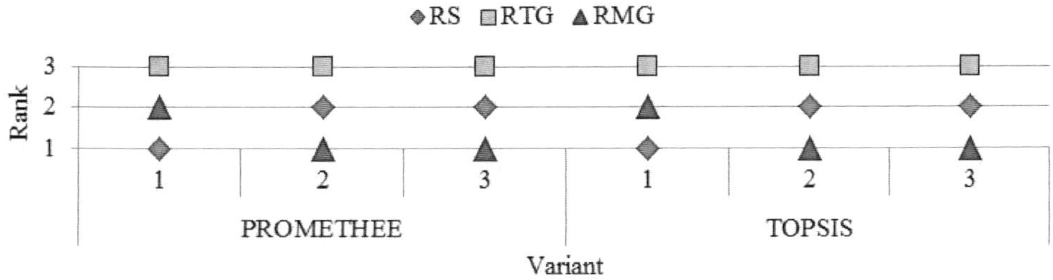

Figure 8. Ranking the handling equipment by PROMETHEE method and TOPSIS method.

4. Conclusion

A methodology for choosing the container handling equipment in a rail-road intermodal terminal has been elaborated. The multi-criteria analysis has been proposed as a tool to make decision. The integration of AHP and PROMETHEE methods have been studied. Three types of container handling equipment – electric Rail Mounted Gantry crane (RMG), diesel driven Rubber Tyred Gantry crane (RTG) and mobile Reach Stacker (RS), have been studied.

The TETE analysis that includes Technical, Economical, Technological and Ecological main groups of criteria has been introduced to study the indices influencing the choice of handling equipment. The sub-criteria for each main group have been determined. A total of twenty quantitative and qualitative criteria have been defined. The weights of main TETE criteria and sub-criteria have been determined by applying AHP method. It was found that the main importance of TETE group has the economics group criteria (39%). The sub-criteria that have main impact of the choice of an alternative are procurement costs C9 (11%), annual operation costs C5 (9%), stacking capacity of the terminal stacking area C14 (9%), annual equipment maintenance costs C6 (6%), flexibility C12 (6%) and CO_2 emissions C16 (6%).

A methodology to determine the values of the quantitative criteria has been developed.

Three types of rail-road terminals have been studied – small, medium and large terminal according the number of TEU per year. The necessary handling equipment from one type has been determined. It was found that the mobile Reach Stacker (RS) is the suitable alternative for small rail-road terminals, and the electric Rail Mounted Gantry crane (RMG) is the best one for the medium and large rail-road terminals.

The results have been verified by applied TOPSIS multi-criteria method. The verification shows that the both methods TOPSIS and PROMETHE give similar results. It can be concluded that the developed methodology and the results obtained are correct.

The elaborated methodology could be applied to investigate real rail-road terminals or to determine the handling equipment when planning a new ones.

The methodology could help to rail-road terminal's operators to make decision when planning the development and the activities of the terminal.

The presented approach, defined criteria and created methodology could be used to investigate and other types of intermodal or freight terminals.

Acknowledgments
This research has been supported by contract №КП-06-Н27/12 of 11.12.2018 "Modelling and elaboration of complex system for selection of transport technology in transport network" funded by the National Science Fund of the Ministry of Education and Science of Bulgaria.

References

[1] Wiegmans B, Masurel E and Nijkamp P 1999 Intermodal freight terminals: an analysis of the terminal market *Transportation Planning and Technol.* **23** 105–28
[2] Wiegmans B and Behdani B 2018 A review and analysis of the investment in, and cost structure of, intermodal rail terminals *Transport Reviews* **38(1)** 33–51
[3] Stahlbock R and Voß S 2007 Operations research at container terminals: a literature update *OR Spectrum* (2008) **30** 1–52
[4] Huang W C and Chu C Y 2004 A selection model for in-terminal container handling systems *J. Marine Science and Technology* **12(3)** 159–70
[5] Moghadam M K and Noori R 2011 Cost function modelling for semi-automated SC, RTG and automated and semi-automated RMG container yard operating systems *Int. J. Business and Development Studies* **3(1)** 85–122
[6] Jachimowski R 2018 Selected aspects of costs shaping in the intermodal terminal *J. KONES Powertrain and Transport* **25** No. 1 151–8
[7] Michele A and Patrizia S 2014 Strategic determinants of terminal operating system choice: An empirical approach using multinomial analysis *Transportation Research Procedia* **3** 592–601
[8] Nooramin A S, Moghadam M K, Jahromi A R M and Jafar S 2012 Comparison of AHP and FAHP for selecting yard gantry cranes in marine container terminal *J. Persian Gulf* **3(7)** 59–70
[9] Wang L and Zhu X 2019 Container loading optimization in rail–truck intermodal terminals considering energy consumption *Sustainability MDPI (Basel, Switzerland)*
[10] Ballis A and Golias J 2002 Comparative evaluation of existing and innovative rail–road freight transport terminals *Transportation Research A* **36** 593–611
[11] Sahin B, Yilmaz H, Ust Y, Guneri A F, Gulsun B and Turan E 2014 An approach for economic analysis of intermodal transportation *The Scientific World J.* **2014** 10
[12] Odu G and Charles-Owaba O 2013 Review of multi-criteria optimization methods – Theory and applications, *IOSR J. Engineering* **3(10)** 1–14
[13] Nooramin A S, Moghadam M K and Sayareh J 2012 Selecting yard cranes in marine container terminals using analytical hierarchy process *J. Maritime Research* **IX(2)** 39–44
[14] Saaty T 2000 *Fundamentals of Decision Making and Priority Theory with the Analytic Hierarchy Process* (New York: RWS Publications 1st ed.) p 477
[15] Brans J P and Mareschal B 2005 *Multiple Criteria Decision Analysis: State of the Art Surveys* (ISOR & Management Science book series New York: Springer Chapter Promethee Methods) **78** pp 163–86
[16] Srinivasa R K 2014 *Multicriterion Analysis in Engineering and Management* (PHI Lerning Private Limited Delhi) p. 266
[17] *EU energy in figures – Statistical pocketbook* 2018 (Luxembourg: Publications Office of the European Union) p. 268

IOP Conf. Series: Materials Science and Engineering **664** (2019) 012032 doi:10.1088/1757-899X/664/1/012032

[18] Zhang Y, Peng Y, Wang W, Gu J, Wu X J and Feng X J 2017 Air emission inventory of container ports' cargo handling equipment with activity-based "bottom-up" method *Advances in Mechanical Engineering* **9(7)** 1–9

[19] Shao Z 2016 Non-road emission inventory model methodology *The international council on clean transportation (Working paper)* 2016–4

[20] Environmental Protection Agency 2004 *Exhaust and Crankcase Emission Factors for Nonroad Engine Modeling - Compression-Ignition*

[21] Saaty T 2004 Fundamentals of the Analytic network process – Dependence and feedback in decision-making with a single network *J. Systems Sci. and Systems Engineering* **13(2)** 129–57

[22] Greenhouse gas emissions in Bulgaria 1988-2016 – National inventory report 2018 The Ministry of Environment and Water – Republic of Bulgaria

[23] National Report on the Status and Environment of the Republic of Bulgaria 2018 The Ministry of Environment and Water – Republic of Bulgaria

[24] Mikalauskiene A, Štreimikis J, Mikalauskas I, Stankūnienė G and Dapkus R 2019 Comparative assessment of climate change mitigation policies in fuel combustion sector of Lithuania and Bulgaria *Energies MDPI (Basel, Switzerland)*

[25] Van Duin J H R and Geerlings H 2011 Estimating CO2 footprints of container terminal port-operations *Int. J. Sus. Dev. Plann.* **6(4)** 459–73

[26] Ecological Transport Information Tool for Worldwide Transports – Methodology report 2018 EcoTransIT World Initiative (Berne – Hannover – Heidelberg)

[27] EcoTransIT: Ecological Transport Information Tool - Environmental Methodology and Data Final report 2003 (Heidelberg)

[28] Green and Effective Operations at Terminals and in Ports 2014 Green EFFORTS FP7 (Erasmus University Rotterdam (EUR) in cooperation with the Technical University Delft)

[29] Clausen U and Poeting M 2017 Allocation of greenhouse gas emissions for containers in multimodal transshipment terminals using simulation *SNE Technical Note* **27(2)** 77–85

[30] Spengler T and Wilmsmeier G 2016 energy consumption and energy efficiency indicators in container terminals – a national inventory *IAME Conf. (Hamburg, Germany)* 27

[31] Rubber Tyred Gantry (RTG) crane load factor study 2009 (Starcrest Consulting Group, Poulsbo WA)

[32] Zhu L, Chen J F and Duan J Y 2019 Air Pollution and control of cargo handling equipments in ports *E3S Web of Conferences* **93** 02001

[33] http://visualpromethee.com (Available online on 03.05.2019)

IOP Conf. Series: Materials Science and Engineering **664** (2019) 012033 doi:10.1088/1757-899X/664/1/012033

Strategic planning and development of transport infrastructures based on agile methodology

Dimitar Dimitrov[1] and Irena Petrova

Todor Kableshkov University of Transport – Sofia, Faculty of Transport Management, 158 Geo Milev Str., 1574 Sofia, Bulgaria

[1] E-mail: ddimitrov@vtu.bg

Abstract. The development of transport infrastructures is an important task for both their strategic planning and development and the provision of quality and safe domestic and international transport services. The analysis of the approaches used to determine transport demand is based on a number of well-established theoretical and practical models, such as the classic four-step transport model. On the other hand, however, the planning of transport needs is of an international affair, requiring multi-state coordination, as is the case with the development of the transport corridors on the Balkan Peninsula. The current publication applies the Agile methodology and approach to defining needs in the design and assessment of infrastructure objects, corridors, plans and strategies. The characteristic of this approach is that it is used to define more precise requirements, using shorter cycles to determine the requirements as well as constant review and evaluation. Based on this approach, a concrete example is presented illustrating the model for the region.

1. Introduction

The planning and development of transport infrastructure are a high priority task determining the level of development of a country or region and reflecting the fundamental problems of society and its development. In this sense, the management, modernization and development of transport infrastructure has a continuous cyclical process that lags behind the expansion of economic relations between different groups of people and countries, even more so that the globalization of society has been already a fact and the way backwards impossible.

In this sense, transport increasingly ensures the daily mobility of people, goods and resources and is becoming more and more important for the production and distribution of goods and services. Transport expenses are part of the cost of production and the main prerequisite for the development of transport systems is the availability of adequate infrastructure.

Transport infrastructure maintains ongoing business connections, but there must be a reserve to ensure the mobility of the economy in the future. Planning and building, transport infrastructure, must reflect fundamental societal problems and issues such as transport safety and security. Every government has problems such as insufficient and inadequate transport infrastructure, missing interconnections, lack of funds for their removal. There are infrastructure problems in all types of transport. The unresolved problems of one type of transport create additional difficulties for other kinds of transport. Infrastructure financing issues have an impact on the quality of infrastructure, with a number of examples being addressed in [1]. Strategic planning and development of transport infrastructure is a complex task related to providing an efficient, effective and sustainable transport system requiring the design and assessment

Content from this work may be used under the terms of the Creative Commons Attribution 3.0 licence. Any further distribution of this work must maintain attribution to the author(s) and the title of the work, journal citation and DOI.
Published under licence by IOP Publishing Ltd

of infrastructure, corridors, plans and strategies as well as support the balanced development of individual regions and countries [2].

In [3, 4] the long experience of a number of researchers in the world has been systematically described and the main principles, models and methods of transport traffic research and analysis are described in relation to the assessment and the need for infrastructure development.

Traditional methods for exploring and analysing models for strategic planning and development of transport infrastructures have the following directions:

- Comparative analysis of needs and requirements for strategic planning and development of transport infrastructure.
- Techno-economic, financial analysis, benefits, costs and efficiency of the transport infrastructure development projects.
- Preparation of model schemes, models and solutions for the design and development of transport infrastructures.

In figure 1 is presented the general model (The classification model is used in the Strategic Planning and Development of Transport Infrastructure, as well as in exploring the results and studies as well as in [3]) for strategic planning and development of transport infrastructure based on the classic Four Stages Transportation / Land Use Model.

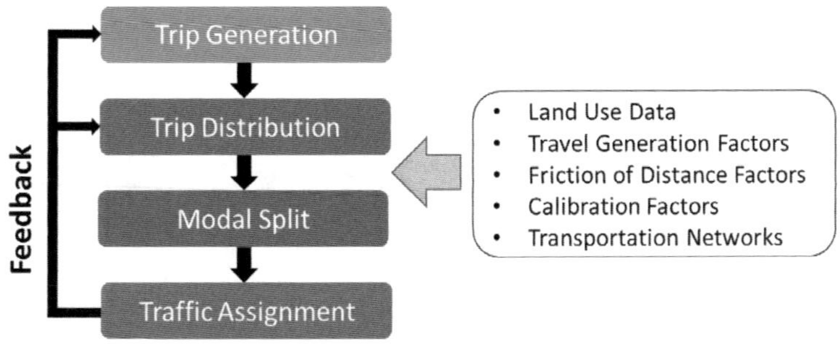

Figure 1. Four Stages Transportation / Land Use Model.

The first stage deals with predictions of travel by zones based on regression analysis to determine the number of trips by zone as a function of independent variables. The second stage deals with the spatial movements and links between travel sources and destinations that are most often judged on the distribution of the journey through a gravity model or multiple regression.

The third stage is a modal split of trips by types of transport based on the source / destination preference probability. The latter stage recognizes traffic patterns and transport links, mainly by using methods to investigate operations aimed at minimizing travel costs or time on a transport network.

This procedure is repeated by looking at the minimum cost of travel for the characteristics of the existing transport network on the basis of available data from studies, assessments and forecasts.

There are lots of applied research in this area and some of them will be mentioned in this publication. The near circle of researchers alone [5-13] shows that these methods are actively used both in research studies and practice in planning the design of transport infrastructures, including building the national strategy of our country [14].

All of this is based on complex analysis and strategic planning of transport infrastructure. This analysis is permanent and necessary for updating and upgrading, as well as coordinating regional and global strategies for the development of transport infrastructure and transport services. In a number of cases, it is established that strategic infrastructure goals are delayed over time and others are not realized on a number of objective (financial) and subjective reasons. Others are transformed to a lesser or more extent, and the end infrastructure product differs from the original design, even more so that sometimes poorly planned actions exist due to the complexity of circumstances and requirements. Another major drawback is the fact that the initial hard-core requirements in the strategy are changing and their updating is labour-intensive and practically impossible. This requires seeking a new approach to planning and managing investment projects to build transport infrastructure.

2. Agile methodology

Agile approach is increasingly being used in the modern software industry to develop sophisticated and time-based projects. Unlike the traditional Waterfall approach, which includes 5 phases (Requirements, Design, Implementation, Verification, Maintenance), Agile Methodology fundamentally differs and has the following features:

- users of the product are actively involved in its development;
- the executive team has decision-making power;
- the requirements change, but the deadlines are fixed;

In figure 2 are shown the differences between Classic and Agile project management methods (Agile approach is successfully used and developed in the software industry, https://www.aoe.com/en/agile.html). With the Classic method Scope, Time and Budget are fixed in advance, while the Agile method is fixed to Time and Budget, and Scope is flexible.

Of course, this can lead to a Scope difference in the project, but it greatly improves the runtime and gaining effective results in the overall iterative development process of the systems. Thus the main task is to get incremental results related to the initial implementation as well as subsequent improvements by building up the next product versions of the project. Fixing Time and Budget aims to close the project implementation cycle by providing the start and end of the project as well as its funding.

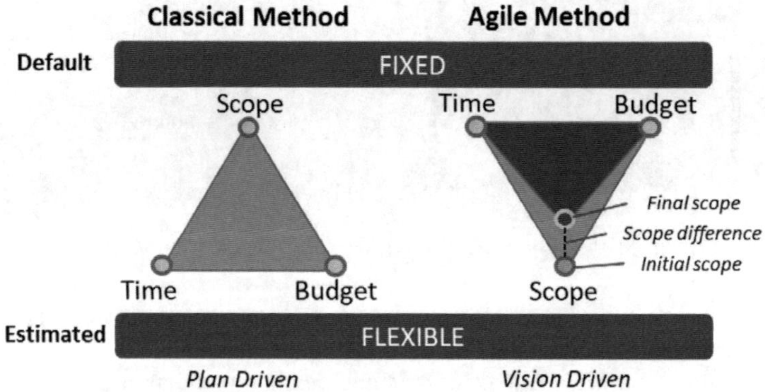

Figure 2. Classical via Agile Methods.

Agile Manifesto (http://agilemanifesto.org/) allows the development of a different management strategy based on the following principles:

- interactive development, allowing review and improvement of the outcome;
- incremental development and consistency of developing versions;
- version control and tracking changes in requirements;
- controlling the team by assigning tasks with shared responsibility.

The idea [15] for evolutionary development of project requirements goes through the following three stages:

- Generic concept of activity (Explain problem, Present rules, etc.);
- Enriching semantic by subtyping (Task 1, Task 2, ... , Task *n*);
- Encrising semantics by machine-readable properties (Property 1, Property 2, ... , Property *m*).

Evolutionary stages should be traced to the same concept, in the same modelling method, while meeting evolving requirements relating to the depth of specialization of the concepts and specifics of the object under consideration.

In a nutshell [16-18], the Agile methodology provides the flexibility to develop sophisticated systems for which traditional design, planning, and management would be dynamic and labour-intensive, moreover, the life of the systems themselves is sometimes shorter than their design and production times. As mentioned, this approach is successfully implemented in one of the most dynamic areas, namely the software industry and the results are present. The accumulated experience on the subject can also be transferred to other dynamic areas, such as the transport sphere, and the features that characterize it would be useful for the development of the Agile approach.

3. Agile Model, Algorithm Example and Results

Combining these two seemingly contradictory approaches can contribute to the qualitative change in strategic planning by taking advantage of the Agile approach to quickly define the initial requirements, and then, with shorter steps, move towards solving the individual tasks of the principle the downstream analysis and iterative solution and construction of the different elements of the transport infrastructure. This will of course result in Infrastructure Flexible Requirements (IFR) for Fixed Time (FxT) and Fixed Budget (FxB) features.

3.1. Agile Methodology Flexible Model

As mentioned above, the basic dependence applied in an area can be represented by the interaction function of the type:

$$IFR_t \rightarrow (FxT_t, FxB_t) \rightarrow \text{Most Important IFR}, \tag{1}$$

where t is a function of time and the matrix comprising the plurality of IFRs:

$$Matrix\ of\ IFRs(t)\ is \begin{pmatrix} ifr_{1,1} & \cdots & ifr_{1,n} \\ \vdots & \ddots & \vdots \\ ifr_{m,1} & \cdots & ifr_{m,n} \end{pmatrix}, \tag{2}$$

where n is type of transport (road, railway, water, air, etc.); m – infrastructure type (pathway, track, bridge, etc.); ifr – infrastructure flexible requirement by transport and infrastructure types.

In general, it is interesting to determine the budget costs R_{IFRs} for a certain period of time, which can be found in the following way:

$$R_{IFRs}(t \in [start2end]) = \sum_{n=1}^{N}\sum_{m=1}^{M} ifr_{n,m}\, c_{n,m}, \text{ money/period}, \tag{3}$$

where $ifr_{n,m}$ is infrastructure flexible requirements for n transport and m infrastructure object; $c_{n,m}$ – the budget value for the design and realization of the site; R_{IFRs} – budget costs; *start2end* – Period of time from the beginning to the end of the project management period considered; N – number of transport; M – number of infrastructures.

The summative semantic dimension of the common differences for a given period $\nabla IFRs$ is the set of:

$$\nabla IFRs\ (t \in [start2end]) = \sum_{n=1}^{N}\sum_{m=1}^{M} \Delta ifr_{n,m}, \text{ differences/period}. \tag{4}$$

where $\nabla IFRs$ is part of a multitude of documents pertaining to a specific transport project.

In this regard, $\nabla IFRs$ have a multi-quantitative dimension, which is measured by the amount of documents (files) containing the semantic descriptions of the Infrastructure Flexible Requirement Differences (IFRD), i.e.:

$$\nabla IFRs \in IFRD. \tag{5}$$

Total Flexible Scope Alternatives for n transport and m infrastructure object can be represented by the following Linear Additive Model:

$$FA(n,m)_j = \sum_i w_i\, fa(n,m)_{i,j};\ w_i > 0;\ fa(n,m)_{i,j} \in \left[fa(n,m)_{i,\min}, fa(n,m)_{i,\max}\right], \tag{6}$$

where $FA(n,m)_j$ is the total flexible scope for alternative j; w_i and $fa(n,m)_{i,j}$ are the weight and flexible scope of criterion i for alternative j; $fa(n,m)_{i,\min}$ and $fa(n,m)_{i,\max}$ define the range of flexible scopes that can be awarded for the performance under criterion i.

The Linear Additive Model is widely used when comparing alternatives to infrastructure projects management [4, 13]. It is simple and easy to use and can also be used as shown in the assessment of agile flexible alternative.

3.2. Agile Flexible Algorithm
In the management of Agile Transport Infrastructure Projects (ATIP), the following step algorithm is used:

3.2.1. Step 1. Defining a matrix of ITFPs on FxT and FxB for n mode of transport and m type infrastructure. Model of the matrix is presented in table 1.

Table 1. ITFPs matrix example.

id(n)	Transport, n	id(m)	Infrastructure, m	FxT, time	FxB, cost
1	road	1	road $_{(1, 1)}$	FxT $_{(1, 1)}$	FxB $_{(1, 1)}$
		2	highway $_{(1, 2)}$	FxT $_{(1, 2)}$	FxB $_{(1, 2)}$
		3	bridge $_{(1, 3)}$	FxT $_{(1, 3)}$	FxB $_{(1, 3)}$
2	rail	1	railway $_{(2, 1)}$	FxT $_{(2, 1)}$	FxB $_{(2, 1)}$
		2	station $_{(2, 2)}$	FxT $_{(2, 2)}$	FxB $_{(2, 2)}$
3	water	1	port $_{(3, 1)}$	FxT $_{(3, 1)}$	FxB $_{(3, 1)}$
		2	port $_{(3,2)}$	FxT $_{(3, 2)}$	FxB $_{(3, 2)}$
4	air	1	airport $_{(4, 1)}$	FxT $_{(4, 1)}$	FxB $_{(4, 1)}$
5	pipe	1	pipeline $_{(5, 1)}$	FxT $_{(5, 1)}$	FxB $_{(5, 1)}$

3.2.2. Step 2. According to the table 2, changes into IFRs and IFRD, are defined, as well as setting priorities for implementation.

Table 2. Range of FxR and Priority.

FxR min	FxR max	ΔFxR	Priority
FxR $_{(1, 1)}$ min	FxR $_{(1, 1)}$ max	ΔFxR $_{(1, 1)}$	low
FxR $_{(1, 2)}$ min	FxR $_{(1, 2)}$ max	ΔFxR $_{(1, 2)}$	high
FxR $_{(1, 3)}$ min	FxR $_{(1, 3)}$ max	ΔFxR $_{(1, 3)}$	medium
FxR $_{(2, 1)}$ min	FxR $_{(2, 1)}$ max	ΔFxR $_{(2, 1)}$	low
FxR $_{(2, 2)}$ min	FxR $_{(2, 2)}$ max	ΔFxR $_{(2, 2)}$	high
FxR $_{(3, 1)}$ min	FxR $_{(3, 1)}$ max	ΔFxR $_{(3, 1)}$	low
FxR $_{(3, 2)}$ min	FxR $_{(3, 2)}$ max	ΔFxR $_{(3, 2)}$	medium
FxR $_{(4, 1)}$ min	FxR $_{(4, 1)}$ max	ΔFxR $_{(4, 1)}$	high
FxR $_{(5, 1)}$ min	FxR $_{(5, 1)}$ max	ΔFxR $_{(5, 1)}$	high

3.2.3. Step 3.
- Define weight w_i and flexible scope $fa(n,m)_{i,j}$ of criterion *i* and alternative *j*;
- Define the range of flexible scopes $fa(n,m)_{i,min}$ and $fa(n,m)_{i,max}$ that can be awarded for the performance under criterion *i*;
- Solve $FA(n,m)_j$.

3.2.4. Step 4. In line with the priorities, an Action Plan is prepared for the phased development of infrastructure projects, taking into account and assessing the progress towards improving the functionality of the transport infrastructure.

3.2.5. Step 5. Scope (Initial, Final, and Difference) is evaluated to what extent the requirements meet the needs and then initiate a subsequent version of the same project to improve transport needs (Traffic Capacity and More Requirements).

Once all the steps have been taken, things are reversed in the starting position again, and all potential projects that have not reached the initial requirements for its scope as it is already involved if necessary in the list of upcoming projects or versions, to achieve new outcomes and priorities for development.

3.3. Results

The Agile Methodology Flexible Model and Agile Flexible Algorithm presented above is used in the analysis of part of the upcoming infrastructure projects in the country. In order to appraise the model and the algorithm, major infrastructure projects embedded in the national transport strategy [14], are chosen and presented in table 3.

Table 3. Major Infrastructure Projects for carrying out.

ID	Project Description	Period, year	Cost, BGN
1	Construction of an intermodal terminal in Ruse	2018-2020	43 055 008
2	Motorway I-1 (E-79) "Vidin-Montana-Vratsa"	2020	2 774 937
3	Rehabilitation of road sections in the direction of Kostinbrod-Berkovitsa (lot 2 – road II-81 Kostinbrod-Buchin pass and lot 3 – road II-81 Buchin pass-Berkovitsa	2020	27 348 581
4	Project "Bypass of Gabrovo" – Stage connection including tunnel under Shipka pass (through Stara Planina mountain)	2017-2019	152 554 740
5	MTWY Hemus from Yablanitsa to Belokopitovo (sections 1 to 7) (section 1 – Yablanitsa road II-35 and Belokopitovo-Shumen)	2017-2022	2 658 152 061
6	Second bridge near Ruse on Danube River	2029-2033	267 759 437
7	Construction of a speed bypass of city of Burgas	2022	30 030 348
8	Modernization of the railway line Karnobat-Sindel (tunnelling Lozarevo-Bat)	2018-2022	338 400 000
9	Rehabilitation of Road II-86 Plovdiv-Asenovgrad	2020	25 392 217
10	Modernization and rehabilitation of the railway section Mezdra-Gorna Oryahovitsa	2018-2022	647 663 250
11	Construction of intermodal terminal Varna	2018-2020	520 000 000

After application of the model, the minimum and maximum requirements Range of FxR are determined and in table 4 the defined priorities for implementation of individual projects are given.

Table 4. Project Implementation Priority.

Project ID	Transport	Priority
1	rail and intermodal	low
2	road	high
3	road	medium
4	road	medium
5	road	high
6	road	low
7	road	low
8	rail	high
9	road	medium
10	rail	high
11	water and intermodal	low

The interactive map of figure 3 graphically illustrates the planned infrastructure projects according to their priority. The map shown dynamically changes depending on the time of the strategic goals for the implementation of current and new infrastructure projects.

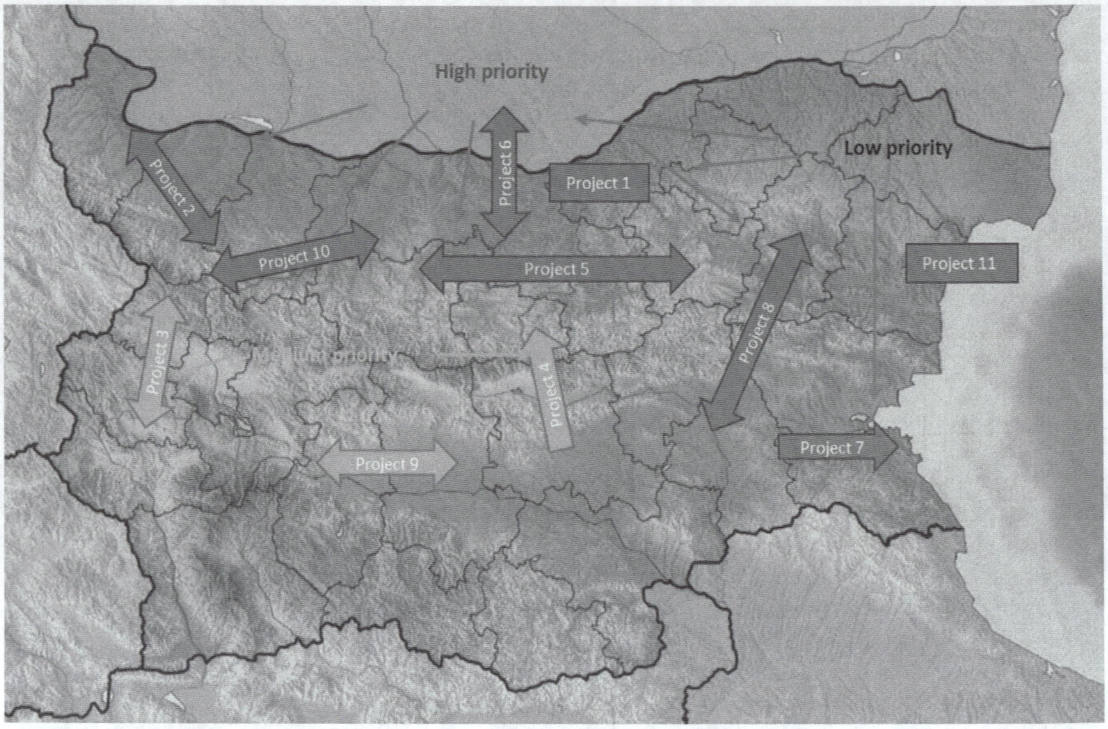

Figure 3. Interactive maps on active projects by priority.

4. Conclusion

Strategic planning of transport infrastructure is an important task which is the basis for the preparation and implementation of investments in applied infrastructure projects and is part of the economic development at the regional and global level [2, 11, 19]. This process has a permanent and cyclical character and is related to the development and modernization of existing and new transport sites.

As it has been presented here, this proves to be a difficult and dynamic task, especially when it comes to complicated objects, such as transport. This puts the traditional approaches to planning, construction and development of these sites in question for their effective implementation, despite the capabilities of the modern tools for data processing and analysis, and therefore their modification and development is necessary. The presented Agile approach allows for a new concept of consideration and problem solving, namely "a faster and more effective solution, than very accurate and precise but out of time". This advantage, which is in fact a modification of the traditional approach, allows better planning and realization of infrastructure projects according to the priorities and the degree of their significance. The release of the Scope parameter while preserving Time and Budget (figure 2) reveals Agile's main advantage over the Classical method, namely the ability to dynamically define and solve the significant problems of managing infrastructure projects in transport.

The approach is presented in a general form and shows its general concepts and results that are the outcome of the work on a scientifically applied project related to research and analysis of models for strategic planning and development of modern transport infrastructures. The presented interactive map is an exemplary model for visualization of infrastructure objects that are included in the Action Plan for Design and Development and which are classified by priority for implementation.

Finally, it can be said that when developing a regional development strategy, both the time horizon and the dynamic changes in transport needs related to general economic and social development must be taken into account. The question of the concrete implementation of the stage for the development of infrastructure projects is up to date and adequate and quick decisions and changes in their scope are of primary importance for their solution.

References

[1] Valkov R and Krumov Z 2010 Infrastructure Project Implementation Within Crisis Conditions, Collection Z 15, *Int. Sci. Conf. Resolving Crisis Situations (Zilina Slovakia)* 789–94
[2] Dimitrov D and Petrova I 2018 Research and analysis of the development of the transport infrastructure in Bulgaria and the region *Mechanics Transport Communications - Academic J.* ISSN 2367-6620
[3] Klyambarski L 2017 *Contemporary State of Research in the Field of Transport Planning and Modeling Studios* (Publishing house of "Todor Kableshkov" Sofia)
[4] Antov A 2017 *Transport Modeling 101 Practical Guide* (ISSN 978-619-188-109-3 Sofia)
[5] Razmov T 2017 Assessment of the current state and the potential development of passenger rail transport in the Republic Of Bulgaria *XV Int. Sci. Conf. "Management and Engineering"* **II** 860–9
[6] Razmov T and Klymbarski L 2016 Modeling of transport traffic – a basic tool for comprehensive evaluation of infrastructure projects and measures to improve the performance of the transport sector *Sci. and Practical Conf. "Mobility for a Connected World" (UNWE Sofia)*
[7] Todorova M and Dzhaleva-Chonkova A 2015 Concept of urban rail in Sofia *RailNewcastle Conference 2015 (Newcastle UK)*
[8] Todorova M 2015 Monitoring on rail transport condition in eastern european countries *Mechanics Transport Communications - Academic J.* ISSN 2367-6620
[9] Todorova M 2016 Survey and forecasting the attendance of Rila Monastery for the purpose of "Project on Construction of Innovative Transport Infrastructure" *XVII Int. Sci.-expert Conf. on Railways (Niš Serbia)*
[10] Stoilova S 2018 An approach for choosing the optimal route and type of transport for freight carriage using Compromise programming *MATEC Web of Conferences* **234** 06002
[11] Stoilova S 2019 Methodology for classification sections of railway network using cluster analysis *Engineering for rural Development (Jelgava)*
[12] Dimitrov D Zlateva P and Velev D 2018 A Methodology for designing an information system for road infrastructure monitoring and traffic management in disasters *IOP Conference Series: Earth and Environmental Science* **167** conf. 1
[13] Antov, A., 2018. Multi-Criteria Analysis for Transport Infrastructure Projects. Bucharest: Infrarch Ltd., Report 02/2018
[14] *Integrated transport strategy for the period until 2030* 2017 (Ministry of Transport, Information Technology and Communications of Republic of Bulgaria 2017)
[15] Karagiannis D 2015 Agile modeling method engineering *Proc. 19th Panhellenic Conf. on Informatics (Athens Greece)* 5–10
[16] Marcos A, Castro V and Olivera S 2015 An analysis of specialized literature about agile method application in the context of technical solution process area included in software process quality models *Revista de Sistemas de Informação da FSMA* **15** 52–9
[17] Heeager L and Nielsen P 2018 A conceptual model of agile software development in a safety-critical context: A systematic literature review *Information and Software Technolody Pages* 22–39
[18] Kumar G and Bhatia K 2012 Impact of Agile methodology on software development process *Int. J. Computer Technology and Electronics Engineering (IJCTEE)* **2**
[19] Tuzkaya U, Onut S and Tuzkaya G 2014 *A* Strategic planning methodology for the multimodal transportation systems: a case study from Turkey *J. Applied Mathematics* **2014** 931456

IOP Conf. Series: Materials Science and Engineering **664** (2019) 012034 doi:10.1088/1757-899X/664/1/012034

Modelling passenger service rate at a transport hub serviced by a single urban bus route as a queueing system

A Asenov[1,3], V Pencheva[1] and I Georgiev[2]

[1] "Angel Kanchev" University of Ruse, Department Transport, 8 Studentska str., 7017, Ruse, Bulgaria
[2] "Angel Kanchev" University of Ruse, Department Applied Mathematics and Statistics, 8 Studentska str., 7017, Ruse, Bulgaria

[3] Corresponding author: asasenov@uni-ruse.bg

Abstract. In this paper the daily service rate at a bus stop from the urban passenger transport network is reviewed. The bus stop is located close to a railway and a bus station. A mathematical model has been developed for the average daily incoming flows of passengers to the monitored bus stop. The service rate has been modelled using pulse Dirac delta functions. The daily irregularity of operation of a particular bus line at this bus stop has been evaluated and the operation of the system has been modelled as a queueing system, in order to assess the capacity of the bus line and the organisation of work. The type of incoming flow, as well as the service rate has been defined as a non-stationary Poisson flow. A system of differential-algebraic equations has been chosen as a model according to Kolmogorov's model for stochastic processes. A specific numerical method for solving system differential equations has been devised with Dirac delta functions in the right part of the equations. Under non-stationary conditions, the basic values of the system parameters have been calculated and an application has been created in a MatLab environment. The results are suitable for accurately determining the routing speed and the average waiting time of the passengers at the bus stop. They can also determine the refusal of passengers to wait at the passenger stop.

1. Introduction

Public transport of passengers in modern urban conditions is a dynamically developing area in which, despite common objectives, many mutually exclusive interests often clash. However, the socio-economic status of the cities significantly depends on it.

In figure 1, the main stakeholders related to mass urban passenger transport in populated areas are indicated where: A is the set of carriers; B – the set of passengers; C – the set of state and local authorities; D – the society.

The combination of the four sets $A \cup B \cup C \cup D$ defines the operation of mass urban passenger transport system and their cross section $A \cap B \cap C \cap D$ represents the common interests of the stakeholders.

The main task of the carriers (set A) is to obtain maximum profit. Passengers (set B) strive for: the shortest time for transportation, the lowest price and maximum comfort and safety. The aspiration of the state and the local government (set C) is to make public transport as accessible and safe as possible, respecting the interests of all parties. The interests of society (set D) are related to providing ecological and energy efficient transport. In these relationships, a number of theoretical and practical problems arise, and the effectiveness of the system of mass urban passenger transport depends on their solution.

Content from this work may be used under the terms of the Creative Commons Attribution 3.0 licence. Any further distribution of this work must maintain attribution to the author(s) and the title of the work, journal citation and DOI.
Published under licence by IOP Publishing Ltd

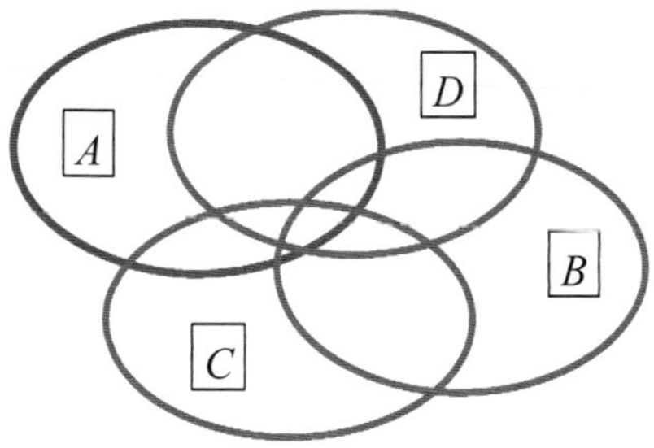

Figure 1. Scheme of key stakeholders.

The use of mathematical modelling in solving these problems allows for modification of the parameters of the system in different combinations, in order to find the optimal one. In resolving such problems, the system is formalised as mathematical models that reflect the fundamental laws of its operation.

The analysis of the situation in question in the given area indicates a significant number of studies on the formation of criteria, indicators and methodologies that allow for assessment of the quality of the transport service provided to the population.

A number of studies have addressed issues related to urban traffic modelling and the impact of the route scheme of mass urban passenger transport [1, 2]. Assessment of the transport mobility of the population for a given populated location [3], determining and exploring areas for pedestrian accessibility [4, 5], modelling and evaluating urban routing schemes [6, 7], improving the quality of the transport service [8] are some studies in such fields. The models developed allow us to solve a number of application problems related to rolling stock, the regularity of transport, ensuring safety, forming multimodal systems, etc.

It is worth noting that the application of the models in question improves safety, reduces costs, ensures high quality of service, improves the environmental performance of cities, ensures energy efficiency of transport, etc.

Special attention shall be paid to the survey of mass urban passenger transport at bus stops providing transportation of passengers on the territory of a given city from a transport hub. Given the nature of the processes at these bus stops at certain periods of time (on arrival of trains and coaches) a large number of passengers gather there. There are also many changing passenger flows, which are serviced at specific moments of time with the ability to master passenger traffic to vehicle capacity.

An examination of the nature of interaction between passenger traffic and urban passenger transport (UPT) is essential for proper organisation of work. A number of studies assess the capacity and operation of mass urban passenger transport at bus stops on a given route. A model of microscopic simulation of bus stops with multiple connections is proposed in [9] while in [10] a simulation of bus stops in mixed traffic is made. Overloads and congestion at bus stops are reviewed in [11].

Each of the transport nodes has its own specific characteristics in terms of location, infrastructure, as well as ensured interaction with mass urban passenger transport. This requires a study of the specific systems of interaction for each of the transport nodes. At the same time, the theoretical development of a common model for servicing passenger flows from UPT for a given bus stop is related to the exploration of all inbound passenger flows oriented towards certain routes. Similarly, the development of a specific inflow pattern to a given route can also be applied to other inbound flows to the specific routes.

The basis for the modelling in this work is a passenger transport hub in the city of Ruse and the need to ensure the movement of passenger traffic with the mass urban passenger transport. The transport hub includes bus and railway transport and the connection to the mass urban passenger transport is provided by trolley and bus transport.

2. Formalizing the system as a single channel queueing system

The number of passengers arriving at a stop from the MUPT can be seen as a stochastic process because it depends on many factors. What is important in this case is the service intensity, the maximum allowable waiting time for the passengers at the bus stop, the number and type of individual routes passing through it. The analysis of passenger flow and the service of the UPT can be done by making a 24-hour model, a queueing system, [12, 13]. For the specific passenger transport unit in the city of Ruse, which includes passenger flow from railway $\lambda_1(t)$, bus $\lambda_2(t)$ terminals and passengers who have arrived at a specific station of the UPT route from the respective district of $\lambda_3(t)$ it can be assumed that the flow of passengers is ordinary (Poisson) and in general from arbitrary periods, the total λ flow is non-stationary $\lambda = \lambda(t)$. Passengers arriving at the bus stop travel on different routes and use vehicles of different capacity. It is, therefore, appropriate that each route passing through the stop is considered as part of multiple independent single-channel systems, servicing corresponding routes from that stop, along with the corresponding inflows. The work analyses and models the total flow of queries on one route passing through one stop [14-16].

Service intensity $\mu(t)$ can be judged by the daily timetable of the respective route at the bus stop, as well as the capacity of the servicing unit (the capacity of the buses/trolleys that service that specific route). It is also known that at a time when the server is busy, the incoming order waits for a certain time T_w to be serviced (waits for the relevant bus/trolley) and then rejects it (the passenger looks for another alternative for transportation). To study the operation of the system, it is necessary to find the probability in the system to have k number of passengers for moment t for the specific server (route), i.e.

$$P_k(t) = ?,\ k = \overline{0, \infty},\ t \in [0, T],\ n = 1, \tag{1}$$

where one full working day $T = 0$ of 24 hours is considered to be unit period T. The start of the working day $T = 0$ coincides with the astronomical start of the day (0 hours and 0 min), the end of the workday $t = T = 24$, coincides with the end of the astronomical day (24 hours and 0 min).

The following summary can be made for the model as a queueing system: a non-stationary flow of queries with density $\lambda(t)$ is entered into a queueing system with $n = 1$ service channel. The service time of the request is a random dimension with an indicative distribution and service rate $\mu(t)$, also non-stationary in time. Number n of servicing channels is one. An order that arrives at a time when the channel is busy, waits for some time T_w then leaves the system as not-serviced, i.e. a denial system. The flow of distressed queries is $v = 1/T_w$, which is a constant over time.

Consequently, the system is type ($M/M/n$) in stationary mode. To describe a system of this type the following system of differential equations of Kolmogorov (Erlang-Kolmogorov) is used [17, 18]:

$$\begin{aligned}
\frac{dP_0(t)}{dt} &= -\lambda(t)P_0(t)+\mu(t)P_1(t);\ldots;\\
\frac{dP_k(t)}{dt} &= \lambda(t)P_{k-1}(t)-\big(\lambda(t)+k\mu(t)\big)P_k(t)+\mu(t)(k+1)P_{k+1}(t);\ldots;\\
\frac{dP_n(t)}{dt} &= \lambda(t)P_{n-1}(t)-\big(\lambda(t)+n\mu(t)\big)P_k(t)+(n\mu(t)+\upsilon)P_{k+1}(t);\ldots;\\
\frac{dP_{n+s}(t)}{dt} &= \lambda(t)P_{n+s-1}(t)-(\lambda(t)+n\mu(t)+sv)P_k(t)+(n\mu(t)+(s+1)v)P_{k+1}(t),
\end{aligned} \tag{2}$$

where s is the number of passengers in the queue when the server is busy.

The differential equation system is infinite and for the numerical solution it is necessary to take a finite number of equations. The number of equations is determined by the number N in the sum $\sum_{i=N+1}^{\infty} P_i(t)$, which must not exceed a predetermined number ε. For that reason, a system of $N = 502$ differential equations with system error less than $\varepsilon = 10^{-8}$ has to be calculated. In this case, the number of probabilities $P_k(t)$ exceeds the number of equations by one and this requires the introduction of an additional algebraic equation for normalisation $\sum_{i=0}^{N} P_i(t) = 1$. In the calculations related to the system (2) after these transformations, it is taken into account that the system has a large dimensionality; it is differential-algebraic one and is of the type "stiff system" (solid system equations). Special numerical methods have been developed to overcome these difficulties. A program of Matlab [19] has been devised to solve system (2), using the built-in "Solver" ode15s, a method of Gir. When entering $\lambda(t)$, $\mu(t)$, N the application gives a numerical solution to the $P_k(t)$. The initial state of the system $P_k(t_0)$ is unknown, but it is known that such type of processes are persistent and after a long enough time go into regular mode of operation. Therefore, a random initial state can be taken. The integration of the system in the interval [0, *T*] needs to be done not once, but a sufficiently large number of times, after each integration the values at the end of the period become initial values for the next integration. In this way, the probability functions $P_k(t)$ begin to tilt to their regular values. After several years of integration, it was found out that only after 5-6 periods do functions $P_k(t)$ enter a regular mode (for two contiguous periods they remain the same). The accuracy in this case is high because the integration is done over 20 periods, the difference of all $P_k(t)$ in the last and penultimate integration is less than 10^{-8} for each *t*.

The input flow $\lambda(t)$ consists of the sum of the three above-mentioned independent streams:

$$\lambda(t)=\lambda_1(t)+\lambda_2(t)+\lambda_3(t). \qquad (3)$$

An appropriate way to model the inflows is by using a sufficiently large statistical database to approximate the least-squares method or some other method, such as the minimax method, [20]. In order to model more precisely the aggregate inflow, it is necessary to study separately all three incoming streams, and then to unite them to model the total inflow $\lambda(t)$, in the absence of a sufficient number of statistical data. The approximate average values, per day, of the number of passengers $A_1, A_2, \dots, A_i, \dots, A_k$ arriving at specific time ranges in the 24-hour day $[t_i, t_{i+1}]$ from the train station, bus station and passengers arriving from nearby areas waiting for the specific route at the station in question. For modelling of $\lambda(t)$ it is appropriate to choose a relatively elementary function with periodicity. The integral amount of $\lambda(t)$ shall correspond to the total average number of requests for the whole day, i.e.:

$$\int_0^{24} \lambda(t)\mathrm{d}t = \sum_{i=1}^{k} A_i\,. \qquad (4)$$

In this case a 24-hour day is divided into time intervals $[t_i, t_{i+1}]$, t_0 corresponds to 0 hours and 0 minutes for which the average number of passengers A_i is known. One such sample allocation is given in table 1.

Table 1. Average number of passengers arriving at a specific time interval.

Interval [a]	A_i [b]	Interval [a]	A_i [b]	Interval [a]	A_i [b]	Interval [a]	A_i [b]
[0, 1]	round (23.4) = 23	[6, 7]	32.5	[12, 13]	36.4	[18, 19]	42.9
[1, 2]	round (19.5) = 20	[7, 8]	44.2	[13, 14]	33.8	[19, 20]	39.0
[2, 3]	15.6	[8, 9]	41.6	[14, 15]	32.5	[20, 21]	32.5
[3, 4]	14.3	[9, 10]	39.0	[15, 16]	32.5	[21, 22]	28.6
[4, 5]	15.6	[10, 11]	42.9	[16, 17]	37.7	[22, 23]	27.3
[5, 6]	22.1	[11, 12]	39.0	[17, 18]	44.2	[23, 24]	23.4

[a] Interval from 24-hour day $[t_i, t_{i+1}]$, with beginning and end respectively at t_i^{-th} and t_{i+1}^{-th}.
[b] Average number of passengers A_i arriving at $[t_i, t_{i+1}]$, rounded to an integer.

For a $\widetilde{\lambda}(t)$ model, a Fourier row is selected with seven members with a 24-hour periodicity:

$$\widetilde{\lambda}(t) = a_0 + a_1 \cos\left(\frac{2\pi t}{24}\right) + b_1 \sin\left(\frac{2\pi t}{24}\right) + a_2 \cos\left(\frac{4\pi t}{24}\right) + b_2 \sin\left(\frac{4\pi t}{24}\right) + a_3 \cos\left(\frac{6\pi t}{24}\right) + b_3 \sin\left(\frac{6\pi t}{24}\right). \qquad (5)$$

Coefficients $a_0, a_1, b_1, a_2, b_2, a_3, b_3$ can be calculated by using a method of least squares or by some other method minimizing approximation errors (e.g. Minimax method). In this case coefficients $a_0, a_1, b_1, a_2, b_2, a_3, b_3$ are calculated by using a method of least squares. The coefficient of determination in the approximation of the model is $R^2 = 0.94$. This forms are very solid, statistically significant relationship, which means that the selected pattern describes the behaviour of the incoming flow well. Coefficients $a_0, a_1, b_1, a_2, b_2, a_3, b_3$ in the model, as well as their confidential intervals, guaranteed with a probability of $\gamma = 0.95$, are given in table 2. For the approximation of the data in table 1 on the abscissa are taken the intervals in the middle $\frac{t_j+t_{j+1}}{2}$, and on the ordinate the values are not rounded.

IOP Conf. Series: Materials Science and Engineering **664** (2019) 012034 doi:10.1088/1757-899X/664/1/012034

Table 2 shows that there is no confidential interval containing 0. This means that all odds are statistically significant (an alternative to the Student's T-test). The approximate flow $\tilde{\lambda}(t)$ and the data are given in figure 2.

Table 2. Values and confidential intervals of the coefficients in a model (5).

Coefficient	Value	Confidential interval of the coefficient guaranteed with probability of $\gamma = 0.95$
a_0	31.69	(30.47, 32.9)
a_1	–8.853	(–10.57, –7.132)
b_1	–4.848	(–6.568, –3.128)
a_2	–2.448	(–4.168, –0.7275)
b_2	–6.777	(–8.497, –5.057)
a_3	2.292	(0.5716, 4.012)
b_3	2.433	(0.7132, 4.154)

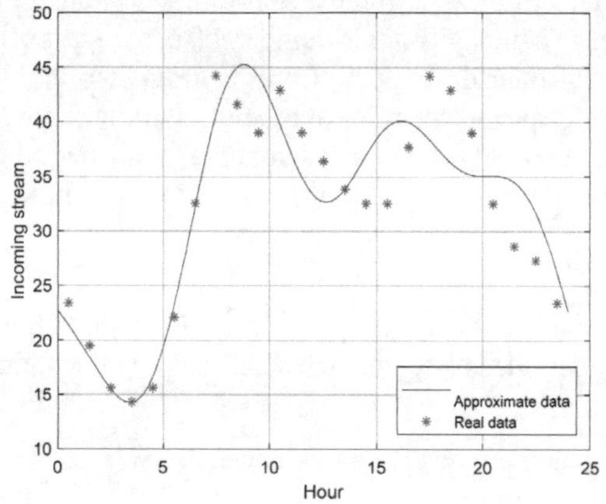

Figure 2. Approximate flow $\tilde{\lambda}(t)$ and average number of passengers A_i, arriving in the interval $[t_i, t_{(i+1)}]$.

Thus, the built-in flow $\tilde{\lambda}(t)$ well reflects the dynamics of change at the individual intervals of the day, but the integral sum of the total volume $\int_0^{24} \tilde{\lambda}(t)\mathrm{d}t$ is not equal to $\sum_{i=1}^{k} A_i$, which necessitates the selection of a norming coefficient $\tilde{C}$ with property:

$$\tilde{C} \int_0^{24} \tilde{\lambda}(t)\mathrm{d}t = \sum_{i=1}^{k} A_i. \quad (6)$$

Then the inflow after normalization takes the form of:

$$\lambda(t) = \tilde{C}\,\tilde{\lambda}(t). \quad (7)$$

In this case $\tilde{C} = 1.0071$.

Finally, for $\lambda(t)$ we have:

$$\lambda(t)=1.0071(31.69 - 8.853\cos\left(\frac{2\pi t}{24}\right) - 4.848\sin\left(\frac{2\pi t}{24}\right) - \\ - 2.448\cos\left(\frac{4\pi t}{24}\right) - 6.777\sin\left(\frac{4\pi t}{24}\right) + 2.292\cos\left(\frac{6\pi t}{24}\right) + 2.433\sin\left(\frac{6\pi t}{24}\right). \quad (8)$$

When modelling service intensity $\mu(t)$, it is necessary to know the traffic schedule and the capacity of the vehicles servicing the route. In this case a route, which is serviced by one type of vehicle with a maximum service capacity of 35 passengers is considered. The service takes place at times t_j^* (time of arrival of the bus/trolley) from the time of service to the C_j^* (maximum vehicle capacity) of the passenger at one time. Namely, for a time interval Δt, theoretically $\Delta t \rightarrow 0$. Normally this interval is never 0, but it is negligible as compared to a 24-hour day (20-30 sec). More precise modelling takes place with the δ function on the right hand side of (2). For time interval $[t_j^*\text{-}\varepsilon, t_j^*\text{+}\varepsilon]$, where ε is negligible (theoretically

$\varepsilon \to 0$), the vehicle services C_j^* passengers at once. For this small amount of time, the integral sum of the service rate must be equal to capacity C_j^*. This is expressed by the basic properties of the δ in the following way:

$$\int_{t_j^*-\varepsilon}^{t_j^*+\varepsilon} \mu(t)\mathrm{d}t = \int_{t_j^*-\varepsilon}^{t_j^*+\varepsilon} C_j^*\delta(t-t_j^*)\mathrm{d}t = \mathrm{C}_j^*. \quad (9)$$

Then the service rate $\mu(t)$ is calculated by the dependence:

$$\mu(t) = \sum_{j=1}^{k} C_j^*\delta(t-t_j^*). \quad (10)$$

To demonstrate the capabilities of the model, specific input data values are adopted, tailored to the work of the transport unit under consideration, including the railway station and the bus station and its interaction with mass urban passenger transport at a particular stop.

In this case, everywhere it is assumed that all vehicles at any time of arrival t_j^* have the same capacity of up to 35 people, ie. $C_j^*=35, \forall j$. An example of the distribution of periods of time in hours of arrival t_i^* after 0:00h in hours, per working day, is as follows: 0,5; 1,5; 2,5; 3,5; 4,5; 5,33; 5,66; 6,25; 6,5; 6,75; 7; 7,25; 7,50; 7,75; 8; 8,2; 8,4; 8,6; 8,8; 9; 9,25; 9,5; 9,75; 10; 10,2; 10,4; 10,6; 10,8; 11; 11,25; 11,50; 11,75; 12; 12,25; 12,50; 12,75; 13; 13,25; 13,50; 13,75; 14; 14,2; 14,4; 14,6; 14,8; 15; 15,25; 15,50; 15,75; 16; 16,2; 16,4; 16,8; 17; 17,2; 17,4; 17,8; 18; 18,2; 18,4; 18,6; 18,8; 19; 19,33; 19,66; 20,33; 20,66; 21,5; 22,5; 23,5.

The presence of δ functions in the part on the right of equation system (2) leads to serious computational difficulties. Standard numerical methods are not appropriate, so specific numerical methods have been constructed. One way to overcome these difficulties is through the so-called immersed interface method (IIM), [21, 22]. Another common approach to the numerical solution of differential equations similar to (2) is using differential schemes in which δ function is replaced by its discrete analogues. Some of these analogues are as follows:

- basic "hat" function

$$d_h^1(x) = \begin{cases} \frac{h-|x|}{h^2}, |x| \le h \\ 0, |x| > h \end{cases}; \quad (11)$$

- wide "hat" function

$$d_h^2(x)=\begin{cases} \frac{2h-|x|}{4h^2}, |x| \le 2h \\ 0, |x| > 2h \end{cases}; \quad (12)$$

- smooth variant

$$d_h^3(x)=\begin{cases} \frac{\cos(\frac{\pi x}{2h})+1}{4h}, |x|\le 2h \\ 0, |x|>2h \end{cases}; \quad (13)$$

- function with a higher approximation level

$$d_h^4(x)=\left(\frac{1}{h}\right)\begin{cases} 1-\left(\frac{|x|}{h}\right)^2, |x| \le h \\ 2-\left|\frac{3x}{h}\right|+\left(\frac{x}{h}\right)^2, h \le |x| \le 2h \\ 0, |x| > 2h \end{cases}, \quad (14)$$

where h is the step of numerical integration.

Other properties and variants of the δ function are discussed in [23]. In the present work, the calculation of the δ function is a variant (12). The waiting time of a passenger before deciding to leave is accepted to be 20 minutes, i.e. $T_w = \frac{1}{3}h$.

IOP Conf. Series: Materials Science and Engineering **664** (2019) 012034 doi:10.1088/1757-899X/664/1/012034

3. Mathematical model application results

The "ode15s" solver was used when integrating system (2) into a Matlab programming environment, taking into account the integration step and constructing the discrete analogue of δ. After integrating system (2), probabilities $P_k(t)$ are calculated. On the basis of these probabilities, the major system characteristics are also calculated. Figure 3 shows the probability of waiting for 1 to 5 passengers; in figure 4 from 6 to 10 passengers; in figure 5 from 11 to 20. As the traffic intervals between 6:30 am to 7:30 pm are short and the likelihood for a large number of passengers to wait at the stop is small, especially with an increase in the number of passengers, in figure 5, the probability is below 20% and bends to zero.

With the so-called waiting times and arrival intervals of the vehicles at the stop, according to figure 6 it follows that most often there will be about 6-8 passengers in the queue. Regarding the average waiting time at the stop, it is in the range of 0.15 to 0.17 hours, mainly in the period after 6.30 to 7.30 pm, as shown in figure 7 and it is higher for the rest of the time for objective reasons. This naturally affects the average number of failures, as shown in figure 8, which are at most 3 and more passengers mainly after 19:30. During the remaining periods the failures are for up to 2 people. The results obtained enable the planning and organization of vehicle traffic on routes and stops from the UPT in the case of the passenger transport unit involving passengers from three railway stations, a bus station and a pedestrian area. For detailed analysis and accurate planning of UPT work in any particular city, it is necessary to conduct surveys to collect real data related to the specifics of population movement.

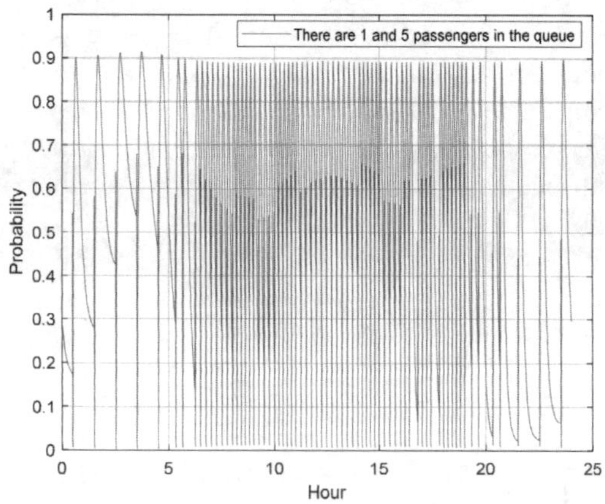

Figure 3. Probability for 1 to 5 passengers in the queue.

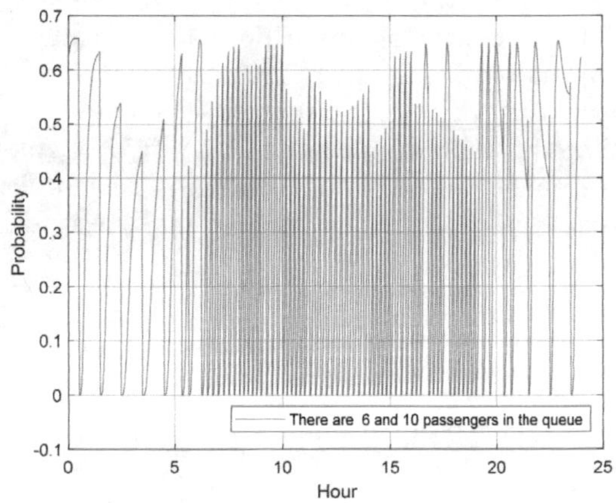

Figure 4. Probability for 6 to 10 passengers in the queue.

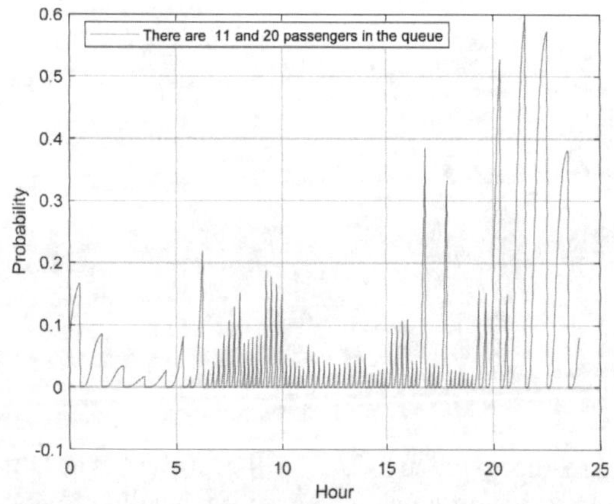

Figure 5. Probability for 11 to 20 passengers in the queue.

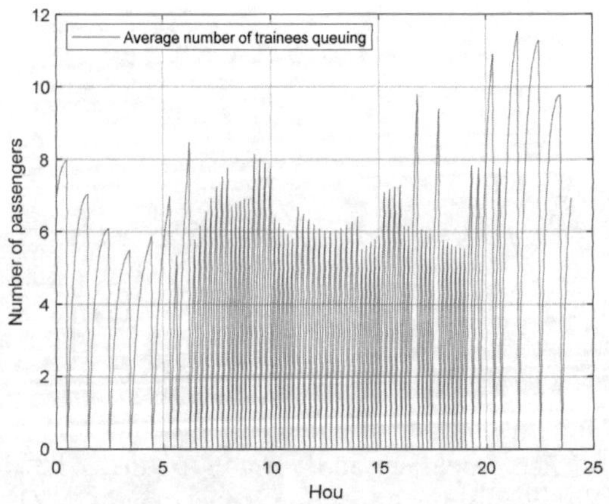

Figure 6. Average number of passengers in a queue.

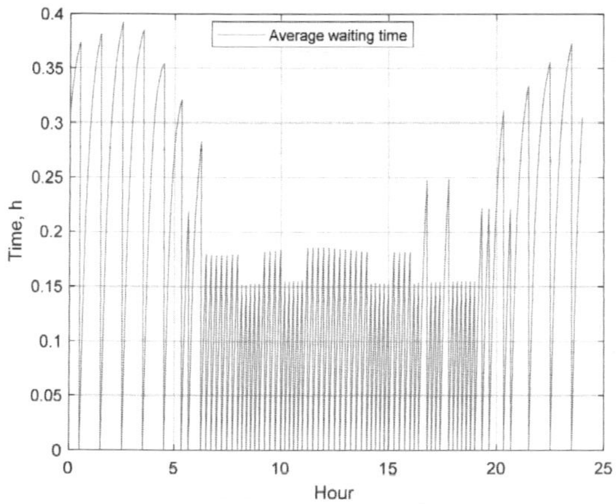

Figure 7. Average waiting time.

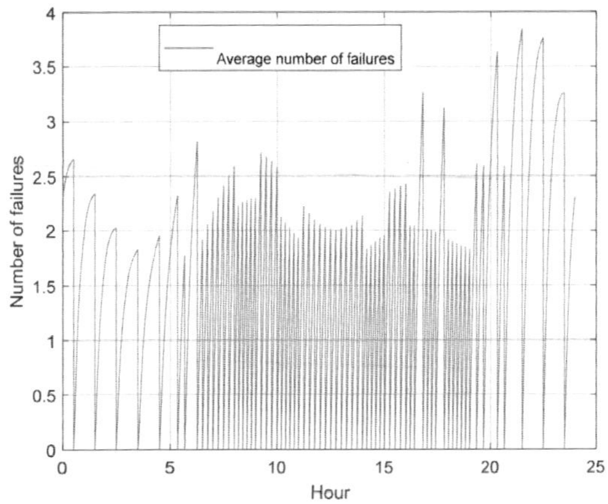

Figure 8. Average number of failures.

4. Conclusion

The developed model allows for real input data (passenger flow rates for a specific route) to describe the input flow density. The model is suitable for application to a bus and railway transport node.

Modelling the speed of service through the Dirac delta functions allows for a detailed description of the standard approximation. While the standard approximation assumes continuity of the entire service, in this case it is assumed that the service is at discrete moments of time. Input data are the service moments and the capacity of the vehicle.

The developed numerical method for solving differential algebraic equation systems takes into account the characteristic of Dirac delta functions. A file function that calculates the probabilities of the model is programmed in Matlab.

The solution of the given model allows us to determine the probabilities of a given state of the system at any point of time. The probability determination also allows us to calculate the system characteristics.

Acknowledgments

This research has been supported by contract No КП-06-Н27/12 of 11.12.2018 "Modelling and elaboration of complex system for selection of transport technology in transport network" funded by the National Science Fund of the Ministry of Education and Science of Bulgaria.

References

[1] Cortes C, Fernandez R and Burgos V 2007 Modelling passengers, buses and stops in traffic microsimulators *In: MISTRANSIT Approach on the Paramics Platform (86th TRB Annual Meeting, Washington DC)* 21–5

[2] Karlsson M, Corazza M and Musso A 2015 Understanding barriers to the diffusion of innovations for accessible and safe bus systems *Proc. 14th Int. Conf. Mobility and Transport for Elderly and Disabled Persons (Transed Lisbon Portugal)* 211–26

[3] Grava S 2003 *Urban Transportation Systems: Choices for Communities* (Sigurd Grava McGraw-Hill Prof Med/Tech) p 840

[4] Corazza M and Favaretto N 2019 A methodology to evaluate accessibility to bus stops as a contribution to improve sustainability in urban mobility *Sustainability* **11(3)**

[5] Ding J, Yi Z and Li L 2018 Accessibility measure of bus transit networks *IET Intell. Transp. Syst.* **12** 682–8

[6] Cervero R, Guerra E and Al S 2017 *Beyond Mobility: Planning Cities for People and Places* (Island Press: Washington DC USA) pp. 109–17

[7] Stoilova S 2020 *Methodology for Multi-criteria Selection of Transportation Technology in Transport Network.* In: Sładkowski A. (eds) Modelling of the Interaction of the Different Vehicles and Various Transport Modes. Lecture Notes in Intelligent Transportation and Infrastructure. (Springer. Cham) 1-103 https://doi.org/10.1007/978-3-030-11512-8_1
[8] Oña J and Oña R 2015 Quality of service in public transport based on customer satisfaction surveys: A review and assessment of methodological approaches *Transportation Sci.* **49(3)**
[9] Fernandez R 2007 Passion 5.0 - A model for microscopic simulation of multiple-berth bus stops *Traffic Engineering and Control* **48(7)** 324–8
[10] Silva P 2000 Simulating bus stops in mixed traffic *Traffic Engineering and Control* **41(4)** 160–7
[11] Gibson I, Baeza N and Willumsen L 1989 Bus-stop, congestion and congested bus-stops *Traffic Engineering and Control* **30(6)** 291–302
[12] Sztrik J 2016 *Basic Queueing Theory* (GlobeEdit, OmniScriptum GmbH, KG, Saarbrucken, Germany)
[13] Zukerman M 2016 *Introduction to Queueing Theory and Stochastic Teletraffic Models* (EE Department, City University of Hong Kong)
[14] Mitrofanova A 2007 *Continuous Times Markov Chains. Poisson Process. Birth and Death Process* (New York University)
[15] Norris J 1997 *Markov Chains* (Cambridge University Press)
[16] Ross S 2007 *Introduction to Probability Models* (Elsevier)
[17] Gautam N 2012 *Analysis of Queues: Methods and Applications* (CRC Press)
[18] Haghighi A and Mishev D 2013 *Difference and Differential Equations with Applications in Queueing Theory* (Wiley)
[19] Shampine L and Reichelt M 1997 The MATLAB ODE Suite *SIAM J. Scientific Computing* **18** 1–22
[20] Grozev D, Pencheva V, Georgiev I and Beloev I 2018 Investigation of the operation mode at Ruse-Danube Bridge border checkpoint considered to be a mass service system with incoming flow of automobiles at a non-stationary mode of operation *MATEC Web of Conferences* **234** 06003
[21] Georgiev I and Kandilarov J 2009 An immersed interface FEM for elliptic problems with local own sources *American Institute of Physics CP* **1186** 335–42
[22] Georgiev I and Kandilarov J 2010 An immersed interface FEM for 1D elastic continuum vibrations under a traveling load *American Institute of Physics CP* **1301** 379–86
[23] Liu Y and Mori Y 2012 Properties of discrete delta functions and local convergence of the immersed boundary method *SIAM J. Numer. Anal.* **50(6)** 2986–3015

IOP Conf. Series: Materials Science and Engineering **664** (2019) 012035 doi:10.1088/1757-899X/664/1/012035

Analysis of the load on the taxi system in a medium-sized city

D Grozev[1,3], M Milchev[1] и I Georgiev[2]

[1] Department of Transport, University of Rousse "Angel Kanchev", 8 Studentska Str., Rousse, Bulgaria
[2] Department of Applied Mathematics and Statistics, University of Rousse "Angel Kanchev", 8 Studentska Str., Rousse, Bulgaria

[3] E-mail: dgrozev@uni-ruse.bg

Abstract. In this research, the carriage of taxi drivers is considered by means of the queue theory. The study of the taxi system found out that the number of servers was too large. This results in low efficiency of taxi cars. The purpose of this study is to analyse the workload of the working servers in terms of their number within one day. An application was created in Matlab environment, which solves the problem and allows to evaluate the effect of this solution. The result of this research recognizes the need for reducing the number of working taxi vehicles by half or keeping their number but reducing their load factor variation between 11% and 19%.

1. Introduction

According to Germany's Taxi Association there are too many taxis on the roads. It means driver's income is reduced in the hunt for fares. Lower revenues mean slower system development, poorer quality of transport service and, most importantly, reduced transportation safety due to poor car maintenance because of lack of finances. This problem exists in other countries as well, as different cities and companies apply different solutions. For example, Berlin plans a minimum wage for the taxi sector – raising the prospect of fare hikes [1].

Many researchers came across issues to find optimal solution - such as leveraging computers and updated computing systems [2]. Previous attempts, to solve these issues, have been made also with other mathematic methods. The most famous way previously used to name the problem was the "traveling salesman problem," in which mathematicians would minimize the total distance a salesman would travel if he had to visit a certain number of locations to stop in a day of sales. The problems described also apply to medium-sized Bulgarian cities such as Rousse.

Another problem with the large number of taxis and shared car services is the generation of too much traffic in cities. There are a total of 2800000 vehicles registered in Bulgaria, which in each group are roughly divided as follows [3]: light cars – 2200000; lightweight cargo vans to 3.5 tons – 50000; buses for urban transport – 6000.

The large number of passenger cars in the country cause problems related to road traffic, which necessitates thc study of the characteristics of traffic flow and, above all, traffic intensity.

In connection with Bulgaria's membership in the European Union and the integration of its transport infrastructure, considerable efforts are being made to introduce European standards and laws for modern, environmentally friendly and safe transport. Insufficient investment in building new infrastructure and maintaining existing infrastructure requires timely planning for its future development [4, 5].

Content from this work may be used under the terms of the Creative Commons Attribution 3.0 licence. Any further distribution of this work must maintain attribution to the author(s) and the title of the work, journal citation and DOI.
Published under licence by IOP Publishing Ltd

Road transport in Bulgaria is developed at fast pace due to its high velocity and great flexibility. The same applies to urban passenger transport and, in particular, to taxi transport. Its advantages over mass city transit of passengers are:

- possibility to carry out a free route;
- possibility to use 24 hours a day;
- the possibility of transporting "from door to door" [5].

One of the main features of traffic flow is traffic intensity. It is expressed in the number of cars passing through a section of the road per unit of time. This intensity changes throughout the day, week, month and year. The intensity of traffic in a given city depends on the amount of vehicles in motion, the links between transport and pedestrian flows, the density of the street network and the city's territory. The intensity of traffic and the parameters of taxi transport strongly influence the number of taxi drivers operating in the particular settlement.

As of January 2019 in the Municipality of Ruse are registered 151 companies with a permit for carrying out taxi activity. The total number of taxi cars operating on the territory of the Municipality of Rousse is 588 [6]. The average number of taxi cars in a company is 3.89, and the number of taxi cars per 1000 people is 4.15. For Rousse, the number of companies providing radio communication for management of taxis is 4, shown in table 1.

Table 1. Taxi companies providing radio communication for taxi management as of 30.05.2019.

Number	Firm
1	Taxi 2222 – to4nite
2	8806 – Taxi 6
3	Orion Taxi 8800
4	Milanov Taxi 8111

Table 2 shows data on registered taxis in five European countries [6]. The selected benchmark "number of taxi cars per 1000 people" shows a number a few times higher than that for Bulgaria.

Table 2. Registered taxis in five European countries and Bulgaria.

	Bulgaria	Holland	Switzerland	Great Britain	Belgium	France
Inhabitants, thousands	7364	15900	8900	59000	10200	60000
Number of taxi cars	30560	19291	14558	70000	4000	44000
Number of taxi cars per 1000 people	4.15	1.2	1.67	1.2	0.4	0.7

On average, for the five European countries, the number of taxi cars per 1000 people is 1.03, while for Bulgaria this index is 4.17.

In the table 3 is a number of taxis per thousand people for five European capitals with an average of 2.22. The table shows that for Rousse this indicator is 1.8 times higher than the average. From the information presented, it seems that the number of taxi cars is much higher than the average for the big European cities. What is needed to achieve efficiency in the taxi system is to establish the optimal number of taxi cars for meeting the needs for this type of transport in society and a transport scheme tailored to the infrastructure of the respective settlement.

Table 3. The density of the taxi network to 10.10.2019.

	Rousse	Stockholm	London	Bruxel	Copenhagen	Paris
Inhabitants	150000	1600000	7000000	960000	1400000	6000000
Taxi cars	588	5337	21000	1283	2513	14900
Number of taxi cars per thousand people	3.89	2.8	2.5	1.3	2.0	2.5

The purpose of this study is to find a proper solution for the large number of taxis in a medium-sized Bulgarian city.

2. Methodology of the research

The chosen method of examining the organization of the taxi system is the mathematical apparatus of a queue theory system. This theory has been chosen because it best describes the process of taxi transportation in its entirety.

To describe the operation mode of a taxi company by means of the queue theory, it is necessary to know the characteristics of the input stream of orders treated as a stochastic process, the service intensity, the maximum length of the queue and the number of service units [7, 8].

For the incoming flow of cargo it could be made the following assumptions for the ordinal flow and flow without consequences. Regarding the stationarity / non-stationarity of the flow – for sufficiently long periods of time – 1 month, 6 months, 1 year, etc. it is possible to assume the steady-state of the incoming stream, i.e. with some provisionallyty. The probability of availability of a certain number of cars in a given, sufficiently long interval depends only on the length of that interval. Generally, at random times, flow λ is non-stationary $\lambda = \lambda(t)$. This non-stationarity stands out well over a 1-day working day (24 hours). The incoming and outgoing stream is modelled [9-11].

For the intensity of service μ, it is known by the company manager that the time for servicing one order is a relatively constant quantity and is 8.5 minutes in the time intervals between 07:00-10:00 h and 16:00-19:30 h, while for the rest of the day it is 5.5 min. The number of serving channels (servers), over the day, is variable and changes between 26 and 120 ($26 \leq n \leq 120$). It is also known that at the moment when all servers are busy, the received order is rejected. To test the operation of the system, it is necessary to find the probability that the system will have k-number of vehicles at time t when n-servers are operating, i.e.

$$P_k(t)=?,\ k=\overline{0,n},\ t\in[0,\ T],\ 26 \leq n(t) \leq 120,$$

where for one period T is taken a full working day $T = 0$ of 24 hours. The beginning of the working day $t = 0$ coincides with the astronomical start of the day (0 hours and 0 min), the end of the working day $t = T = 24$, the end of the astronomical day (24 hours and 0 min).

For the queue theory model, the following can be summarized: a non-stationary flow of densities $\lambda(t)$ flows into a queue theory with n-number service channels. Request processing time is a random variable with an indicative distribution and μ parameter is a partial constant function. The n-number of the service channels also changes overnight, i.e. $n = n(t)$. An order that arrives at a busy time leaves the system unusable, i.e. system with failures.

From what has been said so far, we can say that the system is type ($M/M/n$) in non-stationary mode with variable number of channels. To describe a system of this type, the following system of differential equations of Kolmogorov (Erlang-Kolmogorov) [12,13]:

$$\frac{dP_0}{dt} = -\lambda(t)P_0(t) + \mu(t)P_1(t);\ \dots\ ;$$

$$\frac{dP_k}{dt} = P_{k-1}(t)\lambda(t) - \big(\lambda(t) + k\mu(t)\big)P_k(t) + \mu(t)(k+1)P_{k+1}(t);\ \dots;\ 0 < k < n-1;\ \dots; \tag{1}$$

$$\frac{dP_n}{dt} = P_{n-1}(t) - n\mu(t)P_n.$$

For modelling $\lambda(t)$, it is advisable to choose a relatively elementary function with periodicity. This is appropriate given the daily fluctuations. Approximate squares [14] was used to approximate the average of the received queries over a 24-hour period (figure 1). The model should be as simple as possible but reflect the most characteristic behaviour of the actual stream. For a model, the following trigonometric line (7-point Fourier row) is selected, linear to the quoted odds:

$$\lambda(t)=a_0+a_1\cos\left(\frac{2\pi t}{24}\right)+b_1\sin\left(\frac{2\pi t}{24}\right)+a_2\cos\left(\frac{4\pi t}{24}\right)+b_2\sin\left(\frac{4\pi t}{24}\right)+a_3\cos\left(\frac{6\pi t}{24}\right)+b_3\sin\left(\frac{6\pi t}{24}\right). \tag{2}$$

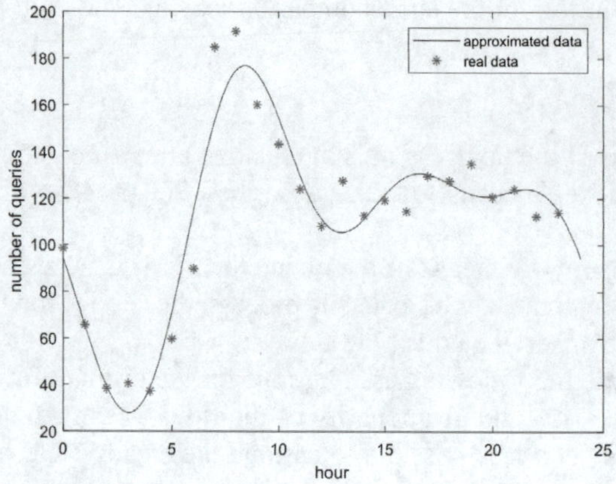

Figure 1. Graph of the density of incoming $\lambda(t)$ stream approximated by least squares method (LSM) and observed midstream input data over a 24-hour period of 24 hours over a week.

3. Result of the research

The data for the received requests (by hour) for 24 hours over a week are given in a table 4.

Table 4. Received requests (by hour) for 24 hours, over a week.

Date/hour	00-01	01-02	02-03	03-04	04-05	05-06	06-07	07-08	08-09	09-10	10-11	11-12	12-13	13-14	14-15	15-16	16-17	17-18	18-19	19-20	20-21	21-22	22-23	23-24	Number of cars in the day
2.4.2019	41	30	18	25	23	24	59	102	137	127	127	119	108	121	108	107	118	118	101	72	76	65	55	49	275
3.4.2019	34	35	20	19	20	27	60	112	142	132	120	126	96	108	110	94	102	118	95	89	81	68	66	50	284
4.4.2019	40	29	21	20	26	31	59	118	136	137	127	106	105	114	93	115	90	108	94	101	81	72	59	51	290
5.4.2019	40	31	19	19	22	29	63	119	133	134	118	108	109	113	100	127	123	118	117	121	102	98	89	84	292
6.4.2019	76	53	44	39	32	42	46	72	78	99	106	94	92	96	92	77	86	99	100	113	107	105	95	100	276
7.4.2019	93	76	58	46	38	49	50	51	57	73	78	78	63	92	80	80	79	82	85	76	73	59	63	68	246
8.4.2019	40	26	17	20	16	35	73	130	157	146	146	116	100	127	107	116	109	111	95	80	65	60	61	67	285

The coefficients a_0, a_1, b_1, a_2, b_2, a_3, b_3 calculated by the LSM, as well as their confidence intervals are given in table 5. All coefficients are statistically significant (the confidence intervals do not contain 0). The determinant coefficient is $R^2 = 0.9067$, also statistically significant, meaning that the chosen model is adequate.

Table 5. Coefficients and their confidence intervals.

Value of the coefficients in the model	Trust confidence interval guaranteed by probability $\gamma = 0.95$
$a_0 = 111.6$	(105.3; 117.8)
$a_1 = -29.73$	(–38.57; –20.89)
$b_1 = -10.64$	(–19.48; –1.803)
$a_2 = -8.927$	(–17.77; –0.0879)
$b_2 = -36.8$	(–45.64; –27.96)
$a_3 = 21.13$	(12.29; 29.97)
$b_3 = -4.83$	(–9.66; –0.009)

For the service $\mu(t)$ speed, it is known that during the more busy daytime, 07:00-10:00 h and 16:00-19:30 h, the service of one The request takes about 8.5 minutes. For one hour, average service speed is 120/17 = 7.0588 queries per hour, and during the rest of the day, queries are served for 5.5 minutes; the speed is 120/11 = 10,909 requests per hour.

$$\mu(t) = \begin{cases} \frac{120}{11}, & 0 \le t < 7;\ 10 \le t < 16;\ 19.5 \le t < 24 \\ \frac{120}{17}, & 7 \le t < 10;\ 16 \le t < 19.5 \end{cases} \tag{3}$$

Statistical data over the course of one week for the number of servers (t) operating overnight are given in table 6.

Table 6. Number of servers (t) operating overnight

Date/Hour	00-01	01-02	02-03	03-04	04-05	05-06	06-07	07-08	08-09	09-10	10-11	11-12	12-13	13-14	14-15	15-16	16-17	17-18	18-19	19-20	20-21	21-22	22-23	23-24	Total applications
2.4.2019	71	54	27	35	39	40	85	184	216	176	157	137	119	157	116	122	133	145	122	90	106	104	85	75	2595
3.4.2019	51	46	24	24	22	37	94	183	229	173	132	153	108	115	126	97	108	136	109	128	99	110	111	84	2499
4.4.2019	71	44	28	30	27	43	83	203	221	173	153	104	112	129	103	123	98	125	120	125	113	110	103	95	2536
5.4.2019	61	48	31	39	48	52	93	212	211	175	138	142	124	117	103	159	139	132	167	162	136	171	120	148	2928
6.4.2019	137	82	65	45	52	81	66	117	108	120	132	105	113	110	120	87	102	120	155	238	176	170	159	193	2853
7.4.2019	230	132	71	69	47	93	89	105	89	108	110	99	71	117	92	119	97	111	103	101	118	104	127	91	2493
8.4.2019	68	54	22	42	25	72	118	290	267	196	181	128	109	148	128	128	124	138	117	100	97	99	81	110	2842

Given the statistics, the number of working servers is approximated as a whole (figure 2). The length of time intervals in which the number of servers is more dynamic is shorter (a shorter time interval of 1 hour cannot be taken due to the specificity of the statistical measurements), those in which the dynamics is smaller have longer intervals.

$$n(t)=\begin{cases} 46, & 0 \le t < 1 \\ 34, & 1 \le t < 2 \\ 26, & 2 \le t < 5 \\ 46, & 5 \le t < 6 \\ 80, & 6 \le t < 7 \\ 120, & 7 \le t < 10 \\ 104, & 10 \le t < 18 \\ 91, & 18 \le t < 21 \\ 68, & 21 \le t < 24 \end{cases} \tag{4}$$

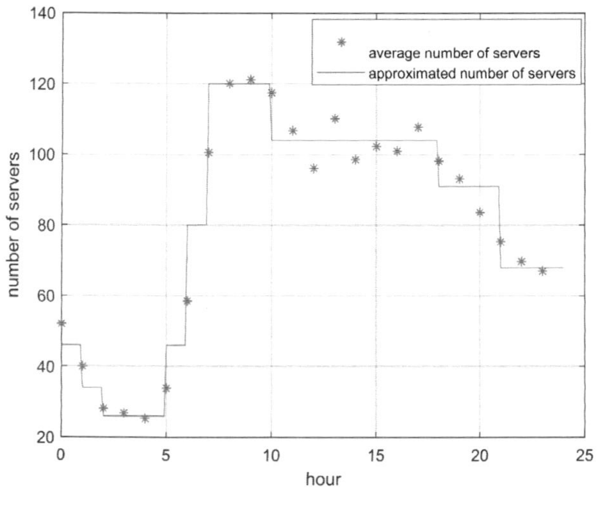

Figure 2. Graph of the approximated average number of measured taxis (servers) over the day and the graph of the function $n(t)$.

The computational features associated with system (1) after the following:

- a large-scale system (generally);
- the system is of the "rigid system" type (rigid system of equations);
- the number of equations is variable at the various integration intervals.

This can be summarized as: a system with varying number of solid, differential equations of large size. To overcome these difficulties, special numerical methods have been developed. A Matlab [14] program was developed to solve a system (2), using the built-in "solver" ode15s, implementing the Gir method. When entering $\lambda(t)$, $\mu(t)$, n the application returns a numerical solution to $P_k(t)$. Artificially, the precision is greater than that implied for the solver – from 10^{-6} to 10^{-9} for absolute error and 10^{-3} to

10^{-5} for a relative error. The integration interval [0, T] is broken down into subfields $t_0, t_1, \ldots, t_{i-1}, t_i, t_{i+1}, \ldots, t_N$ so that in each subinterval $[t_i, t_{i+1}]$, $\mu(t)$ and $n(t)$ are constants.

The initial state of the system $P_k(t_0)$ is unknown. It is known that these types of processes are persistent and after a long period of time they enter into regular mode of operation. Therefore, a random state may be taken arbitrarily. The values $P_k(t)$, at the end of the time interval $[t_{i-1}, t_i]$, become initial values for the subinterval $[t_i, t_{i+1}]$. If there are more equations for a given time interval than the previous interval, then zero values are assumed for the initial conditions of these equations. System integration in the [0, T] interval is required to be done not once but a sufficient number of times, and after each integration the values at the end of the period become initial values for the next integration. In this way, the probability functions $P_k(t)$ begin to bend to their regular values. After multiple integration, it was found out that only after 5-6 periods the functions $P_k(t)$ entered into regular mode (for two adjacent periods, they remain the same). Accuracy is also increased here, with integration being done over 20 periods, with the difference of all $P_k(t)$ in the last and penultimate periods being less than 10^{-8} for each t. Table 7 lists the time intervals $[t_{i-1}, t_i]$, the values of the function μ, and the number of service channels n in the given interval [14].

Table 7. Distribution of servers at time intervals, the values of the function μ, and the number of service channels n in the given interval.

A time interval within 24 hours $[t_{i-1}, t_i]$	[0, 1]	[1, 2]	[2, 5]	[5, 6]	[6, 7]	[7, 10]	[10, 16]	[16, 18]	[18, 19.5]	[19.5, 21]	[21, 24]
Average Service Rate μ	120/11	120/11	120/11	120/11	120/11	120/17	120/11	120/17	120/17	120/11	120/11
Average number of serving servers n	46	34	26	46	80	120	104	104	91	91	68

After integrating a system (2) and calculating the $P_k(t)$, it is interesting to see how the average number of servers are occupied and how the load factor of a server changes.

Figure 3 shows the average number of customer-driven taxi cars for a 24-hour period. Two peaks are observed, the first one is between 7 and 10 hours, and the smaller is between 16-19 hours. The smallest peaks are in the range of 2-3 hours in the morning.

Figure 4 shows the load factor of a server (taxi) over the course of a day. It can be seen that even with the highest peaks, the coefficient is below 0.25, which means that at the busiest intervals, the server occupancy is below 25% of the employed at that particular moment. Average server occupancy is about 15%.

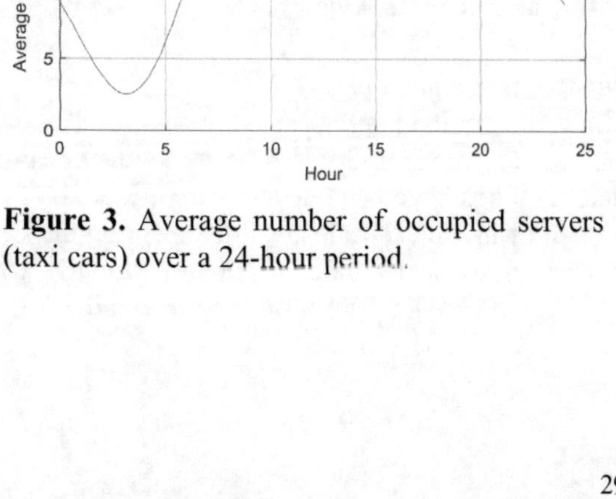

Figure 3. Average number of occupied servers (taxi cars) over a 24-hour period.

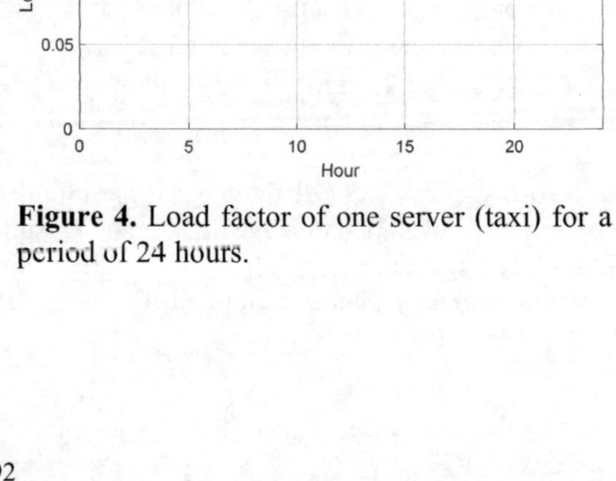

Figure 4. Load factor of one server (taxi) for a period of 24 hours.

IOP Conf. Series: Materials Science and Engineering **664** (2019) 012035 doi:10.1088/1757-899X/664/1/012035

There are two options for distributing the average number of servers in the considered time intervals. The first option is to keep the average number of working cars during the day so the load factor does not change, but it is intended to reduce the limits of that factor over the day. A distribution of servers in the time slots that narrows the boundaries of the load factor is shown in table 8.

Table 8. Distribution of servers at time intervals adjusting the boundaries of the load factor.

A time interval within 24 hours [t_{i-1}, t_i]	[0, 1]	[1, 2]	[2, 5]	[5, 6]	[6, 7]	[7, 10]	[10, 16]	[16, 18]	[18, 19.5]	[19.5, 21]	[21, 24]
Average Service Rate μ	120/11	120/11	120/11	120/11	120/11	120/17	120/11	120/17	120/17	120/11	120/11
Average number of serving servers n	44	34	23	31	69	138	83	101	96	75	61

In this embodiment, the average load factor is again about 15%. Except for the larger but shorter peak at about 10.00 hours, the load factor limits are about 11% and 19%.

According to the second option the number of servers must be reduced in such a way that the average ratio of workload to increase to 40%, but the difference between the smallest / largest and average (40%) of the load factor, is about 10% within the day. A variant of such distribution of servers in the time slot is shown in table 9.

Table 9. Distribution of servers in the time slots increasing the average load factor.

A time interval within 24 hours [t_{i-1}, t_i]	[0, 1]	[1, 2]	[2, 5]	[5, 6]	[6, 7]	[7, 10]	[10, 16]	[16, 18]	[18, 19.5]	[19.5, 21]	[21, 24]
Average Service Rate μ	120/11	120/11	120/11	120/11	120/11	120/17	120/11	120/17	120/17	120/11	120/11
Average number of serving servers n	22	14	8	14	24	58	27	48	42	30	29

The load factor, calculated from the distribution in table 9, is given in figure 5. From the graph of Figure 6 shows that the average load factor is about 49%. Except for the peak of about 10 hours, which is negligibly short and the peak of about 7 hours, which is slightly longer but relatively small, it can be seen that the daytime load factor is between 30% and 50%. In order to achieve this effect, the average number of servers is the day-to-day need to be reduced by a little more than twice.

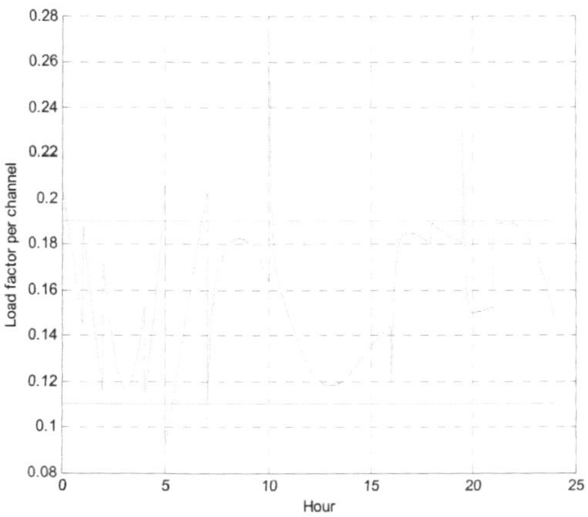

Figure 5. Load factor of one server (taxi) over a 24-hour period with distribution from table 8. Red lines – load factor limits without short peaks.

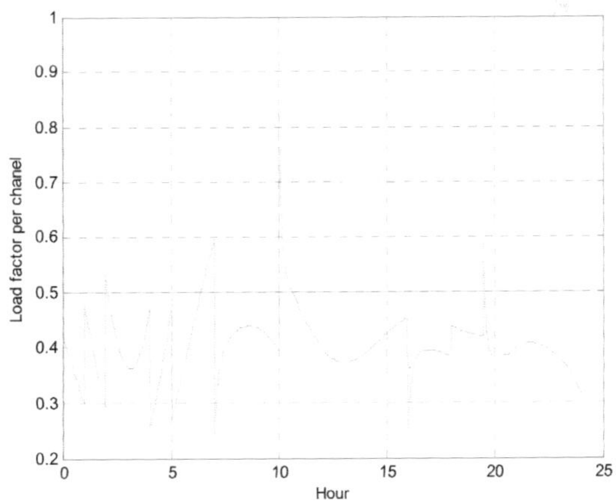

Figure 6. Load factor of one server (taxi) over a 24-hour period with allocation from table 9. Green lines – load factor limits without short peaks; Red line – average values of the load factor.

4. Conclusion

The input flow is modelled on a query stream by means of statistical inventory. The function describing the average number of working servers in the day is created using statistical data.

The statistics given are calculated by the load factor of a server. This factor is too small and leads to inactivity of the servers. In addition, with this server distribution during daytime, the probability range is between 10% and 25%.

A redistribution of servers is proposed overnight, while retaining their number but shrinking the range of load factor variation between 11% and 19%. This provides a relatively uniform workload for servers over the day or the number of orders they serve.

A variant is proposed to reduce and appropriately allocate servers at different times of the day. In this variant, the average server load is about 40%, with the range of variation being between 30% and 50%. The number of servers has been reduced slightly more than twice. This ensures a higher average workload per server.

With a database available for a similar city in size, it is possible to carry out an analogous study of the workload of the taxi-transport system, and determine the relevant conclusions. Based on the findings, it is possible to offer variants for changing the number of working servers in 24 hours.

Acknowledgments

The study was supported by contract of University of Ruse “Angel Kanchev”, № BG05M2OP001-2.009-0011-C01, “Support for the development of human resources for research and innovation at the University of Ruse “Angel Kanchev”. The project is funded with support from the Operational Program “Science and Education for Smart Growth 2014-2020” financed by the European Social Fund of the European Union.

References

[1] Singh S, Singh P and S Samra 2000 Perth Taxi Group’s Submissiontothe ERA inquiry into Microeconomic Reform: Taxi Industry Regulation

[2] Darbéra R 2015 Principles for the regulation of for-hire road passenger transportation services *Discussion paper for the ITF workshop Principles for the regulation of for-hire road transport passenger services (26-27 October 2015, Paris)*

[3] Mitrofanova A 2007 *Continuous Times Markov Chains. Poisson Process. Birth and Death Process* (New York University)

[4] Cooper J 2007 *Taxi Licensing, Regulation and Control An analysis of taxi supply in medium sized UK cities* (Napier University, Edinburgh)

[5] Frazzani S, Grea G and Zamboni A 2016 Study on passenger transport by taxi, hire car with driver and ridesharing in the EU *Final Report for the European Commission* 2015-564

[6] Grozev D 2014 *Efficiency in the Urban Taxi System* (Dissertation work, Bulgaria, Ruse, http://kr.uni-ruse.bg/index.php?r=file/download&id=3586)

[7] Sztrik J 2016 *Basic Queueing Theory* (GlobeEdit, OmniScriptum GmbH, KG, Saarbrucken, Germany)

[8] Zukerman M 2016 *Introduction to Queueing Theory and Stochastic Teletraffic Models* (EE Department, City University of Hong Kong)

[9] Norris J R 1997 *Markov Chains* (Cambridge University Press)

[10] Ross S M 2007 *Introduction to Probability Models* (Elsevier)

[11] Gautam N 2012 *Analysis of Queues: Methods and Applications* (CRC Press)

[12] Haghighi Al and Mishev D 2013 *Difference and Differential Equations with Applications in Queueing Theory* (Wiley)

[13] Grozev D, Pencheva V, Georgiev I and Beloev I 2018 Investigation of the operation mode at Ruse-Danube Bridge border checkpoint considered to be a queue theory with incoming flow of automobiles at a non-stationary mode of operation *MATEC Web of Conferences* **234** 06003

[14] Shampine L F and Reichelt M W 1997 The MATLAB ODE Suite *SIAM J. Scientific Computing* **18** 1–22

IOP Conf. Series: Materials Science and Engineering **664** (2019) 012036 doi:10.1088/1757-899X/664/1/012036

Network model of passenger railway rolling stock scheduling

Mirena MironovaTodorova

Todor Kableshkov University of Transport, 158 Geo Milev Str., Sofia, Bulgaria

E-mail: mirena_todorova@abv.bg

Abstract. Passenger rolling stock scheduling to the passenger train timetable is based on the same type and number of carriages, which is the standard way of passenger trainset assignment. The possibility of train scheduling with a single carriage, a group of carriages or a railcar is topical issue for each carrier engaged in rail passenger transport in order to reduce the required rolling stock fleet. The paper presents an original network model developed for rolling stock scheduling for trainsets. The model makes possible to determine the optimum distribution of rolling stock in trainsets minimizing downtime at stations through the out-of-kilter algorithm. The model developed is applied to the scheduling of trainsets in part of the railway network in Bulgaria served by BDZ Passenger Transport Enterprise at the Plovdiv region. The use of the model allows to save passenger rolling stock and increases the efficiency of passenger rolling stock operation.

1. Introduction

The number of passengers at different times in a 24-hour period is different, so the number of carriages required for each train is determined in compliance of demands. The sample scheduling is performed on the basis of one and the same number and type of carriages. According to the traffic volume served, the so-called Plan of Train Formation (PTF) presets trains that consist of 2-10 carriages, one or two multiple units with different numbers of seats. On the other hand, although two trains can be of one and the same number of carriages, their type could vary depending on the train category. In practice is used train scheduling with one and the same type and number of carriages, which is the standard way to operate passenger carriages and trainsets. This way of train scheduling is not always rational because of long downtimes existing in practice.

It is why train scheduling that considers a carriage, a group of carriages or multiple units is topical for each carrier engaged in passenger transportation by rail in order to reduce the rolling stock fleet needed for train operation.

2. Overview of references

The references on passenger railway transport describe a variety of ways to perform train scheduling. The model in [1] does not consider the number of carriages required while the main problem discussed in [2] is how to reduce the number of trains depending on demands for passenger seats. The method suggested is based on calculating the number of carriages of each type for each train. Another model [3] defines rolling stock scheduling on a single-line for one day aiming to minimize the number of trainsets needed for the Train Timetable (TTT).

The approach using linear programming for scheduling trainsets [4] is complimented in [5] with specificity of train at stations. The two-step optimization model of rolling stock scheduling in [6] concerns the operation of passenger trains with existence of delays and large disturbances in the rail

Content from this work may be used under the terms of the Creative Commons Attribution 3.0 licence. Any further distribution of this work must maintain attribution to the author(s) and the title of the work, journal citation and DOI.
Published under licence by IOP Publishing Ltd

system, which means that it has been designed in a way to deal with delays. The two-step optimization model is formulated as a Mixed Integer Linear Programming model (MILP).

The optimization framework of railway rolling stock scheduling described in paper [7] is based on a two-step approach that combines the scheduling tasks related to train services, short-term maintenance operations and empty runs. Paper [8, 9] presents an analytical comparison between two approaches within the operational and real-time planning phase. The first one is known as Mixed Integer Linear Program (MILP) while the second one is a new method, an extension of a recently introduced MILP, which uses a column and row generation approach. The authors have validated the theoretical methods developed in this paper on the railway networks of Denmark and the Netherlands.

The problem of scheduling in the references examined above is solved by using linear programming. However, this paper examines the rail transport network where passenger train traffic and rolling stock scheduling take place through a structure of respective graphs that give a possibility to use the well-developed mathematic apparatus of the graph theory and especially the so-called out-of-kilter algorithm [10, 11].

3. Model of rolling stock scheduling to trains

There are two mathematical models appropriate to perform rolling stock scheduling to trains: the maximum flow through a network with minimum cost and out-of-kilter algorithm. Both of them allow to find different options depending on the constraints with choosing optimality criteria, which after being matched can determine the optimal solution. The model of rolling stock scheduling developed on the basis of out-of-kilter algorithm passes through the following steps (figure 1):

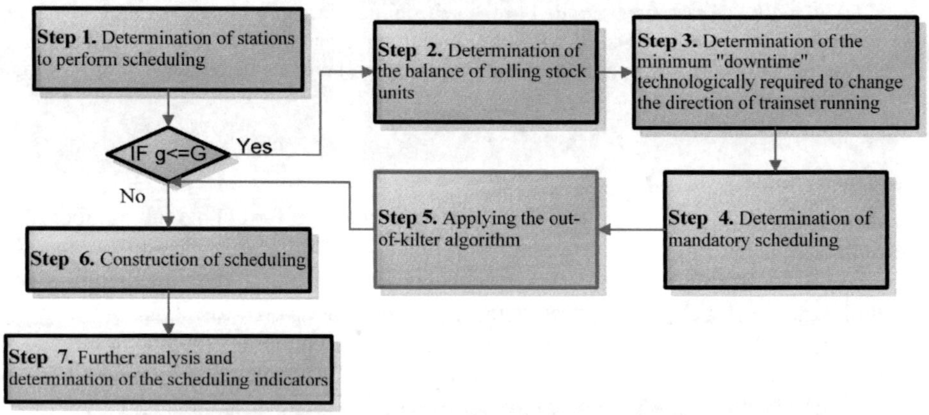

Figure 1. Algorithm of rolling stock scheduling to trains.

The number of stations, where scheduling is performed, is labelled with „G" in figure 1 and determined in step 1. The iterative procedure of scheduling at each current station „g" involving steps 2-5 is performed until all stations included in specified number „G" are considered - g=1…G.

3.1. *Determination of stations to perform scheduling*

The setting of this task in practice requires to allocate rolling stock to trains with the given TTT and PTF. It is therefore necessary to define stations G where rolling stock scheduling is performed. G stations can be considered a subset of the set of all stations in the polygon examined where at least one of the following conditions is met: to be terminal or starting stations for one or more trains or a place where carriages are attached to or detached off one or more trains. In cases of scheduling multiple units (multiple units), the algorithm is applied to the whole network, and in cases of scheduling carriages it is applied to railway lines or sections. By the distribution of trains depending on grouping carriages, the task is reduced to solving a number of subtasks with decreased dimensionality as each of them would give a solution of rolling stock scheduling to the trains that serve a certain section.

3.2. Determination of the balance of rolling stock units

Depending on scheduling, a unit of rolling stock is understood as a railcar, a group of carriages or a carriage. Passenger rolling stock (carriages or multiple units) does not have dead mileage, which is taken into account when designing the TTT and respective PTF, so the number of arriving carriages and multiple units is equal to that of departing ones. Unlike the number of rolling stock units, the number of arriving and departing trains at a given station can be different.

3.3. Determination of minimal "downtime" technologically required to change the direction of trainset running

To ensure the real application of scheduling, it is necessary that the values of technologically needed minimal "downtimes" are greater than any predetermined minimal permissible value depending on the station type and rolling stock handling, namely: the downtime is only for scheduling (according to standards), cleaning and equipment of trains at the station or technical maintenance and arrangement of carriages and multiple units at the technical station. This is achieved by introducing additional restrictive condition assumed in the form of:

$$T_{st_{ij}} \min = T_{ij}(a, T_{st}), \tag{1}$$

where: $T_{st_{ij}} \min$ is the estimated value of required downtime of trainset from the moment of its release from train i to the moment of its inclusion to train j; $T_{ij}(a, T_{st})$ – a predetermined value of the down time of trainset depending on the size of carriage group a and the type of station Tst and technological operations.

3.4. Determination of mandatory scheduling

For example, in relatively short intervals between the arrival and departure of a suburban passenger train to a given destination, it is desirable to use electric multiple units for a number of train pairs, which facilitates the specialization of station tracks and organization of their service. Such a type of scheduling leads to higher technology performance in train timetable implementation and can be achieved by setting zero downtimes in scheduling to certain trains or by artificially raising the values of arrival-departure interval of trains.

3.5. Applying the out-of-kilter algorithm for scheduling trains with carriages

The advantage of this algorithm in comparison to other network algorithms is that it is more generalized, more computerized and allows to solve problems with a large number of variables and constraints. Since the balance of rolling stock is provided in advance for the station, the carriages or multiple units of arriving train i can participate entirely or in groups of carriages (multiple units) in the departing trains j.

The criterion function has the following form:

$$T_k = \sum_{i=1}^{n}\sum_{j=1}^{m} T_{ksi}F_{ksi} + \sum_{i=1}^{n} T_{kij}\,F_{kij} + \sum_{j=1}^{m} T_{kjt}\,F_{kjt} + T_{kts}\,T_{kts} \rightarrow \min \tag{2}$$

$$\min T = \sum_{k=1}^{k} T_k, \tag{3}$$

where T is total carriage hours of downtime of all rolling stock units involved in scheduling;

F_{ksi}, F_{kij}, F_{kjt}, F_{kts} – number of rolling stock units on links s-i, i-j, j-t, t-s;

T_{ksi}, T_{kij}, T_{kjt}, T_{kts} – downtime of rolling stock units on links s-i, i-j, j-t, t-s;

n – number of rolling stock units arriving with trains;

m – number of rolling stock units departing with trains;

k – type of rolling stock unit involved in scheduling at a given station.

In this case, due to the balance provided per unit of rolling stock for each station, $n = m$.

The group of constraints related to keeping the flow at each node:

$$F_{ksi} = \sum_{j=1}^{m} F_{kij}, \text{ for each fixed node } i, \tag{4}$$

$$\sum_{i=1}^{n} F_{kij} = F_{kjt}, \text{ for each fixed node } j, \tag{5}$$

$$F_{kts} = \sum_{i=1}^{n} F_{ksi}, \text{ for node } s, \tag{6}$$

$$\sum_{j=1}^{m} F_{kjt} = F_{kts}, \text{ for node } t. \tag{7}$$

The group of constraints related to capacity of links:

$$L_{ksi} \le F_{ksi} \le U_{ksi}, \text{ with } F_{ksi} > 0, \tag{8}$$

$$L_{kij} \le F_{kij} \le U_{kij}, \text{ with } F_{kij} > 0, \tag{9}$$

$$L_{kjt} \le F_{kjt} \le U_{kjt}, \text{ with } F_{kjt} > 0, \tag{10}$$

$$L_{kts} \le F_{kts} \le U_{kts}, \text{ with } F_{kts} > 0, \tag{11}$$

where U_{ksi}, U_{kij}, U_{kjt}, U_{kts} – maximum capacity of links *s-i, i-j, j-t, t-s*;
L_{ksi} L_{kij}, L_{kjt}, L_{kts} – minimum capacity of links *s-i, i-j, j-t, t-s*.

The constraint related to the downtime of the *k*-th type of rolling stock unit

$$T_{ksi} = 0,\ T_{kjt} = 0,\ T_{kts} = 0,\ T_{kij} \ge T_{kijmin}, \tag{12}$$

for each *i, j* where T_{ijmin} is minimum necessary downtime of a rolling stock unit with scheduling from train *i* to train *j*, min.

The downtime of carriages is defined as a difference between the departure time t_{dti} of any type of rolling stock unit at a given station and the time of arrival t_{ati} of any type of rolling stock unit at the same station by one of the following formulas:

$$T_{ij} = t_{st_{ij}} = t_{dtj} - t_{ati}, \text{ min, with } t_{dtj} > t_{ati}; \tag{13}$$

$$T_{ij} = t_{st_{ij}} = 1440 - (t_{ati} - t_{dtj}), \text{ min, with } t_{dtj} \le t_{ati}. \tag{14}$$

This downtime is graphically shown in figure 2 (a) when the train arrives earlier than the departing one and in figure 2 (b) when the train arrives later than the departing one.

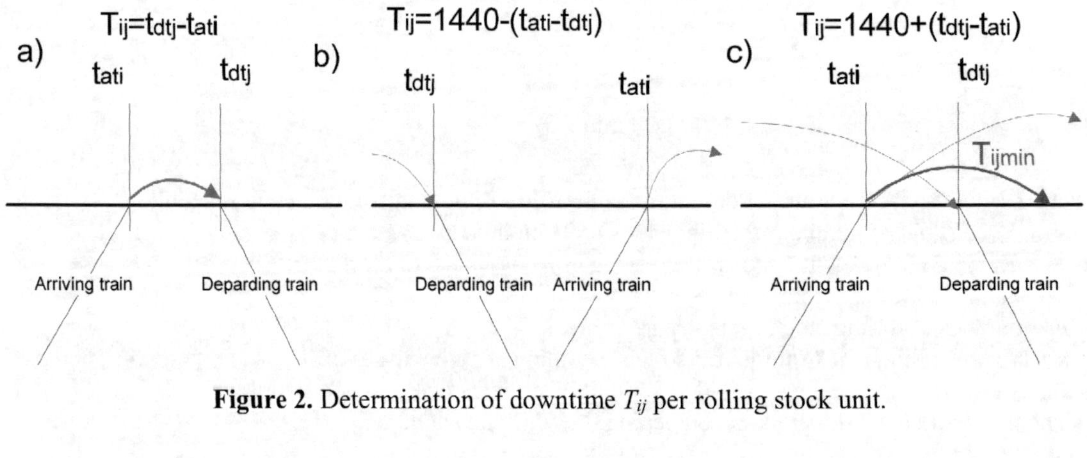

Figure 2. Determination of downtime T_{ij} per rolling stock unit.

While checking if the condition in formula 7 is satisfied, the following cases are possible:

- with $T_{ij} \geq T_{ijmin}$ the so-specified interval remains;
- with $T_{ij} < T_{ijmin}$ the interval for a given train $T_{ij} = 1440 + (t_{dtj} - t_{ati})$, min, figure 2 (c).

The use of out-of-kilter algorithm requires to make a graph structure (figure 3) in the following sequence:

- A fictitious main manufacturer node S, and a fictitious main user node t are introduced. All trains arriving at the station are represented as nodes manufacturers of rolling stock units – P_1, P_2, ... , P_n. All trains departing from the station are represented as nodes users of rolling stock units – K_1, K_2, ... , K_m.
- Directed links are constructed as follows:

1. From node S to all nodes manufacturers with parameters: maximum capacity U_{ksi} equal to the number of the rolling stock units arriving by the train represented with the respective node equal to 1, $L_{ksi} = 0$, $T_{ksi} = 0$ and $F_{ksi} = 0$ (figure 3, link s-P_i);
2. From all nodes manufacturers to all nodes users with parameters: maximum t capacity $U_{kij} = \infty$, minimum capacity $L_{kij} = 0$, downtime T_{kij} equal to the value of T_{ij} for this pair of trains and flow $F_{kij} = 0$;
3. From all nodes users to nodes with parameters: maximum U_{kjt} throughput, equal to the necessary number of carriages to service the corresponding train for the given node equal to 1, $L_{kij} = 0$, $T_{kjt} = 0$ and $F_{kjt} = 0$ (figure 3, link K_1-t);
4. Closing link from node L to node S with parameters: maximum capacity equal to the total production of rolling stock units, minimum capacity L_{kts} equal to the total consumption of a unit of rolling stock, T_{kts} and F_{kts} equal to zero (figure 3, link t-S). Since the task is balanced, the values of the maximum and minimum capacities of closing link must be equal to each other.

With using the out-of-kilter algorithm in excel, (U_{kij}, L_{kij}, T_{kij}) are introduced as input and F_{kij} is determined.

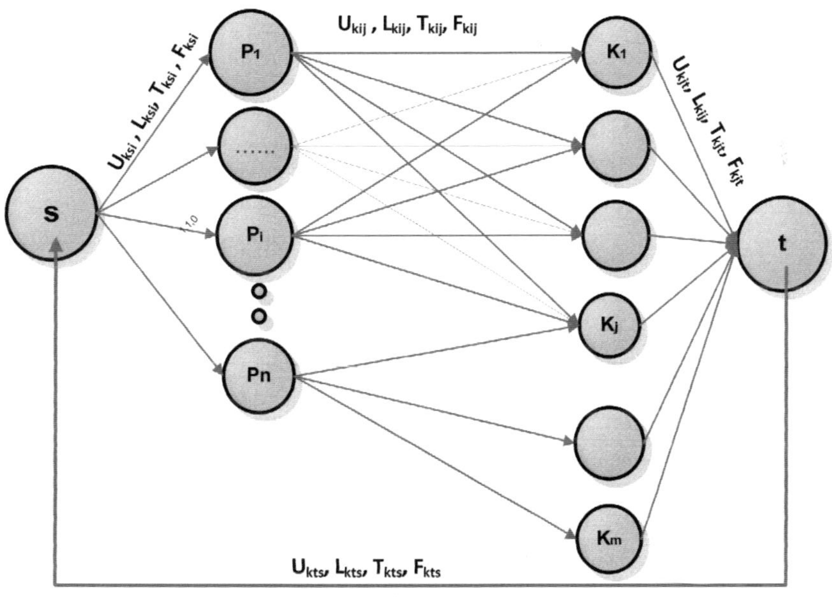

Figure 3. Basic graph structure.

3.6. *Construction of scheduling*

The final result of applying the out-of-kilter algorithm is optimal allocation of carriages between the trains at each station and these are the links where the flows are different from zero. To construct scheduling, it is necessary to walk around all these nodes at different stations in one or several cycles and to determine the sequence of operation of carriages (multiple units) included in these cycles.

3.7. Further analysis and determination of the indicators of scheduling obtained

The problem solution obtained by applying the out-of-kilter algorithm is not the only one possible. In order to enable comparison of the final results obtained to the initial option or of alternative solutions to each other, it is necessary to determine the values of the following quantitative indicators: the total number of rolling stock /carriages or multiple units/ involved in scheduling, the average vehicle mileage and the coefficient of their use in running.

4. Implementation of the scheduling model with diesel multiple unit (DMU) and electric multiple unit (EMU) rolling stock

For comparison, the described methods will be applied to the actual TTT and PTF in the period from December 2018 to December 2019 in the BDZ Passenger Transport Enterprise in Plovdiv, which has available 5 DMU of series 10 with 123 seats, 5 EMU of series 30 with 3 carriages and 190 seats and 5 EMU of series 31 with 4 carriages and 254 seats. The trains set in the TTT and PTF to be provided with multiple units are 119.

The stations where scheduling will take place are sixteen as shown in figure 4. These stations are of different category [12] related to the service of trains and the carriages they consist.

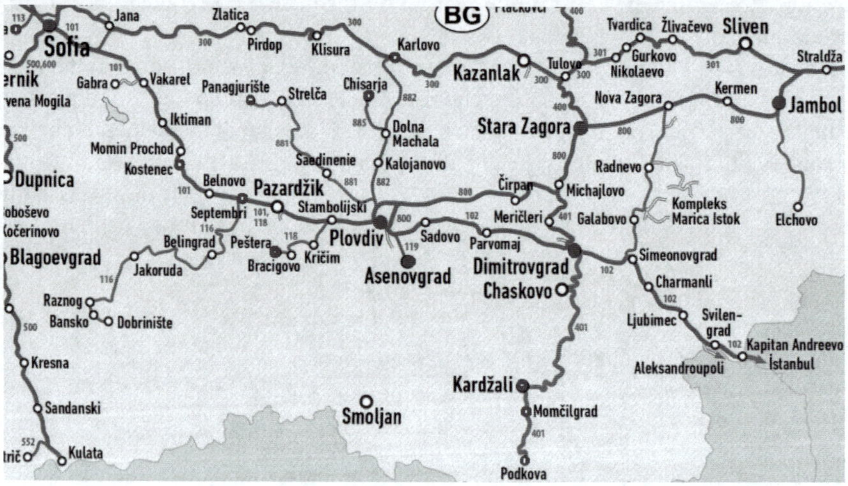

Figure 4. Map of stations where multiple units are scheduled.

The number of arriving and departing trains at most stations is equal and consists of one multiple unit. The station of Plovdiv is an exception because train 10141 departing at 6:10 has two multiple units and train 10144 arriving in Plovdiv at 19:44 is also of 2 multiple units.

The necessary minimal downtime for equipment and cleaning is 50 minutes determined as work [13] and the minimal downtime to turn the direction of running is 5 minutes.

The downtime for obligatory scheduling at Plovdiv station has to be set 0 for the trains serving suburban section Plovdiv-Asenovgrad and this will be the first multiple unit 31 in the samples schedule. The obligatory DMU scheduling will be set for the trains in sections Plovdiv-Peshtera, Plovdiv-Panagyurishte and Dimitrovgrad-Podkova because they are not electrified.

The use of the out-of-kilter algorithm requires to form a graph structure for each station as nodes manufacturers are equal to the number of arriving DMUs and EMUs, and nodes users are equal to the number of departing units. A procedure to create the excel files of graph structure has been developed. These files are used as input data in excel procedure for the out-of-kilter algorithm and the scheduling is built on the basis of the obtained result.

To make scheduling, it is necessary to walk around all trains in several cycles and in succession starting from the earliest departure time from one of the main passenger stations. Each multiple unit of the train is presented as a node of departure and a node of arrival. As shown in figure 5, DMU multiple unit 1 is scheduled to trains 81021, 81202, 81203, 18206 and 18207 and finally to 81206. Its turnover begins and ends in Panagyurishte and the current technical inspection takes place in Plovdiv from 13.00-15.00 PM. In this way, by a number of walking around cycles, the sequence of multiple unit operation is determined on the basis of scheduling developed at different stations and the respective sample schedule is built. The hours of 24-four period are put on the horizontal of sample schedule and multiple units needed to provide scheduling to TTT are shown on the vertical (figure 5). The figure shows the scheduling of DMU 10 and EMU 30 where there is a change in regard to the sample schedules of BDZ PPE. The turnovers of multiple units are given as background in two colours and it can be seen that there are 4 turnovers lasting 24 hours, 2 with a duration of 48 hours.

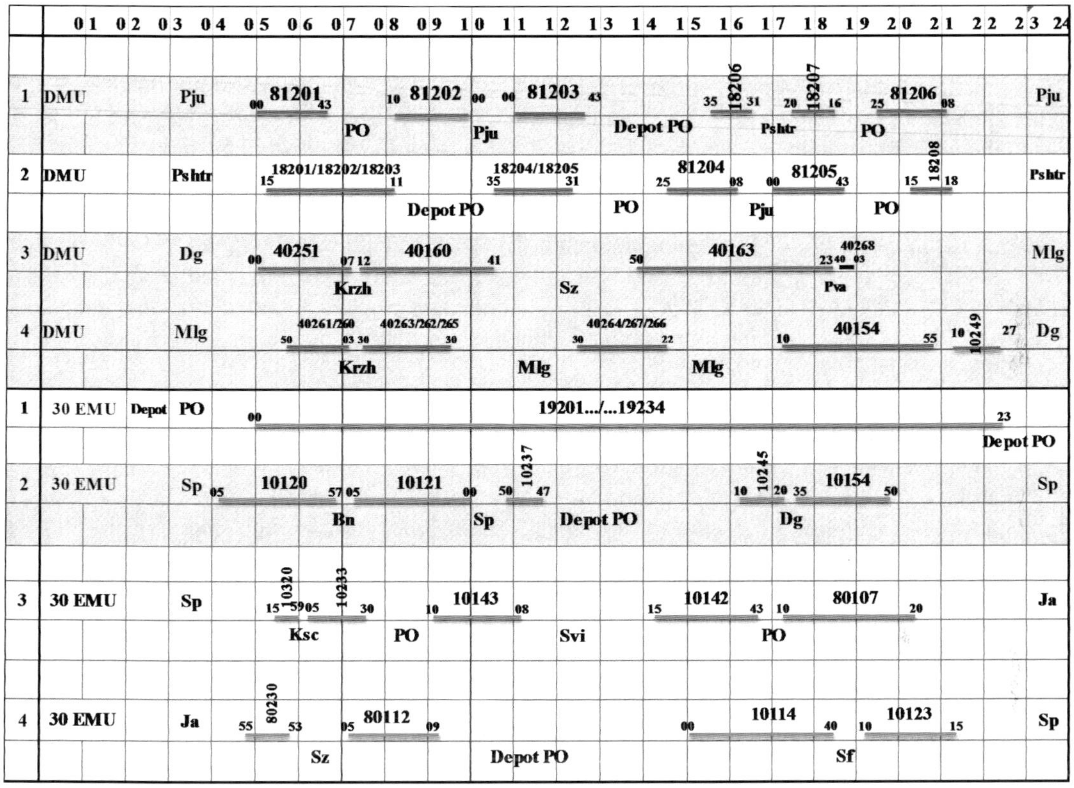

Figure 5. Sample schedule of multiple units.

Using the TTT and sample schedules of the turnover of multiple units – real and optimized, the parameters to evaluate the two models of scheduling are determined: work in km carriage and usage coefficient, which are given in table 1.

As it can be seen from the parameters determined in sample schedule, the optimized model is significantly better since there are 2 spare units left in reserve. The reserve can be used in case of an accident or breakdown, thus reducing train delays due to broken scheduling. There is also an increase of all three indicators used to assess the quality of rolling stock performance – number, mileage and coefficient of use in traffic.

Table 1. Indicators of implementation of MU scheduling.

	Active scheduling	Optimized scheduling
Multiple units, number	15	13
Average mileage, kmwagon^{-1}	357.62	412.63
Usage coefficient in traffic	0.38	0.46

5. Conclusion

The developed model of passenger train scheduling based on the first come, the first served principle where each carriage/multiple unit or main group of carriages is individually assigned gives better results than the standard scheduling of whole trainsets, which has been proved in practice. The model allows to determine the optimal scheduling of trainsets by minimizing the carriage/hours of downtime using the out-of-kilter algorithm.

In order to implement the model, it is necessary to provide database for TTT and PTF for the respective timetable and to predefine the minimal necessary downtime sat stations according to the number of carriages and operations carried out at the station depending on its category. It is also necessary to determine the balance of carriages for each station, to establish scheduling mandatory needed for passenger trains depending on the station track development and destinations served; to build the graph structure and apply the optimized model.

The practical implementation of rolling stock scheduling involves the trains along three main lines and their deviations provided with rolling stock by the BDZ Passenger Transport Enterprise in Plovdiv region. The following options are compared: the actual train scheduling for the TTT being in force from 12.12.2018 to 11.12.2019, which is based on one and the same number and type of carriages, and scheduling performed on principle the first come, the first served using the optimization model based on the out-of-kilter algorithm. The timetable includes 119 trains with scheduling at 16 stations, so the railway carrier needs totally 15 DMUs and EMUs. The application of the proposed model has shown that only 13 multiple units are needed for scheduling of the same number of trains. Thus the model leads to railway rolling stock saving and increasing the indicators of its effective usage.

References

[1] Peeters M and Kroon L 2009 Circulation of railway rolling stock: a branch-and-price approach *Erasmus University Rotterdam(ERIM Research Report)* 32

[2] Schrijver A 1993 Minimum circulation of railway stock *CWI Quarterly* **6** 205–17

[3] Maroti G and Kroon L 2004 Maintenance routing for train units: the transition model *Transportation Science* **39(4)**

[4] Cordeau J, Soumis F and Desrosiers J 2000 A benders decomposition approach for the locomotive and car assignment problem *Transportation Science* **34** 133–49

[5] Cordeau J, Soumis F and Desrosiers J 2006 Simultaneous assignment of locomotives and cars to passenger trains *European J. of Operational Research* **174** 1281–97

[6] Cacchiani V, Caprara A, Galli L, Kroon L, Maróti G and Toth P 2012 Railway rolling stock planning: robustness against large disruptions *Transportation Science* **46**(2) 217

[7] Giacco G, Carillo D, D'Arianob A, Pacciarellib D and Marínc Á 2014 Short-term rail rolling stock rostering and maintenance scheduling *Transportation Research Proc.* **3** 651–9

[8] Haahr J, Wagenaar J, Veelenturfc L and Kroon L 2016 A comparison of two exact methods for passenger railway rolling stock (re) scheduling *Transportation Research Part E: Logistics and Transportation Review* **91** 15–32

[9] Nielsen L 2011 *Rolling Stock Rescheduling in Passenger Railways* (Erasmus Universiteit Rotterdam, ERIM Electronic Series Portal) p 284

[10] Ahuja R, Magnati T and Orlin J 1993 *Network Flows: Theory, Algorithms and Applications* (Prentice Hall Upper Saddle River N.J.)

[11] Phillips D and Garcia-Diaz Al 1981 *Fundamentals of Network Analysis* (Prentice Hall Englewood Cliffs NJ) p 474

[12] Stoilova S. 2019 Methodology for classification sections of railway network using cluster analysis *Proc. 18th Int. Sc. Conf. Engineering for Rural Development (Jelgava Latvia)* 886–97

[13] Todorova M 2019 Gert model of a technical station for determining the passenger trainset delay *Transport Problems* **14(1)** 81–94